MODELING OF IN SITU TECHNIQUES FOR TREATMENT OF CONTAMINATED SOILS

MODELING of IN SITU TECHNIQUES for TREATMENT of CONTAMINATED SOILS

Soil Vapor Extraction, Sparging, and Bioventing

David J. Wilson, Ph.D.

Professor of Chemistry and Environmental Engineering
Vanderbilt University

WITH CONTRIBUTIONS BY:
A. N. Clarke, J. L. Pintenich, K. H. Oma
R. A. Brown, C. Gómez-Lahoz, J. M. Rodríguez-Maroto

TECHNOMIC
PUBLISHING CO., INC.
LANCASTER · BASEL

Modeling of In Situ Techniques for Treatment of Contaminated Soils
a TECHNOMIC publication

Published in the Western Hemisphere by
Technomic Publishing Company, Inc.
851 New Holland Avenue, Box 3535
Lancaster, Pennsylvania 17604 U.S.A.

Distributed in the Rest of the World by
Technomic Publishing AG
Missionsstrasse 44
CH-4055 Basel, Switzerland

Printed in the United States of America
10 9 8 7 6 5 4 3 2 1

Main entry under title: Modeling of In Situ Techniques for Treatment of Contaminated Soils:
  Soil Vapor Extraction, Sparging, and Bioventing

A Technomic Publishing Company book
Bibliography: p.
Includes index p. 563

Library of Congress Catalog Card No. 95-60515
ISBN No. 1-56676-234-0

In loving memory of Alice Olmsted Wilson, who taught her three children to love this planet passionately, and to do something about it.

Table of Contents

Preface

SOIL vapor extraction (SVE) within the last few years has gone from being an innovative, relatively untried technology which was viewed with some distrust, to being one of the most reliable, accepted, and cost-effective of the techniques available for the remediation of hazardous waste sites. (SVE is also known as soil vapor stripping, soil venting, in situ soil aeration, soil vacuum extraction, etc.) As with most of the in situ techniques, its use is restricted; the compounds to be removed must have vapor pressures and Henry's constants that are sufficiently large; SVE is effective for removing only those VOCs that are in the vadose zone or floating on the water table, and the soil at the site must be sufficiently permeable that one can move air through all of the domain to be treated by SVE. The design of optimal, cost-effective SVE installations tends to be highly site-specific, and often at complex sites the technique is deployed in concert with such other technologies as pump-and-treat, in situ air sparging, and fixation/stabilization. SVE may also be operated in such a manner as to optimize the extent to which biodegradation of contaminants occurs, a technique that is called bioventing or bio-assisted SVE, a quite new technology addressed in Chapter 9 of this book.

Recently, another aeration technique has been developed—in situ air sparging (ISAS, sparging, in situ aeration); this is for use in the remediation of aquifers contaminated with VOCs. Virtually all of the points raised above with regard to SVE are also applicable to ISAS. The technique, being newer than SVE, has not yet achieved the wide degree of acceptance that SVE now enjoys, but ISAS has already achieved a number of successes, has a sound basis in science and engineering, has very little environmental impact, and is applicable to situations for which no other technology has proven cost-effective. It is evident that universal acceptance of this technology will be achieved in the near future.

In this book about these techniques, we hope to serve the needs of three groups of readers. First, we have attempted to provide sufficient information on SVE and ISAS to give managers and regulators who may not be familiar with these remediation methods a clear idea of their ranges of applicability. Secondly, we have tried to meet the needs of those engineers who are not experts in the fields of SVE and ISAS, but whose work requires that they read proposals, reports, recommendations, etc., involving the technique with some degree of critical understanding. Thirdly, we have tried to provide a detailed discussion of SVE and ISAS, complete with all the mathematical nuts and bolts, for the environmental engineer whose work is involved with the selection, preliminary planning, design, construction, and operation of SVE and/or ISAS facilities and who may have to use or even to develop computer software in connection with these tasks.

This last objective has necessitated a somewhat longer and much more mathematical book than would be needed by the first two groups, who are herewith encouraged to skip over the heavier mathematical sections without feelings of guilt. These readers will probably find, however, that their intuitive grasp of SVE and ISAS is aided by browsing through the introductory sections and the sections on results and conclusions, which follow the math sections, and by examining the numerous figures that have been included to help the reader develop good intuition about what is going on "down there" when SVE, bioventing, and/or ISAS are being carried out.

Many of the models presented here include the effect of diffusion kinetics, which adds considerably to their complexity but which also makes them much more realistic than the local equilibrium models. As with pump-and-treat operations, in SVE and ISAS one often has contaminants diffused into low-permeability porous structures such as lenses or noncontinuous strata of clay, silt, or till, through which it is virtually impossible to maintain an adequate flux of advecting gas or water. Diffusion transport, thus, provides the only mechanism by which these contaminants can be removed. This may have a profoundly deleterious effect on the rate of remediation, especially during its later stages. Toward the end of the cleanup, slow diffusion processes may result in prolonged "tailing" of the effluent soil gas or groundwater VOC concentrations and in the rebound of, initially, quite low soil gas or groundwater VOC concentrations to frustratingly high values after a period of shutdown of the operation. Failure to be aware of the possibility of diffusion kinetics limits on remediation rates and failure to design pilot studies that ascertain the extent to which these might be a problem at the site under consideration can lead to bitter disillusionment, recriminations, and dramatic scenes of the highest emotional intensity. We hope to provide the reader with sufficient insight into these potential "bottleneck" diffusion processes so that he or she can avoid unpleasant surprises.

This book places heavy emphasis on mathematical modeling, and we believe that there is much to be gained by thoughtful modeling exercises. A word of warning is necessary in connection with this, however. Computer scientists have long had a slogan, "garbage in, garbage out," which applies here. At most sites the quantity and quality of the data available for site characterization are, at best, execrable because of time and budgetary constraints, as well as ignorance on the part of many managers of the savings that result from an optimally designed remediation system. One is fortunate to have enough data to get even a very rough qualitative idea of what is "down there." From these miserable scraps of data, one extracts estimates of the parameters needed in the models and runs the simulations. Under these circumstances, the estimated cleanup times resulting from the model computations are definitely not accurate to five decimal places. No amount of sophisticated mathematics and computation can offset lack of adequate information about the characteristics of the site. There is a tendency on the part of the young, the unsophisticated, and the uneducated to regard numbers coming from a computer as being the revealed "Word of God." It is the duty of engineers commissioning and doing modeling work to enlighten these people before someone is hurt.

I am deeply indebted to a number of people and institutions in connection with this book. Ann Clarke, Jeff Pintenich, and Ken Oma (Eckenfelder, Inc.); Richard A. Brown (Groundwater Technology, Inc.) and César Gómez-Lahoz and José Miguel Rodríguez-Maroto (University of Málaga) authored chapters of the book, bringing to it a level of experience and expertise for which I am most grateful. My wife Marty provided invaluable assistance in editorial work and proofreading. I am indebted to the Spanish government (DGICYT) for a fellowship during which the book was completed, to Vanderbilt University for support during my leave, to Juan J. Rodríguez-Jiménez for making my visit to Málaga possible, and to the faculty and staff of the Departamento de Ingeniería Química of the University of Málaga for making my stay in Spain so enjoyable and rewarding.

Acronyms and Abbreviations Used

AFB	Air Force Base
API	American Petroleum Institute
ARAR	Applicable or relevant and appropriate requirement
BET	Brunauer-Emmett-Teller (adsorption isotherm)
BETX	Benzene, ethylbenzene, toluene, xylenes
BOD_5	5-Day biological oxygen demand
BRA	Baseline risk assessment
CAMU	Corrective action management unit
CAP	Corrective action plan
CERCLA	Comprehensive Environmental Response, Compensation, and Liability Act of 1980 (Superfund)
CMS	Corrective measure study
DCA	Dichloroethane
DCE	Dichloroethylene
DDT	A chlorinated hydrocarbon pesticide, dichlorodiphenyltrichloroethane
DNAPL	Dense nonaqueous phase liquid
DOD	Department of Defense
DOE	Department of Energy
EPA	Environmental Protection Agency
FS	Feasibility study
HEA	Health and environment assessment
HRS	Hazard ranking system
HSWA	Hazardous and Solid Waste Amendments of 1984
ISAS	In situ air sparging
J-A	Jet aircraft fuel for commercial planes
J-A1	Jet aircraft fuel for commercial planes
JP-4	Jet aircraft fuel used by U.S. Air Force
JP-5	Jet aircraft fuel used by U.S. Army
LNAPL	Light nonaqueous phase liquid

MC	Methylene chloride
MEK	Methyl ethyl ketone
MIK	Methyl isobutyl ketone
MTBE	Methyl *t*-butyl ether
NAPL	Nonaqueous phase liquid
NPL	National Priority List (Superfund list)
NVOC	Nonvolatile organic compound
PAH	Polynuclear aromatic hydrocarbon
PA/SI	Preliminary assessment/site investigation
PCB	Polychlorinated biphenyl
PCE	Tetrachloroethylene (= perchloroethylene)
ppmv	Parts per million by volume
PRG	Preliminary remediation goal
PRP	Potentially responsible party
PVC	Polyvinyl chloride
RCRA	Resource Conservation and Recovery Act
RFA	RCRA facility assessment
RFI	RCRA facility investigation
RI/FS	Remedial investigation/feasibility study
ROD	Record of decision
ROI	Radius of influence
SARA	Superfund Amendments and Reauthorization Act
SITE	Superfund Innovative Technology Evaluation
SOB	Statement of basis
SVE	Soil vapor extraction (same as soil vacuum extraction, soil venting, soil vapor stripping)
SVOC	Semivolatile organic compound
SWMU	Solid waste management unit
TBA	*t*-Butyl alcohol
TCA	Trichloroethane
TCE	Trichloroethylene
ThOD	Theoretical oxygen demand
TPH	Total petroleum hydrocarbons
TSDF	Treatment, storage, and disposal facility
TTCA	Tetrachloroethane
UST	Underground storage tank
VOC	Volatile organic compound

CHAPTER 1

Soil Vapor Extraction – Introduction and Background

SOIL vapor extraction [the use of vacuum wells to remove volatile organic compounds (VOCs) from the vadose zone] has become a widely used technology for the remediation of spills, leaks, and hazardous waste sites. Where it is applicable, its effectiveness, low cost, and virtually negligible environmental impact recommend it to responsible parties, environmental engineers, and regulators alike. In this book we present a discussion of this extremely useful technique, and introductions to two somewhat newer related technologies, in situ air sparging and bioventing, now coming into use for the removal of VOCs from contaminated aquifers.

Our first objective is to provide sufficient information on in situ soil vapor extraction (abbreviated as SVE; also called soil vapor stripping, soil vacuum extraction, soil vacuuming, soil venting) to give environmental engineers unfamiliar with the technique a clear idea of the range of its applicability. A second purpose is to describe the screenings and studies needed to determine if soil vapor stripping is suitable for any given site. This book should be useful to engineers who are not experts in the field of soil vapor stripping, but who need to read proposals, reports, and recommendations involving the technique with some critical understanding. Thirdly, we have tried to provide a discussion of soil vapor stripping that is sufficiently detailed so that it will serve the needs of the engineer who is involved in the preliminary planning, design, construction, and operation of soil vacuum extraction facilities. The SVE technique is one in which mathematical modeling is often quite helpful; we therefore discuss the theory of SVE modeling in some detail and include a large number of modeling results.

Conceptually, soil vapor extraction is a simple technology, applicable to the removal of volatile organic compounds (VOCs) and some semivolatile compounds (SVOCs) from the vadose zone. The equipment is shown schematically in Figure 1.1. A well is drilled down through the domain of contamination and a blower is then used to draw air down through the con-

1

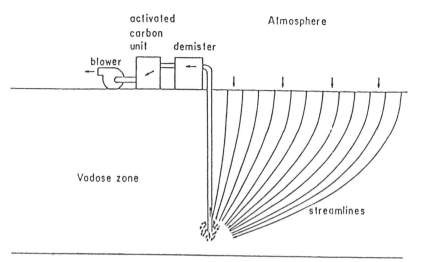

Figure 1.1 Schematic diagram of a soil vapor extraction well.

taminated soil, into the well, up through a demister to remove excess water, and then through a unit such as an activated carbon bed or catalytic combustion unit to remove the VOCs, after which the gas is exhausted through the blower to the atmosphere. For larger scale operations a number of wells may be manifolded and served by a single demister, activated carbon unit, and blower. Several other geometries for SVE (trenches, horizontal drilling, and soil piles) are shown in Figure 1.2.

The environmental impact of the technique is low, and site disturbance is generally minimal. Costs are generally a fraction of those of other technologies. Large volumes of soil can readily be treated. Soil vapor extraction (SVE) systems are relatively easy to install and utilize off-the-shelf equipment. Cleanup times are generally fairly short, and the toxic material is removed from the soil and ultimately destroyed, rather than being sequestered and/or relocated. SVE lends itself well to use in combination with other techniques, such as groundwater pump-and-treat operations and sparging. If the contaminants are biodegradable, SVE may be enhanced by a significant bioremediation component. These advantages make soil vapor stripping a method of choice in those sites at which it is applicable. Pederson and Curtis (1991a,b) have discussed the advantages of SVE in some detail in an EPA handbook on the subject.

The range of applicability of soil vapor stripping is bounded by the following constraints:

(1) The chemicals to be removed must be volatile or at least semivolatile

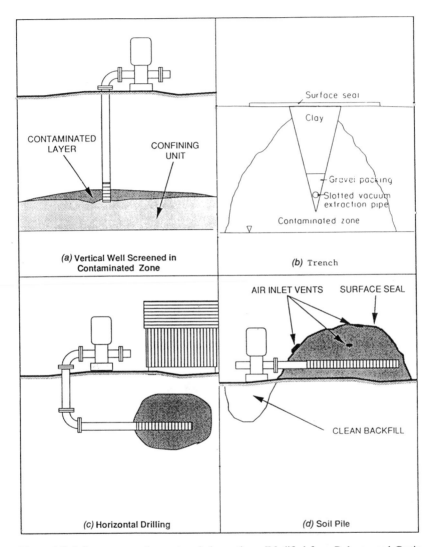

Figure 1.2 Soil vapor extraction system design options. [Modified from Pedersen and Curtis (1991a,b).]

(a vapor pressure of 0.5 torr or greater); it is not feasible to remove metals, most pesticides, oils and greases, and PCBs by vacuum extraction because their vapor pressures are too low.

(2) The chemicals to be removed must have relatively low water solubility, or the soil moisture content must be quite low; generally, it is not feasible to vapor strip chemicals such as acetone or alcohols because their vapor pressures in moist soils are too low. However, many of these compounds biodegrade readily if an adequate oxygen supply (such as is provided by SVE) is available.

(3) The contaminants to be removed must be above the water table or, in the case of light nonaqueous phase liquids (LNAPLs), floating on it. In some situations it is possible to draw down the water table by pumping, thereby extending the domain that can be vapor stripped.

(4) The soil must be sufficiently permeable to permit the vapor extraction wells to draw air through all of the contaminated domains at a reasonable rate; a basic principle in soil vapor stripping is that you must be able to move air through the domain if you are going to clean it up.

Very few standardized design and operating criteria are available for SVE, due mainly to the limitations imposed by site-specific factors. These are (1) contaminant characteristics, including the degree of weathering (particularly important with petroleum products); (2) the vertical and horizontal extent of the contamination; (3) soil characteristics, particularly permeability and heterogeneities in the permeability; (4) depth of the water table; (5) VOC emission control requirements; and (6) criteria set for soil cleanup. Pedersen and Curtis (1991a,b) give a listing of the soil cleanup criteria presently adopted by the states; these range from 1 to 500 ppm, with the most common standard being 100 ppm total petroleum hydrocarbons.

Pedersen and Curtis note that the purpose of the SVE design process is to construct an SVE system that removes the greatest amount of contaminant(s) from the site in the most efficient, timely, and economical manner. This requires understanding of the main factors that determine system effectiveness (Johnson et al., 1989). These are (1) the composition and characteristics of the contaminants present and (2) the delivery of flowing air to all of the contaminated zones at a reasonable rate. SVE system design attempts to maximize the interaction of the vapor flow with the contaminated zone.

Sink (Pacific Environmental Services, Inc., 1989) has provided a good description of current SVE facilities and their operation. Sink's summary of SVE sites is given in Table 1.1, and a listing of the applications of innovative technologies at NPL sites is given in Table 1.2. An SVE system

TABLE 1.1. Representative Soil Vapor Extraction Sites.[a]

Facility/Location	State	Pollutants Identified
Groveland Wells, Groveland	Maine	TCE, PCE, MC, DCE, TCA
Service station, Wayland	Maine	Gasoline
Waldick Aerospace	New Jersey	PCE, petroleum hydrocarbons
Industrial tank farm	Puerto Rico	Carbon tetrachloride
Service station, San Juan	Puerto Rico	Gasoline
Tyson Dumpsite, Tyson's Lagoon	Pennsylvania	TCE, toluene, ethylbenzene, xylene, trichloropropane, 1,1,1,2-TTCA
Service station, Bellview	Florida	Gasoline
AWARE study, Nashville	Tennessee	TCE, acetone, chlorobenzene
Petroleum fuels terminal, Grainger	Indiana	Gasoline
Seymour Facility, Seymour	Indiana	1,2-DCA, benzene, vinyl chloride, 1,1,1-TCA, others
Thomas Solvents Co. (Verona Well Field), Battle Creek	Michigan	DCA, TCA, DCE, TCE, PCE, vinyl chloride, chloroform, carbon tetrachloride, benzene, toluene, xylene, ethylbenzene, MEK, MIK
Kimross Facility, Kimross	Michigan	1,1,1-TCA
Lansing Facility, Lansing	Michigan	TCE
Bangor Facility, Bangor	Maine	Toluene, benzene, xylene, ethylbenzene, styrene, ketones, chloroethane, MC

(continued)

TABLE 1.1. (continued).

Facility/Location	State	Pollutants Identified
Hillside Facility, Hillside	Michigan	TCE
Custom Products, Stevensville	Michigan	PCE
Twin Cities Army Ammunition Plant, New Brighton	Minnesota	TCE, TCA, toluene
Troy Facility, Troy	Ohio	Acetone, MC, TCE, toluene, xylene
Paint storage warehouse, Dayton	Ohio	Acetone, toluene, xylene, ketones
Texas Research Inst., Austin	Texas	Gasoline
Waverly Facility, Waverly	Nebraska	Carbon tetrachloride
Hill Air Force Base, Salt Lake City	Utah	Jet fuel
Dowell Schlumberger, Casper	Wyoming	Chlorinated hydrocarbons, toluene, xylene, benzene, ethylbenzene
LARCO, Casper	Wyoming	Toluene
Southern Pacific spill, Benson	Arizona	Dichloropropene
Electronics Co., Santa Clara	California	1,1,1-TCA
Storage tank, Cupertino	California	TCA, TCE, DCA, DCE
Well 12A, Tacoma	Washington	TCE, PCE, MC, TTCA, DCA, TCA
Ponders Corner	Washington	1,2-DCA, TCE, TTCA

aFrom Pacific Environmental Services (1989).
Pollutant key.

DCE:	dichloroethylene	MC:	methylene chloride	TTCA: tetrachloroethane
TCE:	trichloroethylene	DCA:	dichloroethane	MEK: methyl ethyl ketone
PCE:	perchloroethylene (tetrachloroethylene)	TCA:	trichloroethane	MIK: methyl isobutyl ketone

TABLE 1.2. Applications of Innovative Treatment
Technologies at NPL (Superfund) Sites.

	Number of Times Selected
Soil Vapor Extraction	
VOCs	83
SVOCs	18
Metals	0
Thermal Desorption	
VOCs	20
SVOCs	15
Metals	0
Bioremediation	
VOCs	22
SVOCs	37
Metals	0
In situ Flushing	
VOCs	11
SVOCs	5
Metals	6
Solvent Extraction	
VOCs	4
SVOCs	5
Metals	0
Soil Washing	
VOCs	1
SVOCs	9
Metals	12

At some sites, treatment is for more than one contaminant, and more than one treatment technology may be used. Treatment may be planned, ongoing, or completed. Source: U.S. EPA, Technology Innovation Office, Innovative Treatment Technologies: Semi-Annual Status Report, EPA/542/R-92/011, October, 1992.

consists of (1) vacuum extraction wells, (2) inlet or injection wells (possibly), (3) piping headers, (4) vacuum pumps or blowers, (5) vacuum gauges and flow meters, (6) sampling ports, (7) an air/water separator (usually), (8) a VOC control system (usually), and (9) impermeable caps (possibly). The vacuum wells generally fully penetrate the zone of contamination. The screened sections are constructed of slotted plastic pipe in a permeable packing to facilitate gas flow (see Figure 1.3).

VOCs in the zone of contamination move as vapor through the soil into the vacuum extraction well. They are then generally sent to an air pollution control device. These VOCs typically have molecular weights of 200 g/mol or less; larger molecules are insufficiently volatile to move readily

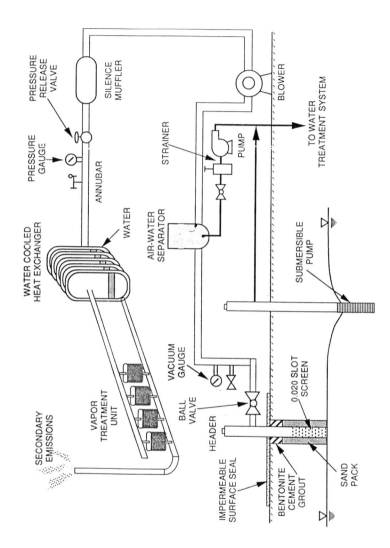

Figure 1.3 Schematic of a typical soil vapor extraction system. [From Pedersen and Curtis (1991a,b), EPA Document EPA/540/2-91/003.]

8

as vapor although they may biodegrade in the presence of oxygen provided by the SVE wells. Inlet wells (passive) or injection wells (in which air is forced into the ground) may be located around the boundaries of the contaminated area to improve air flow in regions in which this may be too slow; injection wells may be used as part of a closed loop SVE system. Piping is usually plastic (often PVC); headers are either plastic or steel. Large sites may require multiple systems, and it is common practice to have one blower serve several wells. Insulation of pipes and headers may be necessary in cold climates. Air/water separators may be needed in the train before off-gas treatment, particularly if VOCs are being removed with activated carbon, since the off-gas generally contains entrained water droplets and is almost always nearly saturated with water vapor. The air/water separator is usually included in the train before the blower to protect it; water removal also increases the efficiency of subsequent VOC removal with activated carbon.

After the system is turned on, a steady-state flow rate is achieved relatively quickly. The gas flow rate is determined by the wellhead vacuum (controlled by the blower), the pipe size and length, the screening and packing of the well, soil pneumatic permeability, the presence of barriers (building foundations, parking lots, impermeable plastic caps, etc.), and flow control devices. If injection wells are employed, the discharge pressure from the blower may be used to provide positive pressure for air injection. The resulting air flow causes VOC vapor movement to the extraction well(s), where the VOCs are drawn up and captured or destroyed in the VOC treatment device. Impermeable plastic caps are occasionally used to reduce the volumes of soil in which air flow is excessively sluggish or wastefully fast. These may also limit fugitive VOC emissions from the soil, particularly important if air is being injected under positive pressure.

Flow rates and off-gas VOC compositions are measured regularly to permit calculations of the rate of VOC removal, which is almost always initially quite high and then decreases gradually with time. If a significant amount of VOC is present in regions such as clay lenses from which it is slowly released by diffusion, the latter portion of the cleanup may be quite prolonged. Initial pilot-scale studies, unless properly designed, will give no warning of this tailing. In the latter stages of cleanup, the blower may be cycled on and off or the well run at a reduced flow rate to energy requirements and costs of off-gas treatment with little reduction in removal rate. Measurements of soil gas VOC concentration during well operation and immediately after shutdown are not reliable indications of the extent of cleanup; one should either obtain soil analyses for VOCs or VOC analyses on gas samples taken after the well has been shut down for an extended period to test the extent to which clean-up is complete.

HISTORICAL BACKGROUND AND LITERATURE REVIEW

The literature on the topic is sufficiently extensive that a complete review would require excessive space; therefore a brief review covering the major topics of interest is presented.

GENERAL REFERENCES, REVIEWS, ETC.

EPA has published a reference handbook on soil vapor extraction (Pedersen and Curtis, 1991a,b), which is extremely useful. This includes a discussion of the principles of soil vapor behavior, including a section on air permeability test methods. The procedures to be followed in carrying out site investigations are then presented. The report discusses system design and equipment selection, system operation and monitoring, techniques for the control of vapor emissions, and costs. The handbook closes with ten research papers on SVE and two appendices on soil cleanup criteria and air discharge criteria.

Another report that presents a broad overview of SVE, along with considerable useful detail, is EPA's guide on SVE treatability studies (Stumbar and Rawe, 1991). The report provides a description of the technology, discussions of the limitations of the technology and of preliminary screening, and a section on the use of treatability tests in the evaluation of SVE as a possible remedy. A detailed discussion on the development of treatability study work plans is also provided.

EPA has published the proceedings of a 1991 symposium on soil venting; these papers give a broad and fairly current overview of the state of SVE art (EPA, 1992). Roy and Griffin (1989) prepared a good introduction to SVE; their report includes a description of a simple model (discussed in Chapter 2 of this book) and tabulations of Henry's constants and adsorption constants for a number of organic solvents. Another relevant report by these authors (Roy and Griffin, 1987) addresses the vapor-phase movement of organic solvents in the unsaturated zone. Pedersen and Fan (1992) have provided a brief overview of SVE development status and trends. Classic works by Millington and Quirk (1961) and Scheidegger (1974) give valuable insight into the behavior of gases and liquids in porous media. A recent EPA report (1993) gives a breakdown of the status of innovative technologies at NPL (Superfund) sites as of October 1992; out of 210 sites, SVE is being used at eighty-three (see Table 1.3). Use of innovative and established treatment technologies is also summarized in Figure 1.4.

Keech (1989) has summarized the research supported by the American Petroleum Institute (API) in soil vapor stripping (subsurface venting), and noted that API would be publishing a manual on this technique.

Hutzler et al. (1989a, 1989b, 1990, 1991) have reviewed soil vacuum ex-

TABLE 1.3. Status of Innovative Technology Projects at NPL Sites as or October 1992.

Technology	Predesign/ in Design	Design Complete/ Being Installed/ Operational	Project Completed	Total
Soil vapor extraction	62	18	3	83
Thermal desorption	19	4	4	27
Ex situ bioremediation	17	7	1	25
In situ bioremediation[a]	14	5	1	20
Soil washing	16	2	0	18
In situ flushing	12	5	0	17
Dechlorination	5	1	1	7
Solvent extraction	5	1	0	6
In situ verification	3	0	0	3
Other innovative treatment	3	0	0	3
Chemical treatment	0	0	1	1
Total	156 (74%)	43 (21%)	11 (5%)	210

[a]Includes in situ groundwater treatment.
Data are derived from RODs for fiscal years 1982–1991 and anticipated design and construction activities.
Source: U.S. EPA, Technology Innovation Office, Innovative Treatment Technologies: Semi-Annual Status Report, EPA/542/R-92/011, October 1992.

traction; their comprehensive articles include a list of representative pilot- and field-scale soil vapor stripping operations with references to detailed descriptions of these sites. Sink (Pacific Environmental Services, Inc., 1989) has also published a list (Table 1.1) of twenty-nine SVE operations, including location, pollutants identified, whether VOC emissions are controlled, and references. The innovative technologies being employed at Superfund sites are listed in Table 1.2. Hutzler and his coworkers include information on the designs of the facilities, a tabulation of Henry's constants for common organic compounds, and a summary of design suggestions. These authors note the importance of diffusion-limited transport; they suggest that intermittent well operation is more efficient than continuous operation when vapor stripping from clays and silts.

Hoag (1989) and Hoag et al. (1991) have reviewed developments in SVE, focusing particularly on mathematical modeling and identifying areas of current and needed future research.

A paper by Danko (1990) discusses the applicability and limitations of SVE in some detail. Criteria that must be evaluated include the identities of the contaminants and their characteristics; the nature, extent, and volume of the contaminated zone; the vadose zone characteristics affecting vapor transport; the depth to the underlying saturated zone; and air emission and vadose zone cleanup standards. Stransky and Blanchard (1989)

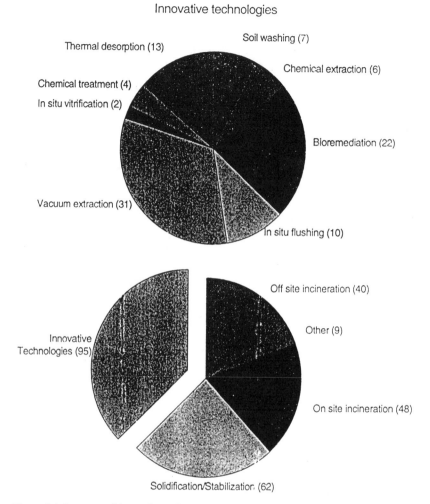

Figure 1.4 Summary of innovative and established treatment technologies at Superfund sites. Data derived from 1982–1989 RODs. [From EPA Document EPA/504/8-91/005 (1991b).]

have described a stepwise methodology for assessing the feasibility of SVE for use at sites involving shallow unconfined aquifers, for design work, and for cleanup monitoring.

Lyman and Noonan (1990) and Lyman et al. (1990) have published guides that identify information about the site subsurface environment and the contaminant characteristics, which is needed for assessment of the suitability of SVE and several other technologies for use at undergound petroleum storage tank sites. Mercer et al. (1989) discuss the processes that control the dynamics of contaminants at a site and note that simple pump-and-treat operations are usually very prolonged; they direct attention to such technologies as SVE, bioremediation, fixation, and in situ heating. Curtis and Noonan (1987) have described a number of remediation technologies and their cost-effectiveness in dealing with releases from underground storage tanks, indicating that, frequently, use of several techniques may be necessary. Ghassemi (1988) has reviewed seven in situ technologies (including SVE) and four ex situ techniques that treat or destroy wastes.

Much of the early work on vacuum extraction was published under the auspices of the American Petroleum Institute (API). In 1980 API published an analysis and evaluation of the ability of venting (SVE in our terminology) to remove subsurface gasoline vapors from contaminated soil (API, 1980). Other early papers on the same topic were published by the Texas Research Institute (1984) and by Thornton and Wootan (1982).

Recently, Dupont (1992) reviewed the soil, site, and waste characteristics that drive the vaporization of VOCs in the vadose zone. In addition, two EPA reports provide rather comprehensive discussions of the practical aspects of the transport and fate of contaminants in the subsurface (Johnson et al., 1989) and of the characterization of sites for subsurface remediation (EPA, 1991a).

BIOREMEDIATION AND SOIL VAPOR EXTRACTION

An early API report demonstrated enhanced microbial degradation of gasoline in the presence of an increased oxygen supply (API, 1982). From the enormous literature on bioremediation, a few entries are selected in which SVE and bioremediation are linked. Staps (1989) has outlined progress in the Netherlands, West Germany and the United States in bioremediation, focusing particularly on the conditions necessary for its use. Recent work on bioremediation has been summarized by Hoeppel and Hinchee (1993). A very helpful brief summary of abiotic and microbiological transformations of chemicals and the principles governing these has been provided by Henson (1991), who provides a discussion of the microbial ecology of the subsurface and the relationship of environmental factors to soil microbiology and chemical biodegradation. He writes a description of microbial metabolism, a brief introduction to the kinetics of biological reactions, and a section on the bioremediation of organic compounds in the subsurface, which includes a summary of types of compounds that are likely candidates for bioremediation. In the same EPA report Boulding and Barcelona (1991) include a table of sources of information on the chemical behavior of natural organics and of contaminants in the subsurface.

Moore and Revell (1989) describe the remediation of an abandoned coal tar site where soil venting and infiltration of nutrients were used to enhance biodegradation of contaminants in the vadose zone. Ostendorf and Kampbell (1990) compare field performance data on light hydrocarbons in a bioreactor with an analytical model that included storage, linear adsorption, advection, and Michaelis-Menton kinetics; the calibrated model was used on an aviation gasoline spill at Traverse City, Mississippi. Wu et al. (1990) developed a mathematical model for in situ biodegradation of contaminants in a soil bed.

Arthur et al. (1991) evaluate aeration techniques for bioremediating fuel-contaminated soils. Over 100 species of organisms were capable of degrading hydrocarbons, and lack of oxygen was probably the most significant limiting factor in biodegradation. Forced air and hydrogen peroxide were both effective electron receptors; forced air was cheaper, making SVE a strong contender in in situ bioremediation work.

Hinchee et al. (1987) compare enhanced bioreclamation, soil venting (SVE), and groundwater extraction with regard to cost-effectiveness and feasibility. SVE was the lowest in cost where it could be used; bioremediation was expensive but was the only technology that cleaned up heavier nonvolatile hydrocarbons. Battelle Columbus has investigated the enhancement of hydrocarbon biodegradation by soil venting; Hinchee et al. report on a study at Hill Air Force Base, Utah (Hinchee, 1989; Hinchee et al., 1989). Results indicated that 25% or more of the contaminant hydrocarbon was destroyed by biodegradion. Bench-scale work showed that supplemental moisture and nutrients increased the rate of biodegradation. This and other work (Dalfonso and Navetta, 1988, for example) indicate a consensus that biodegradation can be made to play a major role in the removal of hydrocarbons.

Hinchee et al. (1991c) have provided further information on the enhancement of petroleum hydrocarbon (jet fuel) biodegradation by SVE at Hill AFB, Utah. Although most of the VOCs were removed by volatilization, 15–25% were biodegraded. Hinchee et al. (1991b) describe the interaction between SVE and bioremediation of petroleum hydrocarbons. Conventional SVE configurations maximize the off-gas hydrocarbon concentrations; air injection into the contaminated zone and its withdrawal up through clean soils maximizes biodegradation and minimizes off-gas treatment costs but requires careful monitoring. This procedure is also described by Hinchee et al. (1991a). They note that air must flow through the contaminated soil at rates and flow patterns that ensure adequate oxygenation and that minimize hydrocarbon concentrations in the off-gas. Added nutrients and water may be needed to optimize biodegradation rates.

Several bioremediation papers were given at EPA's (1992) recent symposium on SVE. Hinchee and Miller (1992) give updates on sites at Hill AFB (UT) and Tyndall AFB (FL), where bioventing is proving to be quite effective. The work at Tyndall is described in more detail by Miller et al. (1992). Pilot-scale bioventing for the remediation of an aviation gasoline spill of some 35,000 gal is described by Kampbell (1992). An SVE project for the remediation of hydrocarbon spills in the Netherlands is discussed by van Eyk and Vreeken (1992). Sorensen and Sims (1992) examine habitat conditions that affect the microbial communities that participate in the bioventing process. Wilson et al.(1992b) explore opportunities for the biotreatment of trichloroethylene (TCE) in the vadose zone.

SVE LABORATORY STUDIES

An early report (Wootan and Voynick, 1984) deals with the vapor stripping of gasoline in a large-scale (3 × 3 × 1.2 m) model aquifer. These workers suggest that vapor stripping wells should be deep and slotted only near the bottom to avoid short-circuiting the air flow. They also suggest that impervious covers at the surface of the vented area might improve efficiency.

Hoag et al (1984) describe SVE experiments, noting that Dalton's and Raoult's laws were followed in the process, so that local equilibrium between vapor and stationary phases could be assumed. They also describe a gas chromatography "fingerprinting" technique for use as a qualitative and quantitative research tool. These workers (Hoag et al., 1987) also measure the evaporation of over fifty gasoline components in soil columns and investigate the SVE of mixtures, the effects of moisture content, particle size, and airflow rate. This work, like their early study, supports the local equilibrium approximation.

Johnson et al. (1987) carried out one-dimensional experiments to investigate mass transport between vapor and aqueous phases in SVE. The advective flow of vapors through damp sand containing residual free product (NAPL) was studied to determine how readily NAPL could be removed by SVE. They conclude that equilibrium was maintained between pore water and pore vapor, that removal of low molecular weight materials would be rapid, and that removal of the higher boiling fractions of mixtures might be troublesome.

However, a lab study by Hutzler et al. (1989a) demonstrates the importance of diffusion-limited transport in certain media, in agreement with the field data of Fall et al. (1988) and a number of other workers. Subsequently, Gierke et al. (1992) carried out laboratory SVE column experiments in connection with the validation of a model for SVE they developed. They found that toluene vapors sorbed to dry soil materials, while sorption on wet soils was negligible. Their findings also imply that SVE in moist, aggregated soils will be affected by nonequilibrium transport.

Rainwater et al. (1990) carried out a study of SVE in large columns (3 m long by a 1 m diameter); hydrocarbon removal was followed for a period of fifteen months. The porous medium was wet sand; these workers found that they could not use the local equilibrium approximation but needed to take diffusion kinetics into account.

Ong et al. (1991) explored the applicability of linear partitioning relationships for sorption of trichloroethylene (TCE) onto soil and soil minerals. Dry soils adsorb TCE much more strongly than do wet soils. The uptake of TCE by wet soils is attributed to (1) solution of TCE in the aqueous phase, (2) sorption at the solid-liquid interface, and (3) condensation and sorption at the water-air interface.

Bouchard et al. (1990) found that the sorption of benzene and naphthalene on soils was markedly increased if the soils had been previously contaminated with high molecular weight residual hydrocarbons from unleaded gasoline. Evidently, contamination of soils with oils, greases, tars, etc., can be expected to substantially decrease SVE cleanup rates. Oja and Kreamer (1992) discussed the effect of moisture on adsorption of TCE vapor on soils; TCE adsorption was greatly increased on dry soils.

Sully (1992) describes the characterization of pneumatic premeabilities in the vadose zone, and Joss et al. (1992) discuss a field technique for determining unsaturated zone air permeabilities.

MATHEMATICAL MODELING

Hoag and his associates have been active in SVE for some years; Hoag and Cliff (1985) describe one of their early soil venting projects for gasoline removal. In a more recent paper (Baehr et al., 1989) this group describes laboratory work, vapor stripping at a field site contaminated with gasoline, and mathematical modeling. They conclude that the technique was extremely effective and that a local equilibrium model was quite adequate for describing vapor stripping at their field site. Other early papers of interest by this group include Marley and Hoag (1984) and Brown et al. (1987). Marley et al. (1989, 1990) have discussed the design of soil vapor extraction systems, and Hoag (1989) has summarized this group's work and indicates the need for additional research. Marley et al. (1990) have discussed the use of airflow modeling both for the evaluation of soil properties (the pneumatic permeability tensor) and for the design of soil vacuum extraction systems. They note that mass transfer limitations may be important and that the distribution of airflow pathways is always of prime importance; cleanup will not occur if soil gas is not moving through the contaminated soil. The air permeability tensor was evaluated in situ in the vadose zone by calibrating a steady-state airflow model with pressure measurements made during the pumping tests. These air permeabilities and steady-state airflow model were then used to determine well spacings, depth of wells, length of screened well sections, and type and size of pumps needed to produce the desired gas flow field. Marley (1992) recently described the development and application of a three-dimensional airflow model in the design of SVE systems.

Johnson et al. (1989a, 1989b, 1990a) at Shell Development have presented a number of useful mathematical models for use as screening tools in assessing the feasibility of soil vapor stripping on a site-specific basis. One of these is a one-dimensional model that assumes radial symmetry and that takes into account the variations in the composition of volatile complex mixtures during the course of a vapor stripping operation. This should be of particular interest in dealing with spills of hydrocarbons, since these normally occur as mixtures of compounds having a relatively wide range of volatilities. The model assumes local equilibrium between the stationary phase and the mobile vapor phase. Another model allows the estimation of the time required for air flow around a vacuum well to approach a steady state. A vapor phase, diffusion-limited model is also given for dealing with low-permeability layers or with free product. These workers (Johnson et al., 1990b, 1991) have also published a practical approach to the design, operation, and monitoring of SVE systems. This paper provides an excellent introduction to the technique; a discussion of the main factors that control SVE (chemical composition of contaminant, gas flow rates through the unsaturated zone, and flowpath of carrier gas relative to the location of the contaminants); and a description of possible pitfalls. Kemblowski and Chowdery (1992) recently discussed the use of models and decision analysis in the design of SVE systems.

Katyal et al. (1992) describe the modeling of the transport of organic compounds in the vapor and liquid phases, providing a rigorous mathematical treatment and then using this to simulate two remediation scenarios.

DiGiulio et al. (1990) have made a number of carefully reasoned recommendations regarding the design of field tests for the assessment of SVE. They note the importance of estimating the effects of diffusion limitation and show that these can be assessed by means of a rather simple field test. A test vapor stripping well is isolated from the bulk of the zone of contamination by surrounding it with passive wells a relatively short distance from it. The vacuum well is then operated for a substantial interval (until the off-gas VOC levels have dropped nearly to zero) and is then shut down. One then monitors the recovery of the VOC concentrations in this well during the shut-down period to obtain information on the extent of diffusion limitation and also on the time constant for this process. These authors provide a formula by means of which effective Henry's constants (or partitioning coefficients) can be related to actual physical and chemical processes. They also point out that one can vapor strip water-soluble organics (such as alcohol, acetone, MEK) if one is able to initiate vapor stripping before these have become highly diluted in the soil water in the vadose zone. These authors give a trenchant criticism of the use of overly simplistic models in designing and predicting the behavior of SVE systems. In a later EPA report DiGiulio (1992) commented in detail on the importance of mass transport limitations in SVE. He also noted that effluent soil gas VOC measurements are meaningless if taken under dynamic conditions (the vacuum well in operation); the well must be shut down and the soil gas allowed to equilibrate with the soil for an extended period for soil gas VOC analyses to give results indicative of the concentrations of VOCs remaining in the soil.

Analysis of data obtained during the remediation of a 1,3-dichloropropene spill near Benson, Arkansas, led Sterrett (1989) to conclude that diffusion kinetics are an important factor in the rate of cleanup of the site. He also noted that high humidity resulted in more rapid release of contaminants (the so-called wet dog effect).

Stephanatos (1990) has described a rather general model for VOC transport in soils in the vadose zone, which includes both fluid flow and vapor transport; the model can be applied to SVE to predict the fate and transport of VOCs. He successfully applied his two-dimensional model to data from the Tyson site near Philadelphia.

Walton et al. (1990) have presented a rather elaborate set of mathematical models for analyzing a very large, deep, complex site in Idaho at which chlorinated VOCs have contaminated a deep, highly structured vadose zone (basalts interspersed with sedimentary layers) and the underlying zone of saturation. The extremely high cost of carrying out an adequate characterization of a site of such size, depth, and complexity has forced heavy reliance on sophisticated modeling. It will be interesting to see if this effort to substitute increased sophistication for a substantial portion of data is successful. This was updated (Walton et al., 1992) in a paper that discusses the model theory and applications of the model to the design of a large-scale SVE system at the Idaho Engineering Laboratory.

Eckenfelder Inc. has published a number of reports and articles on soil vapor stripping. Clarke and Wilson (1988) describe a phased approach to assessing the feasibility of the technique for use at a given site and for subsequent design work. Wilson and his coworkers have developed a two-dimensional (axially symmetric) local equilibrium model for field vapor stripping wells and a one-dimensional local equilibrium model for lab column operation; lab column data can then be used to obtain adsorption isotherm parameters needed in the field model (Wilson et al., 1988; Wilson, 1988). The field-scale model was used to explore the effects of well geometrical parameters (depth, diameter of gravel packing, etc.); the effects of overlying caps and passive wells; and the effects of strata of differing permeabilities and of anisotropic permeabilities, of buried impermeable obstacles, of underlying nonaqueous phase liquid (NAPL), and of low-permeability lenses (Gannon et al., 1989; Wilson et al., 1989; Mutch and Wilson, 1990; Gómez-Lahoz et al., 1991; Mutch et al. 1989).

Major heterogeneities in the porous medium (lenses of high clay content, strata of low permeability, etc.) are of particular interest. Gómez-Lahoz et al. (1991) examined the effects of low permeability lenses on the rate of cleanup by means of buried screened horizontal laterals and developed strategies for reducing the damaging effects of such geological features. Mutch et al. (1989) have used models to interpret piezometer well data in terms of an overlying stratum of low permeability (which greatly extended the range of influence of a vacuum extraction well) and also to interpret the results of pilot vapor stripping operations at a site in Toms River, New Jersey. Vapor stripping can readily be carried out in soils underneath buildings, parking lots, and other impermeable overlying layers; this is modeled by including a no-flow boundary condition over the appropriate portion of the top boundary of the domain of interest (Gannon et al., 1989).

The effects of random variations in the pneumatic permeability of the porous medium in the vicinity of an SVE well on its performance were explored by Roberts and Wilson (1993) and by Bolick and Wilson (1994). Families of permeability functions having a given mean permeability, range, and correlation length and represented as Fourier series were used in calculating SVE cleanup times to get insight into the uncertainties these variations could be expected to produce in the cleanup times. Quite substantial variations in cleanup times resulted. It should be clearly understood that the data sets available for SVE modeling virtually never permit precise, deterministic modeling but only some sort of stochastic estimate of the cleanup time. Environmental engineers need to make it very clear to nontechnical people that the results of SVE modeling with most data sets are definitely not accurate to six decimal places and that the cost and time required to obtain a truly adequate data set for precise SVE modeling are generally prohibitive. Gómez-Lahoz et al. (1991) explored the effects of low-permeability clay lenses on SVE; their results indicate that one may expect substantial uncertainties in the modeling process. On the other hand, Bolick and Wilson (1994) found that even relatively large random uncertainties in initial VOC concentrations resulted in only minor variations in cleanup times. These workers also observed that the extent to which the VOC was allowed to spread in the vadose zone had spectacular effects on the time required for cleanup, suggesting that immediate emergency responses to VOC spills could result in large savings.

Modeling work by Wilson et al. (1989) and by Osejo and Wilson (1991) also indicates that soil vapor stripping should efficiently remove underlying floating light nonaqueous phase liquids. This is in agreement with experimental findings reported by Hoag and Cliff (1985).

A major uncertainty at present is the impact of diffusion and/or desorption kinetics on the rate of cleanup of a site by soil vapor stripping. As mentioned above, some sites can apparently by described by means of a local equilibrium model, while others require that mass transport kinetics limitations be included in the model. Wilson (1990b) has used a lumped parameter approach to develop a soil vapor stripping model for fractured porous bedrock, which includes kinetic effects, and Oma et al. (1990) and Gómez-Lahoz et al. (1993) have examined the impact of these kinetic effects on cleanup times and costs and the advantages to be gained by the use of intermittent or reduced airflow rates with diffusion-limited systems (Gómez-Lahoz et al., 1994b). Megehee and Wilson (1991) and Rodríguez-Maroto and Wilson (1991) have developed a steady-state approach to diffusion desorption–limited transport in soil vapor stripping, which greatly increases the speed with which these computations can be carried out.

More recently, this group has developed models for diffusion-limited and/or NAPL solution-limited SVE in which the lumped parameter simplification is avoided and a rather realistic distributed diffusion picture is used (Gómez-Lahoz et al., 1994c; Rodríguez-Maroto et al., 1994; Wilson et al., 1994a, 1994b). These models show the very rapid initial removal of VOCs, followed by a fairly rapid decrease in effluent soil gas VOC concentration, followed in turn by the prolonged periods of tailing, which are observed at many sites. The models indicate that short-term pilot studies lasting for only a few days cannot provide information about the time that will be required to achieve adequate remediation.

Osejo and Wilson (1991) have carried out lab studies and modeling of the removal of underlying nonaqueous phase liquid floating on the water table; their results indicate that such material can be readily removed by vapor stripping, provided that the wells are screened near the water table. They also measured the diffusivities of several organic solvents in sand and soil.

Gasoline and other petroleum products are usually complex mixtures of compounds having a substantial range of vapor pressures. Kayano and Wilson (1992) have described SVE models for lab column, vertical well, and buried horizontal slotted pipe operation in which nonaqueous phase liquid mixtures obeying Raoult's Law are stripped. The models make the local equilibrium assumption.

Virtually all SVE models assume the validity of Darcy's Law, that the gas flow is directly proportional to the pressure gradient. This is not true in regions in which the Reynolds number is comparable to or larger than unity. This point has been explored by Clarke et al. (1993), who deduced that the wellhead vacuum is a quadratic function of the molar gas flow rate. Experimental measurements on pilot-scale wells supported this. Equations were developed that permit one to use measurements made with pilot-scale systems at one wellhead vacuum to deduce the behavior of a full-scale system of different geometry and operating at a different wellhead vacuum. Models based on Darcy's Law can be used with impunity if the correct value of the molar gas flow rate is used. Attempts to calculate gas flow rates from wellhead vacuum and permeability may be seriously in error, however, and can result in calculated gas flow rates that are substantially larger than those achieved in practice.

Gómez-Lahoz et al. (1994d) developed an SVE model that permits one to explore the effects of changing the SVE well gas flow rate during the course of the remediation. The model allows one to turn the gas flow on and off or to vary it continuously. When the gas flow is turned off, one can follow the "bounce back" of the soil gas VOC concentration in the vicinity of the well as VOCs are gradually released into the vapor phase by diffusion and/or desorption. The model permits one to follow the sort of experiment described by DiGiulio et al. (1990) for the investigation of diffusion/desorption kinetics limitations. García-Herruzo et al. (1993) have also discussed pulsed soil vapor extraction, as well as some aspects of bioremediation. Gómez-Lahoz et al. (1993) and Oma et al. (1990) have discussed the substantial impact of such diffusion limitations on SVE costs. The effects of diffusion kinetics and variable permeabilities on SVE were also discussed recently by Rodríguez-Maroto et al. (1992).

Lingineni and Dhir (1992) have developed a multicomponent, nonisothermal model for the evaporation of VOCs in unsaturated soils during SVE. The model includes the temperature variations produced in the soil by the heat absorbed during evaporation of VOCs and/or condensation of water and also variations produced by provision of heat by, for example, rf heating. SVE column experiments were carried out and verified the assumptions in the model.

Rathfelder et al. (1991) have developed a mathematical model for SVE for use in the design and operation of field-scale venting systems. The model handles two-dimensional flow and transport equations for *n*-component contaminant mixtures and assumes local equilibrium between NAPL, aqueous, gas, and adsorbed phases with regard to VOC transport. Mass transfer rates in volatilization from the aqueous phase and in VOC desorption are potentially rate limiting. VOC vapor pressure and the magnitude and heterogeneity of the soil permeability are controlling variables. They note that proper design of an SVE system can result in quite significant savings.

Kuo et al. (1990) have published a two-dimensional model for estimating the effective radius of influence of an SVE well. Silka et al. (1991) have developed a simple mathematical model for SVE, which includes the effects of a diffusion transport-dominated region. In the model a high porosity, low–soil moisture, middle region is bounded by two low-porosity saturated regions, and a simple one-dimensional mass balance approach was carried out. The system shows an initial period of rapid VOC removal during which VOC is taken from the advective zone. This is followed by a much slower exponential decay in system VOC as the system becomes diffusion-dominated.

Coffin and Glasgow (1992) have developed a steady-state gas flow model and a vapor-phase mass transport model for the modeling of SVE performance, which can be used to optimize well placement and operating pressure, to determine if use of surface caps and/or air injection wells would be advantageous, etc. The model was assessed by comparison with laboratory and field data. Massman (1989) has noted that groundwater flow models can readily be adapted for use in modeling SVE operation; he used analytical groundwater flow models to interpret the results of a field-scale SVE test.

Brusseau (1991) has developed a one-dimensional model, which permits the handling of a structured or heterogeneous porous medium and which also models rate-limited desorption. The model was compared against SVE experimental data from the literature. Shan et al. (1992) have examined analytical solutions for steady-state gas flow to an SVE well in isotropic and anisotropic homogeneous porous media in which the ground surface is open to the atmosphere. Streamlines and streamline travel times were computed and provide a useful tool in SVE system design. Well screen location and porous medium anisotropy have very strong effects on streamline patterns; wells should be screened near the bottom of the vadose zone, and the effective radius of an SVE well in a medium having a permeability with a large anisotropy ratio ($K_{horizontal}/K_{vertical}$) is much larger than that of a well in a medium having an isotropic permeability. Wells screened near the soil surface exhibit short-circuiting, with most of the gas flow occurring in a region very near the well.

Bojd and Nanjundeswar (1992) note that the design of a remediation scheme frequently requires both pneumatic and hydraulic conductivities, which are are generally measured in separate experiments. They develop a relationship between these, which allows one to be calculated from the other, reducing project costs and time requirements.

Bentley and Travis (1992) have described the modeling of in situ biodegradation in unsaturated and saturated soils. Gómez-Lahoz et al. (1994a, 1994c, 1994d) have also published a model for in situ biodegradation during the course of SVE operations.

OFF-GAS TREATMENT

A significant problem in SVE is the removal of VOCs from the off-gas discharged. Pedersen and Curtis (1991a,b) provide a listing of current air discharge standards affecting SVE operations and point out that the states vary widely in their emission regulations, ranging from little or no regulation to contaminant-specific mass discharge rates. Sink (Pacific Environmental Services, Inc., 1989) has prepared a report for EPA on the various options available for VOC control from SVE facilities. The report gives a short summary of remediation techniques available for contaminated soil, followed by a listing of soil vapor extraction systems in the United States, followed by a detailed discussion of the treatment options available for VOC removal from the off-gas from SVE operations. In the past it has often been possible to vent directly to the atmosphere; now this is generally not possible, and VOC control adds substantially to the cost and complexity of SVE operations. At present there are four technologies available for VOC removal: activated carbon adsorption, thermal incineration, catalytic incineration, and condensation. Some characteristics of these are summarized in Table 1.4.

Carbon adsorption is well established for pollution control and solvent recovery. Although it can be used with very dilute VOC streams, its performance is better at concentrations above 700 ppmv (parts per million volume). Although units achieving 99% removal can be designed, actual efficiencies usually range from 60 to 90%, depending on VOC influent

TABLE 1.4. Key Emission Stream Characteristics for Selecting VOC Treatment Systems at SVE Sites.

| Vapor Treatment System | Emission Stream Characteristics | | | VOC Characteristics |
	VOC Concentration	Relative Humidity	Temp. Sensitivity	
Carbon absorption	>700 ppmv	<50%	Insensitive <150°F	VOC mol. wt. between 50–150 g/mol for best performance
Thermal incineration	>100 ppmv	—	Sensitive	Controls most VOCs without difficulty
Catalytic incineration	>100 ppmv	—	Sensitive	P, Bi, Pb, Hg, As, Sn, Zn, Fe oxides, halogenated compounds may poison catalyst
Condensation	>5,000 ppmv	—	Sensitive <200°F	Efficiency limited by vapor pres.-temp. characteristics of VOCs

concentration, temperature, relative humidity, and maintenance. De-humidification is usually necessary if the influent relative humidity exceeds 50% and the stream must be cooled if its temperature exceeds 150°F. The high relative humidity of SVE soil gas effluents is a serious problem if activated carbon is to be used.

Carbon adsorption handles variable VOC concentrations and gas flow rates better than the other three techniques and performs optimally with compounds having molecular weights in the range 50–150 gm/mole (compounds containing four to ten carbon atoms); such compounds are good candidates for SVE. Carbon adsorption has been the most commonly used VOC treatment technique in SVE; it is generally chosen unless the gas stream contains a very high concentration of combustible organics, in which case incineration may be cheaper.

Two methods for using activated carbon are employed for SVE work. The fixed bed regenerative system allows reuse of the carbon; the other system uses carbon canisters that cannot be reused. Fixed bed systems generally have higher capital and annualized costs. They are often used at sites where the expected cleanup time is fairly long, so that the recurring costs of canister replacement exceeds the capital costs of the regenerative system. The costs of the regenerative system are markedly reduced if steam is readily available at the site.

Thermal incineration is able to control a broader range of compounds than the other techniques; it gives efficiencies >99% for concentrations above 200 ppmv and efficiencies >95% at VOC concentrations as low as 50 ppmv. (Carbon adsorption is optimal at VOC concentrations above 700 ppmv.) Thermal incinerators are not well suited to streams with variable flow rates (fluctuations exceeding 10%), since these change mixing and residence times from design values and reduce efficiency. The vapor feed stream must be heated to 1,500–1,800°F for 1–2 sec to oxidize hydrocarbon vapors.

Thermal incineration can be used for streams that are dilute in VOCs, as is the case with most SVE operations during much of their lifetime. Supplemental fuel is generally required, which increases operating costs relative to the other VOC treatment techniques to the point where thermal incineration is rarely used at SVE installations. It is an economical option if little or no supplemental fuel is needed.

Catalytic incinerators are similar to thermal incinerators, except that they use a catalyst to allow the combustion reactions to take place at reduced temperatures, which reduces or eliminates the need for supplemental fuel. Design efficiencies are around 95%, with actual efficiencies running about 90%, depending on operating conditions and maintenance.

Catalytic incinerators are more sensitive to pollutant characteristics and process conditions than are thermal incinerators. Halogen-containing compounds (chlorinated organics, for example), lead, mercury, bismuth, tin, zinc, iron oxides, and phosphorus may poison the catalyst, ruining its performance. Some newer catalysts are able to tolerate higher concentrations of these substances, and it is unlikely that nonvolatile metal compounds will be extracted in SVE operations. Chlorinated organic solvents may present a problem, although Trowbridge and Malot (1990) have described a system that uses a chromium oxide–aluminum oxide catalyst; this system can accept a feed stream the VOCs of which are 100% chlorinated hydrocarbons. The vapor feed stream inlet temperature must be 500–700°F; often, this preheating can be done by means of a heat exchanger. Catalytic incineration can achieve high destruction efficiencies, even at low VOC concentrations, but the technique is sensitive to influent flow rate fluctuations. It is not as commonly used at present as activated carbon, but the development of halogen-resistant catalysts will certainly result in its more widespread use for SVE work. Operating costs are generally lower than those for thermal incineration.

Condensers can be used as preliminary devices for recovering VOCs present in the feed stream at high concentrations; the stream is then sent to another unit for further treatment. Their requirement of inlet VOC concentrations >5,000 ppmv limits their utility in SVE work, since VOC off-gas concentrations at an SVE site are usually quite high initially but drop off relatively quickly. Removal efficiencies of condensers using chilled water are typically about 50 to 80%; values approaching 90% can be achieved if ethylene glycol or freon is used as the refrigerant. (This increases costs significantly.) Sink did not find condensers in use at any of the twenty-nine SVE sites he reviewed.

One interesting off-gas treatment involves the use of soil bioreactors. Miller and Canter (1991) showed that 10–20 min contact times of off-gas in soil columns degraded 8–39% of BETX compounds; the kinetics were first order, with some mass transfer inhibition at long residence times.

Moss (1989) discusses the advantages and disadvantages of several off-gas treatment systems (activated carbon, internal combustion engine, combustion with supplemental fuel, and direct discharge) and gives several case studies. Buck and Smith (1990) examine several off-gas treatment options and conclude that thermal catalytic oxidation, with supplemental fuel if necessary, is one of the better techniques. Buck and Seider (1992), more recently, reviewed the various off-gas treatment techniques, noting that new catalysts have become available which permit the rapid and complete oxidation of chlorinated organics, a most gratifying development. This process was described in detail by Lester (1992); it is capable of destroying C_1 and C_2 chlorohydrocarbons and chlorofluorohydrocarbons and polychlorinated aromatics. Destruction of carbon tetrachloride is better than 99% in moist air at 375°C, and the catalysts have demonstrated stable operation for more than 1,600 hr.

SOIL GAS SURVEYS

The costs of extensive soil sampling and analyses for VOCs can be quite substantial. A number of workers have proposed the use of soil gas surveys as a means to reduce the amount of drilling, soil sampling, and soil analyses needed to characterize a site. Mercer and Spalding (1991) have discussed the use of soil gas analysis for locating areas contaminated by VOCs in the vadose zone. The technique requires the drilling of a shallow hole or the insertion of a sampling tube into the soil; soil gas samples are then withdrawn for collection and analysis in the lab or for on-site analysis by an organic vapor analyzer. Soil gas analysis depends upon the soil being adequately permeable to air and is less reliable in clays and other media of low permeability. It will not detect nonvolatile organics and inorganics (see Table 1.5).

Marrin and Kerfoot (1988) have suggested the use of soil vapor monitoring wells to detect plumes of VOC-contaminated groundwater, and they and Marrin and Thompson (1987) have had some success in this. Mercer and Spalding (1991) list a number of soil gas survey case studies. They note that, if the source of the VOCs is below the water table, the maximum concentration of VOCs in the vadose zone will be in the capillary fringe. Contaminants, on reaching the top of the capillary fringe, will diffuse very rapidly due to the large gas-phase diffusivities in the vadose zone. This will deplete the capillary fringe of VOC, which will be replenished only by very slow aqueous-phase diffusion of VOC from the underlying groundwater. VOC is lost, of course, at the air-soil boundary at the top of the vadose zone, and there may also be biodegradation of VOC in the vadose zone. Under these circumstances, one therfore expects that soil gas VOC concentrations in the vadose zone will be very low and will be most readily detected in samples taken just above the capillary fringe.

On the other hand, if there is residual NAPL in the vadose zone or free product floating on the water table, then mass transfer to the vapor phase is not limited by diffusion of VOC through a thick layer of water, and VOC soil gas concentrations should be much higher. Thus, one expects that the greatest use of shallow soil gas surveys is for locating residues of NAPL in the vadose zone or floating on the water table. The highest gas phase VOC concentrations are expected to be those closest to any residual product. Such surveys would therefore be helpful guides in determining optimal location of SVE wells.

Kerfoot (1991) notes that soil gas surveys have been useful in the delinea-

TABLE 1.5. Contaminant Characteristics in Relation to Soil Gas Surveys.[a]

Group/Contaminants	Applicability of Soil Survey Techniques
Group A: halogenated methanes, ethanes, ethenes	
$CHCl_3$, C_2H_3Cl, CCl_4, CCl_3F, CH_3CCl_3, C_2HCl_3, Ch_2BrCH_2Br	Detectable in soil gas over a wide range of environmental conditions. DNAPLs will sink in aquifer if present as liquid product.
Group B: halogenated propanes, propenes, benzenes	
C_6H_5Cl, $C_6H_4Cl_2$'s, $C_6H_3Cl_3$'s, $CH_2ClCHClCH_3$	Limited value; detectable by soil gas surveys only where probes can sample near contaminated soil or groundwater. DNAPL.
Group C: halogenated polycyclic aromatics	
Aldrin, endrin, DDT, chlordane, heptachlor, PCBs	Do not partition into the gas phase adequately to be detected in soil gas under normal circumstances. DNAPL.
Group D: C_1–C_8 petroleum hydrocarbons	
Benzene, toluene, xylenes, C_1 to C_8 alkanes, cyclohexane, gasoline, JP-4, etc.	Most predictably detected in shallow aquifers or leaking underground storage tanks where probes can be driven near the source of contamination. LNAPLs float as thin film on the water table. Can act as a solvent for DNAPLs, keeping them nearer the ground surface.
Group E: C_9–C_{12} petroleum hydrocarbons	
Trimethylbenzene, naphthalene, decane, diesel and jet A fuels; biphenyl	Limited value; detectable by soil gas surveys only where probes can sample near contaminated soil or groundwater. Mainly LNAPLs.
Group F: polycyclic aromatic hydrocarbons	
Anthracene, benzopyrene, fluoranthene, chrysene, motor oils, coal tars	Do not partition adequately into the gas phase to be detected in soil gas under normal circumstances. DNAPLs.
Group G: low molecular weight oxygenated compounds	
Acetone, ethanol, formaldehyde, methylethylketone	LNAPLs, but dissolve readily in groundwater. May be detected in soil gas if they result from a leak or spill in relatively dry soil.

[a]From Mercer and Spalding (1991).

tion of subsurface VOC contamination and in the design of SVE systems. He notes that gas samples taken under static conditions can be expected to yield quite different results than would be obtained from samples taken during or immediately after SVE well operation, due to diffusion/desorption kinetics limitations, which result in a slow VOC concentration rebound after well shut-down if cleanup is not complete. He describes use of a soil gas survey to design an SVE system in California.

Reisinger et al. (1989) note that soil gas surveys have typically been used in preliminary screening and for monitoring well placement, but that the data so obtained can be used for other purposes as well. They describe active sampling (with vacuum probes) and passive sampling (with traps in flux chambers). In many instances when a rapid response is needed, data from a soil gas survey can form the basis for the design and implementation of the remediation system.

Markley (1988) found soil gas surveys adequate for remediation system design and very much cheaper than collection of soil and groundwater samples by borehole drilling. Henkle et al. (1990) described the use of soil gas surveys in delineating groundwater contaminant plumes at three California sites contaminated with VOCs. This preliminary estimate was then the basis for the siting of monitoring wells. Crouch (1990) has presented a procedure, based on soil gas survey data, for compound-specific mapping of soil contamination, nonaqueous phase liquid product, and severely contaminated groundwater. Gander and Wood (1989) carried out a soil gas survey with a gas chromatograph to ascertain the regions of highest hydrocarbon concentration; this served as the basis for siting borings.

Batterman et al. (1992) have described the design and evaluation of a long-term soil gas flux sampler. Results obtained with the device compared favorably with emission isolation flux chambers; the device is simpler to use and more cost-effective than active samplers. Heins et al. (1992) have pointed out the importance of adequate quality assurance in soil gas surveys, noting a problem they had with false positive acetone analyses.

Walter and Bentley (1992) recently described the use of soil gas measurements in the design of SVE systems, and Siegrist (1992) noted some of the pitfalls involved in soil sampling and VOC analysis.

SVE SITES

An early paper on SVE is an American Petroleum Institute (API) report on a field demonstration of subsurface venting of hydrocarbon vapors from an underground aquifer. Crow et al. (1985) describe data from soil vacuum extraction experiments carried out at a petroleum fuels terminal at which a gasoline spill had taken place. They conclude that the technique removes hydrocarbon vapors from the vadose zone effectively and is also useful in augmenting conventional methods for recovering spilled hydrocarbons from shallow aquifers.

Woodward-Clyde Consultants (1985) report on a pilot soil vacuum extraction study near Tacoma, Washington, and provide data that strongly support their conclusion that vacuum extraction is effective in remediating the site. Roy F. Weston, Inc., report the results of a pilot study of soil vapor stripping for removal of TCE and other VOCs from contaminated sandy soil at the Twin Cities Army Ammunition Plant, Minnesota. They found the technology effective, but TCE removal from soils containing oily deposits was diminished. They note that little was known about adsorption isotherms of contaminants at very low concentrations. They suggest high airflow rates and close spacing of venting wells in domains that are heavily contaminated and note the importance of identifying these domains in the site assessment (Anastos et al., 1985).

Batchelder et al. (1986) describe the use of soil ventilation (synonymous with SVE) for the removal of hydrocarbons from the subsurface and describe the remediation of several former service stations. They include a discussion of a procedure for siting the vacuum wells. Zenobia et al. (1987) describe SVE remediation of soil contaminated by a leaking underground gasolite tank. Instrumentation was unusually complete, permitting estimation of the amount of VOC removed. A mixed SVE-bioremediation cleanup of a site in Vermont contaminated with gasoline is described by Harper and Williams (1987); oxygen and nutrients were provided by means of a reinfiltration gallery.

Connor and Noonan (1988) noted that there are roughly 1.4 million underground storage tanks in the United States, of which between 10 and 30% are thought to have leaked gasoline. He describes the application of SVE to the remediation of a former service station site in Massachusetts. Knieper (1988) describes an SVE operation that remediated a building that was contaminated with gasoline vapor from a nearby underground storage tank. Robitaille and Walen (1988) have described SVE operations at a site in New Jersey and one in Massachusetts, noting that hydrocarbon levels in the underlying groundwater were being substantially reduced. Fawcett (1989) provides a description of an SVE/pump-and-treat facility for the remediation of soil and a shallow groundwater zone that were contaminated with gasoline. The extent of the plume was determined by a soil gas survey.

Johnson et al. (1991) studied the processes controlling the effectiveness of SVE using results from a field study at a service station in California. The water table was at a depth of 60 ft, and there was free floating product. Air permeability measurements indicated good agreement between experiment and model calculations, and predicted and measured contaminant vapor concentrations were also in good agreement. Groundwater extraction was also employed at the site, and dual vapor extraction/groundwater recovery wells were installed after some initial difficulties with airflow rates.

Hill Air Force Base, Utah, was the site of a 27,000 gal JP-4 jet fuel spill, described by Elliott and DePaoli (1989). The highly permeable soil permitted a high gas flow rate (250 cfm) with a vacuum of only 20 in. of water. Initial removal rates were quite rapid. The preliminary results were used to design full-scale systems (DePaoli et al. 1991): a vertical well array at the spill site, a lateral vent system under the area occupied by the new tanks, and a lateral vent system to treat an excavated pile of contaminated soil. Initial removal rates with the full-scale systems were rapid. Downey and Elliott (1990) examined the performance of selected in situ soil remediation technologies employed at Kelly, Elgin, and Hill Air Force Bases. Problems with enhanced biodegradation resulted from low-permeability soils. Radio-frequency (rf) heating appeared promising; further work combining SVE and rf heating and optimizing biodegradation during SVE is planned.

Burgdorf and Lambert (1989) discuss the use of SVE for the removal of VOCs from porous and fractured media at several sites in Pennsylvania. VOC was removed from overlying soil and from unsaturated and dewatered fracture zones and bedding planes in clastic sedimentary bedrock.

The use of horizontal drilling techniques at the Savannah River Laboratory, in connection with SVE and with in situ air stripping of groundwater, has been described by Kaback and Looney (1988). Plans call for two horizontal wells, one below the water table and one in the vadose zone, to test the combination of SVE and in situ air stripping for the removal of VOCs from both soil and groundwater.

Converse Environmental Consultants has presented a description of a vapor extraction system at a Burbank, California, site in which the vented hydrocarbons were destroyed by combustion or catalytic oxidation (Fall et al., 1988). Gasoline had penetrated to depths of 50 ft below grade. The soil consisted of alluvial sand interspersed with silt and clay. Work at this site demonstrated diffusion (or desorption) kinetics limitation; intermittent operation of the facility was used to maintain combustible concentrations of hydrocarbons in the vented gas.

Towers et al. (1989) have presented an analysis of the process of choosing a treatment technique for soil contaminated with volatile halogenated organics at a site in Indiana. Groundwater remediation was necessary at the site. Soil treatment methods considered included excavation and off-site disposal, in situ biodegradation, chemical degradation, soil washing, soil vapor stripping (soil vacuum extraction, SVE), and combined technologies. SVE was substantially cheaper than the other options.

Terra Vac has carried out vapor stripping operations at a number of sites, and descriptions of some of these are available. One of these involved a gasoline leak in Belleview, Florida, which had resulted in contamination of a municipal wellfield. A preliminary report on the vapor stripping operation indicated removal rates averaging about 880 lb/day; removal of this material avoids futher long drawn-out contamination of the underlying aquifer during pump-and-treat operations (Applegate et al., 1987).

The Groveland, Massachusetts, site was used by Terra Vac as a demonstration site; the vadose zone here was contaminated with TCE. An eight-week test run resulted in the removal of 1,300 lb of VOCs and a marked reduction in soil VOC concentrations in the test area. The author of the evaluating report felt that the results of the test were promising and indicated that the process can remove VOCs from soils of both high and low permeability (Michaels, 1989; Stinson, 1989).

Recently, Trowbridge and Malot (1990) have described three SVE sites at which VOC control was accomplished by catalytic incineration. The catalytic oxidizer removed 98.5% or better of the VOCs, which included a number of chlorinated solvents. These authors felt that catalytic oxidation is economical for the majority of SVE systems operating at >25–50 lb/day VOCs and <50 mg/L VOC concentration for most of the operation. These authors note the problem presented by diffusion-limited SVE. They report that SVE works very satisfactorily for removing light nonaqueous phase liquid (LNAPL) floating on top of the water table. The early stages of an SVE operation, when nonaqueous phase liquid product is present, tend to be controlled by Raoult's Law; after this has been removed, an effective Henry's Law becomes controlling.

The third site described by Trowbridge and Malot (1990) is the Thomas Solvents (or Verona Wellfield) site in Battle Creek, Michigan; this has been discussed in more detail by Malmanis et al. (1989) and by Danko et al. (1990). Groundwater extraction (pump-and-treat), started in March 1987, has removed more than 11,000 lb of VOCs, and total groundwater VOC concentrations have dropped from as high as 19,000 μg/L to about 1,500 μg/L as of October 1989. The SVE system included twenty-three wells, a collection manifold, a centrifugal air/water separator, a blower, and an off-gas treatment unit. Initially, activated carbon (nonregenerative) was used; this was later replaced by a catalytic oxidation unit. In this last, the air is preheated to about 430°F in a heat exchanger, after which it enters a burner chamber where it is heated to about 800°F by a natural gas burner. It then passes through the catalyst bed, where oxidation takes place, including the conversion of chlorinated VOCs to carbon dioxide, water vapor, and hydrogen chloride. The gas leaves the catalyst bed at about 820°F and enters the shell side of the heat exchanger, where it is cooled to about 550°F before being discharged through a stack. Advantages of the catalytic incinerator include no down time for carbon changing, destruction of the VOCs on-site, and substantial anticipated cost reductions. The combination of SVE with another technique (in this case groundwater pump-and-treat) will doubtless become more and more common as more complex and difficult sites are addressed.

Fiedler and Shevenell (1990) have used a case study of a cleanup at a gas station by SVE to illustrate estimation of the time required and the cost of an SVE operation. They point out the importance of maintaining high air flow rates through domains of high VOC concentration. One might question their claim that SVE may be effective in remediating contaminated groundwater; diffusion constants in liquids are typically about 10^{-4} times as large as diffusion constants in the gaseous phase, so diffusive mass transport from below the water table into the vadose zone is vanishingly small. The SVE system described in the case study used on-site steam regeneration of activated carbon for VOC control; this was quite satisfactory. These authors use plots of log soil gas VOC concentration versus time to follow the progress of cleanup. This practice, which has been used by others as well, has been trenchantly criticized by DiGiulio et al. (1990); their criticism of this overly simplistic procedure, which fails to take into account diffusion/desorption kinetics limitations and the heterogeneous character of the sorption sites, has a firm basis in fact.

Pope (1988) has given a summary of SVE technology, together with two brief case histories of SVE remediations at service stations. He notes that the long-term operational costs of VOC control with activated carbon are quite high; those of catalytic incineration can be 25–50% less. In one of these cases abatement of explosion hazards in nearby buildings occurred within three weeks after startup of the SVE operation.

Patterson (1989) has described a soil venting pilot study to remove benzene, toluene, ethylbenzene, and xylene from contaminated soil at a construction site and the design of a soil venting remediation facility, which includes an underslab venting system to remove VOCs from below the building to be constructed at the site. Here SVE is combined with groundwater pump-and-treat.

Bailey and Gervin (1988) have provided a rather detailed description of a pilot study on the use of vapor stripping for the removal of 1,1,1-trichloroethane (TCA) and trichloroethylene (TCE) from the vadose zone at a Michigan site. A very substantial quantity of VOC was recovered during the pilot-scale operation. Mutch et al. (1989) have described a pilot SVE study at Toms River, New Jersey, in which a model for gas flow in the vicinity of a vapor extraction well was used to predict soil gas pressures around the well in the presence of an overlying stratum of low-permeability clay fill. A local equilibrium model for vapor stripping was then used to interpret the results of the pilot vapor stripping runs.

A number of Superfund Records of Decision (RODs) have included SVE as a remediation technique. In the future data that will become available from these sites should give considerable insight into the use of this method. The South Valley, New Mexico, site [Superfund ROD: South Valley (PL-83), NM, 1988] involves both VOCs [including tetrachloroethylene (PCE)] and metals; remediation includes both SVE and pump-and-treat. An IBM site in San Jose, California [Superfund ROD: IBM (San Jose), CA, 1988], involves trichloroethane, toluene, xylenes, and other organics. The groundwater is extensively involved, and a complex remediation scheme involving SVE is planned. The Hastings, Nebraska, Colorado Avenue subsite of the Hastings Ground Water Contamination site is extensively contaminated with TCE and PCE (Superfund ROD: Hastings Ground Water/Colorado Avenue, NE, 1988). SVE and groundwater pump-and-treat are planned. The Hastings, Nebraska, FAR-MAR-CO subsite of the Hastings Ground Water Contamination site is contaminated with carbon tetrachloride and ethylene dibromide (Superfund ROD: Hastings Ground Water/Far-Mar, NE, 1988). SVE is planned, and the groundwater will be studied to see if further action is required. The Upjohn Manufacturing Co. site in Puerto Rico is contaminated with carbon tetrachloride and acrylonitrile (Superfund ROD: Upjohn Manufacturing Co., PR, 1988). SVE and pump-and-treat had been used for remediation at this site, and in 1987 EPA asked that remedial activities be continued. Dichloroethane, trichloroethylene, and tetrachloroethylene, presumably from a dry-cleaning establishment, have contaminated water and soil at the Long Prairie, Minnesota, site (Superfund ROD: Long Prairie, MN, 1988); treatment includes SVE and groundwater pump-and-treat.

Ball and Wolf (1990) describe the design of a remediation system for removing PCE from soil at a midwestern federal facility; SVE was used. A model was used to scale up from the laboratory data to predict emission rates for the full-scale SVE system; observed PCE concentrations in monitoring wells were in good agreement with model predictions.

The optimization of the SVE process by means of large physical model studies is described by Johnson (1992). DePaoli and Hutzler (1992) discuss a field test of SVE enhancement by the use of heated air from the system's catalytic oxidation unit; this approach may be economically advantageous if the heat can be more evenly distributed throughout the contaminated soil zone. Hiller (1992) describes the performance characteristics of SVE systems in use in Europe; more than 1,000 systems are in operation in Germany, and SVE is considered to be a standard procedure. Hiller notes that intermittent air flow is often used during the latter portion of a cleanup.

REFERENCES

American Petroleum Institute, 1980, *Examination of Venting for Removal of Gasoline Vapors from Contaminated Soil*, API Publ. No. 4429.

American Petroleum Institute, 1982, *Enhancing the Microbial Degradation of Underground Gasoline by Increasing Available Oxygen*, API Publ. No. 4428.

Anastos, G. J., P. J. Marks, M. H. Corbin, and M. F. Coia, 1985, "In situ Air Stripping of Soils Pilot Study," submitted by Roy F. Weston, Inc., to U.S. Army Toxic and Hazardous Materials Agency, Aberdeen Proving Ground, MD, Oct., Report AMXTH-TE-TR-85026.

Applegate, J., J. K. Genry, and J. J. Malot, 1987, "Vacuum Extraction of Hydrocarbons from Subsurface Soils at a Gasoline Contamination Site," *Proc., 8th Natl. Conf. Superfund*, Washington, D.C., p. 273.

Arthur, M. F., T. C. Zwick, G. K. O'Brien, and S. S. Marsh, 1991, "Evaluation of Aeration Methods to Bioremediate Fuel-Contaminated Soils," in *Innovative Hazardous Waste Treatment Technology Series. Vol. 3: Biological Processes*, Technomic Publishing Co., Inc., Lancaster, PA, p. 185.

Baehr, A. L., G. E. Hoag, and M. C. Marley, 1989, "Removing Volatile Contaminants from the Unsaturated Zone by Inducing Advective Air-Phase Transport," *J. Contam. Hydrology*, 4:1.

Bailey, R. E. and D. Gervin, 1988, "In situ Vapor Stripping of Contaminated Soils: A Pilot Study," *Proc., 1st Ann. Hazardous Materials Management Conf./Central*, Rosemont, IL, March 15–17, p. 207.

Ball, R. and S. Wolf, 1990, "Design Considerations for Soil Cleanup by Soil Vapor Extraction," *Environmental Progress*, 9:187.

Batchelder, G. V., W. A. Panzeri, and H. T. Phillips, 1986, "Soil Ventilation for the Removal of Adsorbed Liquid Hydrocarbons in the Subsurface," *Proc., NWWA/API Conf. on Petroleum Hydrocarbons and Organic Chemicals in Ground Water: Prevention, Detection and Restoration*, Nov. 12–14, Houston, TX, National Water Well Association, Dublin, OH, p. 672.

Batterman, S. A., B. C. McQuown, P. N. Murthy, and A. R. McFarland, 1992, "Design and Evaluation of a Long-Term Soil Gas Flux Sampler," *Environmental Science and Technology*, 26:709.

Bentley, H. W. and B. Travis, 1992, "Modeling in situ Biodegradation in Unsaturated and Saturated Soils," *Proc., Symposium on Soil Venting*, April 29–May 1, 1991, Houston, TX, D. DiGiulio, project officer, U.S. EPA Report No. EPA/600/R-92/174, p. 141.

Bojd, H. G. and B. V. Nanjundeswar, 1992, "Analysis of Soil-Air Permeability and Saturated Hydraulic Conductivity for Remedial System Design," *National Conf. on Environmental Engineering Water Forum '92*, Aug. 2–6, American Soc. of Civil Engineers, Baltimore, MD.

Bolick, J. J. and D. J. Wilson, 1994, "Soil Cleanup by in situ Aeration. XIV. Effects of Random Permeability Variations on Soil Vapor Extraction Cleanup Times," *Separ. Sci. Technol.*, 29:701.

Bouchard, D. C., 1989, "The Role of Sorption in Contaminant Transport," *Workshop on Soil Vacuum Extraction*, April 27–28, U.S. EPA, RSKERL, Ada OK.

Bouchard, D. C., S. C. Mravik, and G. B. Smith, 1990, "Benzene and Naphthalene Sorption on Soil Contaminated with High Molecular Weight Residual Hydrocarbons from Unleaded Gasoline," *Chemosphere*, 21:975.

Bouchard, D. C., A. L. Wood, M. L. Campbell, P. Nkedi-Kizza, and P. C. S. Rao, 1988, "Sorption Nonequilibrium during Solute Transport," *J. Contam. Hydrology*, 2:209.

Boulding, J. R. and M. J. Barcelona, 1991, "Geochemical Variability of the Natural and and Contaminated Subsurface Environment," in *Seminar Publication: Site Characterization for Subsurface Remediation*, U.S. EPA Report No. EPA/625/4-91/026, Nov., p. 103.

Brookman, G. T., M. Flanagan, and J. O. Kebe, 1985, *Literature Survery: Hydrocarbon Solubilities and Attenuation Mechanisms*, American Petroleum Institute, API Publ. No. 4414, August.

Brown, R. A., G. E. Hoag, and R. D. Norris, 1987, "The Remediation Game: Pump, Dig, or Treat," *Water Pollution Control Federation Conf.*, Oct. 5–8.

Brusseau, M. L., 1991, "Transport of Organic Chemicals by Gas Advection in Structured or Heterogeneous Porous Media: Development of a Model and Application to Column Experiments," *Water Resources Research,* 27:3189.

Buck, F. A. M. and E. L. Seider, 1992, "Commercial Vapor Treatment Process," *Proc., Symposium on Soil Venting,* April 29–May 1, 1991, Houston, TX, D. DiGiulio, project officer, U.S. EPA Report No. EPA/600/R-92/174, p. 271.

Buck, F. A. and C. A. Smith, 1990, "Control of Air Emissions from Soil Venting Systems," in *Innovative Hazardous Waste Treatment Technology Series. Vol. 2: Physical/ Chemical Processes,* Technomic Publishing Co., Inc., Lancaster, PA, p. 205.

Burgdorf, G. J. and R. S. Lambert, 1989, "Removal of Volatile Organics from Porous and Fractured Media by Vacuum Extraction Techniques," *Proc. 3rd National Outdoor Action Conf. on Aquifer Restoration, Ground Water Monitoring and Geophysical Methods,* National Water Well Association, Dublin, OH, p. 575.

Clarke, A. N. and D. J. Wilson, 1988, "A Phased Approach to the Development of in situ Vapor Stripping Treatment," *Proc., 1st Ann. Hazardous Materials Management Conf./Central,* Rosemont, IL, March 15–17, p. 191.

Clarke, A. N., M. M. Megehee, and D. J. Wilson, 1993, "Soil Cleanup by in situ Aeration. XII. Effect of Departures from Darcy's Law on Soil Vapor Extraction," *Separ. Sci. Technol.,* 28:1671.

Clarke, A. N., R. D. Mutch, and D. J. Wilson, 1992, "Final Results of a Year-Long USEPA SBIRP/Industry-Funded in situ Vapor Stripping Pilot Scale Study," in *Proc., Symposium on Soil Venting,* April 29–May 1, 1991, Houston, TX, D. DiGiulio, project officer, p. 229.

Clarke, A. N., R. D. Mutch, P. D. Mutch, and D. J. Wilson, 1990, "In situ Vapor Stripping: Results of a Year-Long Pilot Study," *Hazardous Materials Control,* 3(6):25.

Clarke, J. H., A. N. Clarke, and D. J. Wilson, 1988, "Applications of in situ Vapor Stripping Technology to Contaminated Soil," *Proc., 1st Ann. Hazardous Materials Management Conf./Central,* March 15–17., Tower Conference Management Co., Rosemont, IL, p. 379.

Coffin, D. and L. Glasgow, 1992, "Effective Gas Flow Arrangements in Soil Venting," *Water, Air and Soil Pollution,* 62:303.

Connor, J. R. and D. C. Noonan, 1988, "Case Study of Soil Venting," *Pollution Engineering,* 20(7):74.

Crouch, M. S., 1990, "Check Soil Contamination Easily," *Chemical Engineering Progress,* 86(9):41.

Crow, W. L., E. P. Anderson, and E. M. Minugh, 1985a, *Subsurface Venting of Hydrocarbon Vapors from an Underground Aquifer,* American Petroleum Institute, API Publ. No. 4410, submitted by Riedel Environmental Services and Radian Corp.

Crow, W. L., E. P. Anderson, and E. M. Minugh, 1985b, "Subsurface Venting of Vapors Emanating from Hydrocarbon Product on Ground Water," in *Petroleum Hydrocarbons and Organic Chemicals in Ground Water—Prevention, Detection and Restoration—A Conference and Exposition, Proc., NWWA/API Conference,* Nov. 13–15, Houston, TX, p. 536.

Curtis, J. T. and D. C. Noonan, 1987, "Corrective Action Technologies and Their Cost-Effectiveness," *Proc., 4th Ann. Eastern Regional Ground Water Conf.,* July 14–16, Burlington, VT, National Water Well Association, Dublin, OH, p. 217.

Dalfonso, T. J. and M. S. Navetta, 1988, "In situ Treatment of Contaminated Soils Using Vacuum Extraction," *DOE Model Conf. Abstracts,* Oct. 3–7, Oak Ridge, TN, p. 59 (Abstract only).

Danko, J., 1989, "Applicability and Limitations of Soil Vapor Extraction for Sites Contaminated with Volatile Organic Compounds," *Soil Vapor Extraction Technology Workshop,* U.S. EPA RREL, Edison, NJ, June 28–29, 1989.

Danko, J., 1990, "Soil Vapor Extraction Applicability and Limitations," *Proc. 5th Ann. Hazardous Materials Management Conf./West,* Tower Conference Management Co., 800 Roosevelt Rd., Bldg. E, Suite 408, Glen Ellyn, IL 60137-5835.

Danko, J., 1991, "Applicability and Limits of Soil Vapor Extraction for Sites Contaminated with Volatile Organic Compounds," in *Soil Vapor Extraction Technology, Pollution Technology Review No. 204,* Noyes Publications, Park Ridge, NJ, p. 163.

Danko, J. P., M. J. McCann, and W. D. Byers, 1990, "Soil Vapor Extraction and Treatment of VOCs at a Superfund Site in Michigan," CH2M Hill report, Corvallis, OR.

Davies, S. H., 1989, "The Influence of Soil Characteristics on the Sorption of Organic Vapors," *Workshop on Soil Vacuum Extraction,* U.S. EPA, RSKERL, April 27–28, Ada, OK.

DePaoli, D. W. and J. J. Hutzler, 1992, "Field Test of Enhancement of Soil Venting by Heating," *Proc., Symposium on Soil Venting,* April 29–May 1, 1991, Houston, TX, D. DiGiulio, project officer, U.S. EPA Report No. EPA/600/R-92/174, p. 173.

DePaoli, D. W., S. E. Herbes, and M. G. Elliott, 1991, "Performance of in situ Soil Venting Systems at Jet Fuel Spill Sites," in *Soil Vapor Extraction Technology, Pollution Technology Review No. 204,* Noyes Publications, Park Ridge, NJ, p. 260.

DiGiulio, D. C., 1992, *Evaluation of Soil Venting Application,* U.S. EPA Report No. EPA/540/S-92/004, April, 1992.

DiGiulio, D. C., J. S. Cho, R. R. Dupont, and M. W. Kemblowski, 1990, "Conducting Field Tests for Evaluation of Soil Vacuum Extraction Application," *Proc., 4th Natl. Outdoor Action Conf. on Aquifer Restoration, Ground Water Monitoring and Geophysical Methods,* Las Vegas, NV, May 14–17, p. 587.

Downey, D. C. and M. G. Elliott, 1990, "Performance of Selected in situ Soil Decontamination Technologies: An Air Force Perspective," *Environmental Progress,* 9(3):169.

Dupont, R. R., 1992, "Soil, Site and Waste Characteristics Controlling the Volatilization of Organic Contaminants in the Vadose Zone," *Proc., Symposium on Soil Venting,* April 29–May 1, 1991, Houston, TX, D. DiGiulio, project officer, U.S. EPA Report No. EPA/600/R-92/174, p. 11.

Elliott, M. G. and D. W. DePaoli, 1989, "In situ Venting of Jet-Fuel-Contaminated Soil," in *Proc., 44th Purdue Industrial Waste Conf.,* May 9–11, Purdue Univ., West Lafayette, IN, CRC Press, Inc., Boca Raton, FL, p. 1.

EPA, 1988a, "Superfund Record of Decision: Long Prairie, MN," EPA Office of Emergency and Remedial Response, Report No. EPA/ROD/RO5-88/066, June. Available from NTIS, PB89-135339.

EPA, 1988b, "Superfund Record of Decision: Upjohn Manufacturing Co., PR," Office of Emergency and Remedial Response, Report No. EPA/ROD/R02-88/071, Sept. Available from NTIS, PB89-207262.

EPA, 1988c, "Superfund Record of Decision: Hastings Ground Water/Far-Mar, NE," EPA Office of Emergency and Remedial Response, Report No. EPA/ROD-88/017, Sept. Available from NTIS, PB89-182463.

EPA, 1988d, "Superfund Record of Decision: Hastings Ground Water/Colorado Ave., NE," EPA Office of Emergency and Remedial Response, Report No. EPA/ROD/R07-88/018, Sept. Available from NTIS, PB89-182471.

EPA, 1988e, "Superfund Record of Decision: South Valley (PL-83), NM," EPA Office of Emergency and Remedial Response, Report No. EPA/ROD/R06-88/043, Sept. Available from NTIS, PB89-204812.

EPA, 1988f, "Superfund Record of Decision: IBM (San Jose), CA," EPA Office of Emergency and Remedial Response, Report No. EPA/ROD/R09-89/029, Dec. Available from NTIS, PB90-108481.

EPA, 1991a, *Seminar Publication: Site Characterization for Subsurface Remediation,* EPA Report No. EPA/625/4-91/026.

EPA, 1991b, *Superfund Innovative Technology Evaluation (SITE) Program,* Report No. EPA/540/8-91/005.

EPA, 1992, *Proceedings of the Symposium on Soil Venting,* April 29–May 1, 1991, Houston, TX, Report No. EPA/600/R-92/174.

EPA, 1993, *Cleaning Up the Nation's Waste Sites: Markets and Technology Trends,* Report No. EPA/542/R-92/012.

Fall, E. W., et al., 1988, "In-situ Hydrocarbon Extraction: A Case Study," *Southwestern Ground Water Focus Conference,* Albuquerque, NM, March 23–25, 1988; see also *The Hazardous Waste Consultant,* Jan./Feb. 1989. p. 1-1.

Fawcett, J. D., 1989, "Hydrogeologic Assessment, Design and Remediation of a Shallow Ground Water Contaminated Zone," in *Proc., 3rd National Outdoor Action Conf. on Aquifer Restoration, Ground Water Monitoring and Geophysical Methods,* National Water Well Association, Dublin, OH, p. 591.

Fiedler, F. R. and T. C. Shevenell, 1990, "How to Solve the Remediation Twin Dilemmas: How Much? and How Long?—A Case Study Using Vapor Extraction Techniques for Gasoline Contaminated Soils," *Proc., 4th Natl. Outdoor Action Conf. on Aquifer Restoration, Ground Water Monitoring and Geophysical Methods,* Las Vegas, NV, May 14–17, p. 587.

Gander, M. J. and M. R. Wood, 1989, "Preliminary Site Characterization by Soil Gas Surveying with a Portable Gas Chromatograph," in *Proc., 82nd A & WMA Annual Meeting, Vol. 2,* Anaheim, CA, June 25–30, Air and Waste Management Association, Pittsburgh, PA, p. 6.

Gannon, K., D. J. Wilson, A. N. Clarke, R. D. Mutch, and J. H. Clarke, 1989, "Soil Cleanup by in situ Aeration. II. Effects of Impermeable Caps, Soil Permeability, and Evaporative Cooling," *Separ. Sci. Technol.,* 24:831.

García-Herruzo, F., C. Gómez-Lahoz, J. J. Rodriguez, D. J. Wilson, R. A. García-Delgado, and J. M. Rodríguez-Maroto, 1993, "Soil Clean Up: Pulse Soil Vapor Extraction (SVE) and Bioremediation," *6th Mediterranean Congress on Chemical Engineering, Expoquimia '93,* Sociedad Espanola de Quimica Industrial, Barcelona, Spain, 18–20 Oct., p. 639.

Ghassemi, M., 1988, "Innovative On-Site Treatment/Destruction Technologies for Remediation of Contaminated Sites," in *Hazardous and Toxic Materials: Safe Handling and Disposal,* Wiley, New York, NY, p. 345.

Gierke, J. S., N. J. Hutzler, and D. B. McKenzie, 1992, "Vapor Transport in Unsaturated Soil Columns: Implications for Vapor Extraction," *Water Resources Research,* 28:323.

Gómez-Lahoz, C., J. M. Rodríguez-Maroto, and D. J. Wilson, 1991, "Soil Cleanup by in situ Aeration. VI. Effects of Variable Permeabilities," *Separ. Sci. Technol.,* 26:133.

Gómez-Lahoz, C., J. M. Rodríguez-Maroto, and D. J. Wilson, 1994a, "Biodegradation Phenomena during Soil Vapor Extraction: A High-Speed Non-Equilibrium Model," *Separ. Sci. Technol.*, 29:429.

Gómez-Lahoz, C., J. J. Rodríguez, J. M. Rodríguez-Maroto, and D. J. Wilson, 1994b, "Biodegradation Phenomena during Soil Vapor Extraction: Sensitivity Studies for Single Substrate Systems," *Separ. Sci. Technol.*, 29:557.

Gómez-Lahoz, C., J. J. Rodríguez, J. M. Rodríguez-Maroto, and D. J. Wilson, 1994c, "Biodegradation Phenomena during Soil Vapor Extraction: Sensitivity Studies for Two Substrates," *Separ. Sci. Technol.*, 29:1275.

Gómez-Lahoz, C., J. M. Rodríguez-Maroto, D. J. Wilson, and K. Tamamushi, 1994d, "Soil Cleanup by in situ Aeration. XV. Effects of Variable Air Flow Rates in Diffusion-Limited Operation," *Separ. Sci. Technol.* 29:943.

Gómez-Lahoz, C., R. A. García-Delgado, R. García-Herruzo, J. M. Rodríguez-Maroto, and D. J. Wilson, 1993, "Extracción a Vacio de Contaminantes Orgánicos del Suelo. Fenómenos de No-Equilibrio," *III Congreso de Ingeniería Ambiental, Proma '93*, Bilbao, Spain.

Harper, C. J. and L. A. Williams, 1987, "Innovative Aquifer Restoration Techniques at a Site in Northern Vermont Contaminated by Gasoline Hydrocarbons," in *Proc., 4th Annual Eastern Regional Ground Water Conf.*, July 14–16, Burlington, VT, National Water Well Association, Dublin, OH, p. 699.

Heins, T. R., H. B. Kerfoot, D. J. Miller, and D. A. Peterson, 1992, "Quality Assurance in Soil Gas Surveys: False Positive Acetone Identifications at a Codisposal Site," in *Symposium on Current Practices in Ground Water and Vadose Investigations*, Jan. 30–Feb. 1, San Diego, CA, ASTM Special Technical Publication No. 1118, ASTM, Philadelphia, PA, p. 151.

Henkle, W. R., Jr., C. C. Hyde, and B. Cooper, 1990, "Use of Soil Vapor Surveys in Ground Water Contamination Studies. A Technique to Help Minimize Monitoring Well Installation Costs," in *Proc., 1990 Annual Symposium on Engineering Geology and Geotechnical Engineering*, April 4–6, Pocatello, ID, Idaho State University, Pocatello, ID, p. 18.1.

Henson, J. M., 1991, "Characterization is Subsurface Degradation Processes," in *Seminar Publication: Site Characterization for Subsurface Remediation*, U.S. EPA Report No. EPA/625/4-91/026, Nov., p. 193.

Hiller, D. H., 1992, "Performance Characteristics of Vapor Extraction Systems Operated in Europe," *Proc., Symposium on Soil Venting*, April 29–May 1, 1991, Houston, TX, D. DiGiulio, project officer, U.S. EPA Report No. EPA/600/R-92/174, p. 193.

Hinchee, R. E., 1989, "Enhanced Biodegradation through Soil Venting," *Workshop on Soil Vacuum Extraction*, April 27–28, 1989, U.S. EPA RSKERL, Ada, OK.

Hinchee, R. E. and R. N. Miller, 1992, "Bioventing for in-situ Remediation of Petroleum Hydrocarbons," *Proc., Symposium on Soil Venting*, April 29–May 1, 1991, Houston, TX, D. DiGiulio, project officer, U.S. EPA Report No. EPA/600/R-92/174, p. 283.

Hinchee, R. E., D. C. Downey, and E. J. Coleman, 1987, "Enhanced Bioreclamation, Soil Venting and Ground Water Extraction. A Cost-Effectiveness and Feasibility Comparison," in *Proc., NWWA/API Conf. on Petroleum Hydrocarbons and Organic Chemicals in Ground Water—Prevention, Detection and Restoration*, National Water Well Association, Dublin, OH, p. 147.

Hinchee, R. E., D. Downey, and R. Dupont, 1989, "Biodegradation Associated with

Soil Venting," *Soil Vapor Extraction Technology Workshop,* June 28–29, U.S. EPA, RR:EL, Edison, NJ.

Hinchee, R. E., D. C. Downey, and R. N. Miller, 1991a, "In situ Biodegradation of Petroleum Distillates in the Vadose Zone," in *Soil Vapor Extraction Technology. Pollution Technology Review No. 204,* Noyes Publications, Park Ridge, NJ, p. 186.

Hinchee, R. E., R. N. Miller, and R. R. Dupont, 1991b, "Enhanced Biodegradation of Petroleum Hydrocarbons: An Air-Based in situ Process," in *Innovative Hazardous Waste Treatment Technology Series. Vol. 3: Biological Processes,* Technomic Publishing Co., Inc., Lancaster, PA, p. 177.

Hinchee, R. E., D. C. Downey, R. R. Dupont, P. K. Aggarwal, and R. N. Miller, 1991c, "Enhancing Biodegradation of Petroleum Hydrocarbons through Soil Venting," *J. Hazardous Materials,* 27(3):315.

Hoag, G. E., 1989, "Soil Vapor Extraction Research Developments," *Soil Vapor Extraction Technology Workshop,* June 28–29, U.S. EPA RREL, Edison, NJ. Reprinted in *Soil Vapor Extraction Technology, Pollution Technology Review No. 204,* T. A. Pedersen and J. T. Curtis, eds., 1991, Noyes Publications, Park Ridge, NJ, p. 286.

Hoag, G. E. and B. L. Cliff, 1985, "The Use of the Soil Venting Technique for the Remediation of Petroleum-Contaminated Soils," in *Soils Contaminated by Petroleum, Environmental and Public Health Effects,* E. J. Calabrese and P. T. Kostechi, eds., Wiley, New York, NY.

Hoag, G. E., C. J. Bruell, and M. C. Marley, 1984, "Study of the Mechanisms Controlling Gasoline Hydrocarbon Partitioning and Transport in Ground Water Systems," USGS Report No. G832-06. Available from NTIS as PB85-242907/AS.

Hoag, G. E., C. J. Bruell, and M. C. Marley, 1987, "Induced Soil Venting for Recovery/Restoration of Gasoline Hydrocarbons in the Vadose Zone," in *Oil in Freshwater: Chemistry, Biology, Countermeasure Technology,* J. H. Vandermeulen and S. E. Hrudey, eds., Pergamon Press, New York, NY, p. 176.

Hoag, G. E., M. C. Marley, B. L. Cliff, and P. Nangeroni, 1991, "Soil Vapor Extraction Research Developments" in *Hydrocarbon Contaminated Soils and Ground Water: Analysis, Fate, Environmental and Public Health Effects, and Remediation,* Lewis Publishers, Chelsea, MI, p. 187.

Hoeppel, R. E. and R. E. Hinchee, 1993, "Enhanced Biodegradation for On-Site Remediation of Contaminated Soils and Ground Water," in *Hazardous Waste Site Soil Remediation: Theory and Application of Innovative Technologies,* A. N. Clarke and D. J. Wilson, eds., Marcel Dekker, New York, NY, p. 311.

Howe, G. B., M. E. Mullins, and T. N. Rogers, 1986, "Evaluation and Prediction of Henry's Law Constants and Aqueous Solubilities for Solvents and Hydrocarbon Fuel Components Vol. I: Technical Discussion," USAFESE Report No. ESL-86-66. U.S. Air Force Engineering and Services Center, Tyndall AFB, FL, 86 pp.

Hutzler, N. J., J. S. Gierke, and B. E. Murphy, 1990, "Vaporizing VOCS," *Civil Engineering* (ASCE), 60(4):57.

Hutzler, N. J., D. B. McKenzie, and J. S. Gierke, 1989a, "Vapor Extraction of Volatile Organic Chemicals from Unsaturated Soil," *Abstracts, Int. Symposium on Processes Governing the Movement and Fate of Contaminants in the Subsurface Environment,* July 23–26, Stanford, CA.

Hutzler, N. J., B. E. Murphy, and J. S. Gierke, 1989b, "Review of Soil Vapor Extraction System Technology," *Soil Vapor Extraction Technology Workshop,* June 28–29, U.S. EPA RREL, Edison, NJ. Reprinted in *Soil Vapor Extraction Technol-*

ogy. Pollution Technology Review No. 204, 1991, Noyes Publications, Park Ridge, NJ, p. 136.

Hutzler, N. J., B. E. Murphy, and J. S. Gierke, 1991, "State of Technology Review: Soil Vapor Extraction Systems," *J. Hazardous Materials,* 26(2):225.

Johnson, P. C., M. W. Kemblowski, and J. D. Colthart, 1989a, "Practical Screening Models for Soil Venting Applications," *Workshop on Soil Vacuum Extraction,* April 27–28, RSKERL, Ada, OK.

Johnson, P. C., M. W. Kemblowski, and J. D. Colthart, 1990a, "Quantitative Analysis for the Cleanup of Hydrocarbon-Contaminated Soils by in situ Soil Venting," *Ground Water,* 28:413.

Johnson, P. C., M. W. Kemblowski, J. D. Colthart, D. L. Byers, and C. C. Stanley, 1989b, "A Practical Approach to the Design, Operation, and Monitoring of in-situ Soil Venting Systems," *Soil Vapor Extraction Technology Workshop,* June 28–29, U.S. EPA Risk Reduction Engineering Laboratory (RREL), Edison, NJ. Reprinted in *Soil Vapor Extraction Technology. Pollution Technology Review No. 204,* T. A. Pedersen and J. T. Curtis, eds., 1991, Noyes Publications, Park Ridge, NJ, p. 195.

Johnson, P. C., C. C. Stanley, D. L. Byers, D. A. Benson, and M. A. Acton, 1991, "Soil Venting at a California Site: Field Data Reconciled with Theory," in *Hydrocarbon Contaminated Soils and Groundwater: Analysis, Fate, Environmental and Public Health Effects, and Remediation, Vol. 1,* Lewis Publishers, Chelsea, MI, p. 253.

Johnson, P. C., C. C. Stanley, M. W. Kemblowski, D. L. Byers, and J. D. Colthart, 1990b, "Practical Approach to the Design, Operation and Monitoring of in situ Soil Venting Systems," *Ground Water Monitoring Review,* 10:159.

Johnson, R. L., 1992, "Optimization of the Vapor Extraction Process: Large Physical Model Studies," *Proc., Symposium on Soil Venting,* April 29–May 1, 1991, Houston, TX, D. DiGiulio, project officer, U.S. EPA Report No. EPA/600/R-92/174, p. 171.

Johnson, R. L., C. D. Palmer, and J. F. Keely, 1987, "Mass Transfer of Organics between Soil, Water and Vapor Phases: Implications for Monitoring, Biodegradation and Remediation," in *Proc., NWWA/API Conf. on Petroleum Hydrocarbons and Organic Chemicals in Ground Water—Prevention, Detection and Restoration,* National Water Well Association, Dublin, OH, p. 493.

Johnson, R. L., J. F. Keely, C. D. Palmer, J. M. Suflita, and W. Fish, 1989c, *Seminar Publication: Transport and Fate of Contaminants in the Subsurface,* U.S. EPA Report No. EPA/625/4-89/019.

Johnson, S. E., G. H. Emrich, and M. A. Apgar, 1986, "On-Site Removal of Volatile Organic Contaminants from Soils," presented at *9th Ann. Madison Waste Conf.,* Sept. 9–10.

Joss, C. J., A. L. Baehr, and J. M. Fischer, 1992, "A Field Technique for Determining Unsaturated Zone Air Permeability," *Proc., Symposium on Soil Venting,* April 29–May 1, 1991, Houston, TX, D. DiGiulio, project officer, U.S. EPA Report No. EPA/600/R-92/174, p. 149.

Kaback, D. S. and B. B. Looney, 1988, "Final Program Plan: Research Study on Horizontal Well Drilling and in situ Remediation," Report No. DPST-88-346, February 22. Available from NTIS as DE89-000010.

Kampbell, D., 1992, "Subsurface Remediation at a Gasoline Spill Using a Bioventing Approach," *Proc., Symposium on Soil Venting,* April 29–May 1, 1991, Houston, TX, D. DiGiulio, project officer, U.S. EPA Report No. EPA/600/R-92/174, p. 309.

Katyal, A. K., P. K. Patel, and J. C. Parker, 1992, "Modeling Transport of Organic Chemicals in Gas and Liquid Phases," *Proc., Symposium on Soil Venting,* April 29–May 1, 1991, Houston, TX, D. DiGiulio, project officer, U.S. EPA Report No. EPA/600/R-92/174, p. 75.

Kayano, S. and D. J. Wilson, 1992, "Soil Cleanup by in situ Aeration. X. Vapor Stripping of Mixtures of Volatile Organics Obeying Raoult's Law," *Separ. Sci. Technol.,* 27:1525.

Keech, D. A., 1989, "Subsurface Venting Research and Venting Manual by the American Petroleum Research Institute," *Workshop on Soil Vacuum Extraction,* April 17–28, Robert S. Kerr Environmental Research Laboratory (RSKERL), Ada, OK. Most of the work done for API has been carried out by Texas Research Institute, Radian Corp., and Reidel Environmental Services.

Kemblowski, M. W. and S. Chowdery, 1992, "Soil Venting Design: Models and Decision Analysis," *Proc., Symposium on Soil Venting,* April 29–May 1, 1991, Houston, TX, D. DiGiulio, project officer, U.S. EPA Report No. EPA/600/R-92/174, p. 73.

Kerfoot, H. B., 1990, "Soil Venting with Pneumatically Installed 'Shield Screens,' " *Proc., 4th Natl. Conf. on Aquifer Restoration, Ground Water Monitoring and Geophysical Methods,* May 14–17, Las Vegas, NV, p. 571.

Kerfoot, H. B., 1991, "Soil Gas Surveys in Support of Design of Vapor Extraction Systems," in *Soil Vapor Extraction Technology. Pollution Technology Review No. 204,* Noyes Publications, Park Ridge, NJ, p. 171.

Knieper, L. H., 1988, "VES Cleans Up Gasoline Leak," *Plumbing Engineer,* 20(8): 56.

Kuo, J. F., E. M. Aieta, and P. H. Yang, 1990, "A Two-Dimensional Model for Estimating Radius of Influence of a Soil Venting Process," *Proc. Hazardous Materials Conf. '90,* April 17–19, Anaheim, CA, p. 197.

Lester, G. R., 1992, "Catalytic Destruction of Hazardous Halogenated Organic Chemicals," *Proc., Symposium on Soil Venting,* April 29–May 1, 1991, Houston, TX, D. DiGiulio, project officer, U.S. EPA Report No. EPA/600/R-92/174, p. 281.

Lingineni, S. and V. K. Dhir, 1992, "Modeling of Soil Venting Processes to Remediate Unsaturated Soils," *J. Environmental Engineering* (ASCE), 118:135.

Lyman, W. J. and D. C. Noonan, 1990, *Assessing UST Corrective Action Technologies: Site Assessment and Selection of Unsaturated Zone Treatment Technologies,* U.S. EPA Report No. EPA/600/2-90/011.

Lyman, W. J., D. C. Noonan, and P. J. Reidy, 1990, "Cleanup of Petroleum Contaminated Soils at Underground Storage Tanks," in *Pollution Technology Review No. 195,* Noyes Publications, Park Ridge, NJ.

Lyman, W. J., W. F. Reehl, and D. H. Rosenblatt, 1982, *Handbook of Chemical Property Estimation Methods: Environmental Behavior of Organic Compounds,* McGraw-Hill, New York, NY.

Mackay, D. and W. Y. Shiu, 1981, "A Critical Review of Henry's Law Constants for Chemicals of Environmental Interest," *J. Phys. Chem. Ref. Data,* 10(4):1175.

Malmanis, E., D. W. Fuerst, and R. J. Piniewski, 1989, "Superfund Site Soil Remediation Using Large-Scale Vacuum Extraction," *Proc., 6th Natl. Conf. Hazardous Wastes Hazardous Materials,* April 12–14, New Orleans, p. 538.

Markley, D. E., 1988, "Cost Effective Investigation and Remediation of Volatile-Organic Contaminated Sites," in *HWHM 88: Hazardous Wastes and Hazardous Materials. Proc., 5th National Conf.,* April 19–21, Las Vegas, NV, p. 463.

Marley, M. C., 1992, "Development and Application of a Three-Dimensional Air Flow

Model in the Design of a Vapor Extraction System," *Proc., Symposium on Soil Venting,* April 29–May 1, 1991, Houston, TX, D. DiGiulio, project officer, U.S. EPA Report No. EPA/600/R-92/174, p. 125.

Marley, M. C. and G. E. Hoag, 1984, "Induced Venting for the Recovery/Restoration of Gasoline Hydrocarbons in the Vadose Zone," *NWWA/API Conference on Petroleum Hydrocarbons and Organic Chemicals in Groundwater,* Nov. 5–7, Houston, TX.

Marley, M. C., P. E. Nangeroni, B. L. Cliff, and J. D. Polonsky, 1990, "Air Flow Modeling for in situ Evaluation of Soil Properties and Engineered Vapor Extraction System Design," *Proc., 4th Natl. Outdoor Action Conf. on Aquifer Restoration, Ground Water Monitoring and Geophysical Methods,* May 14–17, Las Vegas, NV, p. 651.

Marley, M. C., S. D. Richter, B. L. Cliff, and P. E. Nangeroni, 1989, "Design of Soil Vapor Extraction Systems—A Scientific Approach," *Soil Vapor Extraction Technology Workshop,* June 28–29, U.S. EPA RREL, Edison, NJ. Reprinted in *Soil Vapor Extraction Technology. Pollution Technology Review No. 204,* 1991, Noyes Publications, Park Ridge, NJ, p. 240.

Marrin, D. L., 1987, "Soil Gas Sampling Strategies: Deep vs. Shallow Aquifers," in *Proc., 1st National Outdoor Action Conf. of Aquifer Restoration, Ground Water Monitoring and Geophysical Methods,* National Water Well Association, Dublin, OH, p. 437.

Marrin, D. L. and H. B. Kerfoot, 1988, "Soil Gas Surveying Techniques," *Environ. Sci. Technol.,* 22:740.

Marrin, D. L. and G. M. Thompson, 1987, "Gaseous Behavior of TCE Overlying a Contaminated Aquifer," *Groundwater,* 25:21.

Massmann, J. W., 1989, "Applying Groundwater Flow Models in Vapor Extraction System Design," *J. Environmental Engineering,* 115:129.

Megehee, M. M. and D. J. Wilson, 1991, unpublished work.

Mercer, J. W. and C. P. Spalding, 1991, "Characterization of the Vadose Zone," in *Seminar Publication: Site Characterization for Subsurface Remediation,* U.S. EPA Report No. EPA/625/4-91/026, Nov., p. 59.

Mercer, J. W., D. A. Griffin, J. C. Herweijer, and P. Srinivasan, 1989, "Ground Water Contamination: Processes, Characterization, Analysis, and Remediation," in *Appropriate Methodologies for Development and Management of Ground Water Resources in Developing Countries. Vol. 3. Proc., International Workshop,* Feb. 23–Mar. 4, A. A. Balkema, Rotterdam, Neth., p. 261.

Michaels, P. A., 1989, "Technology Evaluation Report—Terra Vac in situ Vacuum Extraction System, Groveland, Massachusetts," EPA/540/S5-89/003, May.

Miller, D. E. and L. W. Canter, 1991, "Control of Aromatic Waste Streams by Soil Bioreactors," *Environmental Progress,* 10(4):300.

Miller, R. N., R. E. Hinchee, and C. C. Vogel, 1992, "A Field Scale Investigation of Soil Venting Enhanced Petroleum Hydrocarbn Biodegradation in the Vadose Zone at Tyndall AFB, Florida," *Proc., Symposium on Soil Venting,* April 29–May 1, 1991, Houston, TX, D. DiGiulio, project officer, U.S. EPA Report No. EPA/600/R-92/174, p. 293.

Millington, R. J. and J. M. Quirk, 1961, "Permeability of Porous Solids," *Trans. Faraday Soc.,* 57:1200.

Montgomery, J. H. and L. M. Welkom, 1989, *Groundwater Chemicals Desk Reference, Vols. 1 and 2,* Lewis Publishers, Chelsea, MI.

Moore, A. and S. Revell, 1989, "In situ Bioremediation of an Abandoned Coal Tar Site," in *Proc., 3rd National Outdoor Action Conf. on Aquifer Restoration, Ground Water Monitoring and Geophysical Methods,* National Water Well Association, Dublin, OH, p. 491.

Moss, F. J., 1989, "Application of Treatment Techniques to Soil Vapor Extraction System for Remediation of Soils in the Unsaturated Zone," *Proc., FOCUS Conf. on Eastern Regional Ground Water Issues,* Oct. 17–19, Kitchener, Ont., Can., National Water Well Association, Dublin, OH, p. 449.

Mutch, R. D., Jr. and D. J. Wilson, 1990, "Soil Cleanup by in situ Aeration. IV. Anisotropic Permeabilities," *Separ. Sci. Technol.,* 25:1.

Mutch, R. D., Jr., A. N. Clarke, and D. J. Wilson, 1989, "In situ Vapor Stripping Research Project: A Progress Report," *Proc., 2nd Ann. Hazardous Materials Conf./Central,* Rosemont, IL, March 14–16, p. 1.

Oja, K. J. and D. K. Kreamer, 1992, "The Effects of Moisture on Adsorption of Trichloroethylene Vapor on Natural Soils," *Proc., Symposium on Soil Venting,* April 29–May 1, 1991, Houston, TX, D. DiGiulio, project officer, U.S. EPA Report No. EPA/600/R-92/174, p. 13.

Oma, K. H., D. J. Wilson, and R. D. Mutch, 1990, "In situ Vapor Stripping: The Importance of Nonequilibrium Effects in Predicting Cleanup Time and Cost," *Proc., Hazardous Materials Management Conf. and Exhibition/Intern.,* Atlantic City, NJ, June 5–7.

Ong, S. K., S. R. Lindner, and L. W. Lion, 1991, "Applicability of Linear Partitioning Relationships for Sorption of Organic Vapors onto Soil and Soil Minerals," in *Organic Substances and Sediments in Water. Vol. 1: Humics and Soils,* CRC Press, Boca Raton, FL, p. 275.

Osejo, R. E. and D. J. Wilson, 1991, "Soil Cleanup by in situ Aeration. IX. Diffusion Constants of Volatile Organics and Removal of Underlying LNAPL," *Separ. Sci. Technol.,* 26:1433.

Ostendorf, D. W. and D. J. Kampbell, 1990, "Bioremediated Soil Venting of Light Hydrocarbons," *Hazardous Waste and Hazardous Materials,* 7(4):319.

Pacific Environmental Services, Inc., 1989, *Soil Vapor Extraction VOC Control Technology Assessment,* U.S. EPA Report No. EPA-450/4-89-017, Sept.

Patterson, J. H., 1989, "Case History: Soil Venting as a Construction Safety/Remediation Method for Development of Contaminated Property," *Proc., Seminar on Contamination and the Constructed Project,* Conn. Soc. Civil Engineers/Conn. Ground-Water Assoc., Hawthorne Inn, Berlin, Conn., Nov. 2–3.

Pederson, T. A. and J. T. Curtis, 1991a,b, *Soil Vapor Extraction Technology Reference Handbook,* Risk Reduction Engineering Laboratory, U.S. EPA, Report No. EPA/540/2-91/003. Reprinted as *Soil Vapor Extraction Technology. Pollution Technology Review No. 204.* 1991, Noyes Publications (Noyes Data Corp.), Park Ridge, NJ.

Pederson, T. A. and C.-Y. Fan, 1992, "Soil Vapor Extraction Technology Development Status and Trends," in *Proc., Symposium on Soil Venting,* April 29–May 1, 1991, Houston, TX, D. DiGiulio, project officer, U.S. EPA Report No. EPA/600/R-92/174, p. 1.

Peters, M. S. and K. D. Timmerhaus, 1968, *Plant Design and Economics for Chemical Engineers,* 2nd ed., McGraw-Hill, New York.

Pope, J. L., 1988, "Abatement/Remediation of Volatile Organics in the Subsurface Using Soil Vapor Extraction," presented at *11th Ann. Madison Waste Conf.*, Sept. 13–14.

Rainwater, K., M. R. Zaman, B. J. Claborn, and H. W. Parker, 1990, "Experimental and Model Study of Soil Venting," *Proc., 1990 Specialty Conf.*, July 8–11. Arlington, VA, *Natl. Conf. on Environmental Engineering,* ASCE, Environmental Engineering Div., New York, NY, p. 479.

Rathfelder, K., W. W. G. Yeh, and D. Mackay, 1991, "Mathematical Simulation of Soil Vapor Extraction Systems: Model Development and Numerical Examples," *J. Contaminant Hydrology,* 8:263.

Regalbuto, D. P., J. A. Barrera, and J. B. Lisiecki, 1988, "Removal of VOCs by Means of Enhanced Volatilization," in *Proc., NWWA/API Conf. on Petroleum Hydrocarbons and Organic Chemicals in Ground Water—Prevention, Detection and Restoration,* National Water Well Association, Dublin, OH, p. 571.

Reisinger, J. J., J. M. Kerr, R. E. Hinchee, D. R. Burris, and R. S. Dykes, 1989, "Using Soil Vapor Contaminant Assessment at Hydrocarbon Contaminated Sites," in *Petroleum Contaminated Soils. Vol. 2: Remediation Techniques, Environmental Fate, Risk Assessment. Analytical Methodologies,* Lewis Publishers, Chelsea, MI, p. 303.

Roberts, L. A. and D. J. Wilson, 1993, "Soil Cleanup by in situ Aeration. XI. Cleanup Time Distributions for Statistically Equivalent Variable Permeabilities," *Separ. Sci. Technol.,* 28:1671.

Robitaille, L. and S. K. Walen, 1988, "Reduction in Volatile Organic Compound Levels in Soils and Ground Water Using In-Ground Soil Venting," in *Proc., FOCUS Conf. on Eastern Regional Ground Water Issues,* Sept. 27–29, Stamford, CT, National Water Well Association, Dublin, OH, p. 249.

Rodríguez-Maroto, J. M. and D. J. Wilson, 1991, "Soil Cleanup by in situ Aeration. VII. High-Speed Modeling of Diffusion Kinetics," *Separ. Sci. Technol.,* 26:743.

Rodríguez-Maroto, J. M., C. Gómez-Lahoz, and D. J. Wilson, 1991, "Soil Cleanup by in situ Aeration. VIII. Effects of System Geometry on Vapor Extraction Efficiency," *Separ. Sci. Technol.,* 26:1051.

Rodríguez-Maroto, J. M., C. Gómez-Lahoz, and D. J. Wilson, 1992, "Mathematical Modeling of SVE: Effects of Diffusion Kinetics and Variable Permeabilities," *Proc., Symposium on Soil Venting,* April 29–May 1, 1991, Houston, TX, D. DiGiulio, project officer, U.S. EPA Report No. EPA/600/R-92/174, p. 103.

Rodríguez-Maroto, J. M., C. Gómez-Lahoz, and D. J. Wilson, 1994, "Soil Cleanup by in situ Aeration. XVIII. Field Scale Models with Diffusion from Clay Structures," *Separ. Sci. Technol.,* 29:1367.

Roy, W. R. and R. A. Griffin, 1987, *Vapor-Phase Movement of Organic Solvents in the Unsaturated Zone,* Environmental Institute for Waste Management Studies, Univ. of Alabama, Tuscaloosa, open file report no. 16.

Roy, W. R. and R. A. Griffin, 1989, *In-situ Extraction of Organic Vapors from Unsaturated Porous Media,* Environmental Institute for Waste Management Studies, Univ. of Alabama, Tuscaloosa, open file report no. 24.

Scheidegger, A. E., 1974, *The Physics of Flow Through Porous Media,* 3rd ed., University of Toronto Press, p. 306.

Shan, C., R. W. Falta, and I. Javandel, 1992, "Analytical Solutions for Steady State Gas Flow to a Soil Vapor Extraction Well," *Water Resources Research,* 28:1105.

Siegrist, R. L., 1992, "Assessing the Performance of in situ Soil Venting Systems by Soil Sampling and Volatile Organic Compound Measurements," *Proc., Symposium on Soil Venting,* Apr. 29–May 1, 1991, Houston, TX, D. DiGiulio, project officer, U.S. EPA Report No. EPA/600/R-92/174, p. 147.

Silka, L. R., H. D. Cirpili, and D. L. Jordan, 1991, "Modeling Applications to Vapor Extraction Systems," in *Soil Vapor Extraction Technology. Pollution Technology Review No. 204,* Noyes Publications, Park Ridge, NJ, p. 252.

Sorensen, D. L. and R. C. Sims, 1992, "Habitat Conditions Affecting Bioventing Processes," *Proc., Symposium on Soil Venting,* April 29–May 1, 1991, Houston, TX, D. DiGiulio, project officer, U.S. EPA Report No. EPA/600/R-92/174, p. 55.

Staps, S., 1989, "Biorestoration of Contaminated Soil and Ground Water," *Chemistry and Industry (London),* No. 18, Sept. 18, p. 581.

Stephanatos, B. N., 1990, "Modeling the Soil Venting Process for the Cleanup of Soils Containing Volatile Organics," *Proc., 4th Natl. Outdoor Action Conf. on Aquifer Restoration, Ground Water Modeling and Geophysical Methods,* May 14–17, Las Vegas, NV, p. 633.

Sterrett, R. J., 1989, "Analysis of in situ Soil Air Stripping Data," *Workshop on Soil Vacuum Extraction,* April 27–28, U.S. EPA RSKERL, Ada, OK.

Stinson, M. K., 1989, "EPA SITE Demonstration of the Terra Vac in situ Vacuum Extraction Process in Groveland, Massachusetts," *J.A.P.C.A.,* 39:1054.

Stransky, R. and M. W. Blanchard, 1989, "In situ Venting Program for Petroleum Hydrocarbons in Soil and Water," in *Proc., 3rd National Outdoor Action Conf. on Aquifer Restoration, Ground Water Monitoring and Geophysical Methods,* National Water Well Association, Dublin, OH, p. 607.

Stumbar, J. P. and J. Rawe, 1991, *Guide for Conducting Treatability Studies under CERCLA: Soil Vapor Extraction Interim Guidance,* U.S. EPA Report No. EPA/540/2-91/019A.

Sully, M., 1992, "Characterizing Permeability to Gas in the Vadose Zone," *Proc., Symposium on Soil Venting,* April 29–May 1, 1991, Houston, TX, D. DiGiulio, project officer, U.S. EPA Report No. EPA/600/R-92/174, p. 143.

Texas Research Institute, 1984, *Forced Venting to Remove Gasoline Vapors from a Large-Scale Model Aquifer,* American Petroleum Institute, Washington, D.C.

Thornton, J. S. and W. L. Wootan, 1982, "Venting for the Removal of Hydrocarbon Vapors from Gasoline-Contaminated Soil," *J. Environment Sci. Health,* A17:31.

Towers, D., M. J. Dent, and D. G. Van Arnam, 1989, "Part 1: Choosing a Treatment for VHO-Contaminated Soil," *Hazardous Materials Consultant,* March/April, p. 8.

Trowbridge, B. E. and J. J. Malot, 1990, "Soil Remediation and Free Product Removal Using in-situ Vacuum Extraction with Catalytic Oxidation," *Proc., 4th Natl. Outdoor Action Conf. on Aquifer Restoration, Ground Water Monitoring and Geophysical Methods,* May 14–17, Las Vegas, NV, p. 559.

van Eyk, J. and C. Vreeken, 1992, "A Demo-Project for in situ Subsoil and Aquifer Restoration Following Hydrocarbon Spills at a Tankstation," *Proc., Symposium on Soil Venting,* April 29–May 1, 1991, Houston, TX, D. DiGiulio, project officer, U.S. EPA Report No. EPA/600/R-92/174, p. 317.

Verschueren, K., 1983, *Handbook of Environmental Data on Organic Chemicals,* Van Nostrand Reinhold Co., New York, NY.

Walter, G. R. and H. W. Bentley, 1992, "Use of Soil Gas Measurements in the Design of Soil Vapor Extraction Systems," *Proc. Symposium on Soil Venting,* April 29–May 1, 1991, Houston, TX, D. DiGiulio, project officer, U.S. EPA Report No. EPA/600/R-92/174, p. 145.

Walton, J. C., R. G. Baca, J. B. Sisson, and T. R. Wood, 1990, "Application of Soil

Venting at a Large Scale: A Data and Modeling Analysis," *Proc., 4th Natl. Outdoor Action Conf. on Aquifer Restoration, Ground Water Monitoring and Geophysical Methods,* May 14–17, Las Vegas, NV, p. 559.

Walton, J.C., R. G. Baca, J. B. Sisson, A. J. Sondrup, and S. O. Magnusen, 1992, "Application of Computer Simulation Models to the Design of a Large-Scale Soil Venting System and Bioremediation," *Proc., Symposium on Soil Venting,* April 29–May 1, 1991, Houston, TX, D. DiGiulio, project officer, U.S. EPA Report No. EPA/600/R-92/174, p. 91.

Weast, R. C., M. J. Astle, and W. H. Beyer, 1985, *CRC Handbook of Chemistry and Physics,* 65th ed., CRC Press, Inc., Boca Raton, FL, 1985.

Wilson, D. J., 1988, "Mathematical Modeling of in situ Vapor Stripping of Contaminated Soils," *Proc., 1st Ann. Hazardous Materials Management Conf./Central,* March 15–17, Glen Ellyn, IL, p. 194.

Wilson, D. J., 1989, "Soil Vapor Stripping Models: Stripping of Underlying NAPL and Stripping from Fractured Bedrock," *DOE Model Conference Abstracts,* Oct. 3–6, Oak Ridge, TN.

Wilson, D. J., 1990a, "Mathematical Modeling of in situ Vapor Stripping of Contaminated Soils," *Israel J. Chem.,* 30:281.

Wilson, D. J., 1990b, "Soil Cleanup by in situ Aeration. V. Vapor Stripping from Fractured Bedrock," *Separ. Sci. Technol.,* 25:243.

Wilson, D. J., 1993a, "Advances in the Modeling of Several Innovative Technologies," in *Proc., 6th Annual Environmental Management and Technology Conf./Central,* March 9–11, 1993, Rosemont, IL, p. 135.

Wilson, D. J., 1993b, "Soil Cleanup by in situ Aeration. XIII. Effects of Solution Rates and Diffusion in Mass-Transport-Limited Operation," *Separ. Sci. Technol.,* 29:579.

Wilson, D. J. and A. N. Clarke, 1993, "Soil Vapor Extraction," in *Hazardous Waste Site Soil Remediation: Theory and Application of Innovative Technologies,* D. J. Wilson and A. N. Clarke, eds., Marcel Dekker, New York, NY, p. 171.

Wilson, D. J., A. N. Clarke, and J. H. Clarke, 1988, "Soil Cleanup by in situ Aeration. I. Mathematical Modeling," *Separ. Sci. Technol.,* 23:991.

Wilson, D. J., A. N. Clarke, and R. D. Mutch, 1989, "Soil Cleanup by in situ Aeration. III. Passive Vent Wells, Recontamination, and Removal of Underlying Nonaqueous Phase Liquid," *Separ. Sci. Technol.,* 24:939.

Wilson, D. J., C. Gómez-Lahoz, and J. M. Rodríguez-Maroto, 1992a, "Mathematical Modeling of SVE: Effects of Diffusion Kinetics and Variable Permeabilities," in *Proc., Symposium on Soil Venting,* April 29–May 1, 1991, Houston, TX, D. DiGiulio, project officer, U.S. EPA Report No. EPA/600/R-92/174.

Wilson, D. J., C. Gómez-Lahoz, and J. M. Rodríguez-Maroto, 1994a, "Soil Cleanup by in situ Aeration. XVI. Solution and Diffusion in Mass Transport-Limited Operation and Calculation of Darcy's Constants," *Separ. Sci. Technol.,* 29:1133.

Wilson, D. J., J. M. Rodríguez-Maroto, and C. Gómez-Lahoz, 1994b, "Soil Cleanup by in situ Aeration. XIX. Effects of Spill Age on Soil Vapor Extraction Remediation Rates," *Separ. Sci. Technol.* 29:1645.

Wilson, J. T., D. H. Kampbell, and J. Cho, 1992b, "Opportunities for Biotreatment of Trichloroethylene in the Vadose Zone," *Proc. Symposium on Soil Venting,* April 29–May 1, 1991, Houston, TX, D. DiGiulio, project officer, U.S. EPA Report No. EPA/600/R-92/174, p. 71.

Woodward-Clyde Consultants, 1985, "Performance Evaluation Pilot Scale Installation and Operation, Soil Gas Vapor Extraction System, Time Oil Company Site, Tacoma, Washington, South Tacoma Channel, Well 12A Project," work assignment 74-0N14.1, Walnut Creek, CA, Dec. 13.

Wootan, W. L., Jr. and T. Voynick, 1984, *Forced Venting to Remove Gasoline Vapor from a Large-Scale Model Aquifer,* American Petroleum Institute, API Publ. No. 4431.

Wu, J. C., L. T. Fan, and L. E. Erickson, 1990, "Modeling and Simulation of Bioremediation of Contaminated Soil," *Environmental Progress,* 9:47.

Zenobia, K. E., D. K. Rothenbaum, S. B. Chargee, and E. S. Findlay, 1987, "Vapor Extraction of Organic Contamination from the Vadose Zone: A Case Study," in *Proc., NWWA FOCUS Conf. on Northwestern Ground Water Issues,* National Water Well Association, Dublin, OH, p. 625.

CHAPTER 2

Soil Vapor Extraction— A Phased Approach

INTRODUCTION

THIS chapter is mainly concerned with the development of a phased approach to the evaluation and implementation of soil vapor extraction (SVE) on a site-specific basis. A very small investment of time and effort in preliminary screening (the first part of the phased approach) may allow one to determine that SVE is not feasible at a particular site, while the additional work involved in technology selection by means of the phased approach is quite slight if SVE is the technology (or one of the technologies) ultimately selected.

The interconversion of permeabilities in various units is then addressed, since one commonly needs these in m²/atm sec in SVE work, and they are generally not given in these units. The chapter closes with a discussion of some simple models, which can be helpful for preliminary assessments (only) and which do not have the computational and data requirements of the more elaborate models discussed in Chapters 3 to 6.

PRELIMINARY ASSESSMENT OF VAPOR STRIPPING

Site investigations normally break down into the following steps: (1) site history review, (2) preliminary site screening, (3) detailed site characterization, (4) contaminant assessment, and (5) pilot testing (Johnson et al., 1989). The points to be addressed include (1) the types and quantities of contaminants released (gasoline, oxygenated and chlorinated solvents, jet and diesel fuel, etc.); (2) the extent to which the contaminants have migrated and pathways for further contaminant movement; (3) product behavior in the soil (Is it sorbed, present as NAPL, in the vapor phase, dissolved in groundwater?), and (4) the potential for injury to possible receptors (Pedersen and Curtis, 1991).

VAPOR PRESSURE CONSIDERATIONS

Before one invests much effort in assessing the feasibility of SVE at a site, there are some preliminaries. SVE requires that the contaminant be removed as vapor; therefore, the best that the technique can possibly do at a site is limited by the vapor pressures of the contaminants to be removed. The maximum number of kilograms of VOC that can be removed by 1 m³ of air is given by

$$m = \frac{(M.W.) \cdot P_0}{760 \cdot .08206 \cdot T} = 0.01603 \cdot \frac{(M.W.) \cdot P_0}{T} \tag{1}$$

where

m = maximum mass of VOC removed per m³ of air, kg/m
$(M.W.)$ = VOC molecular weight, gm/mole
P_0 = VOC vapor pressure at ambient temperature, torr
T = ambient temperature, deg K (273.15 + deg C)

Thus, for p-xylene (molecular weight = 106 gm/mole, vapor pressure at 20°C = 6.5 torr), one can remove no more than 0.0377 kg of VOC per cubic meter of air (0.0024 lb/ft³). A reasonable airflow rate from a vacuum extraction well is of the order of 100 cfm; such a well could therefore remove no more than about 340 lb/day. This would probably be regarded as quite acceptable, so one would then go on to the next phase of the evaluation. A similar calculation for methylnaphthalene gives 3.2×10^{-4} kg/m³, and a removal rate for a 100 cfm well of 2.9 lb/day. This is rather low, and one might wish to consider other technologies at this point. Vapor pressures for some common organic chemicals at various temperatures are given in Table 2.1. A useful formula for calculating vapor pressures as functions of temperature is

$$\log_{10} P(T) = A - 0.05223 \cdot B/T \tag{2}$$

where

$P(T)$ = VOC vapor pressure at temperature T, torr
T = ambient temperature, deg K
A = constant
B = constant (molar heat of vaporization, j/mol)

Values for the parameters A and B for a number of VOCs are given in Table 2.2. A rough rule of thumb is that vacuum extraction is not likely to be

TABLE 2.1. Vapor Pressures of Selected VOCs.

Compound	Mol. Wt. (gm/mol)	Temperature, °C Vapor Pressures, torr					
		1	10	40	100	400	760
CCl_2F_2	120.9	−118.5	−97.8	−81.6	−68.6	−43.9	−29.8
CCl_3F	137.4	−84.3	−59.0	−39.0	−23.0	+6.8	23.7
CCl_4	153.8	−50.0	−19.6	+4.3	23.0	57.8	76.7
$CHCl_3$	119.4	−58.0	−29.7	−7.1	+10.4	42.7	61.3
CH_2Br_2	173.8	−35.1	−2.4	+23.3	42.3	79.0	98.6
CH_2Cl_2	84.9	−70.0	−43.3	−22.3	−6.3	+24.1	40.7
CS_2	76.1	−73.8	−44.7	−22.5	−5.1	+28.0	46.5
C_2Cl_4 (PCE)	165.8	−20.6	+13.8	40.1	61.3	100.0	120.8
C_2HCl_3 (TCE)	131.4	−43.8	−12.4	+11.9	31.4	67.0	86.7
cis-1,2-DCE	96.9	−58.4	−29.9	−7.9	+9.5	41.0	59.0
trans-1,2-DCE	96.9	−65.4	−38.0	−17.0	−0.2	+30.8	47.8
1,1-DCE	96.9	−77.2	−51.2	−31.1	−15.0	+14.8	31.7
1,1,1,2-TTCA	167.8	−16.3	+19.3	46.7	68.0	108.2	130.5
1,1,2,2-TTCA	167.8	−3.8	+33.0	60.8	83.2	124.0	145.9
1,1,1-TCA	133.4	−52.0	−21.9	+1.6	20.0	54.6	74.1
1,1,2-TCA	133.4	−24.0	+8.3	35.2	55.7	93.0	113.9
1,1-DCA	99.0	−60.7	−32.3	−10.2	+7.2	39.8	57.4
1,2-DCA	99.0	−44.5	−13.6	+10.0	29.4	64.0	82.4
Ethyl chloride	64.5	−89.8	−65.8	−47.0	−32.0	−3.9	+12.3
1,1,1-Trichl. prop.	147.4	−28.8	+4.2	29.9	50.0	87.5	108.2
1,2,3-Trichl. prop.	147.4	+9.0	46.0	74.0	96.1	137.0	158.0
1,2-Dichl. propane	113.0	−38.5	−6.1	+19.4	39.4	76.0	96.8
1-Chloropropane	78.5	−68.3	−41.0	−19.5	−2.5	+29.4	46.4
2-Chloropropane	78.5	−78.8	−52.0	−31.0	−13.7	+18.0	36.5
n-Pentane	72.1	−76.6	−50.1	−29.2	−12.6	+18.5	36.1
2-Methylbutane	72.1	−82.9	−57.0	−36.5	−20.2	+10.5	27.8
2,2-diMepropane	72.1	−102.0	−76.7	−56.1	−39.1	−7.1	+9.5
1,2,3,4-$C_6H_2Cl_4$	215.9	68.5	114.7	149.2	175.7	225.5	254.0
1,2,3,5-$C_6H_2Cl_4$	215.9	58.2	104.1	140.0	168.0	220.0	246.0
1,2,3-$C_6H_3Cl_3$	181.4	40.0	85.6	119.8	146.0	193.5	218.5
1,2,4-$C_6H_3Cl_3$	181.4	38.4	81.7	114.8	140.0	187.7	213.0
1,3,5-$C_6H_3Cl_3$	181.4	—	78.0	110.8	136.0	183.0	208.4
1,2-$C_6H_4Cl_2$	147.0	20.0	59.1	89.4	112.9	155.8	179.0
1,3-$C_6H_4Cl_2$	147.0	12.1	52.0	82.0	105.0	149.0	173.0
1,4-$C_6H_4Cl_2$	147.0	—	54.8	84.8	108.4	150.2	173.9
C_6H_5Cl	112.5	−13.0	+22.2	49.7	70.7	110.0	132.2
Benzene	78.1	−36.7	−11.5	+7.6	26.1	60.6	80.1
Cyclohexane	84.2	−45.3	−15.9	+6.7	25.5	60.8	80.7
Hexane	86.2	−53.9	−25.0	−2.3	+15.8	49.6	68.7
2,2-diMebutane	86.2	−69.3	−41.5	−19.5	−2.0	+31.0	49.7
Toluene	92.1	−26.7	+6.4	31.8	51.9	89.5	110.6

(continued)

TABLE 2.1. (continued).

Compound	Molecular Wt. (gm/mol)	Temperature, °C					
		Vapor Pressures, torr					
		1	10	40	100	400	760
Heptane	100.2	−34.0	−2.1	+22.3	41.8	78.0	98.4
2,2-diMepentane	100.2	−49.0	−18.7	+5.0	23.9	59.2	79.2
Styrene	104.1	−7.0	+30.8	59.8	82.0	—	—
Ethylbenzene	106.2	−9.8	+25.9	52.8	74.1	113.8	136.2
2-Xylene	106.2	−3.8	+32.1	59.5	81.3	121.7	144.4
3-Xylene	106.2	−6.9	+28.3	55.3	76.8	116.7	139.1
4-Xylene	106.2	−8.1	+27.3	54.4	75.9	115.9	138.3
Octane	114.2	−14.0	+19.2	45.1	65.7	104.0	125.0
2,2-diMehexane	114.2	−29.7	+3.1	28.2	48.2	85.6	106.8
Nonane	128.2	+1.4	38.0	66.0	88.1	128.2	150.8
Decane	142.3	+16.5	55.7	85.5	108.6	150.6	174.1

From Weast et al. (1985).

feasible if the vapor pressure of the compound is less than 0.5 torr (Pedersen and Curtis, 1991).

HENRY'S LAW CONSIDERATIONS

If soil vacuum extraction passes this first test, there is a second that should be investigated. One of the forms by which VOCs are held in soils is in solution in the soil moisture. Such aqueous solutions obey Henry's Law, which states that the vapor phase concentration of VOC in contact with a solution of the VOC is proportional to the VOC concentration in the solution. If one assumes that the VOC is held in the soil only in solution in the soil moisture, one can obtain estimates of the cleanup times required, given various limiting assumptions about the vapor stripping process. If this can be regarded as stripping from a column having a very large number of theoretical transfer units (the most efficient possible process), then the time required for cleanup is given by

$$t_{100} = \frac{w \cdot V_s}{Q \cdot K_H} \tag{3}$$

where

w = volumetric soil moisture content, dimensionless
V_s = volume of soil being treated, m³
Q = volumetric airflow rate, m³/sec

TABLE 2.2. Vapor Pressure Parameters of Selected VOCs.

| Compound | Molecular Wt. | $\log_{10} P(T)$ (torr) $= A - .05223B/T$ | |
		A	B
CCl_2F_2	120.9	7.5356	21,670
CCl_3F	137.4	7.6088	26,853
CCl_4	153.8	7.8584	33,259
$CHCl_3$	119.4	7.9412	32,273
CH_2Br_2	173.8	8.0444	36,525
CH_2Cl_2	84.9	7.9629	30,512
CS_2	76.1	7.5228	28,381
C_2Cl_4 (PCE)	165.8	8.2769	40,006
C_2HCl_3 (TCE)	131.4	7.9635	34,747
cis-1,2-DCE	96.9	8.0250	32,610
trans-1,2-DCE	96.9	7.9335	31,029
1,1-DCE	96.9	7.7616	28,466
1,1,1,2-TTCA	167.8	8.1029	39,818
1,1,2,2-TTCA	167.8	8.2868	42,724
1,1,1-TCA	133.4	7.8744	33,025
1,1,2-TCA	133.4	8.2994	39,493
1,1-DCA	99.0	7.8442	31,400
1,2-DCA	99.0	7.9571	34,520
Ethyl chloride	64.5	7.6915	26,265
1,1,1-Trichl. prop.	147.4	8.2175	38,400
1,2,3-Trichl. prop.	147.4	8.5650	46,252
1,2-Dichl. propane	113.0	8.0400	36,075
1-Chloropropane	78.5	7.7875	30,018
2-Chloropropane	78.5	7.4849	27,255
Pentane	72.1	7.6621	28,275
2-Methylbutane	72.1	7.5885	27,095
2,2-diMepropane	72.1	7.0864	22,794
1,2,3,4-$C_6H_2Cl_4$	215.9	8.3860	54,846
1,2,3,5-$C_6H_2Cl_4$	215.9	8.1073	51,402
1,2,3-$C_6H_3Cl_3$	181.4	7.8860	47,279
1,2,4-$C_6H_3Cl_3$	181.4	8.1429	48,555
1,3,5-$C_6H_3Cl_3$	181.4	8.0537	47,418
1,2-$C_6H_4Cl_2$	147.0	8.3849	47,034
1,3-$C_6H_4Cl_2$	147.0	8.1411	44,455
1,4-$C_6H_4Cl_2$	147.0	8.1248	44,721
C_6H_5Cl	112.5	8.2666	41,148
Benzene	78.1	7.8948	33,803
Cyclohexane	84.2	7.7703	33,026
Hexane	86.2	7.9399	32,916
2,2-diMebutane	86.2	7.5709	28,955
Toluene	92.1	8.2910	39,082

(continued)

TABLE 2.2. (continued).

| Compound | Molecular Wt. | $\log_{10} P(T)$ (torr) $= A - .05223B/T$ | |
		A	B
Heptane	100.2	8.1853	37,270
2,2-diMepentane	100.2	7.8432	33,285
Styrene	104.1	7.9923	40,713
Ethylbenzene	106.2	8.3462	42,072
2-Xylene	106.2	8.4295	43,453
3-Xylene	106.2	8.4726	43,169
4-Xylene	106.2	8.4075	42,646
Octane	114.2	8.6481	42,878
2,2-diMehexane	114.2	8.2728	38,522
Nonane	128.2	8.4220	44,251
Decane	142.3	8.3351	46,208

Calculated from data taken from Weast et al. (1985).

K_H = Henry's constant of VOC; c(vapor) $= K_H C$(solution), dimensionless

t_{100} = time required for complete cleanup, sec

A less sanguine assumption for the vapor stripping process is to regard it as analogous to stripping from a well-stirred reactor. This leads to the following expressions for 90%, 99%, and 99.9% cleanup:

$$t_{90} = 2.303 \cdot \frac{w \cdot V_s}{Q \cdot K_H} \tag{4}$$

$$t_{99} = 4.605 \cdot \frac{w \cdot V_s}{Q \cdot K_H} \tag{5}$$

$$t_{99.9} = 6.908 \cdot \frac{w \cdot V_s}{Q \cdot K_H} \tag{6}$$

For example, if one were proposing to vapor strip p-xylene (with a Henry's constant at 15°C of .204) from 5,000 m^3 of soil having a volumetric moisture content of 0.2 by means of a vacuum well operating at 100 cfm, Equation (3) gives t_{100} = 1.2 days, while Equations (4), (5), and (6) yield t_{90} = 2.8 days, t_{99} = 5.5 days, and $t_{99.9}$ = 8.3 days. From this, one would conclude that soil vacuum extraction should still be regarded as a feasible technology for the removal of p-xylene at this site.

Table 2.3 gives the Henry's constants for a number of organic compounds in water at various temperatures. Henry's constants for other com-

TABLE 2.3. Dimensionless Henry's Constants for Selected VOCs.

Compound	10°C	15°C	20°C	25°C
Nonane	17.21	20.98	13.80	16.92
n-Hexane	10.24	17.47	36.71	31.39
2-Methylpentane	30.00	29.35	26.31	33.72
Cyclohexane	4.433	5.329	5.820	7.234
Chlorobenzene	.1050	.1188	.1418	.1471
1,2-Dichlorobenzene	.07015	.06048	.06984	.06417
1,3-Dichlorobenzene	.09511	.09769	.1122	.1696
1,4-Dichlorobenzene	.09124	.09177	.1077	.1296
2-Xylene	.1227	.1527	.1970	.2516
4-Xylene	.1808	.2043	.2681	.3041
3-Xylene	.1769	.2098	.2486	.3041
Propylbenzene	.2445	.3092	.3662	.4414
Ethylbenzene	.1403	.1907	.2498	.3221
Toluene	.1640	.2081	.2307	.2624
Benzene	.1420	.1641	.1879	.2158
Methylethylbenzene	.1511	.1776	.2091	.2281
Vinyl chloride	.6456	.7105	.9021	1.083
1,1-Dichloroethane	.1584	.1920	.2340	.2555
1,2-Dichloroethane	.05035	.05498	.06111	.05763
1,1,1-TCA	.4153	.4864	.6069	.7112
1,1,2-TCA	.01678	.02664	.03076	.03719
cis-1,2-DCE	.1162	.1379	.1497	.1856
trans-1,2-DCE	.2539	.2982	.3563	.3863
Tetrachloroethylene	.3641	.4694	.5861	.6989
Trichloroethylene	.2315	.2821	.3500	.4169
Tetralin	.03228	.04441	.05654	.07643
Decalin	3.013	3.540	4.406	4.782
Chloroethane	.3267	.4052	.4573	.4946
Hexachloroethane	.2552	.2364	.2457	.3413
Carbon tetrachloride	.6370	.8078	.9644	1.206
1,3,5-triMebenzene	.1734	.1945	.2374	.2751
Ethylene dibromide	.01291	.02030	.02536	.02657
1,1-DCE	.6628	.8585	.9062	1.059
Methylene chloride	.06025	.07147	.1014	.1210
Chloroform	.07403	.09854	.1380	.1721
1,1,2,2-TTCA	.01420	.00846	.03035	.01022
1,2-Dichloropropane	.05251	.05329	.07898	.1459
Dibromochloromethane	.01635	.01903	.04282	.04823
1,2,4-$C_6H_3Cl_3$	.05552	.04441	.07607	.07848
2,4-Dimethylphenol	.3568	.2850	.4199	.2015
1,1,2-$C_2Cl_3F_3$	6.628	9.093	10.18	13.04
Methylethylketone	.01205	.01649	.00790	.00531
Methylisobutylketone	.02841	.01565	.01206	.01594
Methyl cellosolve	1.898	1.535	4.822	1.263
CCl_3F	2.307	2.876	3.342	4.128

Adapted from Howe et al. (1986).

pounds can be calculated from the VOC vapor pressure and solubility at the temperature of interest from the following formula:

$$K_H(T) = 0.01603 \cdot \frac{P_0(T) \cdot (M.W.)}{T \cdot c_{sat}} \tag{7}$$

$P_0(T)$ = VOC equilibrium vapor pressure at temperature T, torr
$M.W.$ = VOC molecular weight, gm/mol
T = ambient temperature, deg K
c_{sat} = VOC saturation concentration in water at temperature T, gm/L
$K_H(T)$ = Henry's constant of VOC at temperature T, dimensionless

A number of extensive data compilations are available (Verschueren, 1983; Mackay and Shiu, 1981; Weast et al., 1985; Lyman et al., 1982; Brookman et al., 1985; Howe et al., 1986; Danko, 1989; Montgomery and Welkom, 1989). Generally, soil vacuum extraction is not feasible for compounds having dimensionless Henry's constants less than about 0.01 (Danko, 1989; Pedersen and Curtis, 1991).

OTHER HURDLES FOR SVE—SORPTION AND PERMEABILITY

The above two tests of vapor stripping (the vapor pressure and Henry's constant of the VOC) may clearly eliminate the technique for use at a particular site. They do not, however, demonstrate that vacuum extraction is, in fact, feasible. VOCs are held in soils by other mechanisms than solution in the soil moisture—adsorption on clays and humic materials, for example. These mechanisms may reduce the effective Henry's constant (or linear isotherm parameter) of a VOC by a factor of 1/100 or less. The stratigraphy of the site may be such that an adequate air flow cannot be maintained in some portions of the contaminated zone. EPA notes (Pedersen and Curtis, 1991) that SVE is most effective if the hydraulic conductivity of the soil is above 0.001 cm/sec; contaminant removals have been demonstrated in soils having hydraulic conductivities as low as 10^{-6} cm/sec, however (Danko, 1989). Freeze and Cherry (1979) have provided a convenient summary of the permeabilities associated with a wide variety of soils; this is reproduced as Figure 2.1. A very useful nomograph from EPA's SVE handbook (Pedersen and Curtis, 1991) is shown in Figure 2.2.

DIFFUSION/DESORPTION KINETICS

Diffusion of desorption kinetics may be rate-limiting if the soil contains lumps or lenses of porous but low-permeability clay or porous fractured rock, or some of the VOC may be contained in buried drums. These mat-

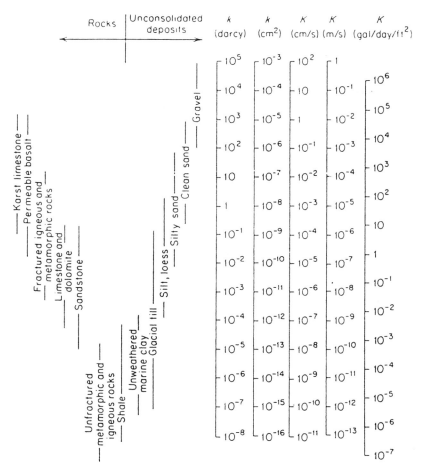

Figure 2.1 Range of values of hydraulic conductivity and permeability. (From Chapter 2 in *Groundwater*, Freeze/Cherry, eds., © 1979, reprinted by permission of Prentice-Hall, Inc., Englewood Cliffs, NJ.)

ters are addressed in the second and third phases of the feasibility assessment.

PHASE DISTRIBUTION OF VOCs IN SOILS

Typically, a VOC in the vadose zone will be present in at least three or four different forms: (1) as vapor in the interstitial soil gas, (2) dissolved in the soil water, (3) adsorbed on sorption sites in the soil, and (possibly) (4) as nonaqueous phase product, in the form of droplets or ganglia. Pedersen and Curtis (1991) report that several workers have found residual

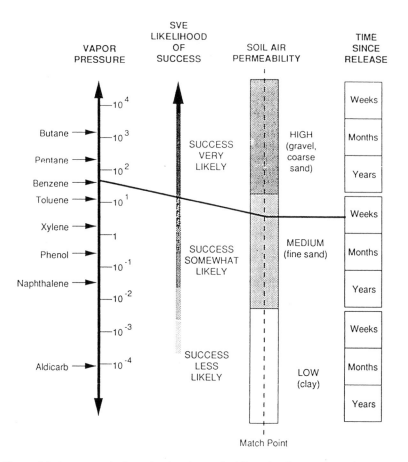

Figure 2.2 A nomograph for estimating the applicability of soil vapor extraction. [From Pedersen and Curtis (1991), EPA Document EPA/540/2-91/003.]

NAPL occupying on the order of 10% of the available pore space after the wetting front moves through the soil. If the adsorption isotherm for the VOC in the soil is known, one can calculate the equilibrium partitioning of the VOC between these phases as follows. We have

$$C_{tot} = \sigma C_v + \omega C_w + C_{ads} + C_{napl} \qquad (8)$$

where

C_{tot} = total VOC concentration in the soil, kg/m³ of bulk soil
C_v = VOC concentration in the soil gas, kg/m³ of soil gas
C_w = VOC concentration in the soil water, kg/m³ of soil water

C_{ads} = adsorbed VOC concentration in the soil, kg/m³ of bulk soil
C_{napl} = concentration of VOC present as nonaqueous phase liquid, kg/m³ of bulk soil
  σ = air-filled porosity, dimensionless
  ω = water-filled porosity, dimensionless

Then C_{ads} and C_w are related to C_v as follows:

$$C_w = C_v/K_H \tag{9}$$

$$C_{ads} = f_v(C_v) \tag{10}$$

Here $f_v(C_v)$ is the adsorption isotherm for the VOC in the soil of interest, which describes the partitioning between the adsorbed and vapor phases. This may be a linear, Langmuir, Freundlich, BET, or other isotherm.

One proceeds by initially assuming that $C_{napl} = 0$, then substituting Equations (9) and (10) into Equation (8), which yields an implicit equation for C_v,

$$C_{tot} = C_v + \omega C_v/K_H + f_v(C_v) \tag{11}$$

This equation is then solved for C_v. The solution is trivial if the adsorption isotherm f_v is a linear function of C_v; if f_v is more complex, one can use a numerical method such as binary search. If C_v is less than the equilibrium vapor concentration of the pure VOC, no NAPL phase VOC is present, the vapor phase VOC concentration is C_v, and the aqueous and adsorbed phase concentrations can be calculated by Equations (9) and (10). If C_v is equal to or greater than the equilibrium vapor concentration of the pure VOC, then one calculates C_{napl} as follows. Substitute Equations (9) and (10) into Equation (8), set $C_v = C_{v0}$, where C_{v0} is the equilibrium vapor concentration of the VOC, and solve for C_{napl}. This yields

$$C_{napl} = C_{tot} - [\sigma C_{v0} + \omega C_{v0}/K_H + f_v(C_{v0})] \tag{12}$$

Then C_w and C_{ads} are given by setting $C_v = C_{v0}$ in Equations (9) and (10). Usually, the available data are insufficient to justify the use of any adsorption isotherm more sophisticated than a simple linear expression, which is why effective Henry's constants are commonly used.

Note that the above analysis assumes that the menu of VOCs present at the site can be adequately approximated by a single representative compound. If one is dealing with complex mixtures such as gasoline, this approach may not be adequate, and a model based upon Raoult's Law and the assumption that the bulk of the VOC mixture is present as NAPL may be more suitable, at least during the early stages of the remediation. EPA's reference handbook on SVE discusses the SVE of gasoline and similar mixtures in some detail (Pedersen and Curtis, 1991), and the modeling of SVE for Raoult's Law mixtures is addressed in Chapters 5 and 6 of this book.

SECOND PHASE OF SVE ASSESSMENT

The second phase of vapor stripping feasibility assessment involves a laboratory study. Soil samples are taken at the site to represent, roughly, the average and the highest levels of contamination and the various media present (sand, silt, till, clay, etc.). Portions of these samples are analyzed for VOCs, and portions are placed in laboratory soil vapor stripping columns. Stripping is carried out until analysis of the gas vented from the columns indicates VOC removal is nearly complete. (This is important if one is to avoid what may be extremely over-optimistic assessments of the rate of SVE, in that during the initial phase of SVE, VOC removal is rather rapid as weakly bound VOC is removed. Later in the run the more strongly adsorbed VOC is removed, and this process is a good deal slower.) The vapor stripped soil samples are then analyzed for residual VOCs and the percent removals calculated. A simple method of obtaining adsorption isotherms from lab column data using a plug flow model is described later in this chapter.

A mathematical model for laboratory column operation [such as is described below and in Wilson et al. (1988)] may be used to process the resulting data to obtain estimates of the effective Henry's constants for the VOCs in the media present at the site. This is most conveniently done by constructing a plot of percent removal versus effective Henry's constant for a system having the same parameters (column length and diameter, gas flow rate, soil moisture content, run duration, etc.) as the laboratory columns, and the effective Henry's constants yielding the observed percent removals are then read off of the plot. Such a plot is shown in Figure 2.3. These effective Henry's constants can then be used to estimate cleanup times by means of Equations (3)–(6). If these are felt to be acceptable, one is ready to carry out the third phase of the feasibility evaluation.

A SIMPLE LAB COLUMN SVE MODEL

A simple laboratory column vapor stripping model is constructed as follows. The differential equation governing the pressure of an ideal gas in steady flow through a porous medium is

$$\nabla \cdot K_D \nabla P^2 = 0 \tag{13}$$

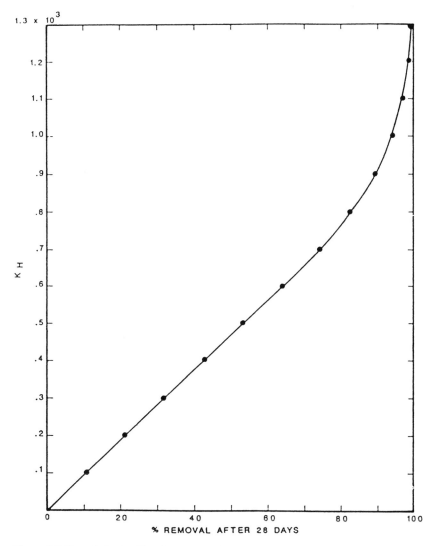

Figure 2.3 Plot of effective Henry's constant (linear adsorption parameter) versus percent removal after twenty-eight days of vapor stripping. A lab column simulation. Column height = 32.1 cm, column radius = 3.15 cm, air flow rate = 5.0 mL/min, air-filled porosity = 0.2, specific moisture content = 0.2. [Reprinted from Wilson et al. (1988), p. 991, by courtesy of Marcel Dekker, Inc.]

which, for a column packed with soil of uniform permeability, gives

$$\frac{d^2(P^2)}{dx^2} = 0 \tag{14}$$

where K_D is the pneumatic permeability of the soil (the Darcy's constant, m²/atm sec), P is the pressure (atm), and x is the distance measured from the inlet end of the column. Integration of Equation (14) and insertion of the boundary conditions at the ends of the column then yields

$$P = \left[P_i^2 - \frac{P_i^2 - P_f^2}{L} \cdot x \right]^{1/2} \tag{15}$$

where

P_i = column inlet pressure, atm
P_f = column outlet pressure, atm
L = column length, m

Darcy's Law,

$$v = -K_D \nabla P \tag{16}$$

then yields

$$v_x = \frac{K_D \cdot (P_i^2 - P_f^2)}{2 \cdot L} \cdot \left[P_i^2 - \frac{(P_i^2 - P_f^2)}{L} \cdot x \right]^{-1/2} \tag{17}$$

The molar gas flow rate (mol/sec) through the column is readily shown to be

$$Q = \frac{\pi r_c^2 \sigma}{RT} \cdot \frac{K_D(P_i^2 - P_f^2)}{2L} \tag{18}$$

where

r_c = column radius, m
R = gas constant, 8.206×10^{-5} m³ atm/mol deg
T = temperature, deg K
σ = soil air-filled porosity

It is not advisable to use the pressure drops across the soil columns to estimate the pneumatic permeabilities of the soil samples from Equation (18). The samples are sufficiently disturbed during sampling and packing into the columns that the measured pneumatic permeabilities are likely to be far different from the permeabilities of the undisturbed media. Permeability must be measured by a pilot-scale field study on the site, which constitutes the third and last phase of the feasibility assessment, or by field tests with a small portable unit. These tests can be done during the course of sampling. EPA's reference handbook on SVE (Pedersen and Curtis, 1991) provides discussions of methods for determining permeabilities; Appendix E of this handbook (Johnson et al., 1989b) gives detailed instructions for this, and we address the problem in Chapter 3 here.

In modeling the elution of a VOC from the soil column, we assume a Henry's Law–type isotherm (with an effective Henry's constant K_H) and local equilibrium. The column is partitioned into a set of N volume elements, and a mass balance is carried out on the VOC being vapor stripped. Axial dispersion is handled by adjusting the number of volume elements into which the column is partitioned; the larger the number of volume elements, the smaller is the axial dispersion. The partial differential equation describing advective transport is

$$\frac{\partial c}{\partial t} = -\frac{\sigma K_H}{\omega + K_H} \cdot \frac{\partial v_x c}{\partial x} \tag{19}$$

which we approximate by a set of ordinary differential equations, one for each volume element; this yields

$$\frac{dc_i}{dt} = \frac{K_H}{\Delta x(\omega + \sigma K_H)} \cdot [(v_x c)_{i-1} - (v_x c)_i], \qquad i = 1, 2, \dots, N \tag{20}$$

Here

c = VOC concentration in the column at the point x, kg/m³
c_i = mean VOC concentration in the ith volume element, kg/m³
ω = soil volumetric moisture content, dimensionless
t = time, sec

An initial VOC concentration is selected and Equation (20) is then integrated forward in time by a simple predictor-corrector method. The total mass of VOC remaining in the column at time t is given by

$$m_{total} = \sum \pi r_c^2 \cdot \Delta x \cdot c_i(t) \tag{21}$$

Generally $\omega \gg \sigma K_H$, so that removal rates are essentially proportional to the effective Henry's constant. This makes it quite easy to scale results, since one can make a plot using a reduced time $K_H \cdot t$.

One should note that the soil structure is severely disturbed in the course of preparing a lab column experiment. Porous structures of low permeability (clay, till, or silt lenses or layers) are likely to be disrupted and mixed in with material of higher permeability. Therefore, lab column experiments will not identify possible limitations in the rate of SVE due to slow diffusion of VOC from these low-permeability structures; this can *only* be done by field experiments, although the possibility that there may be a problem can often be ascertained by inspection of the well logs.

SVE PILOT STUDIES

In the pilot study, one or more vapor stripping wells are placed in or near regions of high contaminant concentration and/or at points where stratigraphic data (well logs, etc.) indicate potential problems may exist (regions of low permeability, buried obstacles, etc.). Preliminary site data can be used in mathematical models to optimize the design of the pilot study. Piezometer wells can be used around a vacuum well to determine its range of influence and the effects of clay lenses, strata of low and high permeabilities, perched water tables, etc. Well vacuum and gas flow rate measurements can be used to obtain a mean value for the pneumatic permeability in the vicinity of the well. VOC removal rates can be used to verify the effective Henry's constant values obtained in the lab study.

The set of parameters resulting from completion of the third phase of the feasibility assessment is then used in the models to develop an optimized design for the facility and to simulate its operation. From these simulations one can estimate cleanup times and costs, at which point one is in position to make a final assessment of the feasibility of soil vapor stripping at the site. One can also use the results of the simulations to locate those portions of the domain of interest that will be cleaned up most slowly; particular attention may be paid to these during post-cleanup monitoring. It should be noted, however, that prediction of cleanup times from model calculations is fraught with considerable uncertainty, since the data set on which the model calculations are based is generally quite sparse because of time and budget constraints.

SOIL CHARACTERISTICS

Soil characteristics have a very major effect on the initial transport of VOCs in the soil and on their removal by SVE. Soil porosity and grain size, stratigraphy, and soil moisture content will determine the ability of a vacuum well to deliver air flow to the contaminated soil. The effect of soil

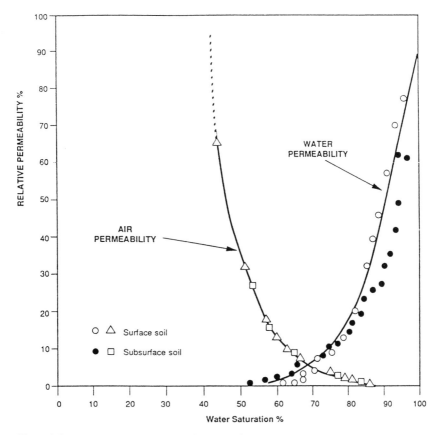

Figure 2.4 Air and water permeability as functions of water content. [From Pedersen and Curtis (1991), EPA Document EPA/540/2-91/003, adapted from Corey (1957).]

moisture content on air and water permeabilities is shown in Figure 2.4 (Corey, 1957) and is seen to be quite large; SVE cannot be carried out on saturated soils. Soil moisture content and the presence of adsorption sites (on clay or on natural humic materials) will affect the efficiency with which the advecting air will be able to move VOC from the soil. Very low soil moisture content results in increased binding of VOCs to soil surfaces, thereby reducing SVE efficiency—the so-called wet dog effect (Davies, 1989).

MULTIPLE-TECHNOLOGY APPROACHES

Frequently, soil vapor stripping will be used in conjunction with other remediation techniques. Pump-and-treat may be necessary to remove (or at least to contain) contaminants already in the groundwater, with vacuum extraction being used to remove VOCs in the vadose zone, which would slowly recontaminate the groundwater were they not removed. The aquifer may be suitable for remediation by sparging or bioremediation. One may need to remove and treat a surface layer of soil contaminated with a slow-moving nonvolatile compound such as a heavy metal, PCBs, or chlorinated pesticides after VOCs were removed by vapor stripping. A complex site may require the deployment of several sequential and/or simultaneous techniques for effective cleanup.

The logic of the selection process for SVE is illustrated in Table 2.4, which provides a decision tree that can be used as a guide. EPA's *Guide for Conducting Treatability Studies under CERCLA: Soil Vapor Extraction* gives an excellent discussion of preliminary screening, treatability tests, and data interpretation in connection with SVE; the guide includes a treatability flow chart (Figure 2.5) and an example project schedule for an SVE treatability study (Figure 2.6).

INTERCONVERSION OF INTRINSIC PERMEABILITIES AND PNEUMATIC PERMEABILITIES

Often, one can use well log information and tables showing the permeability ranges of various porous media (sand, silt, etc.), such as that provided by Freeze and Cherry (1979) to make a first estimate of the intrinsic permeabilities of the media present at a site. Most SVE models, however, use pneumatic permeabilities. In this section we develop the relationship between pneumatic permeability and the intrinsic permeability. We define the pneumatic permeability by the equation

$$v = -K_D \nabla P \tag{22}$$

where

v = superficial gas velocity, m/sec
K_D = pneumatic permeability tensor, m²/sec atm, assumed to be a scalar in the subsequent treatment (i.e., isotropic)
P = gas pressure, atm

From Freeze and Cherry's Equation (2.25), the magnitude of the pressure gradient is given by

$$|\nabla P| = \varrho g \frac{dh}{dl} \tag{23}$$

TABLE 2.4. **Soil Vapor Stripping Decision Tree.**

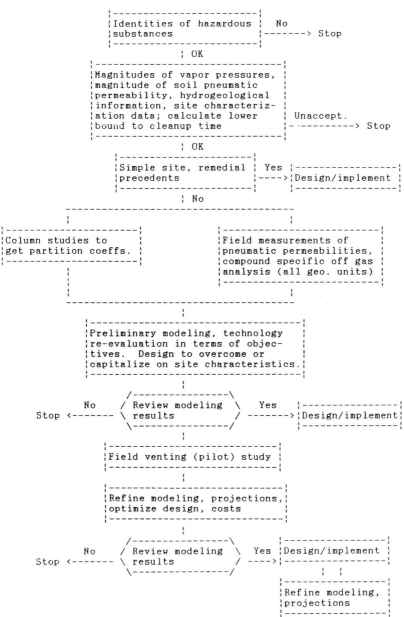

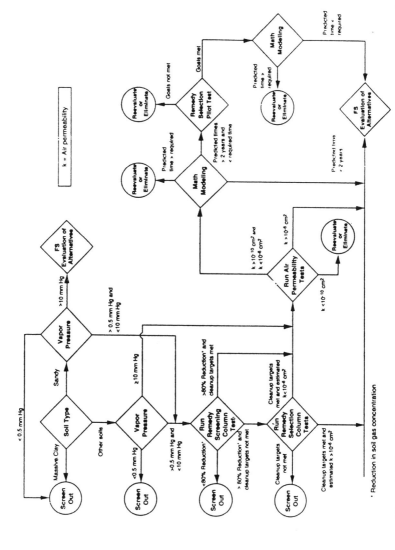

Figure 2.5 Treatability flowchart for evaluating SVE. [From Stumbar and Rawe (1991), EPA Document EPA/540/2-19/019A.]

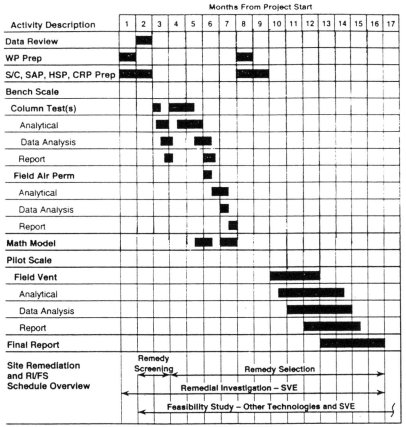

Months From Project Start

Activity Description	1	2	3	4	5	6	7	8	9	10	11	12	13	14	15	16	17
Data Review		■															
WP Prep	■							■									
S/C, SAP, HSP, CRP Prep	■■■							■■■									
Bench Scale																	
Column Test(s)			■ ■■														
Analytical			■ ■														
Data Analysis			■ ■														
Report			■	■													
Field Air Perm				■													
Analytical						■											
Data Analysis						■											
Report						■											
Math Model				■ ■													
Pilot Scale																	
Field Vent									■■■								
Analytical										■■■							
Data Analysis										■■■							
Report											■■■						
Final Report												■■■					

| Site Remediation and RI/FS Schedule Overview | Remedy Screening ⟶ | | Remedy Selection ⟶ | | | | | | | | | | | | | | |

Remedial Investigation – SVE ◀————————————▶

Feasibility Study – Other Technologies and SVE ◀————————————

Note: Abbreviations defined in Abbreviations, Acronyms, and Symbols (p. Ix)

Figure 2.6 Example project schedule for a full-tier SVE treatability study. [From Stumbar and Rawe (1991), EPA Document EPA/540/2-91/019A.]

where

ϱ = density of fluid (water), kg/m³
g = gravitational constant, 9.8 m/sec²
h = hydrostatic head, m
l = distance in the direction of maximum increase of hydrostatic head, m

From this we see that

$$K_D = Cd^2/\mu \tag{24}$$

in Freeze and Cherry's notation, where μ is the dynamic viscosity of the fluid. The intrinsic permeability, k, is defined [Freeze and Cherry's Equation (2.27)] as

$$k = Cd^2 \qquad (25)$$

So

$$K_D = k/\mu \qquad (26)$$

However, some unit conversions are necessary before one can use this simple relationship; k is commonly given in units of cm^2, so we need to express K_D in c.g.s. units, K_D'. This is done as follows:

$$K_D \left[\frac{m_2}{atm\ sec}\right] \cdot \frac{1\ atm}{1.01325 \times 10^5\ kg\ m\ sec^{-2}\ m^{-2}} \cdot \frac{10^6\ cm^3}{1\ m^3} \cdot \frac{1\ kg}{10^3\ gm} = K_D'$$

$$(27)$$

Then

$$k = K_D' \left[\frac{cm^3\ sec}{gm}\right] \cdot \left[\frac{gm}{cm\ sec}\right] \qquad (28)$$

so

$$k\ (cm^2) = K_D \left[\frac{m^2}{atm\ sec}\right] \times \frac{10^3}{1.01325 \times 10^5} \times 1.8 \times 10^{-4}$$

or

$$k\ (cm^2) = 1.776 \times 10^{-6} \times K_D (m^2/atm\ sec) \qquad (29)$$

and

$$K_D\ (m^2/atm\ sec) = 5.631 \times 10^5 \times k (cm^2) \qquad (30)$$

A SIMPLE EXPONENTIAL-REMOVAL MODEL FOR SVE

Roy and Griffin (1989) have published a simple model for SVE for use as a screening tool and for conducting sensitivity analyses. We follow their discussion and notation here.

Assume that the total mass of contaminant per unit volume of soil is present in three phases:

$$M = M_v + M_d + M_a \qquad (31)$$

where

M = total contaminant mass per unit volume of soil
M_v = contaminant mass per unit volume of soil in the vapor phase
M_d = contaminant mass per unit volume of soil dissolved in interstitial water
M_a = contaminant mass per unit volume of soil adsorbed to solid phases

Also assume that the equilibrium distribution of the vapor phase VOC concentration C_v and the aqueous phase VOC concentration C_d is described by Henry's Law,

$$C_v/C_d = K_H \tag{32}$$

Further, assume that the adsorption-desorption of VOCs can be described by a linear reversible isotherm with an adsorption constant K_d, where

$$C_a/C_d = K_d \tag{33}$$

Then if Equation (31) is written on a mass VOC per unit volume basis, it becomes

$$M = v_a C_v + v_w C_d + K_d \varrho_b C_d \tag{34}$$

where

v_a = air-filled porosity
v_w = water-filled porosity
ϱ_b = dry bulk density of the soil

Define

C_t = total VOC concentration
v_t = total porosity

Then substitution of Equation (32) into Equation (34), use of these definitions, and rearranging yields

$$C_v/C_t = [(K_d \varrho_b/K_H) + (v_w/K_H) + (v_t - v_w)]^{-1} \tag{35}$$

Lyman et al. (1982) have defined Equation (35) as the strippability relationship, which describes the partitioning of a VOC between vapor, liquid, and adsorbed phases in the vadose zone. (Note that the model omits the possible presence of an NAPL phase.)

As an SVE well draws air through the contaminated zone, the organic vapors are removed, and, provided that the partitioning of the VOC between phases remains at equilibrium, passage of a volume of air dV through the system results in a change in total mass of VOC in the system of

$$dM = -C_v dV \tag{36}$$

Define the right side of Equation (35) as β; then Equation (36) can be rewritten as

$$\frac{dM}{M} = -\frac{\beta(\nu_t - \nu_w)}{V_a} \cdot dV \tag{37}$$

since the total volume of the soil may be written as $V_a/(\nu_t - \nu_2)$, where V_a is the total volume of air in the pore spaces.

Equation (37) is then integrated from initial conditions $M(0) = M_0$ to $M(V)$, and $V' = 0$ to $V' = V$ to yield, after some rearrangement,

$$M(V) = M_0 \exp[-\beta(\nu_t - \nu_w)(V/V_a)] \tag{38}$$

or, on noting that $V_a = \nu_a V_{soil}$ and that $\nu_a = \nu_t - \nu_w$,

$$M(V) = M_0 \exp(-\beta V/V_{soil}) \tag{39}$$

Equation (39) provides a very simple way to make a rather rough estimate of the volume V of gas that will be required to remediate a site. The assumptions of (1) a completely mixed system and (2) no diffusion/desorption kinetics limitations must be kept in mind in interpreting the results, which may be far too optimistic. For preliminary screening, however, Equation (39) can be quite useful.

A SIMPLE PLUG FLOW REACTOR MODEL FOR SVE LAB COLUMN OPERATION

A local equilibrium, linear isotherm model for SVE lab column operation can be constructed by treating the column as a simple plug flow reactor. This is done as follows.

Let

$C(x,t) =$ total VOC concentration at a distance x from the entrance to the column at time t

C_0 = initial total VOC concentration in the soil sample
$C^v(x,t)$ = vapor phase VOC concentration, mass/volume of vapor phase
$C^a(x,t)$ = adsorbed VOC concentration, mass/volume of soil
K_a = VOC adsorption coefficient
σ = air-filled porosity of soil
Q = volumetric airflow rate through column
L = length of column
A = cross-sectional area of column

A mass balance on an arbitrary volume element in the column then yields

$$\frac{\partial C}{\partial t} = -\frac{K_a Q}{A(1 + \sigma K_a)} \cdot \frac{\partial C}{\partial x} \tag{40}$$

The general solution to this equation is

$$C = F(t - x/r) \tag{41}$$

where

$$r = KQ/[A(1 + \sigma K_a)] \tag{42}$$

The initial condition is

$$C(x,0) = 0, \quad x < 0$$

$$= C_0, \quad 0 < x < L \tag{43}$$

At a later time t this then gives

$$C(x,t) = 0, \quad x < rt$$

$$C(x,t) = C_0, \quad rt < x < L \tag{44}$$

The total mass of VOC remaining in the column at time t is then given by

$$M(t)/M_0 = (1 - rt/L) \tag{45}$$

Let the absolute value of the slope of a plot of $M(t)/M_0$ (determined experimentally) be S; then from Equation (45)

$$S = r/L \tag{46}$$

Substitution of Equation (42) into Equation (46) and solution for K_a then yields

$$K_a = \frac{LAS}{Q - LA\sigma S} \qquad (47)$$

for the adsorption coefficient of the VOC. The time required for 100% removal of the VOC from the column is readily obtained from Equations (42) and (45); it is

$$t_{100} = \frac{AL(1 + \sigma K_a)}{K_a Q} \qquad (48)$$

This analysis is based on the assumption that diffusion/desorption rates are fast, so that local equilibrium is maintained and so that the adsorption isotherm is linear. One can test as to whether the assumption of local equilibrium is true by making runs at markedly different gas flow rates Q. If these yield quite similar values of K_a, then the assumption is valid. If not, it will be necessary to use an SVE lab column model that includes diffusion/desorption mass transport rates and/or a more realistic adsorption isotherm. The approximation of neglecting dispersion is probably not a major source of error in the calculations.

REFERENCES

Brookman, G. T., M. Flanagan, and J. O. Kebe, 1985, *Literature Survey: Hydrocarbon Solubilities and Attenuation Mechanisms,* American Petroleum Institute, API Publ. No. 4414, August.

Clarke, A. N. and D. J. Wilson, 1988, "A Phased Approach to the Development of in situ Vapor Stripping Treatment," *Proc., 1st Ann. Hazardous Materials Management Conf./Central,* March 15–17, Rosemont, IL, p. 191.

Corey, A. T., 1957, "Measurement of Water and Air Permeability in Unsaturated Soils," *Soil Sci. Soc. Am. Proc.,* 21:7.

Danko, J., 1989, "Applicability and Limitations of Soil Vapor Extraction for Sites Contaminated with Volatile Organic Compounds," *Soil Vapor Extraction Technology Workshop,* June 28–29, U.S. EPA RREL, Edison, NJ.

Danko, J., 1990, "Soil Vapor Extraction Applicability and Limitations," *Proc., 5th Ann. Hazardous Materials Management Conf./West,* Tower Conference Management Co., 800 Roosevelt Rd., Bldg. E, Suite 408, Glen Ellyn, IL 60137-5835.

Danko, J., 1991, "Applicability and Limits of Soil Vapor Extraction for Sites Contaminated with Volatile Organic Compounds," in *Soil Vapor Extraction Technology, Pollution Technology Review No. 204,* Noyes Publications, Park Ridge, NJ, p. 163.

Davies, S. H., 1989, "The Influence of Soil Characteristics on the Sorption of Organic

Vapors," *Workshop on Soil Vacuum Extraction,* April 27–28, U.S. EPA, RSKERL, Ada, OK.

Freeze, R. A. and J. A. Cherry, 1979, *Groundwater,* Prentice-Hall, Englewood Cliffs, NJ.

Howe, G. B., M. E. Mullins, and T. N. Rogers, 1986, "Evaluation and Prediction of Henry's Law Constants and Aqueous Solubilities for Solvents and Hydrocarbon Fuel Components Vol. I: Technical Discussion," USAFESE Report No. ESL-86-66. U.S. Air Force Engineering and Services Center, Tyndall AFB, FL, 86 pp.

Johnson, P. C., M. W. Kemblowski, and J. D. Colthart, 1989a, "Practical Screening Models for Soil Venting Applications," *Workshop on Soil Vacuum Extraction,* April 27–28, RSKERL, Ada, OK.

Johnson, P. C., M. W. Kemblowski, and J. D. Colthart, 1990a, "Quantitative Analysis for the Cleanup of Hydrocarbon-Contaminated Soils by in situ Soil Venting," *Ground Water,* 28:413.

Johnson, P. C., M. W. Kemblowski, J. D. Colthart, D. L. Byers, and C. C. Stanley, 1989b, "A Practical Approach to the Design, Operation, and Monitoring of in-situ Soil Venting Systems," *Soil Vapor Extraction Technology Workshop,* June 28—29, 1989, U.S. EPA Risk Reduction Engineering Laboratory (RREL), Edison, NJ. Reprinted in *Soil Vapor Extraction Technology. Pollution Technology Review No. 204,* T. A. Pedersen and J. T. Curtis, eds., 1991, Noyes Publications, Park Ridge, NJ, p. 195.

Johnson, P. C., C. C. Stanley, M. W. Kemblowski, D. L. Byers, and J. D. Colthart, 1990b, "Practical Approach to the Design, Operation and Monitoring of in situ Soil Venting Systems," *Ground Water Monitoring Review,* 10:159.

Lyman, W. J. and D. C. Noonan, 1990, "Assessing UST Corrective Action Technologies: Site Assessment and Selection of Unsaturated Zone Treatment Technologies," U.S. EPA Report No. EPA/600/2-90/011.

Lyman, W. J., D. C. Noonan, and P. J. Reidy, 1990, "Cleanup of Petroleum Contaminated Soils at Underground Storage Tanks," in *Polluation Technology Review No. 195,* Noyes Publications, Park Ridge, NJ.

Lyman, W. J., W. F. Reehl, and D. H. Rosenblatt, 1982, *Handbook of Chemical Property Estimation Methods: Environmental Behavior of Organic Compounds,* McGraw-Hill, New York, NY.

Mackay, D. and W. Y. Shiu, 1981, "A Critical Review of Henry's Law Constants for Chemicals of Environmental Interest," *J. Phys. Chem. Ref. Data,* 10(4):1175.

Montgomery, J. H. and L. M. Welkom, 1989, *Groundwater Chemicals Desk Reference, Vols. 1 and 2,* Lewis Publishers, Chelsea, MI.

Pacific Environmental Services, Inc., 1989, "Soil Vapor Extraction VOC Control Technology Assessment," U.S. EPA Report No. EPA-450/4-89-017, Sept.

Pedersen, T. A. and J. T. Curtis, 1991, "Soil Vapor Extraction Technology Reference Handbook," Risk Reduction Engineering Laboratory, U.S. EPA, Report No. EPA/540/2-91/003. Reprinted as *Soil Vapor Extraction Technology. Pollution Technology Review No. 204,* 1991, Noyes Publications (Noyes Data Corp.), Park Ridge, NJ.

Roy, W. R. and R. A. Griffin, 1989, "In-situ Extraction of Organic Vapors from Unsaturated Porous Media," Environmental Institute for Waste Management Studies, Univ. of Alabama, Tuscaloosa, open file report no. 24.

Stumbar, J. P. and J. Rawe, 1991, *Guide for Conducting Treatability Studies under*

CERCLA: Soil Vapor Extraction Interim Guidance, U.S. EPA Report No. EPA/540/2-91/019A.

Verschueren, K., 1983, *Handbook of Environmental Data on Organic Chemicals,* Van Nostrand Reinhold Co., New York, NY.

Weast, R. C., M. J. Astle, and W. H. Beyer, 1985, *CRC Handbook of Chemistry and Physics,* 65th ed., CRC Press, Inc., Boca Raton, FL.

Wilson, D. J., A. N. Clarke, and J. H. Clarke, 1988, "Soil Cleanup by in situ Aeration. I. Mathematical Modeling," *Separ. Sci. Technol.,* 23:991.

CHAPTER 3

Gas Flow in Soil Vapor Extraction

INTRODUCTION

THE feasibility of SVE at any particular site crucially depends on being able to deliver an adequate flux of air to the domains that are contaminated with volatile organic compounds (VOCs) or semivolatile organics (SVOCs). It is therefore common practice in assessing the feasibility of SVE at a site to make measurements of the pneumatic permeability of the soil. In this chapter, we first explore the range of validity of Darcy's Law,

$$v = -K_D \nabla P \tag{1}$$

and the implications of the results of this investigation. The calculation of soil gas pressure distributions and velocity fields in the vicinity of vapor stripping well (or a horizontal slotted pipe) by the method of images is then examined. This is followed by the development of a numerical over-relaxation method, which, although requiring considerably more computation than the method of images, permits one to compute soil gas velocity fields when the pneumatic permeability tensor is anisotropic and a function of position, when buried or surface impermeable obstacles are present, and when passive vent wells are used in addition to the soil vapor stripping well. The chapter closes with a section in which soil gas stream-lines and transit times are calculated for a number of cases to illustrate the impact of various soil features on the dynamics of the soil gas in the vicinity of a soil vapor extraction well.

DARCY'S LAW IN SOIL VAPOR EXTRACTION

In SVE treatability studies one commonly estimates the pneumatic permeability of the soil at the site by measuring the gas flow rate through a well that results from a given pressure difference (wellhead vacuum) (Stumbar and Rawe, 1991). If the soil gas is in the viscous flow regime throughout the entire domain of interest, Darcy's Law applies, the gas flow rate is proportional to the wellhead vacuum, and K_D can readily be estimated by making measurements at a single wellhead vacuum.

If, on the other hand, there are regions in the domain of interest in which the Reynolds Number of the air is of the order of one or larger, the simple linear form of Darcy's Law no longer holds, and one must consider the effects of momentum as well as those of viscosity (Freeze and Cherry, 1979). The Reynolds Number is given by

$$\text{Re} = \frac{\varrho v d}{\mu} \tag{2}$$

where

ϱ = fluid density
v = fluid velocity
d = characteristic length, of the order of a pore diameter
μ = fluid dynamic viscosity

de Marsily (1986) gives an equation relating the hydraulic gradient to the fluid velocity for the case of an incompressible fluid as one moves into the turbulent flow regime; it is

$$h = \alpha' U + \beta' U^2 \tag{3}$$

where

h = hydraulic head
α' = constant
β' = constant
U = fluid velocity

Perry (1969) gives a similar relationship for incompressible fluids,

$$\frac{P_1 - P_2}{L} = \frac{\alpha \mu V}{g} + \frac{\beta \varrho V^2}{g} \tag{4}$$

where

P_1 = absolute upstream pressure
P_2 = absolute downstream pressure

L = length through porous medium
V = superficial velocity of liquid (based on total cross section)
ϱ = fluid density
μ = fluid dynamic viscosity, mass/length time
g = gravitational constant
α = viscous resistance coefficient, length^{-2}
β = inertial resistance coefficient, length^{-1}

as well as an expression pertaining to ideal gases, which we shall examine later.

One may expect that near the screened section of the well, where there are high pressure gradients, there may be departures from the simple linear form of Darcy's Law, and that gas flow rate through the well may therefore not be simply proportional to wellhead vacuum. In the following sections this nonlinear effect is explored for an incompressible fluid (i.e., we assume that all pressures are sufficiently close to 1 atm that the compressibility of air can be ignored) and for an ideal compressible gas. The incompressible approximation permits permits analytical treatment, and we close this section with its application to some experimental data and calculation of some Reynolds Numbers for conditions appropriate to SVE well operation [see Wilson et al. (1993); Wilson (1993)].

INCOMPRESSIBLE FLUID APPROXIMATION

Equation (4) is readily rewritten in the limit as $L \rightarrow 0$ as

$$-\frac{dP}{dL} = (A + BV)V \tag{5}$$

where A and B are constants determined by the characteristics of the fluid and the porous medium and L is the distance measured in the direction of positive flow. If we consider the flow field of an incompressible fluid moving toward a constant point sink at the origin we may work in spherical coordinates (r, θ, ϕ), and readily have

$$-V = v_r = -\frac{Q}{4\pi r^2} \tag{6}$$

and

$$-\frac{dP}{dL} = \frac{dP}{dr} \tag{7}$$

Then Equation (5) becomes

$$\frac{dP}{dr} = \frac{AQ}{4\pi r^2} + \frac{BQ^2}{(4\pi)^2 r^4} \tag{8}$$

One then integrates Equation (8) between r_1, the radius of the well's gravel packing, and r_2, a point at some distance from the well where the pressure $P(r_2)$ is essentially 1 atm. The result is

$$P(r_2) - P(r_1) = \frac{AQ}{4\pi} [(r_1)^{-1} - (r_2)^{-1}] + \frac{BQ^2}{3(4\pi)^2} [(r_1)^{-3} - (r_2)^{-3}] \tag{9}$$

For a given well, A, B, r_1, and r_2 may be regarded as constants (r_2 could generally be set equal to infinity), so that the wellhead vacuum is just a quadratic function of the gas flow rate. Note that A and B are independent of well geometry and operating conditions; they depend only on the properties of the porous medium and those of air. Later, this will permit the scaling up of the results of small test well measurements.

The smaller the value of r_1 (the radius of the well gravel packing), the larger is the coefficient of Q^2, and the greater is that component of the flow resistance associated with turbulence.

We now turn to the case of axial geometry—where the well is screened over a substantial length but where the top of the screened section is well below the surface of the soil. Here, we use cylindrical coordinates (r,θ,z). We assume radial flow of the fluid to a length h of the z-axis. For this case it is readily shown that

$$-V = v_r = -\frac{Q}{2\pi rh} \tag{10}$$

and

$$-\frac{dP}{dL} = \frac{dP}{dr} \tag{11}$$

Equation (5) becomes

$$\frac{dP}{dr} = (A + BV)V \tag{12}$$

as before; substitution of Equation (10) into Equation (12) then yields

$$\frac{dP}{dr} = \frac{AQ}{2\pi hr} + \frac{BQ^2}{(2\pi)^2 h^2 r^2} \tag{13}$$

Integration of Equation (13) between r_1 (the packed radius of the well) and some large value r_2 of the radius at which the pressure is essentially 1 atm then gives

$$P(r_2) - P(r_1) = \frac{AQ}{2\pi h} \log_e (r_2/r_1) + \frac{BQ^2}{(2\pi)^2 h^2} [(r_1)^{-1} - (r_2)^{-1}] \quad (14)$$

As before, for fixed r_1, r_2, A, and B, we get a quadratic expression in Q for the wellhead vacuum.

IDEAL GAS APPROXIMATION

If the wellhead vacuum is an appreciable fraction of the ambient pressure (of the order of 0.1 atm or more), the incompressible fluid approximation for air becomes questionable, and it becomes necessary to develop an approach that takes into account the compressibility of air. At pressures of the order of an atmosphere and below, air behaves as an ideal gas to an excellent approximation. Perry (1969) gives the following equation for the isothermal flow of an ideal gas through a porous medium:

$$\frac{P_1^2 - P_2^2}{L} = \frac{2\alpha RT\mu G}{Mg} + \left[\beta + \frac{1}{L} \log_e (P_1/P_2) \right]\left[\frac{2RTG^2}{Mg} \right] \quad (15)$$

Here

P_1 = absolute upstream pressure (mass length/time^2 area)
P_2 = absolute downstream pressure
L = thickness of the porous medium
G = superficial mass velocity of gas, mass/sec area
g = gravitational constant, length/sec^2
μ = dynamic viscosity of air, mass/length time
M = molecular weight of gas, mass/mol
R = gas constant, mass length^2/sec^{2}mol deg
T = temperature, deg K
α = viscous resistance coefficient, length^{-2}
β = inertial resistance coefficient, length^{-1}

In Equation (15) we replace L by dL, which will be allowed to approach zero, and P_2 by $P_1 + dP$, where $dP \to 0$ as $dL \to 0$. The log term then becomes

$$-(P^{-1})dP/dL$$

and the left-hand side becomes

$$-(dP^2/dL) = -2P(dP/dL)$$

With these substitutions it becomes possible to write Equation (15) as

$$-2P\frac{dP}{dL} = aF + bF^2 - cP^{-1}\frac{dP}{dL}F^2 \tag{16}$$

where

F = molar gas flux, moles/time area
a, b, c = constants dependent on gas characteristics, temperature, and porous medium characteristics

If we consider the case where the well is screened only along a short distance at the bottom, the flow field is that associated with a point sink at the origin, and we work in spherical coordinates. The flux in the direction of flow is readily seen to be

$$F = Q/(4\pi r^2) \tag{17}$$

If we look at the case where the well is screened along a substantial portion h of its length and the top of the screened section is well below the surface of the soil, the problem can be viewed as axially symmetric. Cylindrical coordinates are appropriate, and the molar gas flux in the direction of flow is

$$F = Q/(2\pi hr) \tag{18}$$

We then replace dP/dL by $-dP/dr$ in Equation (16) and solve for dP/dr to get

$$\frac{dP}{dr} = \frac{(aF + bF^2)P}{2P^2 - cF^2} \tag{19}$$

where F is defined as a function of r and of the molar gas flow rate Q by Equations (17) (spherical geometry) or (18) (cylindrical geometry). One then integrates Equation (19) from some large initial value of r_2 at which $P(r_2) = 1$ atm, to a value of r_1 equal to the radius of the well gravel packing, at which point $P = P(r_1)$, the wellhead pressure. Note that if the terms in F^2 are neglected in Equation (19), one recovers Darcy's Law for an ideal compressible gas, which suggests that the constant a can be obtained fairly easily by making measurements at small values of Q. The values of b and c could be obtained by a numerical least squares fit of experimental values of the wellhead pressure to calculated values of $P(r_1)$, which would be rather laborious.

FITTING TO EXPERIMENTAL RESULTS

We here use the incompressible fluid model to illustrate the calculation of model parameters from experimental data. For a particular well, Equations (9) or (14) can be written as

$$V_w = A_1 Q + B_1 Q^2 \tag{20}$$

where A_1 and B_1 are the corresponding coefficients in Equations (9) or (14) and V_w is the wellhead vacuum in atm. The coefficients A_1 and B_1 can be calculated by the method of least squares, with the following results. Let

$$\sum_{i=1}^{i_{expts}} Q_i^n V_{wi}^m = SQ^n V_w^m \tag{21}$$

Define

$$D = \begin{vmatrix} SQ^2 & SQ^3 \\ SQ^3 & SQ^4 \end{vmatrix} \tag{22}$$

Then

$$A_1 = \begin{vmatrix} SQ^1 V_w^1 & SQ^3 \\ SQ^2 V_w^1 & SQ^4 \end{vmatrix} D^{-1} \tag{23}$$

and

$$B_1 = \begin{vmatrix} SQ^2 & SQ^1 V_w^1 \\ SQ^3 & SQ^2 V_w^1 \end{vmatrix} D^{-1} \tag{24}$$

Equation (20) is easily solved for the gas flow rate Q; the result is

$$Q = \frac{A_1}{2B_1} [-1 + (1 + 4B_1 V_w/A_1^2)^{1/2} \tag{25}$$

In the limit of small V_w this yields

$$Q = A_1^{-1}[1 - B_1/A_1^2)V_w]V_w \qquad (26)$$

which shows the relationship of this approach to Darcy's Law.

EXPERIMENTAL RESULTS

In this section we discuss the interpretation of some experimental data provided by Eckenfelder, Inc., in terms of the theory developed above. Data sets from four wells will be examined. The first three data sets were from an industrial waste landfill site in Pennsylvania at which VOCs were present. Monitoring wells were installed with a hand auger to a depth of about 5–6 ft; these were equipped with a 1-ft screened section with 0.020-in. openings. Two-inch diameter PVC pipe was used. The screened section was surrounded by a pea gravel packing approximately 4 in. in diameter and 20 to 24 in. in length, above which a 6-in. bentonite seal was placed. Each of the wells was operated as an extraction well at three or more airflow rates, and each run was conducted until the wellhead vacuum and airflow rate had reached steady state.

Another data set was obtained at a site in the southern United States with a long history of chemicals-related activity. The soils were relatively poorly draining silt loam and fine sand loam; subsoils were silty clay, silty clay loam, and sandy clay. The two wells, 14.3 ft apart, were essentially identical. Well pipes of 2-in. PVC extended to a depth of 10 ft and were screened (0.010-in slots) for a length of 5 ft. A sand packing 6.25 in. in diameter extended from a depth of 4 ft to a depth of 16 ft. The wells were sealed with bentonite.

Testing at both sites was carried out with a portable in situ vapor stripping unit developed by Eckenfelder, Inc. This unit is capable of flow rates in the 1 to 10 SCFM range and vacua in the 0 to 150 in. of water column (in W.C. range). The five data sets were fitted to Equation (20) by the least squares procedure described above. The coefficients of determination r^2 obtained were 0.9964, 0.9923, 0.9986, 0.9982, and 0.9944, indicating quite good fits. (Fits to the linear relationship required by Darcy's Law gave r^2 values of 0.9076, 0.9446, 0.9026, 0.9122, and 0.9028.) A plot of wellhead vacuum versus gas flow rate is shown in Figure 3.1. Evidently, this approach appears to provide good fits to within the limits of the experimental data. The data show substantial departures from the linear relationship required by Darcy's Law. On the other hand, there is no evidence from these data to suggest that the incompressible fluid model is inadequate. One therefore has little incentive to pursue the significantly greater complexities of the ideal compressible gas model.

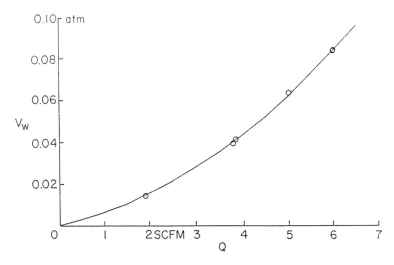

Figure 3.1 Plot of wellhead vacuum versus airflow rate for a data set from the southern United States site. The curve is the calculated dependence, given by $V_w = 0.004883 \cdot Q + 0.001513 \cdot Q^2$ with $r^2 = 0.9993$. [Reprinted from Clarke et al. (1993) p. 1671, by courtesy of Marcel Dekker, Inc.]

REYNOLDS NUMBERS

Here, we estimate values of the Reynolds Numbers for conditions roughly corresponding to the permeability tests discussed above and also for some corresponding to typical SVE well operation. The Reynolds Number is given by Equation (2),

$$\text{Re} = \frac{\varrho v d}{\mu} \qquad (2)$$

The gas density ϱ is calculated from

$$\varrho = \frac{0.3515(1 - V_w/406.9)}{T} \qquad (27)$$

where

ϱ = gas (air) density (gm/cm³)
T = temperature, deg K
V_w = wellhead vacuum, in W.C. (inches of water column)

The dynamic gas viscosity is calculated from

$$\mu(T) = (10^{-6})[0.3008 + 0.072082T - (3.7131 \times 10^{-6})T^2] \qquad (28)$$

where μ is the dynamic gas viscosity at temperature T in gm/cm sec (poise). Equation (28) results from a least squares quadratic fit to viscosity data for air between 100 and 500°K, taken from the *Handbook of Chemistry and Physics* (Lide, 1990); for this fit $r^2 = 0.999934$.

The gas velocity v is given by

$$v = \frac{471.9 Q_{SCFM}}{2\nu(1 - V_w/406.93)r_w h} \qquad (29)$$

where

$$
\begin{aligned}
v &= \text{linear gas velocity at the edge of the well gravel packing, cm/sec} \\
Q_{SCFM} &= \text{gas flow rate, standard cubic feet per min} \\
V_w &= \text{wellhead vacuum, in W.C.} \\
r_w &= \text{radius of well gravel packing, cm} \\
h &= \text{length of screened section of well, cm} \\
\nu &= \text{soil porosity, dimensionless}
\end{aligned}
$$

The characteristic length d must be estimated from the soil characteristics; this would be essentially the diameter of the pores that are most important in contributing to the conductivity of the soil to gas.

Reynolds Numbers calculated with parameters roughly corresponding to some of the test well runs are given in the first three cases presented in Table 3.1. Parameters for the last case in Table 3.1 were selected to correspond to a typical full-size SVE well. All yield Reynolds Numbers on the order of one or larger. Given the uncertainty in the characteristic length d, it would be imprudent to interpret these Reynolds Numbers too closely, but they certainly do not indicate that these wells are being operated in the viscous flow regime. This is consistent with the experimental results, as shown by comparison of the coefficients of determination of the Darcy's Law fits with those of the fits to the quadratic equation, and also as illustrated in Figure 3.1.

SCALING UP FROM FIELD TEST DATA

Inspection of Equations (9) and (14) permits us to obtain values for the constants A and B in these equations from the values of A_1 and B_1 in Equation (20), which are obtained from small-scale tests. The values of A and B can then, in turn, be used to calculate the values of A_1 and B_1 appropriate for wells having different parameters (i.e., radius of packing, length of

TABLE 3.1. Reynolds Numbers for SVE Wells.

Common parameters, all wells	
Temperature	15°C
Soil pore diameter	0.05 cm
Soil porosity	0.3
Common parameters, small-scale test wells	
Radius of well gravel packing	2 in.
Length of well screened section	12 in.
Small-scale test # 1	
Well head vacuum	4.07 in. of water
Gas flow rate of well	2 SCFM
Reynolds Number near the well	1.096
Small-scale test # 2	
Wellhead vacuum	24 in. of water
Gas flow rate of well	5 SCFM
Reynolds Number near the well	2.741
Small-scale test # 3	
Wellhead vacuum	28.48 in. of water
Gas flow rate of well	6 SCFM
Reynolds Number near the well	3.289
Large-scale run	
Radius of well gravel packing	6 in.
Length of well screened section	36 in.
Wellhead vacuum	61 in. of water
Gas flow rate of well	100 SCFM
Reynolds Number near the well	6.091

screened section). The relevant expressions are Equations (30) and (31) for point sinks (for which $h = r_w$) and Equations (32) and (33) for line sinks (for which $h \gg r_w$).

$$A_1 = \frac{A}{4\pi} [(r_1)^{-1} - (r_2)^{-1}] \qquad (h = r_w) \qquad (30)$$
$$\text{(point sinks)}$$

$$B_1 = \frac{B}{3(4\pi)^2} [(r_1)^{-3} - (r_2)^{-3}] \qquad (h = r_w) \qquad (31)$$

and

$$A_1 = \frac{A}{2\pi h} \log_e (r_2/r_1) \qquad (h \gg r_w) \qquad (32)$$
$$\text{(line sinks)}$$

$$B_1 = \frac{B}{(2\pi)^2 h^2} [(r_1)^{-1} - (r_2)^{-1}] \qquad (h \gg r_w) \qquad (33)$$

This permits calculation of the behavior of a large, field-scale well from small test well data within the framework of non-Darcian flow, which appears to be generally applicable to SVE wells.

CONCLUSIONS REGARDING DARCY'S LAW AND SVE

Several conclusions can be drawn from the above analysis. These are as follows.

First, soil pneumatic permeability measurements should be carried out over a substantial range of wellhead vacua and gas flow rates, so that the effect of transition and turbulent flow can be taken into account. Failure to consider this factor will usually result in the serious over-estimation of gas flow rates when wellhead vacua are increased above the value used in estimating the pneumatic permeability of the soil.

Second, these effects are by far the most severe in the immediate vicinity of the screened section of the well, where Reynolds Numbers are large. One therefore expects that if the correct molar gas flow rate values are used in SVE models relying on Darcy's Law (as most do), these models should yield correct results, in that gas flow is certainly Darcian at any appreciable distance from the well.

Third, Darcian models should not be relied upon to calculate molar gas flow rates from the wellhead vacua and pneumatic permeability constants because of the strong dependence of the latter on molar gas flux under conditions commonly occurring in soil vapor extraction.

This non-Darcian approach permits scale-up to larger systems almost as easily as does the simpler approach, which assumes the validity of Darcy's Law for these systems. One simply uses Equations (30) and (31) or (32) and (33) to calculate A and B from the test well data, then uses these equations to calculate A_1 and B_1 for the proposed full-scale well(s). These new values are then used in Equations (20) or (25) to calculate wellhead vacuum as a function of flow rate [Equation (20)] or flow rate as a function of wellhead vacuum [Equation (25)] for the proposed wells.

CALCULATION OF SOIL GAS FLOW FIELDS

Here, the equations governing the flow of a compressible gas in a porous medium are obtained first. This is followed by the development of two methods for calculating the soil gas velocities in the vicinity of a soil vapor extraction well. The first involves the use of the method of images, borrowed from electrostatics. The second employs over-relaxation, a numerical method well-adapted to deal with variable pneumatic permeabilities and complex boundary conditions, both of which are outside the scope of the method of images. Simple vertical wells and long buried horizontal slotted pipe wells are examined. The section closes with the calculation of soil gas streamlines and transit times for a number of examples. Such calculations are helpful in identifying domains in the contaminated zone that may be particularly slow to clean up.

EQUATIONS GOVERNING THE FLOW OF A COMPRESSIBLE GAS IN A POROUS MEDIUM

The continuity equation for a gas in a porous medium may be taken as

$$\partial c / \partial t = -\nu^{-1}(\nabla \cdot \mathbf{v}c) \tag{34}$$

where

ν = porosity of medium
c = concentration of the gas, mol/m^3
$\mathbf{v}$ = superficial velocity of gas, m/sec

The ideal gas law can be written as

$$P = cRT \tag{35}$$

where

P = pressure, atm
R = 8.206 × 10^{-5} m^3 atm/mol sec
T = temperature, deg K

From Darcy's Law (Freeze and Cherry, 1979) we have

$$v = -K_D \nabla P \tag{36}$$

where K_D is Darcy's constant (actually a tensor), m^2/atm sec.
For steady flow,

$$\partial c / \partial t = 0 \tag{37}$$

This, plus Equations (34), (35), and (36) then yields

$$0 = [\nabla \cdot K_D(\nabla P)P/RT] \tag{38}$$

which can be rewritten as

$$0 = (1/2)\nabla \cdot [(K_D/RT)\nabla P^2] \tag{39}$$

If the medium is isothermal and isotropic with constant permeability and porosity, Equation (39) simplifies to

$$\nabla^2(P^2) = 0 \tag{40}$$

It is thus apparent that suitably chosen solutions to Laplace's equation, $\nabla^2 u = 0$, may be used together with Equation (36) to construct velocity fields for compressible gases in simple porous media.

For a first case we examine the pressure distribution in a cylindrical laboratory aeration column. For this one-dimensional case Equation (40) becomes

$$\frac{d^2(P^2)}{dx^2} = 0 \tag{41}$$

where x is the distance measured from the inlet end of the column. Integration of this equation twice and introduction of the boundary conditions on the pressures at the ends of the column then give

$$P(x) = \left[P_i^2 - \frac{P_i^2 - P_f^2}{L} x \right]^{1/2} \tag{42}$$

From Equation (39),

$$v_x = -K_D dP/dx \tag{43}$$

which yields

$$v_x = \frac{K_D(P_i^2 - P_f^2)}{2L} \left[P_i^2 - \frac{(P_i^2 - P_f^2)}{L} x \right]^{-1/2} \tag{44}$$

for the superficial velocity of the gas at point x in the column.

THE METHOD OF IMAGES

Application to a Single Vertical Extraction Well

The velocity field of an ideal gas in the vicinity of a sink (the vent pipe) in a homogeneous isotropic porous medium can be calculated by the method of images as follows (see Figure 3.2 for the geometry and notation). We use cylindrical coordinates $(r,z,0)$, with the origin under the

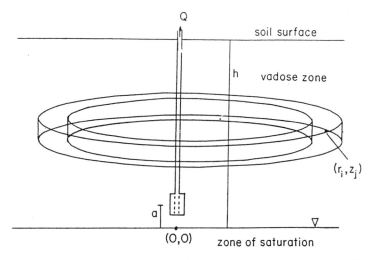

Figure 3.2 Geometry and notation used in describing a soil vapor extraction well.

well and at the bottom of the domain of interest. The boundary conditions that must be satisfied by the solution to Laplace's equation are as follows. (1) At the bottom of the domain of interest (the water table or other impermeable layer) one has a no-flow boundary, giving

$$\frac{\partial P^2(r,0)}{\partial z} = 0 \tag{45}$$

(2) At the top of the domain of interest, at which, $z = h$, the soil gas pressure is 1 atm,

$$P^2(r,h) = 1 \text{ atm}^2 \tag{46}$$

(3) There must be a sink of magnitude Q_a mol/sec at the point $(0,a)$, to represent the molar flow of gas to the vacuum well, represented as a point sink a distance $h - a$ below the surface of the soil.

Construction of a potential function satisfying these requirements is readily carried out by the method of images (Smythe, 1950); one obtains

$$W = 1 - P^2 = -\frac{RTQ_a}{2\pi K_D} \sum_{n=-\infty}^{\infty} \left[-\frac{1}{\{r^2 + [z - 4nh - a]^2\}^{1/2}} \right.$$

$$-\frac{1}{\{r^2 + [z - 4nh + a]^2\}^{1/2}} + \frac{1}{\{r^2 + [z - (4n - 2)h - a]^2\}^{1/2}}$$

$$\left. + \frac{1}{\{r^2 + [z - 4n - 2)h + a]^2\}^{1/2}} \right] \tag{47}$$

The scale factor in front of the sum here is obtained by integrating the molar gas flux over the surface of a small sphere centered about the well at $(0,a)$, noting that all of the terms in the summation [except the one with the singularity at $(0,a)$] contribute nothing, and setting the result equal to the molar gas flow to the well, Q_a. Then

$$P(r,z) = [1 \text{ atm}^2 - W]^{1/2} \qquad (48)$$

Darcy's Law then gives us the relationship between soil gas pressure gradient and the superficial gas velocity v; this is

$$v = -K_D \nabla P \qquad (36)$$

The components of the superficial gas velocity in the vicinity of the well are then

$$v_r = \frac{K_D}{2P} \frac{\partial W}{\partial r} \qquad (49)$$

and

$$v_z = \frac{K_D}{2P} \frac{\partial W}{\partial z} \qquad (50)$$

Note that these formulas are not valid in the immediate vicinity of the well, where, as seen in the first section of this chapter, we may have significant departures from Darcy's Law because of the very high gas velocities that are found as one approaches quite closely to the well.

The derivatives are given by

$$\frac{\partial W}{\partial r} = \frac{RTQ_a r}{2\pi K_D} \sum_{n=-\infty}^{\infty} \left[-\frac{1}{\{r^2 + [z - 4nh - a]^2\}^{3/2}} \right.$$

$$-\frac{1}{\{r^2 + [z - 4nh + a]^2\}^{3/2}} + \frac{1}{\{r^2 + [z - (4n - 2)h - a]^2\}^{3/2}}$$

$$\left. + \frac{1}{\{r^2 + [z - (4n - 2)h + a]^2\}^{3/2}} \right] \qquad (51)$$

and

$$\frac{\partial W}{\partial z} = \frac{RTQ_a}{2\pi K_D} \sum_{n=-\infty}^{\infty} \left[-\frac{z - 4nh - a}{\{r^2 + [z - 4nh - a]^2\}^{3/2}} \right.$$

$$-\frac{z - 4nh + a}{\{r^2 + [z - 4nh + a]^2\}^{3/2}} + \frac{z - (4n - 2)h - a}{\{r^2 + [z - (4n - 2)h - a]^2\}^{3/2}}$$

$$\left. + \frac{z - (4n - 2)h + a}{\{r^2 + [z - (4n - 2)h + a]^2\}^{3/2}} \right] \tag{52}$$

Note that while W and its derivatives are directly proportional to Q_a; P, v_r, and v_z are not. If we denote by primes ($'$) values of W, $\partial W/\partial r$, and $\partial W/\partial z$ calculated with $Q_a = 1$ mol/sec, we can then express the soil gas pressure and soil gas velocity components at other values of Q_a by Equations (53), (54), and (55).

$$P(r,z,Q_a) = [1 \text{ atm}^2 - Q_a W'(r,z)]^{1/2} \tag{53}$$

$$v_r(r,z,Q_a) = \frac{K_D Q_a \partial W'/\partial r}{2[1 \text{ atm}^2 - Q_a W'(r,z)]^{1/2}} \tag{54}$$

$$v_z(r,z,Q_a) = \frac{K_D Q_a \partial W'/\partial z}{2[1 \text{ atm}^2 - Q_a W'(r,z)]^{1/2}} \tag{55}$$

This permits us to evaluate W' and its derivatives at the mesh points of the SVE model initially and then to use the much simpler Equations (54) and (55) to calculate the gas velocities as functions of $Q_a(t)$ during the course of the SVE simulation.

If one wishes to obtain an estimate of the wellhead pressure of the extraction well at a given flow rate, this can be done within the framework of the above approach, although one should bear in mind the limitations to Darcy's Law mentioned earlier. One simply evaluates the pressure at the point (r_w,a) or at $(0,a \pm r_w)$, where r_w is the radius of the gravel packing of the well and a is the distance of the screened section of the well above the water table.

Application to a Single Buried Horizontal Slotted Pipe

The geometry of the system is defined in Figure 3.3. Let

h = thickness of vadose zone, m

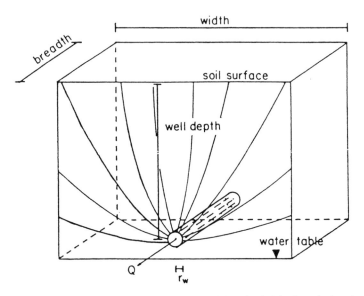

Figure 3.3 Configuration for soil vapor extraction with a buried horizontal slotted pipe. [Reprinted from Kayano and Wilson (1992) p. 1527, by courtesy of Marcel Dekker, Inc.]

$h - a$ = depth of horizontal slotted pipe, m

r_w = radius of gravel packing around the horizontal slotted pipe, m

l = length of horizontal slotted pipe, m

We assume that l is sufficiently large that we can neglect end effects, so that the problem can be regarded as two-dimensional and Laplace's equation for the system can be written as

$$\frac{\partial^2 P^2}{\partial x^2} + \frac{\partial^2 P^2}{\partial y^2} = 0 \tag{56}$$

It is then readily shown by symmetry arguments that the following function $W(x,y)$ satisfies the boundary conditions

$$W(x,h) = 0 \tag{57}$$

and

$$\partial W(x,0)/\partial y = 0 \tag{58}$$

These are required to satisfy the requirements that the pressure at the surface of the vadose zone be 1 atm and that there be no flow of gas through the bottom of the vadose zone.

$$W(x,y) = 1 \text{ atm}^2 - P^2(x,y)$$

$$= A \sum_{n=-\infty}^{\infty} [\log_e \{x^2 + [y - (4n - 2)h - a]^2\}$$

$$+ \log_e \{x^2 + [y - (4n - 2)h + a]^2\}$$

$$- \log_e \{x^2 + y - 4nh - a]^2\}$$

$$- \log_e \{x^2 + [y - 4nh + a]^2\}] \tag{59}$$

The constant A in Equation (59) is evaluated by calculating the surface integral of the molar gas flux into the sink at $(0,a)$ and setting this equal to Q_a; the result is

$$A = \frac{Q_a RT}{2\pi l K_D} \tag{60}$$

As before, if the gas flow rate is sufficiently slow that Darcy's Law can be assumed to be valid right up to the gravel packing of the slotted pipe, one can calculate the wellhead pressure as a function of flow rate. The point (r_w, a) lies at the boundary of the packing, within which we may assume a negligible pressure drop. The wellhead pressure P_w is then given by

$$P_w = [1 - W(r_w, a)]^{1/2} \tag{61}$$

The point $(0, a \pm r_w)$ may also be used for the calculation.

The soil gas superficial velocity components v_x and v_y are obtained from

$$\mathbf{v} = K_D \nabla P = K_D \nabla W / 2P \tag{62}$$

The derivatives are given by

$$\frac{\partial W}{\partial x} = \frac{Q_a RT}{\pi l K_D} \sum_{n=-\infty}^{\infty} \left[\frac{1}{x^2 + [y - (4n - 2)h - a]^2} \right.$$

$$+ \frac{1}{x^2 + [y - (4n - 2)h + a]^2} - \frac{1}{x^2 + [y - 4nh - a]^2}$$

$$\left. - \frac{1}{x^2 + [y - 4nh + a]^2} \right] \tag{63}$$

and

$$
\frac{\partial W}{\partial y} = \frac{Q_a RT}{\pi l K_D} \sum_{n=-\infty}^{\infty} \left[\frac{y - (4n - 2)h - a}{x^2 + [y - (4n - 2)h - a]^2} \right.
$$

$$
+ \frac{y - (4n - 2)h + a}{x^2 + [y - (4n - 2)h + a]^2} - \frac{y - 4nh - a}{x^2 + [y - 4nh - a]^2}
$$

$$
\left. - \frac{y - 4nh + a}{x^2 + [y - 4nh + a]^2} \right] \tag{64}
$$

If it is desired to model a horizontal slotted pipe SVE system for which the molar gas flow rate varies with time, the same procedure as was used to obtain Equations (53)–(55) can be used.

RELAXATION METHODS FOR CALCULATING GAS FLOW FIELDS: A SINGLE WELL

The method of images can be generalized to deal with media having a constant anisotropic permeability tensor, with different horizontal and vertical permeabilities, but runs into difficulty for any more complex permeability. In practice, one often has permeabilities that vary from point to point in the soil; there may be strata of different permeabilities; there may be clay lenses; there may be other types of heterogeneities; there may be buried or overlying impermeable obstacles; there may be passive vent wells. All of these will affect the pattern of gas movement in the vicinity of a vapor extraction well. To handle these situations one requires techniques that are more general than the method of images. In this section we explore one such approach: relaxation methods (Shaw, 1953; Freeze and Cherry, 1979).

We shall illustrate the relaxation method here by applying it to a system having axial symmetry, with permeability components $K_r(z)$ and $K_z(z)$. The equation governing the gas pressure in the vicinity of a single vapor stripping well is

$$
\frac{1}{r} \frac{\partial}{\partial r} \left[r K_r \frac{\partial P^2}{\partial r} \right] + \frac{\partial}{\partial z} \left[K_z \frac{\partial P^2}{\partial z} \right] = 0 \tag{65}
$$

The numerical integration of Equation (65) is done as follows. See Figure 3.4 for the geometry; the cylindrical domain of interest is parti-

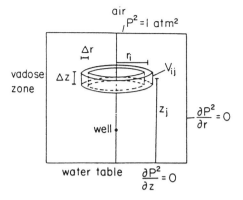

Figure 3.4 Geometry, notation, and boundary conditions for a vapor stripping well; partitioning of the domain of interest. [Reprinted from Mutch and Wilson (1990) p. 14, by courtesy of Marcel Dekker, Inc.]

tioned into ring-shaped volume elements as indicated. For the ijth volume element

$$r_i = (i + 1/2)\Delta r \tag{66}$$

$$z_j = (j + 1/2)\Delta z \tag{67}$$

$$V_{ij} = \pi(2i + 1)(\Delta r)^2 \Delta z \tag{68}$$

$$\text{inner surface} = 2\pi i \Delta r \Delta z \tag{69}$$

$$\text{outer surface} = 2\pi(i + 1)(\Delta r \Delta z \tag{70}$$

$$\text{upper and lower surfaces} = \pi(2i + 1)(\Delta r)^2 \tag{71}$$

We next calculate the net molar gas flux into the ijth volume element, as in electrostatics or diffusion calculations, and set this equal to zero, as required by Equation (55). This gives

$$0 = K_r[(j + 1/2)\Delta z]2\pi i \Delta r \Delta z(P^2_{i-1,j} - P^2_{i,j})$$

$$+ K_r[(j + 1/2)\Delta z]2\pi(i + 1)\Delta r \Delta z(P^2_{i+1,j} - P^2_{i,j})$$

$$+ K_z[j\Delta z]\pi(2i + 1)(\Delta r)^2(P^2_{i,j-1} - P^2_{i,j})$$

$$+ K_z[j + 1)\Delta z]\pi(2i + 1)(\Delta r)^2(P^2_{i,j+1} - P^2_{i,j}) \tag{72}$$

We set $\Delta r = \Delta z = \delta$, and then solve Equation (72) for P_{ij}^2:

$$P_{i,j}^2 = \frac{\begin{array}{c}(2iK_r[(j + 1/2)\delta]P_{i-1,j}^2 + 2(i + 1)K_r[(j + 1/2)\delta]P_{i+1,j}^2 + \\ (2i + 1)K_z[j\delta]P_{i,j-1}^2 + (2i + 1)K_z[(j + 1)\delta]P_{i,j+1}^2\end{array}}{(4i + 2)K_r[(j + 1/2)\delta] + (2i + 1)\{K_z[j\delta] + Kz[(j + 1)\delta]\}}$$

(73)

The boundary conditions are

$$P^2 = 1 \text{ atm}^2 \tag{74}$$

on top of the zone of influence,

$$\partial P^2/\partial r = 0 \tag{75}$$

around the periphery of the zone of influence, and

$$\partial P^2/\partial z = 0 \tag{76}$$

on the base of the zone of influence. Equation (75) is an appropriate approximate boundary condition for the situation in which the well is symmetrically surrounded by other vacuum wells. If the vacuum well is surrounded by passive vent wells screened along their entire length, Equation (75) is replaced by $P^2(b,z) = 1$, where b is the radius of the domain of influence. This configuration, with passive wells, is often useful for pilot studies in which one wishes to isolate a well-defined volume of soil for investigation of diffusion/desorption kinetics effects. We carry out flux balances on each of the volume elements on the surface of the domain of interest, which are consistent with the boundary conditions, with the following results.

At the top of the boundary, where $P = 1$ atm,

$$0 = K_r[(n_z + 1/2)\delta]2\pi i(P_{i-1,nz}^2 - P_{i,nz}^2)$$

$$+ K_r[n_z + 1/2)\delta]2\pi(i + 1)(P_{i+1,nz}^2 - P_{i,nz}^2)$$

$$+ K_z[j\delta]\pi(2i + 1)(P_{i,nz-1}^2 - P_{i,nz}^2)$$

$$+ 2K_z[(j + 1)\delta]\pi(2i + 1)(1 - P_{i,nz}^2) \tag{77}$$

Along the outer boundary, where $\Delta P^2/\Delta r = 0$:

$$0 = K_r[(j + 1/2\delta)2\pi]n_r(P^2_{nr-i,j} - P^2_{nr,j})$$

$$+ K_z[j\delta]\pi(2n_r + 1)(P^2_{nr,j-1} - P^2_{nr,j})$$

$$+ K_z[(j + 1)\delta]\pi(2n_r + 1)(P^2_{nr,j+1} - P^2_{nr,j}) \tag{78}$$

Along the bottom boundary, where $\Delta P^2/\Delta z = 0$, we have

$$0 = K_r[(1/2)\delta]2\pi i(P^2_{i-1,0} - P^2_{i,0})$$

$$+ K_r[(1/2)\delta]2\pi(i + 1)(P^2_{i+1,0} - P^2_{i,0})$$

$$+ K_z[\delta]\pi(2i + 1)(P^2_{i,1} - P^2_{i,0}) \tag{79}$$

Along the left boundary (actually the axis of the domain, $r = 0$) one can use Equation (73) since the coefficient of the term in $P^2_{i-1,j}$ vanishes when $i = 0$. In the upper left corner (on the axis of the system) we have

$$0 = K_r[(n_z + 1/2)\delta]2\pi(P^2_{i,nz} - P^2_{0,nz})$$

$$+ K_z[n_z\delta]\pi(P^2_{0,nz-1} - P^2_{0,nz})$$

$$+ 2K_z[(n_z + 1)\delta]\pi(1 - P^2_{0,nz}) \tag{80}$$

In the upper right corner

$$0 = K_r[(n_z + 1/2)\delta]2\pi n_r(P^2_{nr-1,nz} - P^2_{nr,nz})$$

$$+ K_z[n_z\delta]\pi(2n_r + 1)(P^2_{nr,nz-1} - P^2_{nr,nz})$$

$$+ 2K_z[(n_z + 1)\delta]\pi(2n_r + 1)(1 - P^2_{nr,nz}) \tag{81}$$

In the lower left corner

$$0 = K_r[(1/2)\delta]2\pi(P^2_{1,0} - P^2_{0,0}) + K_z[\delta]\pi(P^2_{0,1} - P^2_{0,0}) \tag{82}$$

And in the lower right corner

$$0 = K_r[(1/2)\delta]2n_r\pi(P^2_{nr-1,0} - P^2_{nr,0})$$

$$+ K_z[\delta](2n_r + 1)\pi(P^2_{nr,1} - P^2_{nr,0}) \tag{83}$$

Equations (77)–(83) are solved for the central value of P^2, as was done in obtaining Equation (73) from Equation (72). The results, along with Equation (73), are the relaxation equations. An over-relaxation procedure provides more rapid convergence than simple relaxation; the algorithm used is as follows. Equations (73) and (77)–(83) can be abbreviated as

$$P^2_{i,j} = f(\{P^2_{k,l}\}) \tag{84}$$

One starts with an initial set of values for the $P^2_{i,j}$ such as $P^2_{i,j} = 1$ generally, and $P^2_{0,J}$ given by Equation (101) below, and then iterates Equation (84):

$$P^{*2}_{i,j} = f(\{P^2_{k,l}\}) \tag{85}$$

is the relaxation step. Then over-relaxed values of the $P^2_{i,j}$ are calculated from

$$P^{*2}_{i,j} = \omega P^{*2}_{i,j} + (1 - \omega)P^2_{i,j} \qquad 1 < \omega < 2 \tag{86}$$

These are then used in the next iteration. Actually, the newly computed values $P^{*2}_{i,j}$ are immediately put back into the array for the $P^2_{i,j}$; this speeds convergence somewhat and reduces computer memory requirements. Convergence of the calculations requires roughly a minute for a 10×20 array on a microcomputer operating at 20 MHz and using an 80286 microprocessor and a math chip.

We next turn to the inclusion of the boundary condition that represents the well. The procedure to be followed is similar to that employed in the method of images above, but included here is the possibility that the porous medium is anisotropic, with a permeability tensor given by

$$K = \begin{pmatrix} K_r & 0 \\ 0 & K_z \end{pmatrix} \tag{87}$$

In cylindrical coordinates and with the assumption of cylindrical symmetry with the vapor stripping well on the z-axis of the coordinate system, Equation (39) becomes

$$\frac{1}{r}\frac{\partial}{\partial r}\left[r\frac{\partial P^2}{\partial r}\right] + \varkappa^2 \frac{\partial P^2}{\partial z^2} = 0 \tag{88}$$

where $\varkappa^2 = K_z/K_r$. We shall carry out the analysis for the case where the length of the screened section of the well is comparable to the diameter of the gravel packing.

A trial solution for Equation (88),

$$U = A/(r^2 + z^2/\varkappa^2)^{1/2} \tag{89}$$

is easily shown to satisfy Equation (88), as does the more general trial solution

$$U = A/\{r^2 + [(z - a)/\varkappa]^2\}^{1/2} \tag{90}$$

A still more general solution is a linear combination of terms of the type shown in Equation (90),

$$U = \sum_i \frac{A_i}{\{r^2 + [(z - a_i)/\varkappa]^2\}^{1/2}} \tag{91}$$

This result indicates that one can use the method of images to construct solutions of Equation (52) satisfying the desired boundary conditions for the case of an anisotropic medium by procedures similar to the analysis of the isotropic medium presented above.

We let the first term in Equation (91) represent the actual sink (the vacuum well), located at $(0,a_1) = (0,a)$. The constant A_1 is evaluated as follows. Recall that, since the soil gas is assumed to be ideal, $P^2 = U + \text{constant}$. Let Q_a be the magnitude of the molar gas flow rate at the sink. Then

$$Q_a = -\int_0^{2\pi} \int_{-\infty}^{\infty} c\mathbf{v} \cdot d\mathbf{S} \tag{92}$$

Here the surface S is taken to be an infinitely long cylinder of radius r' coaxial with the z-axis, and

$c = $ concentration of gas, mol/m³ $= P/RT$
$\mathbf{v} = $ superficial velocity of gas, m/sec
$R = 8.206 \times 10^{-5}$ m³ atm/mol deg

Then

$$Q_a = \frac{1}{RT} \int_0^{2\pi} \int_{-\infty}^{\infty} PK_r \frac{\partial P}{\partial r} r' d\theta dz \tag{93}$$

$$= \frac{K_r 2\pi r'}{RT} \int_{-\infty}^{\infty} \frac{1}{2} \frac{\partial P^2}{\partial r} dz \tag{94}$$

We are concerned only with the sink at $(0,a)$, so we take

$$P^2 = 1 \text{ atm}^2 - \frac{A}{\{r^2 + [(z - a)/x]^2\}^{1/2}} \tag{95}$$

[dropping all terms that do not represent a source at $(0,a)$]. Then

$$\frac{\partial P^2}{\partial r} = \frac{Ar}{\{r^2 + [(z - a)/x]^2\}^{1/2}} \tag{96}$$

and

$$Q_a = \frac{\pi K_r (r')^2 A}{RT} \int_{-\infty}^{\infty} \frac{dz'}{[(r')^2 + (z'/x)^2]^{3/2}} \tag{97}$$

where $z' = z - a$. The substitution $\zeta = z'/(r' x)$ transforms Equation (97) into

$$Q_a = \frac{2\pi K_r A}{RT} \int_{-\infty}^{\infty} \frac{1}{(1 + \zeta^2)^{3/2}} \, d\zeta \tag{98}$$

The definite integral in Equation (98) is equal to 1 (Dwight, 1949), so we finally obtain (on substituting for x and solving for A)

$$A = \frac{RTQ_a}{2\pi(K_r K_z)^{1/2}} \tag{99}$$

To calculate the wellhead pressure we approximate P^2 in the vicinity of the well by Equation (95), substitute for A from Equation (99), and set $z = a$, $r = r_w$, the packed radius of the well. This gives

$$P_w = \left[1 - \frac{RTQ_a}{2\pi(K_r K_z)^{1/2} r_w}\right]^{1/2} \tag{100}$$

To calculate the pressure in the volume element containing the sink, replace r_w in Equation (100) by Δr. This then provides the value of $P_{0,J}^2$ to be used as the boundary condition at the well, given by Equation (101).

$$P_w = \left[1 - \frac{RTQ_a}{2\pi(K_r K_z)^{1/2} \Delta r}\right]^{1/2} \tag{101}$$

The molar flow rate of the well in terms of the wellhead pressure P_w is easily obtained from Equation (100):

$$Q_a = (1 \text{ atm}^2 - P_w^2)\frac{2\pi(K_r K_z)^{1/2} r_w}{RT} \qquad (102)$$

GAS FLOW STREAMLINES AND TRANSIT TIMES

Analysis

The delivery of an adequate flow of air to the contaminated regions of the vadose zone is crucial to the success of soil vapor extraction. Soil gas streamlines and transit times are helpful in quickly identifying regions in the domain of interest, which will clean up slowly because of insufficient air flux. Soil gas pressure measurements (or, in the case of modeling, calculations) are not sufficient to determine whether or not the gas flow will be adequate, since soil gas velocities are not determined by pressure but by pressure gradient and permeability. In this section we describe the calculation of streamlines and transit times for soil gas in the vicinity of a long horizontal slotted pipe; the calculation for a single vertical well is virtually identical. The soil gas pressures required may either be calculated by the method of images or by the relaxation process described above.

The streamlines are calculated by integration of the equations

$$\frac{dx}{dt} = -\nu^{-1}K_x(x,y)\frac{\partial P}{\partial x} \qquad (103a)$$

$$\frac{dy}{dt} = -\nu^{-1}K_y(x,y)\frac{\partial P}{\partial y} \qquad (103b)$$

If the pressures are calculated by the method of images, one has formulas for the pressure derivatives in Equations (103a) and (103b) for both a single vertical well and for a horizontal slotted pipe, but use of these leads to rather slow calculations because of the time required to calculate these derivatives from their infinite series at each point. It is much faster to evaluate the pressures at a set of mesh points covering the domain of interest and then to use finite difference representations for the derivatives, which is the method one must use if the soil gas pressures have been calculated by the relaxation method described above.

For interiors points (not in volume elements) at the borders of the domain, these derivatives are calculated by differentiating the Taylor's series for the pressure. In the ijth volume element this is given by

$$P(x,y) = P_{i,j} + \frac{\partial P}{\partial x_{i,j}} (x - x_i) + \frac{\partial P}{\partial y_{i,j}} (y - y_j) + \frac{1}{2} \frac{\partial^2 P}{\partial x_{i,j}^2} (x - x_i)^2$$

$$+ \frac{1}{2} \frac{\partial^2 P}{\partial y_{i,j}^2} (y - y_j)^2 + \frac{\partial^2 P}{\partial x \partial y_{i,j}} (x - x_i)(y - y_j) \qquad (104)$$

where

$$\partial P / \partial x_{i,j} = (P_{i+1,j} - P_{i-1,j})/2\Delta x \qquad (105)$$

$$\partial P / \partial y_{i,j} = (P_{i,j+1} - P_{i,j-1})/2\Delta y \qquad (106)$$

$$\partial^2 P / \partial x_{i,j}^2 = (P_{i+1,j} - 2P_{i,j} + P_{i-1,j})/(\Delta x)^2 \qquad (107)$$

$$\partial^2 P / \partial y_{i,j}^2 = (P_{i,j+1} - 2P_{i,j} + P_{i,j-1})/(\Delta y)^2 \qquad (108)$$

$$\partial^2 P / \partial x \partial y_{i,j} = (P_{i+1,j+1} - P_{i-1,j+1} - P_{i+1,j-1} + P_{i-1,j-1})/4\Delta x \Delta y \qquad (109)$$

These formulas can also be used on the left, right, and bottom borders by replacing $P_{-1,j}$ by $P_{0,j}$ along the left border, $P_{i,-1}$ by $P_{i,0}$ along the bottom border, and $P_{nx+1,j}$ by $P_{nx,j}$ along the right border, which causes the no-flow boundary conditions to be satisfied. At the top boundary, Taylor's series expansions yield

$$\partial P / \partial y_{i,ny} = -[1/3 \cdot P_{i,ny-1} + P_{i,ny} - 4/3 \cdot P_a]\Delta y \qquad (110)$$

$$\partial^2 P / \partial y_{i,ny}^2 = 4[P_{i,ny-1} - 3P_{i,ny} + 2P_a]/3(\Delta y)^2 \qquad (111)$$

$$\partial^2 P / \partial x \partial y_{i,ny} = \frac{(1/3)(P_{i+1,ny-1} - P_{i-1,ny-1}) + (P_{i+1,ny} - P_{i-1,ny})}{2\Delta x \Delta y}$$

$$(112)$$

where P_a is the ambient atmospheric pressure. These formulas appear to be somewhat more accurate in the border regions than some used earlier; with these, streamlines are well-behaved right out to the edges of the domain. Transit times are calculated simply by keeping track of the accumulating value of t in the numerical integrations generating the streamlines.

Results

Here we examine the results of a number of computations of soil gas streamlines and transit times in the vicinity of a soil vapor extraction well.

Figures 3.5, 3.6, and 3.7 (Table 3.2) demonstrate the effect of well depth on the streamlines and transit times. The permeability of the soil is isotropic and constant. As the well becomes more shallow, the transit times of gas entering the soil out near the edge of the domain become much larger, and the streamlines do not extend nearly as deeply as is the case with the deeper well. The results suggest that wells should be drilled at least through the contaminated zone and that in an isotropic soil transit times for gas entering the soil at a distance of more than one and a half times the well depth from the well are likely to be unacceptably long, resulting in correspondingly long cleanup times.

Often, at hazardous waste sites one encounters buried impermeable obstacles—concrete rubble, buried scrap metal, etc. One would like to have some idea as to how the presence of these will affect soil vapor ex-

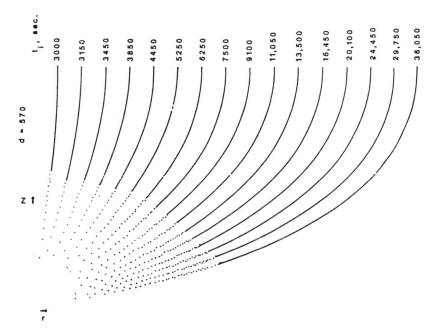

Figure 3.5 Streamlines and gas transit times in the vicinity of a vacuum well at a depth of 5.7 m; t_i is the time required for air to move from the surface of the ground at the indicated point down along the streamline to the vacuum well. See Table 3.2 for parameter values. [Reprinted from Wilson et al. (1988) p. 991, by courtesy of Marcel Dekker, Inc.]

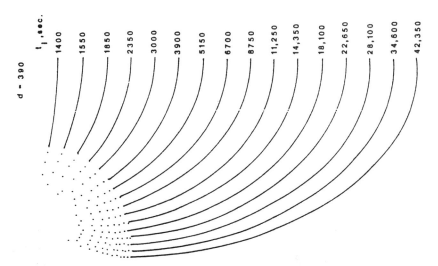

Figure 3.6 Streamlines and gas transit times in the vicinity of a vacuum well at a depth of 3.9 m. See Table 3.2 for parameter values. [Reprinted from Wilson et al. (1988) p. 991, by courtesy of Marcel Dekker, Inc.]

traction. Certainly, the presence of significant amounts of contaminated medium in "dead end zones" such as a ruptured drum will result in extremely long cleanup times, since there will be no advective transport of VOC from such zones. However, the presence of simple obstacles does not present serious difficulties. In Figures 3.8(a), (b), and (c) we see the im-

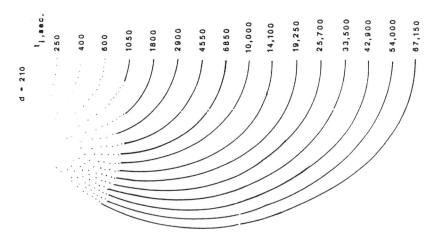

Figure 3.7 Streamlines and gas transit times in the vicinity of a vacuum well at a depth of 2.1 m. See Table 3.2 for parameter values. [Reprinted from Wilson et al. (1988) p. 991, by courtesy of Marcel Dekker, Inc.]

TABLE 3.2. Well Parameters (Figures 3.5–3.7).

Depth of water table	6.1 m
Depth of vacuum well	5.7, 3.9, 2.1 m
Radius of well gravel packing	0.127 m
Wellhead pressure	0.866 atm
Gas flow rate	23.6 L/sec
Soil porosity	0.2

pact of the presence of a horizontally placed, infinitely long obstacle of finite width on the transit times of gas flowing around it. Some increases in transit times do occur, but they are not extremely large, and the size of the "dead volumes" above and below the center of the strip is not large.

Somewhat the same thing is seen in Figures 3.9(a), (b), and (c), in which the impact of a horizontally placed circular disk is shown. The results do indicate that, if one knows that impermeable obstacles are present at a site, one should drill wells in that immediate area, so as to have the maximum possible gas flow rate in their vicinity, thereby minimizing their damaging effect on the rate of cleanup. If such obstacles are out near the periphery of the domain of influence of an SVE well, where the airflow rate would be slow even in the absence of obstacles, they could result in substantial increases in cleanup times.

The presence of low-permeability domains such as lenses high in clay has a very marked effect on streamlines and transit times. As we shall see later, this, in turn, can very substantially increase remediation times, so that failure to identify the presence of such structures by examination of an adequate number of well logs can result in some very unpleasant surprises in soil vapor extraction. Here, the effects of long low-permeability lenses on the soil gas flow patterns around a long horizontal slotted pipe vacuum well are explored. In Chapter 6 we shall see how these effects are reflected in the rates of SVE remediation of these systems. We represent the occurrence of low-permeability lenses in the domain of interest by means of Equation (113).

$$K_x(x,y) = K_{x0} - \sum_{i=1}^{m} A_i \exp\left\{ -\left[\left(\frac{x - x^1}{r_i}\right)^2 + \left(\frac{y - y_i}{s_i}\right)^2 \right]^n \right\}$$

(113)

B_i replaces A_i in a similar expression for K_y.

The default parameters for the streamline plots shown in Figures 3.10 through 3.16 are given in Table 3.3. The numbers at the top ends of the streamlines are the transit times in units of 1,000 sec. In Figure 3.10 we see

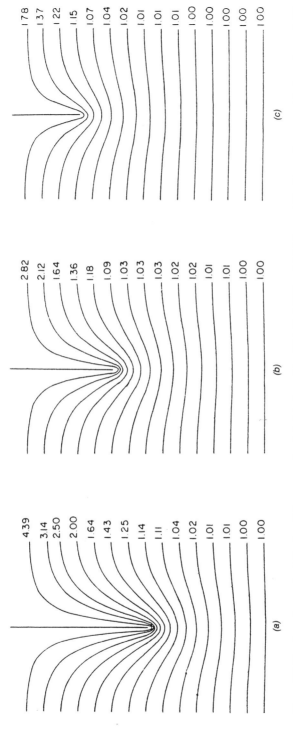

Figure 3.8 Gas streamlines and relative transit times around an infinitely long strip of (a) 40, (b) 60, and (c) 80 cm width. The region mapped in these figures is 80 × 80 cm. [Reprinted from Gannon et al. (1989) pp. 844, 848, by courtesy of Marcel Dekker, Inc.]

108

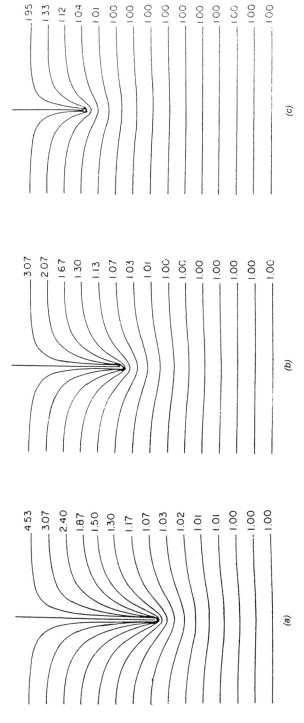

Figure 3.9 Gas streamlines and relative transit times around a circular disk of radius (a) 40, (b) 60, and (c) 80 cm diameter. [Reprinted from Gannon et al. (1989) pp. 849, 850, by courtesy of Marcel Dekker, Inc.]

109

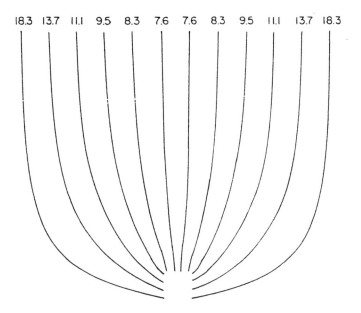

18.3   13.7   11.1   9.5   8.3   7.6   7.6   8.3   9.5   11.1   13.7   18.3

Figure 3.10 Streamlines in the vicinity of a horizontal slotted pipe. No low-permeability lenses are present in this run. The numbers at the tops of the streamlines are the gas transit times in units of 1,000 sec. See Table 3.3 for the run parameters. [Reprinted from Gómez-Lahoz et al. (1991) p. 143, by courtesy of Marcel Dekker, Inc.]

the streamlines and transit times for a well in an isotropic medium having a constant permeability throughout; this is the standard of reference. The transit times out at the edges of the domain are about two and a half times larger than those in the center, and we see that there are relatively small stagnation zones (regions in which there is little or no air flow) in the lower corners of the domain. This is Run 1.

TABLE 3.3. Standard Parameter Set for the Runs Shown in Figures 3.10–3.16.

Domain width	13 m
Domain depth	8 m
Δx, Δy	1 m
Coordinates of well (origin in lower left corner)	(6.5, 0.5) m
Packed radius of well	0.2 m
Wellhead pressure	0.85 atm
Temperature	14°C
Soil gas-filled porosity	0.3
K_{x0}, K_{y0}	0.1 m²/atm sec
Initial soil contaminant concentration	100 mg/kg
Soil density	1.7 gm/cm³
Specific moisture content	0.2
Effective Henry's constant (dimensionless)	0.005

In Figure 3.11 a low permeability lens is located a short distance over the vapor extraction well. This results in a marked increase in the transit times along the streamlines near the center of the domain but has virtually no effect on those out at either side. The size of the stagnation zones in the lower corners appears to have decreased somewhat, and the longest transit times in Figures 3.10 and 3.11 are exceedingly similar, leading us to expect that this clay lens will have very little effect on the overall cleanup time of the system. This is Run 2.

In Figure 3.12 the low-permeability lens has been moved over to the left side of the domain, which is Run 4. The outermost streamline on the left side now has a transit time four times that of the streamlines at the center, and the zone of stagnation on the left has markedly increased in size. These results lead us to expect a very marked increase in the cleanup time of this domain, as compared to that described in Figure 3.10.

The transit times in Figure 3.13, which is Run 5, depict the disastrous results of screening an SVE well in the middle of a low-permeability lens. This has resulted in a greatly reduced airflow rate through the well and has increased gas transit times roughly tenfold. Evidently, if the well log indicates that one is about to screen an SVE well in such a low-permeability structure, it would be advisable to continue drilling until one came out underneath it into a region of higher permeability.

Figure 3.14 shows the effect of two low-permeability lenses in the upper right and left corners of the domain (Run 6). The transit times of the outermost streamlines have been greatly increased, as compared to those in

18.7 14.7 12.7 12.2 13.1 17.2 17.2 13.1 12.2 12.7 14.6 18.6

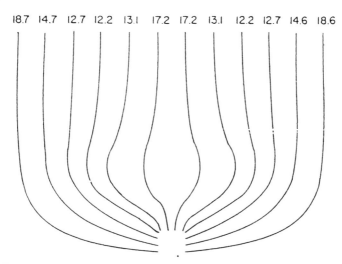

Figure 3.11 Streamlines in the presence of a lens centered over the vapor extraction well. The lens is centered at the point (6.5, 3). $A_i = B_i = 0.095$ m²/atm sec. $r_i = 3$ m, $s_i = 1$ m, $n = 2$. [Reprinted from Gómez-Lahoz et al. (1991) p. 144, by courtesy of Marcel Dekker, Inc.]

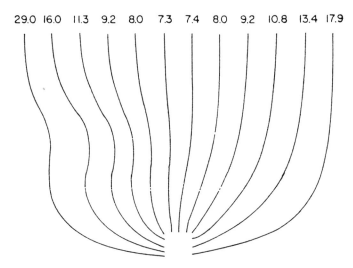

29.0 16.0 11.3 9.2 8.0 7.3 7.4 8.0 9.2 10.8 13.4 17.9

Figure 3.12 Streamlines in the presence of a lens in the far left portion of the domain, centered at (1,4). $r_i = 4$, $s_i = 1$ m, other parameters as in Figure 3.11. [Reprinted from Gómez-Lahoz et al. (1991) p. 145, by courtesy of Marcel Dekker, Inc.]

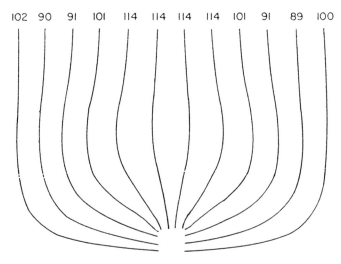

102 90 91 101 114 114 114 114 101 91 89 100

Figure 3.13 Streamlines in the presence of a lens surrounding the horizontal slotted pipe, centered at (6.5,2). $r_i = s_i = 4$ m, other parameters as in Figure 3.11 (Run 5). [Reprinted from Gómez-Lahoz et al. (1991) p. 146, by courtesy of Marcel Dekker, Inc.]

112

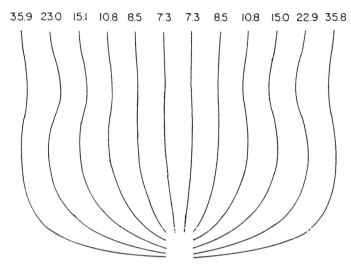

35.9 23.0 15.1 10.8 8.5 7.3 7.3 8.5 10.8 15.0 22.9 35.8

Figure 3.14 Streamlines in the presence of low-permeability lenses in the upper right and upper left corners of the domain, centered at (1,6) and (12,6). $r_i = 4$, $s_i = 1$ m, $A_i = B_i = 0.095$ m²/atm sec ($i = 1,2$), $n = 1$. [Reprinted from Gómez-Lahoz et al. (1991) p. 147, by courtesy of Marcel Dekker, Inc.]

Figure 3.10, so one would expect correspondingly slower cleanup of the soil in the outer portions of the domain, in which cleanup is slow even under the best of circumstances.

Low-permeability lenses are located in the bottom left and right corners of the domain in Figure 3.15. There is some increase in transit times, but of particular concern to us are the large zones of stagnation in the lower corners of the domain. Contaminant in these will be removed quite slowly (Run 7).

The bottom left and upper right corners of the domain of interest contain low-permeability lenses for the run shown in Figure 3.16. Increased transit times and a fairly large zone of stagnation in the lower left corner lead us to expect a decrease in cleanup rate, especially for contaminant in the lower left corner of the domain.

A potentially major factor affecting the pneumatic permeability of the soil is its moisture content. A formula proposed by Millington and Quirk (1961) appears to be in reasonable agreement with experimental results (Pedersen and Curtis, 1991); it can be written as

$$K(w) = K_0 \left[\frac{\nu - \omega}{\nu} \right]^{10/3} = K_0 (1 - R_h)^{10/3} \qquad (114)$$

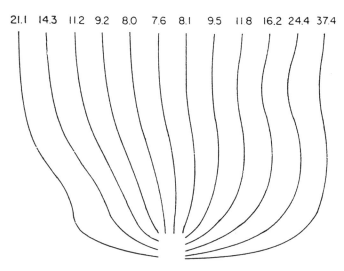

Figure 3.15 Streamlines in the presence of low-permeability lenses in the lower right and lower left corners of the domain, centered at (1,2) and (12,2). Other parameters as in Figure 3.14. [Reprinted from Gómez-Lahoz et al. (1991) p. 148, by courtesy of Marcell Dekker, Inc.]

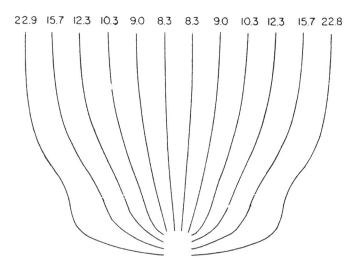

Figure 3.16 Streamlines in the presence of low-permeability lenses in the upper right and lower left corners of the domain, centered at (1,2) and (12,6). Other parameters as in Figure 3.14. [Reprinted from Gómez-Lahoz et al. (1991) p. 149, by courtesy of Marcel Dekker, Inc.]

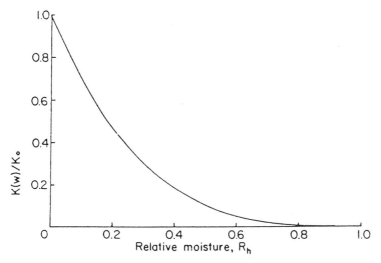

Figure 3.17 Plot of $K(w)/K_0$ versus relative moisture content R_h according to Equation (114). [Reprinted from Gómez-Lahoz et al. (1991) p. 155, by courtesy of Marcel Dekker, Inc.]

where

K_0 = (pneumatic) permeability of dry soil
ν = total voids fraction
ω = volumetric water content
$R_h = \omega/\nu$

A plot of $K(w)/K_0$ versus R_h is shown in Figure 3.17; a significant moisture content is seen to result in a very marked decrease in the pneumatic permeability. A plot of experimental results is shown in Figure 3.18.

Generally, the underlying boundary to a region being vapor stripped is the water table, at a depth b below the surface of the soil. We therefore consider a model in which R_h varies from 1 (saturation) at the water table to a value R_{ha} at the surface of the soil, where R_{ha} is determined by the atmospheric relative humidity and the recent history of rainfall events. We assume that R_h is given by the following function, which satisfies these limiting conditions:

$$R_h(y) = 1 - (1 - R_{ha})(y/b)^g \qquad (115)$$

where

y = distance above the water table, m
g = adjustable parameter; $0 < g < \infty$

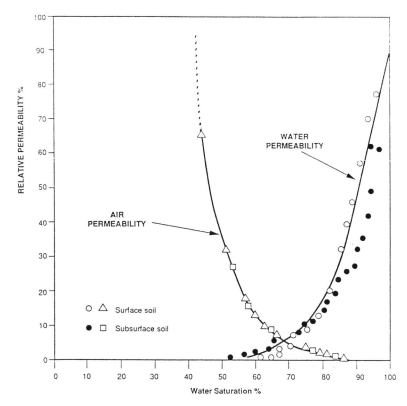

Figure 3.18 Air and water permeability as a function of water content. [Reprinted from Pedersen and Curtis (1991), EPA Document EPA/540/2-91/003.]

Then the equation

$$K(y) = K_0[(1 - R_{ha})(y/b)^g]^{10/3} \qquad (116)$$

can be used to calculate K_x and K_y, the horizontal and vertical components of the permeability tensor, by substituting $K_0 = K_{0x}$ or K_{0y}. Plots of y/b versus R_h for various values of g are given in Figure 3.19.

The effects of soil moisture content are illustrated in Figures 3.20 and 3.21. In Figure 3.20 we see the transit times and streamlines for a homogeneously rather dry soil; in Figure 3.21 the lower portion of the vadose zone is substantially wetter than the upper portion. The transit times for the outer streamlines are substantially larger in Figure 3.21 than they are in Figure 3.20, and in Figure 3.21 there are two quite large zones of stagnation in the lower corners of the domain. We conclude that some effort to prevent infiltration and, perhaps, to lower the water table may be war-

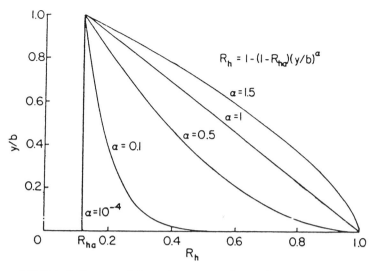

Figure 3.19 Plot of y/b versus relative moisture content R_h: y = height above the water table, b = distance between the water table and the surface of the soil, R_{ha} = 0.12. See Equation (115). [Reprinted from Gómez-Lahoz et al. (1991) p. 156, by courtesy of Marcel Dekker, Inc.]

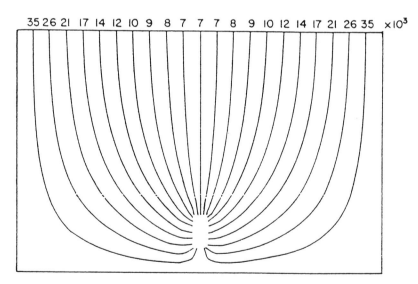

Figure 3.20 Streamlines in a homogeneously rather dry soil (g = 10⁻⁴). See Figure 3.19. [Reprinted from Gómez-Lahoz et al. (1991) p. 157, by courtesy of Marcel Dekker, Inc.]

56 31 20 14 10 7 5 4 4 3 3 3 4 4 5 7 10 14 20 31 56 ×10⁵

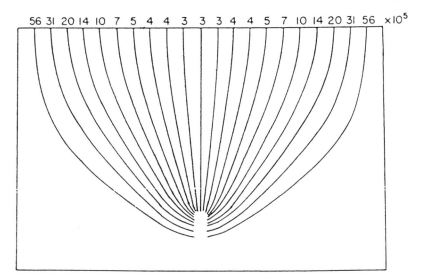

Figure 3.21 Streamlines in a soil that becomes progressively more wet at increasing depths ($g = 1$). See Figure 3.19. [Reprinted from Gómez-Lahoz et al. (1991) p. 158, by courtesy of Marcel Dekker, Inc.]

ranted if there is evidence that the moisture content of the contaminated region is high. In Chapter 6 we shall see the disastrous effect of excessive soil moisture on the rate of VOC removal by SVE.

SOIL GAS PRESSURES: RESULTS

Piezometer measurements taken during the course of pilot-scale SVE operation are commonly available in SVE feasibility studies and can provide a good deal of useful information. In this section we present some soil gas pressure results that illustrate the possibilities of using piezometer data taken in the vicinity of an extraction well to estimate anisotropy in the permeability and to discover or confirm the presence of spatially variable permeabilities. The results reported here were computed by the methods described previously in this chapter.

Default parameters for the plots shown in Figures 3.22–3.25 are given in Table 3.4. Figure 3.22 shows the effect of vacuum well depth on piezometer reading (the piezometer is at a depth of 4 m in all these runs) for various horizontal distances between the vacuum and piezometer wells. The permeability here is isotropic. The influence of the vacuum well is detectable by a piezometer 10 m away provided that the vacuum well is drilled to a depth of at least 4 m.

The effect on piezometer readings of the depth of the water table is

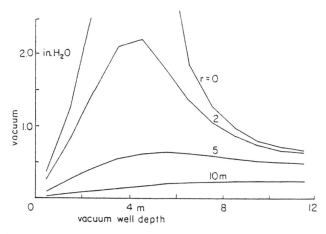

Figure 3.22 Effect of vacuum well depth on soil vacuum at constant water table depth. Distance r between vacuum well and piezometer well $= 0, 2, 5,$ and 10 m as indicated; other parameters as in Table 3.4. All piezometer readings are in inches of water vacuum. [Reprinted from Mutch and Wilson (1990) p. 11, by courtesy of Marcel Dekker, Inc.]

shown for various distances between the vacuum well and the piezometer well in Figure 3.23. The depth of the vacuum well is 11.5 m, and that of the piezometer well is 4 m. Somewhat surprisingly, the influence of the water table on the piezometer readings is seen at all the piezometer stations until the depth to the water table is somewhat over 20 m. The effect is larger, the shorter the distance is between the vacuum well and the piezometer well.

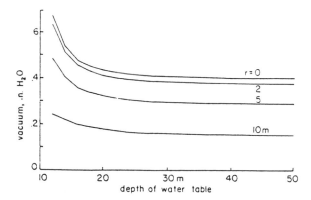

Figure 3.23 Effect of water table depth on piezometer readings: $r = 0, 2, 5,$ and 10 m as indicated. Vacuum well depth $= 11.5$ m. Other parameters as in Table 3.4. [Reprinted from Mutch and Wilson (1990) p. 12, by courtesy of Marcel Dekker, Inc.]

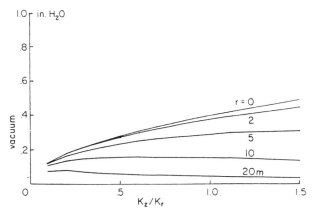

Figure 3.24 Effect of permeability anisotropy K_z/K_r on piezometer readings with a water table depth of 50 m. Depth of vacuum well = 11.5 m; r = 0, 2, 5, 10, and 20 m as indicated. Other parameters as in Table 3.4. [Reprinted from Mutch and Wilson (1990) p. 12, by courtesy of Marcel Dekker, Inc.]

Figures 3.24 and 3.25 exhibit the effect of permeability anisotropy on piezometer readings. The vacuum wells are 11.5 m deep, the piezometer wells are 4 m deep, and the depth to the water table is 50 m (Figure 3.24) and 12 m (Figure 3.25). Experimentally, one typically finds K_z (vertical component) substantially smaller than K_r (horizontal component). The effect of changes in the ratio of the permeability components, K_z/K_r, is

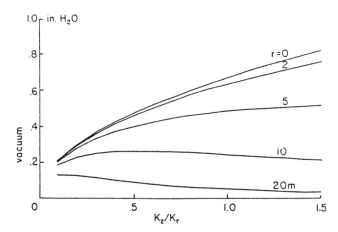

Figure 3.25 Effects of permeability anisotropy K_z/K_r on piezometer readings with a water table depth of 12 m. Depth of vacuum well = 11.5 m; r = 0, 2, 5, 10, and 20 m as indicated. Other parameters as in Table 3.4. [Reprinted from Mutch and Wilson (1990) p. 13, by courtesy of Marcel Dekker, Inc.]

TABLE 3.4. Default Parameters for the
Runs Shown in Figures 3.22–3.25.

Depth of water table	12 m
Piezometer well depth	4 m
Vacuum well molar flow rate	0.1 mol/sec
Screened radius of vacuum well	0.12 m
Soils voids fraction	0.2
Wellhead pressure at vacuum well	0.866 atm
Temperature	12 °C
K_r/K_z	1.0
K_r	0.0623 m²/atm sec

seen to be greatest when the vacuum well and the piezometer well are relatively close together in both Figures 3.24 and 3.25. Comparison of readings taken from piezometers 2 m and 10 or 20 m from the vacuum well should yield a reasonable accurate value of the anisotropy ratio. This is a matter of some interest because the smaller the ratio K_z/K_r, the wider the useful radius of the vacuum well and the more widely apart the vacuum wells can be spaced without adversely affecting the cleanup time.

We next turn to the effects of strata of differing permeability, a very common situation in practice. The geometry of the systems described in Figures 3.27–3.29 is given in Figure 3.26, and the parameters used in the calculations are given in Table 3.5. The vapor extraction well is screened in the lower stratum. In Runs 1 and 2 of this set, the permeability of the overlying layer is greater than that of the lower layer. In Run 3 the permeabilities are identical, and in Runs 4 and 5 the permeability of the overlying layer is less than that of the lower layer. In all cases the permeability is isotropic. For both shallow (1.5 m, Figure 3.27) and deeper (4.5 m, Figure 3.28) piezometer wells, there is a much greater vacuum at

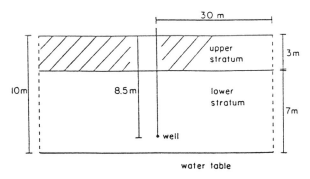

Figure 3.26 The stratified domain of interest for Runs 1 to 5 (Figures 3.27–3.29). [Reprinted from Mutch and Wilson (1990) p. 18, by courtesy of Marcel Dekker, Inc.]

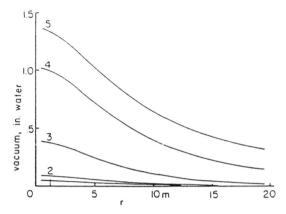

Figure 3.27 Plots of soil vacuum (inches of water) versus radial distance from the vacuum well (Runs 1 through 5). Depth of piezometer = 1.5 m. Other parameters as in Table 3.5. [Reprinted from Mutch and Wilson (1990) p. 20, by courtesy of Marcel Dekker, Inc.]

all vacuum-piezometer well distances when an overlying low-permeability layer is present. We also see that the influence of the vacuum well extends out to much greater distances when an overlying low-permeability layer is present. The impact of an overlying low-permeability layer is also seen in Figure 3.29, in which soil vacuum is plotted as a function of piezometer well depth for the five runs. Even before the low-permeability stratum has been completely penetrated, we see a much higher soil vacuum. From this figure we see that a cluster of piezometers at various depths provides an

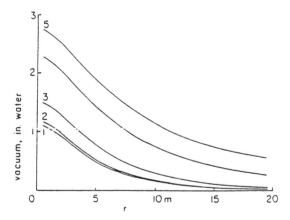

Figure 3.28 Plots of soil vacuum versus radial distance from the vacuum well (Runs 1 through 5). Depth of piezometer = 4.5 m. Other parameters as in Table 3.5. [Reprinted from Mutch and Wilson (1990) p. 20, by courtesy of Marcel Dekker, Inc.]

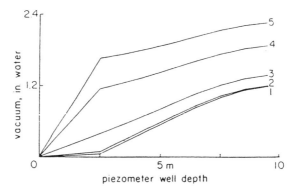

Figure 3.29 Plots of soil vacuum versus piezometer depth. Radial distance from the vacuum well to the piezometer well = 5.5 m. See Table 3.5 for other parameters. [Reprinted from Mutch and Wilson (1990) p. 21, by courtesy of Marcel Dekker, Inc.]

effective method for locating variations in soil permeability with depth. The cluster should be fairly close to the vacuum well, but not so close that the soil in the vicinity of the cluster has been disturbed by the drilling of the vacuum well. The presence of an overlying low-permeability stratum, like an anisotropy ratio substantially less than one, permits one to utilize a substantially greater SVE well spacing than would otherwise be possible, with corresponding reductions in cost.

We next examine the situation illustrated in Figure 3.30, in which the lower stratum lies below the soil vapor extraction well. Parameters for the

TABLE 3.5. Parameters for Runs 1–5 in Figures 3.25–3.29.

Radius of zone of influence	30 m
Depth of water table	10 m
Depth of vacuum well	8.5 m
Screened radius of well	0.12 m
Soil voids fraction	0.2
Temperature	13°C
Depth of discontinuity in the permeability	3 m
Wellhead pressure	0.866 atm

Run	Q_a (mol/sec)	K_z (upper)[a]	K_r (upper)	K_z (lower)	K_r (lower)
1	0.016	0.1	0.1	0.01	0.01
2	0.032	0.1	0.1	0.02	0.02
3	0.16	0.1	0.1	0.1	0.1
4	0.16	0.02	0.02	0.1	0.1
5	0.16	0.01	0.01	0.1	0.1

[a]Units of the permeabilities are m^2/atm sec.

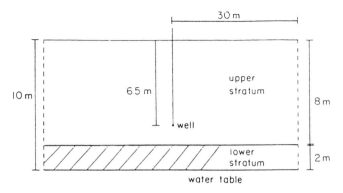

Figure 3.30 The stratified domain of interest for Runs 6–8 (Figures 3.31–3.33). [Reprinted from Mutch and Wilson (1990) p. 22, by courtesy of Marcel Dekker, Inc.]

three runs of this type that were made are given in Table 3.6. In the runs shown in Figures 3.31 and 3.32, the piezometer wells are at a depth of 4.5 m in the upper stratum. Variations in the permeabilities of the two strata have very little effect on the piezometer readings, and it would be impossible to deduce the presence of the underlying stratum from these measurements.

The situation is not so unfavorable if the piezometer is screened at a depth of 9.5 m, down in the underlying stratum, provided that the piezometer well is relatively close to the vacuum well. We see, however, that, in a situation in which the underlying stratum lies underneath the vacuum extraction well, its effect on the radius of influence of the vacuum well is very slight, so that being able to deduct its presence is not particularly important. Examination of Figure 3.33, which shows the dependence

TABLE 3.6. Parameters for Runs 6–8 in Figures 3.31–3.33.

Radius of zone of influence	30 m
Depth of water table	10 m
Depth of vacuum well	6.5 m
Screened radius of well	0.12 m
Soil voids fraction	0.2
Temperature	13°C
Wellhead pressure	0.866 atm

Run	Q_a (mol/sec)	K_z (upper)[a]	K_r (upper)	K_z (lower)	K_r (lower)
6	0.032	0.02	0.02	0.1	0.1
7	0.16	0.1	0.1	0.1	0.1
8	0.16	0.1	0.1	0.02	0.02

[a]Units of the permeabilities are m^2/atm sec.

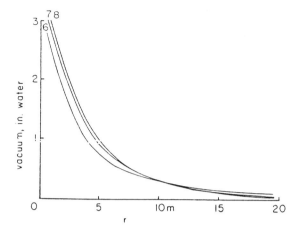

Figure 3.31 Plots of soil vacuum versus radial distance from the vacuum well (Runs 6–8). Depth of piezometer = 4.5 m. Other parameters as in Table 3.6. [Reprinted from Mutch and Wilson (1990) p. 23, by courtesy of Marcel Dekker, Inc.]

of soil vacuum on piezometer well depth for these three runs, does not provide us with a very useful tool for detecting these underlying strata either.

It is apparent from these two sets of runs that the sensitivity of the piezometer readings to the permeability distribution depends markedly on the relative positions of the vacuum well, the permeability discontinuity, and the piezometer wells. Evidently, one would be well advised to make

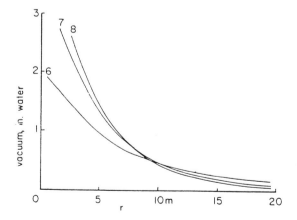

Figure 3.32 Plots of soil vacuum versus radial distance from the vacuum well (Runs 6–8). Depth of piezometer = 9.5 m. Other parameters as in Table 3.6. [Reprinted from Mutch and Wilson (1990) p. 23, by courtesy of Marcel Dekker, Inc.]

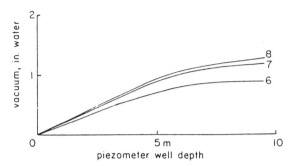

Figure 3.33 Plots of soil vacuum versus piezometer depth (Runs 6–8). Radial distance from the vacuum well = 5.5 m. Other parameters as in Table 3.6. [Reprinted from Mutch and Wilson (1990) p. 24, by courtesy of Marcel Dekker, Inc.]

use of well logs and test boring results in designing a vacuum well–piezometer well array for determining permeabilities, radii of influence, etc., with optimal sensitivity.

The effect of the boundary conditions employed around the periphery of the cylindrical zone of vacuum well influence depends very markedly on the ratio of the radius of the zone of influence to the depth of the vacuum well. The over-relaxation model described earlier in the chapter and used for the calculations involving low-permeability lenses, strata of differing permeabilities, underground obstacles, and overlying impermeable caps (to be discussed shortly) utilizes at the cylindrical periphery of the domain of interest one of two possible boundary conditions. If the well is one in an array of vacuum wells, one uses a no-flow boundary condition, $\partial P / \partial r = 0$. If, on the other hand, the well is surrounded by a set of passive vent wells screened along their entire length, the appropriate boundary conditions is $P = 1$ atm. We next explore some aspects of these points.

Figure 3.34 shows plots of soil gas vacuum versus radial distance for a well depth of 8.5 m and a radius of influence of 30 m. The curves labeled A(9,10) pertain to a piezometer well at 4.5 m depth. Those labeled B(9,10) pertain to a piezometer well at at 9.5 m depth. At this scale the curves for the two types of boundary conditions appear identical. Run parameters are given in Table 3.7. Evidently, the boundary of the domain is sufficiently far from the well that the nature of the boundary condition has virtually no effect on the soil gas pressure distribution.

On the other hand, Figures 3.35 and 3.36 (Table 3.8) show plots of soil gas vacuum versus radial distance for a well depth of 8.5 m and a radius of influence of only 10 m. In these figures, the curve labeled 11 is for the no-flow boundary condition, and that labeled 12 is for the $P = 1$ atm boundary condition. We see, as expected, that the soil vacuum is less for

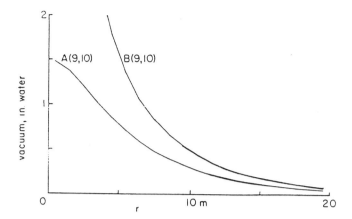

Figure 3.34 Plots of soil vacuum versus radial distance from the vacuum well (Runs 9 and 10). Depth of the piezometer wells are 4.5 m (curve A) and 9.5 m (curve B). At this scale the curves for the two boundary conditions [$P(r = 30$ m) = 1 atm and $\partial P/\partial r(r = 30$ m) = 0] appear identical; four curves are shown on this plot. Other parameters as in Table 3.7. [Reprinted from Mutch and Wilson (1990) p. 24, by courtesy of Marcel Dekker, Inc.]

the 1 atm (passive well) boundary condition at all points at a depth of 4.5 m (Figure 3.35) and also at a depth of 9.5 m (Figure 3.36). These results illustrate the importance of using the correct boundary conditions in modeling calculations, since these can have a significant impact on the soil gas pressure distribution.

An alternative way to get insight into the gas flow patterns around an SVE well is to plot gas velocities throughout the domain of influence. We use this approach to examine the gas flows near a well in which an impermeable circular cap covers the soil in the vicinity of the well (see

TABLE 3.7. **Parameters and Boundary Conditions for Runs 9 and 10 in Figure 3.34.**

Radius of zone of influence	30 m
Depth of water table	10 m
Depth of vacuum well	8.5 m
Screened radius of well	0.12 m
Soil voids fraction	0.2
Temperature	13°C
Permeability components K_r and K_z	0.1 m²/atm sec
Molar airflow rate	0.16 mol/sec
Boundary conditions	
Run 9 $\partial P/\partial r(r = 30$ m) = 0	
Run 10 $P(r = 30$ m) = 1 atm	

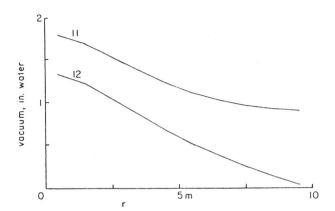

Figure 3.35 Plots of soil vacuum versus radial distance from the vacuum well. Depth of the piezometer wells are 4.5 m. Boundary conditions are $\partial P/\partial r(r = 10 \text{ m}) = 0$ for Run 11; $P(r = 10 \text{ m}) = 1$ atm for Run 12. See Table 3.8 for parameters. [Reprinted from Mutch and Wilson (1990) p. 26, by courtesy of Marcel Dekker, Inc.]

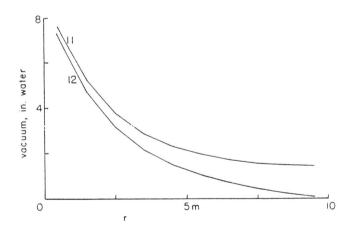

Figure 3.36 Plots of soil vacuum versus radial distance from the vacuum well. Piezometer well depth = 9.5 m. Boundary conditions are $\partial P/\partial r(r = 10 \text{ m}) = 0$ for Run 11; $P(r = 10 \text{ m}) = 1$ atm for Run 12. See Table 3.8 for parameters. [Reprinted from Mutch and Wilson (1990) p. 26, by courtesy of Marcel Dekker, Inc.]

TABLE 3.8. Parameters and Boundary Conditions
for Runs 11 and 12 in Figures 3.35 and 3.36.

Radius of zone of influence	10 m
Depth of water table	10 m
Depth of vacuum well	8.5 m
Screened radius of well	0.12 m
Soil voids fraction	0.2
Temperature	13°C
Permeability components K_r and K_z	0.1 m²/atm sec
Molar airflow rate	0.16 mol/sec
Boundary conditions	
Run 11 $\partial P/\partial r(r = 10\ m) = 0$	
Run 12 $P(r = 10\ m) = 1$ atm	

Figure 3.37). This system can readily be handled by the over-relaxation method; one has a mixed boundary condition at the top of the domain, as follows:

$$\partial P^2(r,h)/\partial z = 0, \qquad r < b \tag{117}$$

$$P^2(r,h) = 1, \qquad b < r < d \tag{118}$$

where

b = radius of impermeable cap
d = radius of domain of interest
h = depth of water table

This is represented by suitable modifications in the finite difference equations along the lines described earlier. In Figure 3.38 we see the gas velocity vectors; the well is located at the lower left of the region shown, which is a plane drawn through the right half of the domain of interest. Comparison of velocity vectors in the lower right corner of the domain with those for a similar system without an impermeable cap, at higher magnification than shown here, shows somewhat increased gas velocities in the zone of stagnation at the lower periphery and a small, centrally located zone of stagnation immediately under the impermeable cap.

These results indicate that there might be advantages in some cases in using impermeable caps to improve the gas flow pattern around an SVE well, particularly if one needed faster removal of contaminants from the lower peripheral portion of the domain of influence. Perhaps more important, they indicate the feasibility of carrying out vapor stripping operations under pre-existing impermeable surface barriers such as concrete floors, asphalt parking lots, etc.

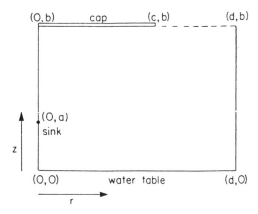

Figure 3.37 SVE with an impermeable cap. The vacuum well is evacuating air from the soil at $(0,a)$, at a height a meters above the water table. The impermeable circular cap at the surface of the soil is of radius c and is located b meters above the water table. The radius of the domain of interest is d meters. [Reprinted from Gannon et al. (1989) p. 835, by courtesy of Marcel Dekker, Inc.]

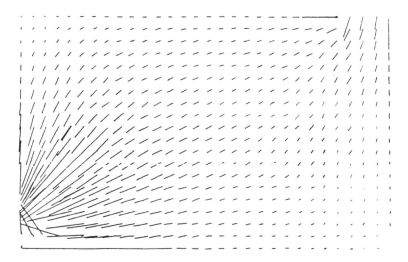

Figure 3.38 Velocity field for a soil vapor extraction well with a circular impermeable cap. The screened radius of the well = 0.12 m; wellhead pressure = 0.866 atm; soil voids fraction = 0.2. $K_r = K_z = 0.654$ m²/atm sec; molar gas flow rate = 0.988 mol/sec; depth of well = 16.5 m; depth of water table = 20 m; radius of domain of interest = 30 m; radius of impermeable cap = 25 m. [Reprinted from Gannon et al. (1989) p. 842, by courtesy of Marcel Dekker, Inc.]

130

CONCLUSIONS

The delivery of an adequate flux of soil gas to all of the contaminated portion of the vadose zone is essential to the success of soil vapor extraction. A clear understanding of the movement of gas through soil in the vicinity of an SVE well is therefore necessary. In this chapter we have explored Darcy's Law, commonly used in analyzing fluid flow in porous media; some limitations to Darcy's Law in its application to SVE wells were found, but ways of circumventing these were also developed. The general problem of gas flow in the vicinity of a vacuum well was then addressed, and two methods of calculating the soil gas pressures were developed. One, the method of images, permits the use of algebraic formulas and requires relatively little computation time but suffers from an inability to deal with many of the complex boundary conditions that may arise. The second, over-relaxation, is a numerical method of great generality but involves the writing of more code and requires somewhat more computing time.

Three methods of visualizing the situation in the vicinity of an SVE well were then used in exploring the application of these methods; these were the plotting of streamlines and listing of gas transit times, plots of soil gas pressures, and plots of soil gas velocity vectors. These can be used to investigate either single vertical wells or buried long horizontal slotted pipes. The effects of well depth, presence of buried obstacles, low-permeability lenses, strata of differing permeability, peripheral boundary conditions (free, well in an array of wells, well surrounded by passive wells), and impermeable caps were investigated.

The extension of the over-relaxation technique and the method for calculating streamlines and transit times to a full three-dimensional model is very straightforward. Computer speed and memory requirements for such calculations are substantially increased but now (1994) are within the capabilities of the more powerful microcomputers and work stations that are available. Unfortunately, the data sets necessary to justify the use of such a sophisticated model are generally not available, so that the two-dimensional simplifications presented here will probably be in use for some time.

The calculations of the soil gas velocities is the first step in the development of mathematical models for the removal of volatile organic compounds by soil vapor extraction. Development of such models will be addressed in later chapters.

REFERENCES

Cho, J. S., 1991, "Forced Air Ventilation for Remediation of Unsaturated Soils Contaminated by VOC," U.S. EPA Report No. EPA/600/2-91/016.

Clarke, A. N., M. M. Megehee, and D. J. Wilson, 1993, "Soil Cleanup by in situ Aeration. XII. Effects of Departures from Darcy's Law on Soil Vapor Extraction," *Separ. Sci. Technol.*, 28:1671.

de Marsily, G., 1986, *Quantitative Hydrogeology: Groundwater Hydrology for Engineers,* Academic Press, San Diego, CA, pp. 73–74.

Dwight, H. B., 1949, *Tables of Integrals and Other Mathematical Data,* MacMillan, New York, NY.

Freeze, R. A. and J. A. Cherry, 1979, *Groundwater,* Prentice-Hall, Englewood Cliffs, NJ.

Gannon, K., D. J. Wilson, A. N. Clarke, R. D. Mutch, and J. H. Clarke, 1989, "Soil Cleanup by in situ Aeration. II. Effects of Impermeable Caps, Soil Permeability, and Evaporative Cooling," *Separ. Sci. Technol.*, 24:831.

Gómez-Lahoz, C., J. M. Rodríguez-Maroto, and D. J. Wilson, 1991, "Soil Cleanup by in situ Aeration. VI. Effects of Variable Permeabilities," *Separ. Sci. Technol.*, 26:133.

Gómez-Lahoz, C., J. M. Rodríguez-Maroto, D. J. Wilson, and K. Tamamushi, 1993, "Soil Cleanup by in situ Aeration. XIII. Effects of Variable Air Flow Rates in Diffusion-Limited Operation," *Separ. Sci. Technol.*, 29:579.

Johnson, P. C., M. W. Kemblowski, and J. D. Colthart, 1989, "Practical Screening Models for Soil Venting Applications," presented at the *Workshop on Soil Vacuum Extraction,* April 27–29, U.S. EPA, Robert S. Kerr Environmental Research Laboratory (RSKERL), Ada, OK.

Johnson, P. C., M. W. Kemblowski, and J. D. Colthart, 1990, "Quantitative Analysis for the Cleanup of Hydrocarbon-Contaminated Soils by in situ Soil Venting," *Ground Water,* 28:413.

Johnson, P. C., M. W. Kemblowski, J. D. Colthart, D. L. Byers, and C. C. Stanley, 1989, "A Practical Approach to the Design, Operation, and Monitoring of in situ Soil Venting Systems," presented at the *Soil Vapor Extraction Technology Workshop,* June 28–29, U.S. EPA Risk Reduction Engineering Laboratory (RREL), Edison, NJ.

Kayano, S. and D. J. Wilson, 1992, "Soil Cleanup by in situ Aeration. X. Vapor Stripping of Mixtures of Volatile Organics Obeying Raoult's Law," *Separ. Sci. Technol.*, 27:1525.

Lide, D. R., ed., 1990, *Handbook of Chemistry and Physics,* 71st ed., CRC Press, Boca Raton, FL.

Marley, M. C., 1991, "Development and Application of a Three-Dimensional Air Flow Model in the Design of a Vapor Extraction System," *Proceedings of the Symposium on Soil Venting,* April 29–May 1, Houston, TX, U.S. EPA Report No. EPA/600/R-92/174, p. 125.

Marley, M. C., S. D. Richter, B. L. Cliff, and P. E. Nangeroni, 1989, "Design of Soil Vapor Extraction Systems – A Scientific Approach," Presented at the *Soil Vapor Extraction Workshop,* June 28–29, U.S. EPA Risk Reduction Engineering Laboratory, Edison, NJ.

Millington, R. J. and J. P. Quirk, 1961, "Permeability of Porous Solids," *Trans. Faraday Soc.,* 57:1200.

Mutch, R. D. and D. J. Wilson, 1990, "Soil Cleanup by in situ Aeration. IV. Anisotropic Permeabilities," *Separ. Sci. Technol.*, 25:1.

Pedersen, T. A. and J. T. Curtis, 1991, *Soil Vapor Extraction Technology Reference Handbook,* U.S. EPA Report No. EPA/540/2-91/003.

Perry, J. H., 1969, *Chemical Engineer's Handbook,* 5th ed., McGraw-Hill, New York, NY, pp. 5–52.

Rodríguez-Maroto, J. M., C. Gómez-Lahoz, and D. J. Wilson, 1991, "Mathematical Modeling of SVE: Effects of Diffusion Kinetics and Variable Permeabilities," *Proceedings of the Symposium on Soil Venting,* April 29–May 1, Houston, TX, U.S. EPA Report No. EPA/600/R-92/174, p. 103.

Shaw, F. S., 1953, *Relaxation Methods: An Introduction to Approximational Methods for Differential Equations,* Dover, New York, NY.

Smythe, W. R., 1950, *Static and Dynamic Electricity,* McGraw-Hill, New York, NY.

Stumbar, J. P. and J. Rawe, 1991, "Guide for Conducting Treatibility Studies under CERCLA: Soil Vapor Extraction Interim Guidance," U.S. EPA Report No. EPA/540/2-91/019A.

Wilson, D. J., 1993, "Advances in the Modeling of Several Inovative Technologies," *Proc., 6th Ann. Environmental Management and Technology Conf./Central (HAZMAT/CENTRAL 93),* March 9–11, Rosemont, IL, p. 135.

Wilson, D. J., A. N. Clarke, and J. H. Clarke, 1988, "Soil Cleanup by in situ Aeration. I. Mathematical Modeling," *Separ. Sci. Technol.,* 23:991.

CHAPTER 4

Equilibria and Mass Transport of VOCs in Soils

INTRODUCTION

IN the last chapter we addressed the movement of soil gas in the vadose zone in the vicinity of a soil vapor extraction well. In this chapter we deal with the equilibrium and kinetic processes that govern the distribution of VOCs between the various phases present in the soil and the rates of the processes by which these compounds migrate between phases. Equilibria are controlled by thermodynamic quantities—vapor pressures, Henry's constants, water solubilities, adsorption isotherm parameters, chemical composition, soil moisture content. Rate processes are controlled by kinetic parameters—diffusion constants, desorption rate constants, etc.—as well as by the thermodynamic quantities. In concert, all of these determine the characteristics of the VOC source on which the advecting soil gas acts in bringing about soil vapor extraction. These characteristics, in turn, control the design and operating parameters of the SVE system or may, in fact, force one to the conclusion that the site is not suitable for SVE.

In the following sections the various equilibria that are operative for VOCs in solids are addressed. What phases are present? How are the masses of VOC in these various phases related at equilibrium? What adsorption isotherms are operative? We then turn to the kinetic processes by which VOCs are transferred from one phase to another. These include desorption of VOCs that are relatively strongly bound to an adsorbate in the soil (such as humic materials or some clays) and diffusion processes, both through a gas phase in the porous medium and through aqueous or possibly nonaqueous layers such as occur in oily or greasy soils.

EQUILIBRIUM

PHASES PRESENT

The first question we must ask is what phases are appropriate for modeling VOCs in soils. Certainly, one must include the vapor phase. Not only does this provide a reservoir (albeit a small one) for the storage of VOCs, it is the only phase where the advective motion can be readily controlled.

A second phase, which is virtually always present and which can contain VOC, is the soil moisture. VOC, dissolved in the aqueous phase, present in the soil and at equilibrium with the vapor phase will be governed by Henry's Law—that is, the vapor pressure of the VOC will be proportional to the concentration of VOC in the aqueous phase. Henry's constants vary in size enormously, from indetectably small (for soluble, nonvolatile, or nearly nonvolatile compounds) to quite large values that are associated with compounds of low aqueous solubility and high vapor pressure.

Often, nonaqueous phase liquid (NAPL) may be present. Most hydrocarbons are light nonaqueous phase liquids (LNAPLs) and will not penetrate the zone of saturation to any great extent as NAPL, although the groundwater may be contaminated by dissolved material, and there may be a zone in and around the capillary fringe in which LNAPL is smeared by rises and drops in the level of the water table. Dense nonaqueous phase liquids (DNAPLs; mostly chlorinated solvents), on the other hand, can sink down through the zone of saturation if the pressure they exert is greater than the capillary pressure resulting from the DNAPL-water interface in the porous medium. DNAPLs in the zone of saturation are not remediable by SVE, since one does not deliver air below the level of the water table in this technique. Residual NAPL, trapped interstitially in the vadose zone, may be present at quite high concentrations—as much as $3-30$ L/m^3 of soil according to Schwille (1988).

VOC is also generally present in the vadose zone as adsorbed material. This is commonly described as an adsorbed phase; however, the strength of the binding of the VOC molecules to the highly heterogeneous adsorption sites is highly variable, and it is perhaps misleading to think of this material as in a single surface phase.

Occasionally, soil at hazardous waste sites may contain substantial quantities of oil, grease, or tarry materials. Many hydrophobic VOCs (BETX, aliphatics, chlorinated solvents, etc.) are much more soluble in such greasy or tarry phases than they are in water, so even comparatively small amounts of soil heavily contaminated with these materials can hold substantial quantities of VOCs, which may cause difficulties in the implementation of SVE.

To get some idea of the bulk quantitative aspects of this distribution of contaminant(s) among the various phases, let us look at some figures provided by Brown and Norris (1986) for a site at which an 80,000-gallon gasoline spill had occurred five years previously. The phase distribution of gasoline at the site is reported in Tables 4.1 (sand and gravel) and 4.2 (fractured bedrock). That portion of the contaminant that is in the aqueous

TABLE 4.1. Phase Distribution of Gasoline in Sand and Gravel.

Phase	Extent of Contamination		Mass Distribution		
	Volume cu yd	% of Total	lb	Conc. ppm	% of Total
Free phase	780	5.3	126,800[a]	—	90.9
Adsorbed (soil)	2,670	18.3	11,500	2,000	8.2
Dissolved (water)	11,120	76.3	390	15	0.3

[a]Actual value recovered from site.

phase is only a small fraction of the total contaminant mass, and that portion in the vapor phase is too small a fraction even to be considered. Evidently, the rates of the mass transfer processes between phases must play a major role in determining the rate of any cleanup technique in which contaminant is removed in the aqueous or vapor phase.

MASS BALANCE

One relationship, which is independent of the details of the phases that are present, is the mass balance. This states that the total mass of VOC in a unit volume of soil is equal to the sum of the masses of VOC present in the vapor phase as NAPL or in the aqueous phase as adsorbed material, and dissolved in any greasy or tarry phase that may be present. This gives

$$C_{total} = \sigma C_{vap} + C_{NAPL} + \omega C_w + C_{ads} + g C_{gr} \qquad (1)$$

where

C_{total} = total concentration of VOC, kg/m³ of soil

TABLE 4.2. Phase Distribution of Gasoline in Fractured Bedrock.

Phase	Extent of Contamination		Mass Distribution		
	Volume cu yd	% of Total	lb	Conc. ppm	% of Total
Free phase	44	0.5	12,390[a]	—	72.8
Adsorbed (soil)	1,630	18.9	4,600	2,400	28.9
Dissolved (water)	6,960	80.6	70	25	0.2

[a]Actual value recovered from site.

C_{vap} = concentration of VOC in the soil gas, kg/m³ of vapor phase

σ = air-filled porosity of soil, dimensionless

C_w = concentration of VOC in the soil moisture, kg/m³ of aqueous phase

ω = water-filled porosity of soil, dimensionless

C_{NAPL} = VOC concentration as NAPL, kg/m³ of soil

C_{ads} = adsorbed VOC concentration, kg/m³ of soil

C_{gr} = VOC concentration in grease phase, kg/m³ of greasy phase

g = grease-filled porosity of soil, dimensionless (m³ of greasy material per m³ of soil)

HENRY'S LAW

We next turn to the partitioning of VOC between the vapor phase and the aqueous phase in the soil. This is generally adequately described by Henry's Law, which states that the vapor pressure of a VOC above a solution of that VOC is proportional to the concentration of the VOC in the solution. That is,

$$P_{VOC} = K_H C_w'$$ (2)

where

P_{VOC} = vapor pressure of VOC, atm

K_H = Henry's constant of VOC, atm L/mg

C_w' = VOC concentration, mg/L of water

Henry's constants can be reported in a variety of units, which can lead to confusion. In our subsequent modeling it will be convenient to express Henry's Law as

$$C_{vap}(\text{kg/m}^3) = K_H \cdot C_w(\text{kg/m}^3)$$ (3)

in which the Henry's constant K_H is dimensionless. We relate the dimensionless K_H to several other forms of Henry's constant in Table 4.3. Table 2.3 in Chapter 2 provides a list of dimensionless Henry's constants for a number of organic solvents at temperatures between 10 and 25°C. The Henry's constants for other compounds can be calculated from the VOC vapor pressure and solubility at the temperature T (deg K) of interest by means of Equation (4).

$$K_H(T) = 0.016034 \left[\frac{P_{vap}^0(T) \cdot (MW)}{TC_{sat}} \right]$$ (4)

TABLE 4.3. Conversion of Henry's Constants.

$$
\begin{aligned}
K_H \text{ (dimensionless)} &= K_H' \text{ (atm L/mg)} \times 1.2186 \times 10^4 \times (MW)/T \\
&= K_H' \text{ (torr L/mg)} \times 16.034 \times (MW)/T \\
&= K_H' \text{ (atm L/gm)} \times 12.186 \times (MW)/T \\
&= K_H' \text{ (torr L/gm)} \times 0.016034 \times (MW)/T \\
&= K_H' \text{ (atm L/mol)} \times 12.186/T \\
&= K_H' \text{ (torr L/mol)} \times 0.016034/T \\
&= K_H' \text{ (atm m}^3\text{/mol)} \times 1.2186 \times 10^4/T
\end{aligned}
$$

(MW) = VOC molecular weight, gm/mol.
T = temperature, deg K.

where

$P_{vap}^0(T)$ = pure VOC equilibrium vapor pressure at temperature T, torr
C_{sat} = VOC saturation concentration in water at temperature T, gm/L
(MW) = molecular weight of compound, gm/mol

NONAQUEOUS PHASE LIQUIDS, RAOULT'S LAW

If NAPLs are present in the soil, the vapor pressures and water solubilities of these compounds become important. The vapor pressures of a number of VOCs are given in Table 2.1, Chapter 2, and vapor pressure constants are provided in Table 2.2 of that chapter for use in calculating equilibrium vapor pressures of pure compounds at arbitrary temperatures from Equation (5).

$$\log_{10} P(T) = A - 0.05223(B/T) \tag{5}$$

where P is the vapor pressure of the VOC in torr and A and B are constants given in Table 2.2 of Chapter 2.

If one has a mixture of VOCs in the NAPL phase Raoult's Law may provide a convenient approximation for calculating the vapor pressures of the various components. Raoult's Law states that the vapor pressure of a VOC component i in an ideal solution mixture is given by Equation (6).

$$P_i = P_i^0 X_i \tag{6}$$

where

P_i = vapor pressure of component i above the mixture
P_i^0 = vapor pressure of pure component i
X_i = mole fraction of component i in the NAPL phase

The mole fractions are given in terms of concentrations by

$$X_i = \frac{C_i/(MW_i)}{\sum\limits_{j=1}^{n} C_j/(MW_j)} \tag{7}$$

where

C_k = concentration of component k in the NAPL, mass/volume
$(MW)_k$ = molecular weight of component k, gm/mol

The partial pressures P_i are related to the vapor concentrations C_{vapi} by

$$C_{vapi} = (MW)_i P_i / RT \tag{8}$$

where the vapor concentration is in kg/m^3, molecular weight $(MW)_i$ is in kg/mol, $R = 8.206 \times 10^{-5}$ m^3 atm/mol deg, and T is in deg K.

Raoult's Law is approximate but generally provides sufficient accuracy for SVE calculations. If desired, more accurate (and complex) calculations can be done utilizing activity coefficients, but the accuracy of the site data rarely warrants such refinements. See Thibodeaux (1979) for discussion of how this is done.

THE KELVIN EQUATION

In porous media that contain VOC, one may have air/NAPL interfaces involving VOC in small pores or as small droplets. If the radii of curvature of these interfaces are sufficiently small, there can be a significant impact on the vapor pressure of the VOC. The Kelvin Equation, derived in Adamson (1976), is

$$RT \log_e [P(r)/P^0] = 2\,\gamma \bar{V}/r \tag{9}$$

where

P^0 = equilibrium vapor pressure of the liquid above a flat air/liquid interface
$P(r)$ = equilibrium vapor pressure of the liquid above a spherical surface of radius of curvature r cm, arbitrary units
γ = surface tension of the air/liquid interface, dyne/cm (erg/cm^2)
$\bar{V}$ = molar volume of the liquid, cm^3/mol
R = gas constant, 8.315×10^7 erg/mol deg
T = temperature, deg K

If the NAPL is in the form of a droplet, the value of r is taken as positive, and the vapor pressure of the liquid is increased over that of the plane surface. If the NAPL is in a capillary pore that it wets, the NAPL/air surface is concave, the value of r is taken as negative, and the vapor pressure of the liquid is decreased. For benzene in wetted capillary pores (r negative), the calculation is as follows:

$$\bar{V} = (78.11 \text{ gm/mol})/(0.877 \text{ gm/cm}^3) = 89.1 \text{ cm}^3$$

$$= 28.88 \text{ erg/cm}^2$$

$$T = 20°C = 293 \text{ deg K}$$

Substitution of these figures into Equation (9) gives

$$\log_e [P(r)/P^0] = -2.11 \times 10^{-7}/r \qquad (10)$$

so a 1% reduction in vapor pressure would require pores 2.11×10^{-5} cm in radius.

AQUEOUS PHASE VOC

The relationship between aqueous phase VOC concentration and the vapor phase has been described above in connection to Henry's Law. If NAPL is present as a single volatile organic compound, the aqueous phase concentration of the VOC at equilibrium is simply the aqueous solubility of the compound. A number of these are listed in Table 4.4. We note that the accuracy of aqueous solubility data for these compounds is not high.

The aqueous solubilities of components in a solution can be readily estimated if the solution obeys Raoult's Law [Equation (6)]. The aqueous solubility of component i is given by

$$C_{wi} = C_{w0i}X_i \qquad (11)$$

where

C_{wi} = aqueous solubility of component i present in a NAPL
C_{w0i} = aqueous solubility of pure component i
X_i = mole fraction of component i in the NAPL solution, calculated from Equation (7)

If the solution is nonideal, one can use activity coefficients, as described by Thibodeaux (1979).

TABLE 4.4. Water Solubilities of Selected VOCs, mg/L.

Compound	T, °C					Mol Wt (gm/mol)
	10	15	20	25		
CCl_2F_2	39	63	99	153		120.9
CCl_3F	966	1,030	1,096	1,164		137.4
$CHCl_2F$				9,500	1 atm	102.9
C_2ClF_5				60	1 atm	154.5
$CHClF_2$				3,000	1 atm	86.5
$1,1,1,2-C_2Cl_4F_2$			100			203.8
$1,1,2,2-C_2Cl_4F_2$			100	120		203.8
CCl_2FCClF_2	136		200	170		187.4
CCl_4	830		800	825		153.8
CH_2Cl_2	19,880	18,660	17,550	16,540		84.9
$CHCl_3$	9,246	8,795	8,380	7,997		119.4
CS_2	2,110	2,040	1,985	1,930		76.1
C_2Cl_4	6	71	150	309		165.8
C_2HCl_3	1,075	1,137	1,157	1,175		131.4
cis-1,2-DCE						96.9
trans-1,2-DCE				8,300		96.9
1,1-DCE	2,362	2,370	2,377	2,383		96.9
Vinyl chloride	8,950	9,030	9,110	9,180		62.5
1,1,1,2-TCA						167.8
1,1,2,2-TCA	2,808	2,966	2,924	2,981		167.8
1,1,1-TCA	1,511	1,200	961	775		133.4
1,1,2-TCA			4,500			133.4
1,1-DCA			5,500	5,060		99.0
1,2-DCA	8,700	8,600	8,600	8,500		99.0
Ethyl chloride	4,800	5,200	5,500	5,800		64.5
1,2-Dichloropropane			2,650	2,800		113.0
n-Pentane	36	38	39	41		72.1
2-Me butane	61	56	52	48		72.1
2,2-diMe propane				33.2		72.1
$1,2,3,4-C_6H_2Cl_4$				5.92		215.9
$1,2,3,5-C_6H_2Cl_4$				5.19		215.9
$1,2,3-C_6H_3Cl_3$				18		181.4
$1,2,4-C_6H_3Cl_3$	0.13	4.2	12	36		181.4
$1,3,5-C_6H_3Cl_3$			5.8	6.0		181.4
1,2-Dichlorobenzene	100	114	130	148		147.0
1,3-Dichlorobenzene	42	63	92	132		147.0
1,4-Dichlorobenzene	37	47	59	74		147.0
Chlorobenzene	405	418	431	445		112.5
Benzene	1,630	1,700	1,770	1,840		78.1
Cyclohexane	40	46	52	59		84.2
n-Hexane	7.5	8.1	8.7	9.3		86.2
2-Me pentane	14	15	15	15		86.2
2,2-diMe butane	30	27	24	21		86.2
Toluene	588	568	549	531		92.1

TABLE 4.4. (continued).

Compound	T, °C				Mol Wt (gm/mol)
	10	15	20	25	
n-Heptane	2.1	2.3	2.4	2.6	100.2
2-Me hexane				2.5	100.2
3-Me hexane				3.8	100.2
2,2-diMe pentane				5.3	100.2
2,4-diMe pentane	5.6	5.3	4.9	4.6	100.2
Styrene	250	270	295	320	104.1
Ethylbenzene	169	170	171	171	106.2
2-Xylene	156	164	172	180	106.2
3-Xylene	179	172	165	159	106.2
4-Xylene	167	171	174	177	106.2
n-Octane	0.55	0.59	0.62	0.66	114.2
3-Me heptane				0.79	114.2
n-Nonane			0.07	0.12	128.2
n-Decane			0.009	0.021	142.3

Values taken or calculated from date given by Montgomery and Welkom (1989) and by Hughes et al. (1985).

VOC DISSOLVED IN OIL, GREASE, OR TAR

A reasonable assumption, with regard to the equilibrium partitioning of a VOC between water and oil, grease, or tar, is that this obeys the simple distribution law

$$C_w = K_{gr} C_{gr} \qquad (12)$$

where K_{gr}, the partition coefficient, depends on the identity of the VOC, the characteristics of the greasy or tarry material, and the temperature.

A study by Bouchard et al. (1990) showed that the presence in soil of tetradecane or high-boiling residual hydrocarbons from gasoline resulted in increased adsorption of neutral organic compounds.

ADSORBED VOC

VOCs may be adsorbed on the various surfaces available in the soil medium. Such adsorption is related to the specific surface area of the medium, which can range from as little as a small fraction of a square meter per gram for coarse sands and gravels up to values of several hundred square meters per gram for fine clays (van Olphen, 1977). Adsorption is also related to the type of substrate that is available; naturally occurring organic material in the soil is quite effective in providing adsorption sites

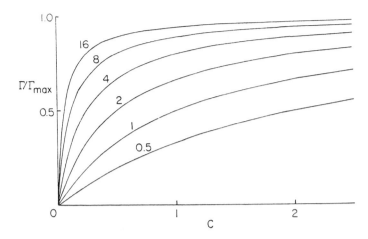

Figure 4.1 Representative Langmuir isotherm plots. Γ_{max} is the VOC surface concentration at maximum coverage. The numbers labeling the plots are the reciprocals of the aqueous VOC concentrations at which the surface sites are 50% occupied.

for VOCs, and in many instances one can adequately calculate the adsorption characteristics of a soil simply from its organic carbon content.

Several adsorption isotherms have been proposed for the description of the adsorption of VOCs on soils; among these are the linear, the Langmuir, the Freundlich, and the Brunauer-Emmett-Teller (or BET). See Figures 4.1, 4.2, and 4.3 for some representative plots of each type. Generally, the experimental data available are sufficiently sparse and of sufficiently poor accuracy that use of a highly sophisticated, parameter-laden isotherm is not justified. In most modeling work, in fact, a simple linear isotherm analogous to Henry's Law is used. One can write these isotherms either in terms of the aqueous VOC concentration in equilibrium with the surface or in terms of the gaseous VOC concentration in equilibrium with it; Henry's Law [Equation (3)] provides the relationship between the two. We shall express the adsorption isotherms here in terms of the aqueous VOC concentration.

The linear isotherm is simply

$$C_{ads} = K_a C_w \tag{13}$$

where

C_{ads} = concentration of adsorbed VOC, kg/m³ of soil
C_w = aqueous concentration of VOC, kg/m³ of aqueous phase
K_a = adsorption isotherm

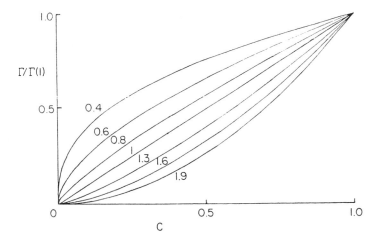

Figure 4.2 Representative Freundlich isotherm dimensionless plots. The numbers labeling the plots are the values of $1/n$.

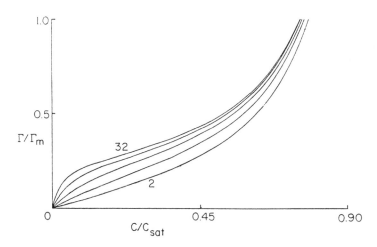

Figure 4.3 Representative BET isotherm plots. Γ_m is the VOC surface concentration corresponding to monolayer coverage. C/C_{sat} is the ratio of the aqueous VOC concentration to the aqueous solubility of the VOC.

Mass balance [Equation (1)], together with the various partitioning equations [Equations (3), (6), and (8)], gives us the relationship

$$C_{total} = [(\nu - g - \omega)K_H + gK_{gr} + K_a + \omega]C_w + C_{NAPL}$$
(14)

where

ν = total porosity, dimensionless
ω = water-filled porosity
g = grease-filled fraction of soil, grease content of soil (kg/m³)/density of greasy phase (kg/m³)

One then sets $C_{NAPL} = 0$ and solves Equation (12) for C_w, which gives

$$C_w = \frac{C_{total}}{[(\nu - g - \omega)K_H + \omega + gK_{gr} + K_a]}$$
(15)

If the resulting value of C_w is less than or equal to C_{sat}, the aqueous saturation concentration of the VOC, then the value of C_w is, in fact, given by Equation (15), and the concentrations in the other phases are given by

$$C_{vap} = K_H C_w$$
(16)

$$C_{ads} = K_a C_w$$
(17)

$$C_{gr} = K_{gr} C_w$$
(18)

and

$$C_{NAPL} = 0$$
(19)

If Equation (15) gives a value of C_w that is larger than C_{sat}, one proceeds as follows. The aqueous solution is evidently saturated with VOC, so

$$C_w = C_{sat}$$
(20)

Then

$$C_{vap} = K_H C_{sat}$$
(21)

$$C_{ads} = K_a C_{sat}$$
(22)

and

$$C_{gr} = K_{gr}C_{sat} \tag{23}$$

Lastly, one uses the material balance equation to calculate C_{NAPL}:

$$C_{NAPL} = C_{total} - [(\nu - g - \omega)K_H + gK_{gr} + K_a + \omega]C_w \tag{24}$$

The Langmuir isotherm introduces a second parameter and is slightly more complex than the linear isotherm; it is

$$C_{ads} = \frac{K_L C_w}{1 + C_w/C_w'} \tag{25}$$

where the Langmuir parameter K_L is the slope of the adsorption isotherm at very low values of C_w, and the Langmuir parameter C_w' is the value of C_w at which the surface sites are half saturated. One expects strongly adsorbed compounds to have small values of C_w' (see Figure 4.1).

We use the material balance equation [Equation (1)], along with Equations (3), (6), (12), and (25), to get an equation in C_w alone:

$$C_{total} = \left[(\nu - g - \omega)K_H + gK_{gr} + \omega + \frac{K_L}{1 + C_w/C_w'}\right]C_w \tag{26}$$

This can be rearranged to give a quadratic equation in C_w,

$$0 = aC_w^2 + bC_w + c \tag{27}$$

where

$$a = \frac{[(\nu - g - \omega)K_H + gK_{gr} + \omega]}{C_w'} \tag{28}$$

$$b = \left[(\nu - g - \omega)K_H + gK_{gr} + \omega - \frac{C_{total}}{C_w'}\right] \tag{29}$$

$$c = -C_{total} \tag{30}$$

The root of the quadratic which lies in the range [0, C_{total}/ω], is the one that is taken. If C_w is less than or equal to C_{sat}, one then substitutes C_w into Equations (16), (18), and (25) to calculate C_{vap}, C_{gr}, and C_{ads}; $C_{NAPL} = 0$.

Alternatively, one may solve Equation (26) by a simple binary search technique, or one can rearrange it to give

$$C_w = \frac{C_{total}}{\left[(\nu - g - \omega)K_H + gK_{gr} + \omega + \dfrac{K_L}{1 + C_w/C'_w} \right]} \tag{31}$$

which can then be solved iteratively. In any case, if the resulting value of C_w is larger than C_{sat}, one calculates C_w from Equation (22), C_{vap} from Equation (23), and C_{gr} from Equation (24). C_{ads} is given by

$$C_{ads} = \frac{K_L C_{sat}}{1 + C_{sat}/C'_w} \tag{32}$$

The NAPL concentration is then calculated from

$$C_{NAPL} = C_{total} - (\nu - g - \omega)C_{vap} - \omega C_w - C_{ads} - gC_{gr} \tag{33}$$

The Freundlich isotherm can be written as

$$C_{ads} = K_F(C_w)^{1/n} \tag{34}$$

where K_F and n are the adjustable parameters in the isotherm. See Figure 4.2 for its behavior. One proceeds exactly as above with the Langmuir isotherm to obtain Equation (35) from mass balance:

$$C_{total} = [(\nu - g - \omega)K_H + gK_{gr} + \omega]C_w + K_F C_w^{1/n} + C_{NAPL} \tag{35}$$

As before, we set $C_{NAPL} = 0$ and solve the resulting equation for C_w. Unless $n = 0.5$, 1, or 2, this must be done numerically, and we have normally used a binary search method for this, starting with the interval [0, C_{total}/ω]. This is simply done, since the right-hand side of Equation (35), like that of Equation (26), is montonically increasing with C_w.

The Brunauer-Emmett-Teller (BET) isotherm can be written as

$$C_{ads} = \frac{C_{ads,mon} \cdot c \cdot (C_w/C_{sat})}{(1 - C_w/C_{sat})[1 + (c - 1)C_w/C_{sat}]} \tag{36}$$

where the isotherm parameters are

$C_{ads,mon}$ = adsorbed VOC concentration, which produces a monolayer on the medium, kg/m^3

    c = parameter measuring the strength of the binding of VOC in the monolayer; the stronger the binding, the larger c

In the derivation of the BET isotherm, one makes the assumption that multilayer adsorption is taking place—that the VOC wets the porous medium. This means that one considers both NAPL and adsorbed VOC in the equation, so one does not calculate a separate C_{NAPL} concentration to describe nonaqueous phase liquid. See Figure 4.3 for some examples of the BET isotherm. The material balance equation gives

$$C_{total} = \frac{C_{ads,mon} \cdot c \cdot (C_w/C_{sat})}{(1 - C_w/C_{sat})[1 + (c - 1)C_w/C_{sat}] +} \qquad (37)$$
$$[\nu - g - \omega)K_H + gK_{gr} + \omega]C_w$$

as the equation to be solved for C_w. Again, since the right-hand side of the equation increases monotonically with C_w, a simple search algorithm is quite satisfactory. Initial bounds are $0 < C_w <$ the least of C_{sat}, C_{total}/ω.

    These adsorption isotherms and a number of others are discussed in more detail in most surface chemistry texts, such as Shaw (1966), Adamson (1976), Popiel (1978), or Vold and Vold (1983). The use of other isotherms follows along the lines discussed above for the Freundlich and BET isotherms. These are generally sufficiently complex that the mass balance equation must be solved numerically for C_w, but this is easily done.

EFFECTS OF ORGANIC CARBON

    The importance of organic carbon (humic and fulvic acids) and of surface area on the adsorption of organic compounds has been discussed by Thibodeaux (1979) and others. The impact of organic carbon is seen in Figure 4.4, which shows Freundlich plots for a polychlorinated biphenyl on illite clay, a natural soil, and on humic acid itself. Freundlich isotherm parameters for the herbicide 2,4-D on a number of substrates are shown in Table 4.3 and illustrate the effects of organic carbon and of a high specific surface area.

    Jury (1986) has summarized the effects of soil organic content on adsorption coefficients. He notes that natural soil organic matter is very complex, variable, and poorly defined. Its adsorption characteristics vary with its state of decomposition, so the adsorption isotherm of a surface soil can be expected to show considerable variation for a given total organic matter content.

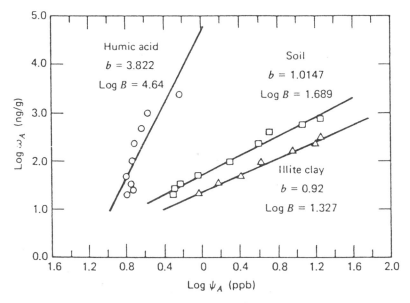

Figure 4.4 Freundlich plot for the adsorption of 2,4,2′,4′-tetrachlorophenol on illite, humic acid, and Woodburn soil surface. [Reprinted from R. Hague and V. H. Freed (1975), "Role of Adsorption in Studying the Dynamics of Pesticides," *Environmental Dynamics of Pesticides,* with permission from Plenum Publishing Corporation.]

Nevertheless, he notes, one observes a positive linear correlation between soil organic matter content and adsorption of organic chemicals. This correlation is generally strongest when the organic matter content is high, but may still be significant in soils with an organic carbon content as low as 0.1% (Hamaker and Thompson, 1972; Lyman, 1982). It is therefore common practice to define an organic carbon partition coefficient K_{oc} as

$$K_{oc} = K_a/f_{oc} \qquad (38)$$

where f_{oc} is the organic carbon fraction of the soil. Hamaker and Thompson (1972) have shown that the coefficient of variability of K_{oc} is considerably less than that of K_a for a given chemical adsorbed on different soils, particularly for nonpolar compounds. Calvet (1980) has noted that the variability in K_{oc} is still fairly large; however, Jury (1986) remarks that it is, nevertheless, commonly used [see Rao and Davidson (1980); Kenaga and Goring (1980); Mingelgrin and Gerstl (1983)]. A number of correlations of K_{oc} with octanol/water partition coefficient K_{ow} and with water solubility have been made; these have been summarized by Palmer and Johnson (1991).

Roy and Griffin (1987, 1989) have published considerable information on adsorption of VOCs on soils. Figure 4.5 shows adsorption isotherms for benzene and four chlorinated benzenes, as well as that for water, on a dry soil; note that water is the most strongly adsorbed. Roy and Griffin (1989) have provided an extensive table of values of K_{oc} and K_H from which the information in Table 4.5 has been taken. Montgomery and Welkom (1989) also provide values of K_{oc}. Table 4.6 provides Freundlich isotherm parameters for three common organic solvents.

EFFECTS OF WATER

The presence of soil water has a strong inhibiting effect on the adsorption of VOCs. This so-called "wet dog effect" has been known to agricultural scientists for many years; Wade's (1955) data on the effect of relative humidity on the adsorption of the fumigant ethylene dibromide show the magnitude of the effect, as seen in Figure 4.6. Jury (1986) comments that, when soil is dry, the preferential coverage of water molecules on the adsorbing surfaces is removed and solute adsorption increases dramatically. Spencer and Cliath (1973), Ehlers et al. (1973), and Harper et al. (1976) have explored the point in connection with agricultural chemicals.

The effect of relative humidity on the adsorption of 1,2,4-trichlorobenzene by topsoil is shown in Figure 4.7 (Chiou and Shoup, 1985). Sterrett

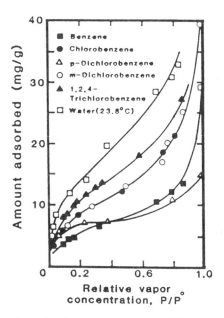

Figure 4.5 Adsorption of organic solvent vapors and water by Woodburn soil at 20°C. [Reprinted with permission from Chiou and Sharp (1985), © 1985 American Chemical Society.]

TABLE 4.5. Values of Soil Organic Carbon/Water Partition Coefficients and Henry's Constants for a Number of Volatile Organics.

Compound	K_{oc} (cm³/gm)	K_H (atm m³/mol)
Benzene	97	5.6×10^{-3}
Bromomethane	131	6.4×10^{-3}
Carbon disulfide	63	1.2×10^{-4}
Carbon tetrachloride	232	3.0×10^{-2}
Chlorobenzene	318	3.9×10^{-3}
Chloroethane	42	1.1×10^{-2}
Chloroform	34	3.7×10^{-3}
Dichlorodifluoromethane (F-12)	269	0.43
1,1-Dichloroethane	43	5.6×10^{-3}
1,2-Dichloroethane	36	1.3×10^{-3}
1,1-Dichloroethylene	58	2.6×10^{-2}
trans-1,2-Dichloroethylene	169	9.4×10^{-3}
cis-1,3-Dichloropropylene	23	2.3×10^{-3}
trans-1,3-Dichloropropylene	26	1.8×10^{-3}
Ethylbenzene	622	8.7×10^{-3}
Methylene chloride	25	3.2×10^{-3}
Styrene	260	2.3×10^{-4}
1,1,2,2-Tetrachloroethane	88	4.5×10^{-4}
1,1,1-Trichloroethane	155	1.7×10^{-2}
1,1,2-Trichloroethane	48	9.1×10^{-4}
Trichloroethylene	152	9.7×10^{-3}
Trichlorofluoromethane (F-11)	479	6.0×10^{-2}
Toluene	242	6.3×10^{-3}
Vinyl chloride	66	2.8×10^{-2}
o-Xylene	363	5.7×10^{-3}
m-Xylene	1,580	6.7×10^{-3}
p-Xylene	204	6.7×10^{-3}

TABLE 4.6. Freundlich Isotherm Parameters at 16% (gm/gm) Soil Moisture.

Compound	Sorption Parameters		
	$K_f \times 10^{-3}$ (mg/gm)/(mg/L)$^{1/n}$	$1/n$	r^2
Benzene	3.64	0.79	0.76
Trichloroethylene	3.02	0.85	0.86
o-Xylene	6.57	1.10	0.90

From English and Loehr (1992).

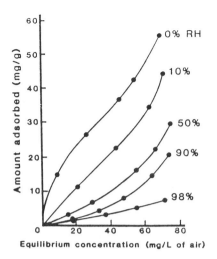

Figure 4.6 Adsorption of ethylene dibromide vapor by Black Fen soil at 20°C as a function of relative humidity (RH). [Reprinted from P. Wade (1955), "Soil Fumigation, III. The Sorption of Ethylene Dibromide," *J. Science of Food & Agriculture,* with permission from Elsevier Science.]

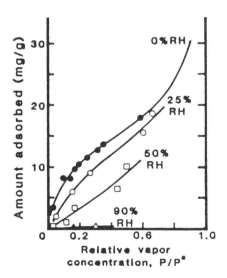

Figure 4.7 Adsorption of 1,2,4-trichlorobenzene by Woodburn soil as a function of relative humidity (RH). [Reprinted from Chiou and Shoup (1985), © 1985 American Chemical Society.]

(1989) reported that the vapor stripping of 1,3-dichloropropane at a site near Benson, Arizona, was enhanced by moisture.

Davies (1989) noted that the interaction between mineral surfaces and nonionic organics is weak in high-humidity environments because water is preferentially adsorbed on the mineral surfaces. In connection with a study on the adsorption of chlorobenzenes, he commented that dehydrated soils are very strong adsorbers for VOCs but that, when water is present, it displaces the organics. Although chlorobenzene is very strongly adsorbed by dry soils, its adsorption on moist soils is no greater than one finds in soil slurries. Fortunately, at most sites the relative humidity of the soil gas is of the order of 98% or more, even a short distance below the surface, so the strong binding observed with dry soils is not likely to be a problem. Davies noted the importance of not drying the soil excessively in SVE operations in order to avoid strong adsorption of VOCs, which could markedly impede SVE. More recently, Thibaud et al. (1992) published adsorption isotherms for chlorobenzene at several relative humidities; their results are shown in Figure 4.8.

Gierke et al. (1992) reported, in connection with a laboratory column SVE study, that sorption of toluene was negligible when the medium was wet but did occur when the medium was dry.

A study by Ong et al. (1991) found that adsorption of trichloroethylene (TCE) on both humic-coated aluminum oxide and Gila silt loam increased dramatically as the solid was made progressively drier. Oja and Kreamer (1991) also investigated the effect of moisture on the adsorption of TCE in a silty sand (estimated surface area 620 cm^2/gm), a sandy loam (900 cm^2/gm), and a clay (1,360 cm^2/gm). K_a' values at various water contents are given in Table 4.7. $K_a' = $ [mg TCE adsorbed/gm of soil]/[mg TCE/cm^3 in gaseous phase/mg TCE/cm^3 in gaseous phase at saturation].

THE LOCAL EQUILIBRIUM APPROXIMATION

In the modeling of soil vapor extraction, one frequently finds the term *local equilibrium* used. This is an approximation in which it is assumed that the rate of advection of the soil gas in the vicinity of a vapor extraction well is sufficiently slow that it does not perturb the equilibria of VOC between the various phases. If this is an adequate approximation, one may use the equations above, unncoupled from advective transport, to calculate the VOC concentrations in the vapor phase, which simplifies the modeling of an SVE well significantly. Use of the local equilibrium approximation always gives one an upper bound to the rate of cleanup (a lower bound to the cleanup time), i.e., an optimistic estimate. In cases where diffusion and/or desorption rates are slow, the estimate provided by local equilibrium calculations may be so optimistic as to be utterly useless.

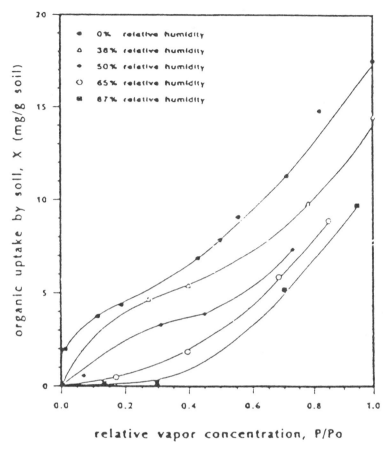

Figure 4.8 Adsorption isotherms for the system chlorobenzene-water-soil-helium at different relative humidities. [Reprinted with permission from Thibaud et al. (1992), © 1992 American Chemical Society.]

MASS TRANSPORT KINETICS

We next turn to the rate processes that affect soil vapor extraction. The literature here is somewhat confusing because there has been disagreement as to whether or not it is necessary to consider the effects of rate processes. Some of the earlier publications (and some later ones, as well) reported that the advective transport of VOCs in the gas phase was sufficiently slow that mass transport processes of VOCs between phases were essentially at equilibrium. See, for example, Hoag et al. (1984); Johnson et al. (1987); Hoag (1991); Rathfelder et al. (1991). If this is true, it would substantially simplify mathematical modeling, since one could

TABLE 4.7. K_a' Values for TCE on Three Natural Soils as a Function of Water Content.

Soil	Water Content (wt %)	K_a' (mg/gm)
Silty sand	Air-dried	1.4
	5	0.34
	10	0.42
	15	0.60
	20	0.41
Sandy loam	Air-dried	4.4
	5	0.96
	10	1.0
	15	0.75
	20	0.45
Clay	Air-dried	3.2
	5	0.71
	10	0.60
	20	0.45

use the local equilibrium approximation to calculate all concentrations in the various phases by algebraic equations as described above and then adjoin to this the differential equations describing advection of the vapor phase. At some sites and in some laboratory column experiments, local equilibrium does, indeed, appear to be an adequate approximation.

However, other sites have been reported at which initial high off-gas VOC concentrations drop off rapidly to relatively low levels, which then persist for an extended period of time (tailing), and it has also been observed that at some sites the soil gas concentration, low during the operation of the vapor extraction well, increases quite markedly within hours after the well is shut down (rebound). Both of these effects suggest mass transfer limitations of some sort, by either diffusion or desorption processes. At present there seem to be enough data to indicate that a substantial fraction of SVE sites are mass transport–limited, so that local equilibrium models are not adequate for their analysis [see Rainwater et al. (1990); Gierke et al. (1992); Brusseau et al. (1991), Brusseau (1992)]. At present, EPA's position appears to be that the possibility of mass transport–limited kinetics should be considered in designing SVE pilot tests [see DiGiulio et al. (1990); DiGiulio (1992)].

DESORPTION

The two mass transfer processes that can impact upon SVE are desorption and diffusion. In desorption one is looking at the rate of a chemical (or quasi-chemical) reaction, in which VOC sorbed at a site on the porous medium is released into aqueous solution or into the vapor phase. Desorption is generally an endothermic process, and the stronger the binding, the more endothermic is the desorption. Also, the rates of virtually all chemical reactions increase dramatically with increasing temperature. This suggests that one can facilitate desorption by increasing the temperature of the medium, perhaps by steam or hot air injection, radio-frequency heating, etc. Heating also increases VOC vapor pressures and Henry's constants, which further accelerates the SVE process [see Downey and Elliott (1990); Lingineni and Dhir (1992); Phelan (1992); Lord et al. (1988)].

One can say little more about desorption kinetics because relatively little has been done to measure the rates of desorption of the VOCs of interest from sorption sites in moist natural soils into the aqueous or the vapor phase.

DIFFUSION

Several diffusion processes appear in SVE. One is the diffusion of VOC from a layer or droplet of water in which it is dissolved into an adjoining vapor phase. A second is diffusion of VOC from the surface of an NAPL droplet/blob through a surrounding layer of water into the vapor phase. A third is diffusion of VOC from NAPL droplets in a lump of low-permeability porous medium out through the gas-filled pores of the lump into a domain through which soil gas is able to advect.

In all cases, these diffusion processes are governed by Fick's Laws of Diffusion [see Freeze and Cherry (1979)]. The first law states that the flux of a diffusing solute is given by

$$F = D\nabla C \tag{39}$$

where

F = flux of solute, kg/m^2 sec
D = diffusion constant, m^2/sec
C = solute concentration, kg/m^3

The second law is given by Equation (40):

$$\frac{\partial C}{\partial t} = \nabla \cdot D\nabla C \tag{40}$$

VAPOR PHASE DIFFUSION OF VOC THROUGH A POROUS MEDIUM

In examining gaseous diffusion of VOCs in soil, one must take into account the fact that the diffusion is taking place in a porous medium. Roy

TABLE 4.8. Equations Relating Porous-Media Diffusion Coefficients D_s to Fick's Law Diffusion Coefficients D_F.

$D_s = D_F \nu_a^{3.3}/\nu_t^2$	Millington and Quirk (1961)
$D_s = D_F(0.9\nu_a - 0.1)$	Wesseling (1962)
$D_s = D_F \cdot 5.25 \cdot \nu_a^{3.36}$	Grable and Siemer (1968)
$D_s = D_F(\nu_a/\nu_t)^4 \nu_a^{3/2}$	Currie (1970)
$D_s = D_F \nu_a^{5/3}$	Lai et al. (1976)
$D_s = D_F \cdot 0.777(\varrho_s/\nu_t) - 0.274$	Albertson (1979)
$D_s = D_F \cdot (Tor)\nu_a + (\nu_t - \nu_a)\varrho_w K_w + (1 - \nu_t)\varrho_s K_s$	Weeks et al. (1982)

where

Tor = soil tortuosity factor (dimensionless)
ϱ_w = density of water (gm/cm^3)
K_w = liquid-gas partition coefficient [volume (gas)/weight (solid)]
ϱ_s = particle density of the porous medium, gm/cm^3
K_s = $K_w K_d$, where K_d is a liquid-solid partition coefficient [volume (gas)/weight (solid)]
ν_a = air-filled porosity
ν_t = total porosity

and Griffin (1987) have reviewed most of the formulas that have been proposed for the calculation of porous media diffusion coefficients. A number of these are given in Table 4.8. They note that diffusivities in soils depend on a variety of factors (such as moisture content, adsorption coefficient, etc.) and that there is no really satisfactory general theory for calculating diffusivities in porous media. They present a table of Fick's Law gas phase diffusion coefficients and gas phase porous-media diffusion coefficients for thirty-seven organic solvents; these vary between 0.0658 (ethylbenzene) to 0.159 cm^2/sec (methanol), with the great bulk of the values clustered between 0.070 and 0.090. Porous-media diffusion coefficients for these VOCs vary from 0.0053 (ethylbenzene) to 0.0130 cm^2/sec (methanol), with most values clustered between 0.006 and 0.008 cm^2/sec.

Osejo and Wilson (1991) measured diffusivities of several VOCs in air-dried sand of porosity 0.42 at room temperature; results are given in Table 4.9. The theoretical values were calculated by using an equation for

TABLE 4.9. Diffusivities of VOCs in Unscreened Washed Sea Sand.

Compound	Experimental D_s cm^2/sec	Theoretical D_s
CH_3CCl_3	0.0178	0.0224
Toluene	0.0193	0.0234
Hexane	0.0218	0.0216
C_2Cl_4	0.0242	0.0223
C_2Cl_3	0.0244	0.0234

the Fick's Law diffusivity from the kinetic theory of gases [see Levine (1988), for example] and Millington and Quirk's formula (1961). The agreement between theoretical and experimental values is not perfect, but it is certainly adequate for engineering estimates.

One can estimate the rate of diffusion transport from an area A of NAPL through a thickness a of porous medium to a region in which the VOC is removed rapidly by advection from Fick's First Law as follows. Let

D_s = diffusivity of the VOC in the porous medium, m²/sec

a = thickness of porous layer through which diffusion must take place, m

C_{sat} = saturation vapor concentration of VOC, kg/m³

We assume that the VOC vapor concentration at a distance a from the NAPL surface is essentially zero due to VOC removal by advection. Then the concentration gradient of the VOC is simply C_{sat}/a, and the mass of VOC evaporated per second through the porous medium is given by

$$\frac{dM}{dt} = \frac{AD_sC_{sat}}{a} \tag{41}$$

DIFFUSION OF AQUEOUS VOC INTO THE VAPOR PHASE

Let us first use Equation (40) to calculate the rate constants for diffusion of VOC from a thin plane layer of aqueous solution into an adjacent vapor phase (see Figure 4.9). We assume that advective transport in the vapor phase is sufficient to maintain a low constant VOC concentration of C_0 at

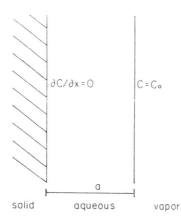

Figure 4.9 Geometry and notation, diffusion of aqueous VOC from a plane layer of solution.

the boundary between the aqueous and the vapor phases. The boundary conditions for the system are

$$\partial C(0,t)/\partial x = 0 \tag{42}$$

(no diffusion of VOC into the solid at the left of the layer of liquid), and

$$C(a,t) = C_0 \tag{43}$$

For this system Equation (40) becomes

$$\partial C/\partial t = D\partial^2 C/\partial x^2 \tag{44}$$

Equation (40) is readily solved by the method of separation of variables; use of the boundary conditions determines the eigenvalues, and we obtain

$$\lambda_n = \frac{(2n + 1)^2 \pi^2 D}{4a^2} \tag{45}$$

and

$$C(x,t) = C_0 + \sum_{n=0} B_n \cdot \cos\left[(2n + 1)\pi x/a\right] \cdot \exp\left[-\frac{(2n + 1)^2 \pi^2 D}{4a^2} \cdot t\right] \tag{46}$$

The decay of this solution to C_0 is evidently determined by the smallest eigenvalue,

$$\lambda_1 = \frac{\pi^2 D}{4a^2} \tag{47}$$

We can regard λ_1 as an approximate rate constant for this type of diffusion process.

Let us next carry out the same calculation for a droplet of aqueous solution of VOC of radius a in contact with the vapor phase. Spherical coordinates are appropriate here, since the system is spherically symmetric, and Equation (40) becomes

$$\frac{\partial C}{\partial t} = D\frac{1}{r^2}\frac{\partial}{\partial r}r^2\frac{\partial C}{\partial r} \tag{48}$$

Boundary conditions are

$$C(a,t) = C_0 \tag{49}$$

and

$$\partial C(0,t)/\partial r = 0 \tag{50}$$

This is solved by separation of variables and use of the substitution $R(r) = U(r)/r$; the eigenvalues that result from the boundary conditions are given by

$$\lambda_n = \frac{(n\pi)^2 D}{a^2} \tag{51}$$

The general solution is

$$C(r,t) = C_0 + \frac{1}{r} \sum_{n=1}^{\infty} A_n \cdot \sin(n\pi x/a) \cdot \exp\left[-\frac{(n\pi)^2 D}{a^2} \cdot t\right] \tag{52}$$

As before, the rate of decay of the solution to C_0 is controlled by the smallest eigenvalue, which is

$$\lambda_i = D(\pi/a)^2 \tag{53}$$

In these equations the value of D will be that for diffusion of the VOC in water (typically about 2×10^{-10} m²/sec), and the value of a will be that appropriate to the thickness of the water films in moist soils, probably a few thousandths of a centimeter. If the film thickness is 0.001 cm, the above value of D in Equation (47) gives a diffusion rate constant of 4.93 sec⁻¹. If, on the other hand, the soil is sufficiently wet that the dimensions of water-saturated regions are of the order of 0.1 cm, Equation (53) gives a diffusion rate constant of only 1.97×10^{-3} sec⁻¹, which would probably result in some interference with SVE.

These rate constants and others, which one could calculate with identical mathematics for the diffusion of dissolved VOC from lumps or slabs of porous medium that are saturated with water, can be used in a lumped parameter approximation in the modeling of SVE. This approximation, which will be used in the next chapter, is as follows. We wish to represent the mass transfer of VOC between the stationary liquid and the moving gaseous phase. Let the aqueous VOC concentration that is in equilibrium with the given vapor concentration be C_e, given by Henry's Law. Then, for the case of a spherical lump at times after the terms involving the higher eigenvalues have become negligible, we have

$$C(r,t) - C_e = A_1 \sin(\pi x/a) \cdot \exp(-\lambda_1 t) \tag{54}$$

Differentiate this with respect to t and substitute Equation (54) in the result to obtain

$$\partial[C(r,t) - C_e]/\partial t = -\lambda_1[C(r,t) - C_e] \tag{55}$$

One then integrates over the volume of the lump and then adds up the dissolved VOC contributions from all the lumps in 1 m³ of medium to obtain an equation for the rate of change of the bulk aqueous VOC concentration, C_w (kg of VOC/m³ of aqueous phase), with time due to mass transport to (or from) the vapor phase; this is

$$\frac{\partial C_w}{\partial t} = -\lambda_1(C_w - C_e) \tag{56}$$

One can readily calculate the coefficients in the Fourier series in Equation (41), which gives the concentration of VOC in the plane layer of water of thickness a as a function of time and position; if we assume that the initial concentration $C(x,t)$ is a constant C_1, this yields

$$C(x,t) = C_0 + \frac{4\Delta C}{\pi} \sum_{n=0}^{\infty} \frac{(-1)^n}{2n+1} \cdot \cos\left[\frac{(2n+1)\pi x}{2a}\right]$$
$$\cdot \exp\left[-\left(\frac{(2n+1)\pi}{2a}\right)^2 Dt\right] \tag{57}$$

where $\Delta C = C_1 - C_0$.

To get the bulk concentration of VOC in the water layer, we integrate Equation (57) with respect to x from 0 to a and divide by a; this results in

$$C_w = C_0 + \frac{8\Delta C}{\pi^2} \sum_{n=0}^{\infty} \frac{1}{(2n+1)^2} \cdot \exp\left[-\left(\frac{(2n+1)\pi}{2a}\right)^2 Dt\right] \tag{58}$$

Our simple lumped parameter approach gives

$$C_w = C_0 + \Delta C\{1 - \exp[-(\pi/2a)^2 Dt]\} \tag{59}$$

Plots of $\Delta C/\Delta C_{initial}$ versus the reduced time $\tau = Dt/a^2$ are given in Figure 4.10 for Equations (58) and (59). The lumped parameter approximation appears to be fairly good, except in one regard. The initial slopes of the two curves are quite different, with the series giving a much larger slope than the lumped parameter method. Therefore, if one measures the rate of VOC removal during only the initial stages to calculate the lumped parameter rate constant for diffusion, one will very substantially overestimate the rate of VOC removal in the later stages of the remediation. Evidently, estimations of lumped parameter diffusion constants should be made utilizing SVE runs in which a substantial fraction of the total dissolved VOC has been removed if one wishes to avoid quite substantial underestimations of cleanup times. Pilot-scale test runs lasting only a few hours or days can give grossly overoptimistic estimates of diffusion kinetics limitations along toward the end of the remediation, with correspondingly bad underestimations of cleanup times. A common statement in reports on field-scale SVE operations is that VOC removal was extremely rapid the first few days but then dropped off quite substantially, as explained by this model.

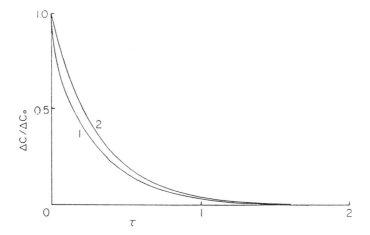

Figure 4.10 Plots of $\Delta C/\Delta C_0$ versus reduced time τ. In plot 1 the quantity

$$\Delta C/\Delta C_0 = \frac{8}{\pi^2} \sum_{n=0}^{\infty} \frac{1}{(2n + 1)^2} \cdot \exp\left\{ -\left[\frac{(2n + 1)\pi}{2} \right]^2 \tau \right\}$$

is graphed. The simple lumped parameter approach gives plot 2,

$$\Delta C/\Delta C_0 = 1 - \exp[-(\pi/2)^2\tau]$$

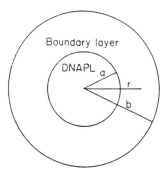

Figure 4.11 Geometry and notation, mass transport by diffusion from an NAPL droplet of radius a through an aqueous boundary layer to the advecting vapor phase. [From Wilson (1990) p. 243, by courtesy of Marcel Dekker, Inc.]

SOLUTION AND DIFFUSION FROM NONAQUEOUS PHASE LIQUID DROPLETS IN WATER

We next address the problem of mass transport from an NAPL droplet through a surrounding water boundary layer into an advecting vapor phase (see Figure 4.11). We shall assume that the solubility of the VOC in water is sufficiently low that we may use a steady-state approximation for the diffusion process in the aqueous phase. We shall also assume that the DNAPL droplet is sufficiently small that its solution and eventual transport to the vapor phase does not affect the radius of the surrounding water boundary layer.

The steady-state diffusion equation for the system shown in Figure 4.11 is

$$\frac{1}{r^2}\frac{d}{dr}\left[Dr^2\frac{dC}{dr}\right] = 0 \tag{60}$$

where C is the local concentration of VOC in the aqueous boundary layer. The boundary conditions are

$$C(a) = C_{sat} \tag{61}$$

and

$$C(b) = C_\infty \tag{62}$$

where C_∞ is the VOC concentration at the outer edge of the aqueous boundary layer. A general solution to Equation (60) is

$$C(r) = A/r + B \tag{63}$$

where A and B are arbitrary constants. Use of Equations (61) and (62) gives

$$A = \frac{\Delta Cab}{b - a}, \qquad \Delta C = C(a) - C(b) \tag{64}$$

so

$$\frac{dC}{dr} = -\frac{\Delta Cab}{b - a} \cdot \frac{1}{r^2} \tag{65}$$

Fick's first law of diffusion then gives the following equation for the rate of change of mass m of the NAPL droplet:

$$\frac{dm}{dt} = 4\pi r^2 D \cdot \frac{dC}{dr} = -4\pi D\Delta C \frac{ab}{b - a} \tag{66}$$

Now

$$m = \frac{4\pi \varrho a^3}{3} \tag{67}$$

so

$$\frac{dm}{dt} = 4\pi \varrho a^2 \frac{da}{dt} \tag{68}$$

Substitution of Equation (68) into Equation (66) and rearranging then yields

$$(ab - a^2)\frac{da}{dt} = -\frac{D\Delta Cb}{\varrho} \tag{69}$$

which integrates to give

$$ba^2/2 - a^3/3 = ba_0^2/2 - a_0^3/3 - (D\Delta Cb/\varrho)t \tag{70}$$

where a_0 = initial NAPL droplet radius, which relates the drop radius to the time. To calculate the time required for complete solution of the drop, set $a = 0$ in Equation (70) and solve for $t = t_{100}$:

$$t_{100} = \frac{\varrho}{bD\Delta C} \cdot \left[\frac{ba_0^2}{2} - \frac{a_0^3}{3} \right] \tag{71}$$

In the limit of large boundary layer thickness b, this simplifies to give

$$t_{100} = \frac{\varrho a_0^2}{2D\Delta C} \tag{72}$$

Also in this limit, substitution of Equation (67) into Equation (70) and rearrangement yields

$$m(t) = \left\{ m_0^{2/3} - \left[\frac{4\pi\varrho}{3}\right]^{2/3} \frac{2D\Delta Ct}{\varrho} \right\}^{3/2} \tag{73}$$

where m_0 is the initial mass of the NAPL droplet.

For use in the next chapter, we would like to have a differential equation describing NAPL solution in terms of $m/m_0 = M/M_0$, where M is the mass of NAPL per m³ and M_0 is the initial value of this quantity. This is obtained as follows.

It is readily seen that the droplet radius at time t is given by

$$a(t) = a_0(M/M_0)^{1/3} \tag{74}$$

and that

$$m = m_0 M/M_0 \tag{75}$$

Substitution in Equation (66) then yields

$$\frac{dM}{dt} = -\frac{3DM_0^{2/3}\Delta CbM^{1/3}}{\varrho a_0^2[b - a_0(M/M_0)^{1/3}]} \tag{76a}$$

which in the limit of large aqueous boundary layer thickness b gives

$$\frac{dM}{dt} = -\frac{3DM_0^{2/3}\Delta CM^{1/3}}{a_0^2\varrho} \tag{76b}$$

Equations (76a) or (76b) then provide a way to estimate mass transport from NAPL droplets that are surrounded by an aqueous phase boundary layer through which the VOC must diffuse to move into the advecting vapor phase. Generally, one expects that the available data on a site would not be sufficient to permit the estimation of b in Equation (75), and we shall therefore generally rely on Equation (76b) to describe this mass transport process.

MASS TRANSPORT–LIMITED SOIL VAPOR EXTRACTION

If the gas flow rate in an SVE system is maintained at a sufficiently high level, advection is not rate limiting, and the bottleneck in the process lies entirely in the desorption/diffusion processes. One can therefore calculate upper bounds to SVE cleanup rates (and lower bounds to cleanup times) by simply examining the equations for these processes under the condition that the ambient gaseous VOC concentration outside the aqueous layer is equal to zero.

For lumped parameter models we have

$$C(t) = C_{initial}[1 - \exp(-\lambda_1 t)] \tag{77}$$

where

$$\lambda_1 = D(\pi/a)^2 \text{ (spherical aqueous droplet of radius } a) \tag{78}$$

$$\lambda_1 = D(\pi/2a)^2 \text{ (planar aqueous layer of thickness } a) \tag{79}$$

A more exact approach to diffusion from the planar aqueous layer gives

$$C(t) = C_{initial} \cdot \frac{8}{\pi^2} \sum_{n=0}^{\infty} \frac{1}{(2n + 1)^2} \cdot \exp\left\{ -\left[\frac{(2n + 1)\pi}{2a} \right]^2 Dt \right\} \tag{80}$$

If the remediation involves the solution of NAPL droplets and diffusion of VOC through the surrounding aqueous boundary layer, we have

$$\frac{dM}{dt} = - \frac{3DM_0^{2/3}\Delta C M^{1/3}}{a_0^2 \varrho} \tag{76b}$$

which integrates and rearranges to give

$$M(t) = \left[M_0^{2/3} - \frac{2DM_0^{2/3}C_{sat}}{\varrho a_0^2} \cdot t \right]^{3/2} \tag{81}$$

where we have replaced ΔC by C_{sat}.

Equations (77), (80), or (81) provide the most favorable removal rates possible for an SVE system involving VOC held in the soil in such a manner as to be controlled by diffusion processes in these ways. One would generally expect that the airflow rates that would be required to operate at such rates would be quite high. Additionally, the off-gas VOC concentrations would be quite low, so off-gas treatment would be costly.

REFERENCES

Adamson, A. W., 1976, *Physical Chemistry of Surfaces,* 3rd ed., Wiley-Interscience, New York, NY.

Albertson, M., 1979, "Carbon Dioxide Balance in the Gas-Filled Part of the Unsaturated Zone, Demonstrated at a Podzol," *Zeitschrift für Pflanzenernahrung und Bodenkunde,* 142:39.

Bouchard, D. C., 1989, "The Role of Sorption in Contaminant Transport," *Workshop on Soil Vacuum Extraction,* April 27–28, Robert S. Kerr Environmental Research Laboratory, U.S. EPA, Ada, OK.

Bouchard, D. C., S. C. Mravik, and G. B. Smith, 1990, "Benzene and Naphthalene Sorption on Soil Contaminated with High Molecular Weight Residual Hydrocarbons from Unleaded Gasoline," *Chemosphere,* 21:975.

Brown, R. A. and R. D. Norris, 1986, "An In-Depth Look at Bioreclamation," *Proc., 4th Ann. Hazardous Materials Management Conf.,* Tower Conference Management Co., Wheaton, IL, p. 138.

Brusseau, M. L., 1992, "Rate-Limited Mass Transfer and Transport of Organic Solutes in Porous Media That Contain Immobile Immiscible Organic Liquids," *Water Resources Research,* 28:33.

Brusseau, M. L., T. Larsen, and T. H. Christensen, 1991, "Rate-Limited Sorption and Nonequilibrium Transport of Organic Chemicals in Low Organic Carbon Aquifer Materials," *Water Resources Research,* 27:1137.

Calvet, R., 1980, "Adsorption-Desorption Phenomena," in *Interactions between Herbicides and Soil,* R. J. Hance, ed., Academic Press, New York, NY.

Chiou, C. T. and T. D. Shoup, 1985, "Soil Sorption of Organic Vapors and Effects of Humidity on Sorption Mechanism and Capacity," *Environmental Science and Technology,* 19:1196.

Currie, J. A., 1970, "Movement of Gases in Soil Respiration," in *Sorption and Transport Processes in Soils,* Rothamsted Experimental Station Monograph, V. 37, Rothamsted, U.K.

Davies, S., 1989, "The Influence of Soil Characteristics on the Sorption Behavior of Organic Vapors," *Workshop on Soil Vacuum Extraction,* April 27–28, Robert S. Kerr Environmental Research Laboratory, U.S. EPA, Ada, OK.

DePaoli, D. W. and N. J. Hutzler, 1992, "Field Test of Enhancement of Soil Venting by Heating," in *Proc. of the Symposium on Soil Venting,* April 29–May 1, 1991, Houston, TX, U.S. EPA Report No. EPA/600/R-92/174, p. 173.

DiGiulio, D. C., 1992, *Evaluation of Soil Venting Application,* U.S. EPA Report No. EPA/540/S-92/004, April, 1992.

DiGiulio, D. C., J. S. Cho, R. R. Dupont, and M. W. Kemblowski, 1990, "Conducting Field Tests for Evaluation of Soil Vacuum Extraction Application," *Proc., 4th Natl. Outdoor Action Conf. on Aquifer Restoration, Ground Water Monitoring and Geophysical Methods,* May 14–17, Las Vegas, NV, p. 587.

Downey, D. C. and M. G. Elliott, 1990, "Performance of Selected in situ Soil Decontamination Technologies: An Air Force Perspective," *Environmental Progress,* 9:169.

Ehlers, W., W. J. Farmer, W. F. Spencer, and J. Letey, 1973, "Lindane Diffusion in Soils. II. Water Content, Bulk Density and Temperature Effects," *Soil Sci. Soc. Amer. Proc.,* 33:504.

English, C. W. and R. C. Loehr, 1992, "Degradation of Volatile Organic Compounds in Unsaturated Soils," *Emerging Technologies for Hazardous Waste Management: 1992 Book of Abstracts for the Special Symposium,* Atlanta, GA, American Chemical Society, Washington, D.C.

Freeze, R. A. and J. A. Cherry, 1979, *Groundwater,* Prentice-Hall, Englewood Cliffs, NJ.

Gierke, J. S., N. J. Hutzler, and D. B. McKenzie, 1992, "Vapor Transport in Unsaturated Soil Columns: Implications for Vapor Extraction," *Water Resources Research,* 28:323.

Grable, A. R. and E. G. Siemer, 1968, "Effects of Bulk Density, Aggregate Size, and Soil Water Suction on Oxygen Diffusion, Redox Potentials, and Elongation of Corn Roots," *Soil Science Society of America Proceedings,* 32:180.

Hamaker, J. W. and J. M. Thompson, 1972, "Adsorption," in *Organic Chemicals in the Soil Environment,* C. A. I. Goring and J. W. Hamaker, eds., Marcel Dekker, New York, NY.

Harper, L. A., A. W. White, R. R. Bruce, A. W. Thomas, and R. A. Leonard, 1976, "Soil and Microclimate Effects on Trifuluralin Volatilization," *J. Environ. Qual.,* 5:236.

Hoag, G. E., 1991, "Soil Vapor Extraction Research Developments," in *Soil Vapor Extraction Technology, Review No. 204,* T. A. Pederson and J. T. Curtis, eds., Noyes Publications, Park Ridge, NJ, p. 286.

Hoag, G. E., C. J. Bruell, and M. C. Marley, 1984, "Study of the Mechanisms Controlling Gasoline Hydrocarbon Partitioning and Transport in Groundwater Systems," Inst. of Water Resources, Univ. of Connecticut, Storrs, CT, NTIS PB85-242907/AS.

Hoag, G. E., C. J. Bruell, and M. C. Marley, 1987, "Induced Soil Venting for Recovery/Restoration of Gasoline Hydrocarbons in the Vadose Zone," in *Oil in Freshwater: Chemistry, Biology, Countermeasure Technology,* J. H. Vandermeulen and S. E. Hrudey, eds., Pergamon Press, New York, NY, p. 176.

Hughes, T. H., K. E. Brooks, B. W. Norris, B. M. Wilson, and B. N. Roche, 1985, "A Descriptive Survey of Selected Organic Solvents," Open File Report No. 1, Environmental Institute for Waste Management Studies, University of Alabama, Tuscaloosa, AL.

Johnson, R. L., C. D. Palmer, and J. F. Keely, 1987, "Mass Transfer of Organics between Soil, Water, and Vapor Phases," *Proc., NWWA/API Conf. on Petroleum Hydrocarbons and Organic Chemicals in Ground Water—Prevention, Detection, and Restoration,* Nat'l. Well Water Association, Dublin, OH, p. 493.

Jury, W. A., 1986, "Adsorption of Organic Chemicals onto Soil," in *Vadose Zone Modeling of Organic Pollutants,* S. C. Hern and S. M. Melancon, eds., Lewis Publishers, Chelsea, MI, p. 177.

Kenaga, E. E. and C. A. I. Goring, 1980, "Relationship between Water Solubility, Soil Sorption, Octanol-Water Partitioning, and Bioconcentration of Chemicals in Biota," *Proc., 3rd ASTM Symposium on Aquatic Toxicology,* ASTM Special Technical Publication, 707:78.

Lai, S. H., T. M. Tiedje, and A. E. Erickson, 1976, "In situ Measurement of Gas Diffusion Coefficients in Soils," *Soil Science Society of America Journal,* 40:3.

Levine, I. N., 1988, *Physical Chemistry,* 3rd ed., McGraw-Hill, Inc., New York, NY.

Lingineni, S. and V. K. Dhir, 1992, "Modeling of Soil Venting Processes to Remediate Unsaturated Soils," *J. Environ. Engineering,* 118:135.

Lord, A. E., Jr., R. M. Koerner, V. P. Murphy, and J. E. Brugger, 1988, "Laboratory Studies of Vacuum-Assisted Steam Stripping of Organic Contaminants from Soil," *Land Disposal, Remedial Action, Incineration, and Treatment of Hazardous Waste, Proceedings, 14th Annual Research Symposium,* May 9–11, Cincinnati, OH, EPA Report No. EPA/600/9-88/021.

Lyman, W. J., 1982, "Adsorption Coefficients in Soil and Sediments," in *Handbook of Chemical Property Estimation Methods,* W. J. Lyman, ed., McGraw-Hill, New York, NY.

Millington, R. J. and J. P. Quirk, 1961, "Permeability of Porous Solids," *Trans. Faraday Soc.,* 57:1200.

Mingelgrin, U. and Z. Gerstl, 1983, "Reevaluation of Partitioning as a Mechanism of Nonionic Chemical Adsorption in Soil," *J. Environ. Qual.,* 12:1–11.

Montgomery, J. H. and L. M. Welkom, 1989, *Groundwater Chemicals Desk Reference, Vols. 1 and 2,* Lewis Publishers, Chelsea, MI.

Oja, K. J. and D. K. Kreamer, 1991, "The Effect of Moisture on Adsorption of Trichloroethylene Vapor on Natural Soils," in *Proc. of the Symposium on Soil Venting,* April 29–May 1, Houston, TX, U.S. EPA Report No. EPA/600/R-92/174, p. 13.

Ong, S. K., S. R. Lindner, and L. W. Lion, 1991, "Applicability of Linear Partitioning Relationships for Sorption of Organic Vapors onto Soil and Soil Minerals," in *Organic Substances and Sediments in Water, Vol. 1: Humics and Soils,* CRC Press, Boca Raton, FL, p. 275.

Osejo, R. E. and D. J. Wilson, 1991, "Soil Cleanup by in situ Aeration. IX. Diffusion Constants of Volatile Organics and Removal of Underlying Liquid," *Separ. Sci. Technol.,* 26:1433.

Palmer, C. D. and R. L. Johnson., 1991, "Physical and Chemical Processes in the Subsurface," in *Seminar Publication: Site Characterization for Subsurface Remediation,* U.S. EPA Report No. EPA/625/4-91/026, p. 155.

Phelan, J. M., 1992, "Thermal Enhanced Vapor Extraction System for VOC's in Soil," *Emerging Technologies for Hazardous Waste Management: 1992 Book of Abstracts for the Special Symposium,* Atlanta, GA, Industrial and Engineering Chemistry Div., American Chemical Society, Washington, D.C., p. 544.

Popiel, W. J., 1978, *Introduction to Colloid Science,* Exposition Press, Hicksville, NY.

Rainwater, K., M. R. Zaman, B. J. Claborn, and H. W. Parker, 1990, "Experimental and Model Study of Soil Venting," *Proc., 1990 Specialty Conf.,* July 8–11, Arlington, VA, ASCE, Environmental Engineering Div., p. 479.

Rao, P. S. C. and J. M. Davidson, 1980, "Estimation of Pesticide Retention and Transformation Parameters," in *Environmental Impact of Nonpoint Source Pollution,* M. R. Overcash and J. M. Davidson, eds., Ann Arbor Science Publications, Ann Arbor, MI.

Rathfelder, K., W. W. G. Yeh, and D. Mackay, 1991, "Mathematical Simulation of Soil Vapor Extraction Systems: Model Development and Numerical Examples," *J. Contaminant Hydrology,* 8:263.

Roy, W. R. and R. A. Griffin, 1987, "Vapor-Phase Movement of Organic Solvents in the Unsaturated Zone," Open File Report No. 16, Environmental Institute for Waste Management Studies, University of Alabama, Tuscaloosa, AL.

Roy, W. R. and R. A. Griffin, 1989, "In-situ Extraction of Organic Vapors from Unsaturated Porous Media," Open File Report No. 24, Environmental Institute for Waste Management Studies, University of Alabama, Tuscaloosa, AL.

Schwille, F., 1988, *Dense Chlorinated Solvents in Porous and Fractured Media*, J. F. Pankow, trans., Lewis Publishers, Chelsea, MI.

Shaw, D. J., 1966, *Introduction to Colloid and Surface Chemistry*, Butterworths, London, U.K.

Spencer, W. F. and M. M. Cliath, 1973, "Desorption of Lindane from Soil as Related to Vapor Density," *Soil Sci. Soc. Amer. Proc.*, 34:574.

Sterrett, R., 1989, "Analysis of in situ Soil Air Stripping Data," *Workshop on Soil Vacuum Extraction*, April 27–28, Robert S. Kerr Environmental Research Laboratory, U.S. EPA, Ada, OK.

Thibaud, C., C. Erkey, and A. Akgerman, 1992, "Investigation of Adsorption Equilibria of Volatile Organics on Soil as a Function of Relative Humidity," *Emerging Technologies for Hazardous Waste Management: 1992 Book of Abstracts for the Special Symposium*, Atlanta, GA, Industrial and Engineering Chemistry Div., American Chemical Society, Washington, D.C., p. 197.

Thibodeaux, L. J., 1979, *Chemodynamics: Environmental Movement of Chemicals in Air, Water, and Soil*, Wiley, New York, NY.

van Olphen, H., 1963, *An Introduction to Clay Colloid Chemistry*, 2nd ed., Wiley-Interscience, New York, NY.

Vold, R. D. and M. J. Vold, 1983, *Colloid and Interface Chemistry*, Addison-Wesley, Reading, MA.

Wade, P., 1955, "Soil Fumigation. III. The Sorption of Ethylene Dibromide by Soils at Low Moisture Contents," *J. Science of Food and Agriculture*, 6:1.

Weeks, E. P., D. E. Earp, and G. M. Thompson, 1982, "Use of Atmospheric Fluorocarbons F-11 and F-12 to Determine the Diffusion Parameters of the Unsaturated Zone in the Southern High Plains of Texas," *Water Resources Research*, 18:1365.

Wesseling, J., 1962, "Some Solutions of the Steady State Diffusion of Carbon Dioxide through Soils," *Netherlands Journal of Agricultural Science*, 10:109.

Wilson, D. J., 1990, "Soil Cleanup by in situ Aeration. V. Vapor Stripping of Fractured Bedrock," *Separ. Sci. Technol.*, 25:243.

CHAPTER 5

Soil Vapor Extraction Modeling — Analysis

INTRODUCTION—WHY DOES ONE MODEL?

THE nature of the soil vapor extraction (SVE) technique is such that assessment of its feasibility and the design of an SVE system in any particular application are rather site-specific. These depend on the site geology (depth to water table, pneumatic permeability of vadose zone soils, presence of overlying impermeable structures such as floors or parking lots, heterogeneity of soil, presence of natural or other nonvolatile organics) and on contaminant properties (vapor pressure, water solubility, partition coefficient on organic carbon, and Henry's constant, all at ambient soil temperature). This has led to considerable interest in the mathematical modeling of SVE for feasibility studies, data interpretation, and system design. Johnson et al. (1989a, 1989b, 1990) have published extensively on this. Hoag, Marley, Cliff, and their associates at Vapex (Hoag and Cliff, 1985; Marley, 1991; Marley et al., 1989) were among the first to use mathematical modeling techniques in SVE. Cho (1991) has carried out a quite detailed study in which modeling work was supported by extensive experimental verification. Our group has published a number of papers on the mathematical modeling of SVE under a variety of conditions (Wilson et al., 1988; Kayano and Wilson, 1992; Roberts and Wilson, 1993a, 1993b; and other papers in this series).

In the preceding chapters the two principal factors governing SVE were examined. These are (1) the flow pattern of soil gas induced in the vicinity of an SVE well and (2) the equilibrium and kinetic factors governing the partitioning of a volatile organic compound (VOC) between stationary phases and the advecting vapor phase. In this chapter we merge these two in the construction of mathematical models for the operation of SVE in various configurations and at various levels of approximation. This chapter is concerned with the mathematical details of SVE modeling, and it is intended to be of help to those who are confronted with the development,

173

coding, and/or modification of SVE models. Chapter 6 presents a variety of results obtained with the various models and can be read independently of Chapter 5.

Mathematical simulations of SVE have a number of uses, which we shall discuss briefly. These are as follows:

- *design of laboratory column experiments and interpretation of the results*—There are enough data presently available in the literature to permit one to make rough first estimates of the parameters needed to describe SVE of the contaminated soil from the site of interest in a lab column. These data can be used in lab column models to design lab column experiments—appropriate column dimensions, gas flow rates, durations of experiments, etc. The same lab column models can then be used after the experiments are complete to extract equilibrium (adsorption isotherm) and kinetic parameters from the data for use in subsequent modeling. Lab column permeability measurements, on the other hand, must be interpreted with caution unless the experiments are carried out on undisturbed, intact cores. Generally, collecting soil samples and packing them into lab-scale SVE columns so disturbs the soil structure as to make permeability measurements quite unreliable.
- *design of pilot-scale field studies*—Isotherm data from lab column experiments or from similar field situations and geological data from well logs and other test borings can be used in models of single SVE wells to simulate operation of a pilot-scale SVE test facility. How large will be the effective domain of influence of the SVE well? Should passive vent wells be used to isolate a well-defined domain for the pilot-scale SVE test? Given the distribution of contaminant VOC and the stratigraphy of the soil, how deep should the well be drilled, and how long a section of well should be screened? What should be the diameter of the gravel packing around the screened section of the well? What gas flow rates should be used? If it is feared that diffusion/desorption kinetics may be a problem, over what time period after well shut-down should one take soil gas samples to estimate the time constant of the diffusion/desorption processes?
- *interpretation of pilot-scale field studies*—The results of the pilot-scale field studies can be used in the models to obtain improved values of the various SVE parameters—impact of soil stratigraphy (i.e., presence of low- or high-permeability layers, clay lenses, other heterogeneities, anisotropy) on soil gas flow field, diffusion/desorption processes, adsorption isotherm parameters, etc. The pilot-scale field study will be particularly useful in assessing any problems with medium permeability. Pilot-scale

SVE runs should be made in one or more domains representative of each type of situation as characterized by the nature of the medium and the distribution and type(s) of contaminants, and parameters then calculated for each situation.

- *design of full-field scale SVE facility* – Once adequate sets of parameters have been obtained, one can begin designing the full field-scale facility. Modeling permits one to explore a large number of design options quickly and cheaply, so that a design may be chosen based on the information available on the site, which will result in reasonably rapid, complete, and economical remediation. Questions to be answered include how far apart the wells should be spaced, how deeply they should be drilled and over what length they should be screened, what airflow rates are both efficient and economical, whether passive wells and/or impermeable caps should be used, etc.

- *modeling to follow the progress of the remediation* – Sometimes there are unpleasant surprises in SVE work. There may be a hot spot with very high VOC concentrations out near the periphery of a well that was not identified in the remedial investigation. A well may have been screened in a low-permeability clay layer so that its airflow rate is extremely low. The location of low-permeability lenses with respect to the well(s) may be such that gas flow through some zones of contamination is very slow. If the soil gas effluent VOC concentrations are monitored from time to time and compared with the results of the modeling, discrepancies can permit one to deduce relatively early in the remediation the existence of one or more of these unpleasant surprises and to take steps to deal with the problem.

- *estimation of projected cleanup time and of costs* – Mathematical modeling of SVE operations permits one to make the most realistic estimations of cleanup time possible and, along with this, estimates of capital and operating costs. Modeling, for example, permits one to examine the trade-offs involved in drilling shallow wells or wells that are spaced rather far apart, which reduces capital costs but increases operating costs and cleanup times. Sensitivity analyses can be used with these models to get a rough idea of the uncertainty in the estimates of cleanup times and costs.

SVE BY MEANS OF A SINGLE VERTICAL WELL, LINEAR ADSORPTION ISOTHERM, LOCAL EQUILIBRIUM AND CONSTANT ISOTROPIC PERMEABILITY

The simplest model one can construct for a single SVE well is that in which one assumes that the adsorption isotherm of the VOC on the porous

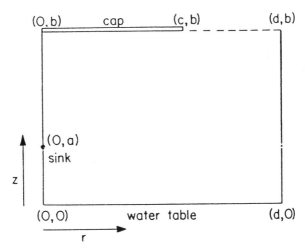

Figure 5.1 The geometry of the SVE model for a vertical well screened at its bottom. The vent pipe is drawing air from the soil at the point $(0,a)$ at a height a meters above the water table. The impermeable circular cap at the soil surface is located b meters above the water table and is of radius c. The radius of the zone of influence is taken as d meters. [Reprinted from Rodríguez et al. (1994) p. 1367, by courtesy of Marcel Dekker, Inc.]

medium is linear (an effective Henry's Law), that there is local equilibrium between the stationary condensed VOC and the VOC in the advecting vapor phase, and that the permeability of the porous medium is constant in space and isotropic [see Wilson et al. (1988), for example]. We use cylindrical coordinates, and represent the vacuum well by a point sink at $(0,0,a)$. The top of the vadose zone is at the plane $z = b$, and the bottom of the vadose zone (or an air-impermeable layer, in any event) is at $z = 0$. Since the problem is axially symmetric, we henceforth drop the coordinate θ. See Figure 5.1 for the geometry and notation used.

Let

$C(t,r,z)$ = mass of VOC per unit volume of soil, kg/m³
   σ = soil air-filled porosity, dimensionless
   K_A = VOC adsorption isotherm on the moist soil, dimensionless; $K_A = C^v/C^a$
   C^v = VOC vapor concentration, kg/m³ of vapor phase
   C^a = sorbed VOC concentration, kg/m³ of soil
   v = soil gas superficial velocity, m/sec

Then, a mass balance on an arbitrary volume element leads in the usual way to

$$\sigma \frac{\partial C}{\partial t} = -\nabla \cdot vC^v \qquad (1)$$

For this case we are assuming that VOC is present only in two forms – in the vapor phase and adsorbed in some way, so

$$C = \sigma C^v + C^a \tag{2}$$

The assumption of a linear adsorption isotherm and local equilibrium gives

$$C^v = K_A C^a \tag{3}$$

By using Equations (2) and (3) to eliminate C^v from the right-hand side of Equation (1), we obtain

$$\frac{\partial C}{\partial t} = -\frac{K_A}{\sigma K_A + 1} \nabla \cdot vC \tag{4}$$

as the partial differential equation governing the change of the total concentration $C(r,z)$ with time.

Note that no dispersion term has been introduced. We shall approximate the physical dispersion in the system by the numerical dispersion in the system of ordinary differential equations to be used to represent Equation (4) in the cylindrical coordinates r and z.

Let

$$r_i = (i + 1/2)\Delta r, \quad i = 1,2,3,\ldots,n_r \tag{5}$$

$$z_j = (j + 1/2)\Delta z, \quad j = 1,2,3,\ldots,n_z \tag{6}$$

specify the mesh points to be used in representing the system for numerical analysis. The volume of the ijth annular-shaped volume element of the system is given by

$$V_{ij} = (2i + 1)\pi(\Delta r)^2 \Delta z \tag{7}$$

The area of the inner face of this volume element is

$$A^I_{ij} = 2\pi i \Delta r \Delta z \tag{8}$$

The area of the outer face is

$$A^O_{ij} = 2\pi(i + 1)\Delta r \Delta z \tag{9}$$

The areas of the top and bottom faces are

$$A^T_{ij} = A_{Bij} = (2i + 1)\pi(\Delta r)^2 \tag{10}$$

The velocity $\mathbf{v}$ is given in terms of its components as

$$\mathbf{v} = v_r e_r + v_z e_z \tag{11}$$

where $v_r(r,z)$ and $v_z(r,z)$ are calculated from Equations (49) and (50) of Chapter 3 if we are dealing with a single well in an unbounded medium with a constant isotropic permeability. For more complex systems the velocity components may be calculated by the numerical relaxation procedures described in Chapter 3.

Then the velocity components normal to the midpoint of each of the four faces of the volume element are given by

$$v_{ij}^I = v_r[i\Delta r, (j + 1/2)\Delta z] \tag{12}$$

$$v_{ij}^O = v_r[(i + 1)\Delta r, (j + 1/2)\Delta z] \tag{13}$$

$$v_{ij}^B = v_z[(i + 1/2)\Delta r, j\Delta z] \tag{14}$$

$$v_{ij}^T = v_z[(i + 1/2)\Delta r, (j + 1)\Delta z] \tag{15}$$

One then approximates Equation (6) in terms of a set of ordinary differential equations by carrying out a mass balance on the ijth volume element, which yields

$$\frac{\partial C_{ij}}{\partial t} = \frac{K_A}{(\sigma K_A + 1)\Delta V_{ij}} \{A_{ij}^I[v_{ij}^I S(v^I)C_{i-1,j} + v_{ij}^I S(-v^I)C_{i,j}]$$

$$+ A_{ij}^O[-v_{ij}^O S(-v^O)C_{i+1,j} - v_{ij}^O S(v^O)C_{i,j}]$$

$$+ A_{ij}^B[v_{ij}^B S(v^B)C_{i,j-1} + v_{ij}^B S(-v^B)C_{i,j}]$$

$$+ A_{ij}^T[-v_{ij}^T S(-v^T)C_{i,j+1} - v_{ij}^T S(v^T)C_{i,j}]\} \tag{16}$$

where

$$S(v) = 0, v > 0$$

$$= 1, v > 0 \tag{17}$$

and

$$S(v^I) = S(v_{ij}^I), \text{ etc.}$$

The rather elaborate expression on the right-hand side of Equation (16) is necessary to correctly calculate the advective transport for both positive and negative values of v_r and v_z; simpler approximations do not conserve mass.

The boundary conditions normally used are that the VOC concentrations around the periphery of the domain of interest are zero, so all terms involving nonexistent volume elements are simply dropped. For volume elements on the axis of the well, the radial velocities in the centers are equal to zero, so these nonphysical terms are dropped automatically. VOC entering the volume element containing the sink representing the well is assumed to be annihilated; i.e., the VOC concentration in that volume element is kept at zero at all times.

In simulating a run, one inputs the parameters needed, initializes the total VOC concentrations in the various volume elements, evaluates the ΔV_{ij}'s, the areas of the faces, and the velocity components, all needed to calculate the coefficients in Equation (16). One then integrates the set of Equation (16) forward in time. We have generally used a predictor-corrector method for doing this; the algorithm is as follows.

Starter predictor

$$y^*(\Delta t) = y(0) + \frac{dy(0)}{dt} \Delta t \tag{18}$$

Starter corrector

$$y(\Delta t) = y(0) + (1/2)\Delta t \cdot \left[\frac{dy(0)}{dt} + \frac{dy^*(\Delta t)}{dt}\right] \tag{19}$$

General predictor

$$y^*[(n + 1)\Delta t] = y[(n - 1)\Delta t] + 2\Delta t \cdot \frac{dy(n\Delta t)}{dt} \tag{20}$$

General corrector

$$y[(n + 1)\Delta t] = y(n\Delta t) + (1/2)\Delta t \cdot \left[\frac{dy(n\Delta t)}{dt} + \frac{dy^*[(n + 1)\Delta t]}{dt}\right] \tag{21}$$

where Equation (16) is used to calculate the derivatives.

The progress of the remediation is conveniently followed by calculating the total residual mass of VOC left in the domain of interest. This is given by

$$M_{total} = \sum_{i=0}^{n_r} \sum_{j=0}^{n_z} \Delta V_{ij} C_{ij} \qquad (22)$$

The off-gas VOC concentration can be determined by printing out the value of C^v_{1q}, where q is chosen so that the volume element is immediately adjacent to the volume element containing the sink representing the vacuum well. This is given by

$$C^v_{1q} = K_A C_{1q} \qquad (23)$$

We implemented this model in TurboBASIC and have run it on a number of MS-DOS microcomputers. The resulting code can be run on computers using 8088 or 8086 microprocessors; on such a machine a typical run may take an hour or two. On machines equipped with an 80286 or 80386 chip and a math coprocessor, a run requires only a few minutes. A word of warning: if the value of Δt selected for the numerical integration is too large, the integration algorithm becomes extremely unstable within a few cycles. If this occurs, simply reduce the value of Δt until the solution is well-behaved.

SVE WITH A SINGLE BURIED HORIZONTAL SLOTTED PIPE, LINEAR ADSORPTION ISOTHERM, LOCAL EQUILIBRIUM, AND CONSTANT ISOTROPIC PERMEABILITY

If the domain to be vapor stripped is shallow, it may be advantageous to use buried horizontal slotted pipes, as indicated in Figure 5.2, which shows the notation to be used. We develop an SVE model for this configuration here under the same assumptions as used in the previous section. From symmetry we need model only the right-hand side of the rectangular domain being vapor stripped, and we neglect end effects so as to be able to work the problem in two dimensions.
   Let

l = length of the slotted pipe, m
$x_i = (i + 1/2)\Delta x$ (horizontal coordinate), m
$y_j = (j + 1/2)\Delta y$ (vertical coordinate), m
$\Delta V = l\Delta x \Delta y$, m³

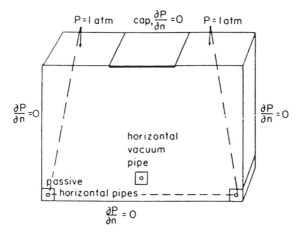

Figure 5.2 The geometry of the SVE model for a buried horizontal slotted pipe. An impermeable strip cap may be present at the surface of the soil, and passive horizontal slotted pipes may be present in the lower edges of the domain.

The areas of the volume element are given by

$$A^L = A^R = l\Delta y$$

$$A^B = A^T = l\Delta x$$

The velocity is given by

$$\mathbf{v} = v_x \mathbf{i} + v_y \mathbf{j} \tag{24}$$

where $v_x(x, y)$ and $v_y(x, y)$ are calculated from Equations (59)–(64) of Chapter 3 for a medium having a constant isotropic permeability and here $\mathbf{i}$ and $\mathbf{j}$ are unit vectors. For variable or anisotropic permeabilities one may use the relaxation method described in Chapter 3 to calculate the velocity components. The normal velocities at the centers of the faces of the ijth volume element are

$$v_{ij}^L = v_x[i\Delta x, (j + 1/2)\Delta y] \tag{25}$$

$$v_{ij}^R = v_x[(i + 1)\Delta x, (j + 1/2)\Delta y] \tag{26}$$

$$v_{ij}^B = v_y[(i + 1/2)\Delta x, j\Delta y] \tag{27}$$

$$v_{ij}^T = v_y[(i + 1/2)\Delta x, (j + 1)\Delta y] \tag{28}$$

Then a mass balance on the ijth volume element yields

$$
\begin{aligned}
\frac{dC_{ij}}{dt} = \frac{K_A}{(\sigma K_A + 1)\Delta V} \cdot \{ & A^L[v_{ij}^L S(v^L)C_{i-1,j} + v_{ij}^L S(-v^L)C_{i,j}] \\
& + A^R[-v_{ij}^R S(-v^R)C_{i+1,j} - v_{ij}^R S(v^R)C_{i,j}] \\
& + A^B[v_{ij}^B S(v^B)C_{i,j-1} + v_{ij}^B S(-v^B)C_{i,j}] \\
& + A^T[-v_{ij}^T S(-v^T)C_{i,j+1} - v_{ij}^T S(v^T)C_{i,j}]
\end{aligned}
\tag{29}
$$

The progress of the remediation is followed by calculating the total mass of residual VOC, which is given by

$$
M_{total} = 2\Delta V \sum_{i=0}^{n_x} \sum_{j=0}^{n_y} C_{ij}
\tag{30}
$$

The factor of 2 is needed to take account of the fact that only the right half of the domain is being simulated. The off-gas VOC concentration can be calculated by the procedure leading to Equation (23). Gas entering the volume element that contains the right half of the horizontal slotted pipe is assumed to be removed; the boundary condition used is that the total VOC concentration in this volume element is kept equal to zero throughout the run.

ONE-DIMENSIONAL MODELS WITH CONSTANT ISOTROPIC PERMEABILITY, LOCAL EQUILIBRIUM, AND A LINEAR ADSORPTION ISOTHERM

Two one-dimensional models may be of practical interest: a simple SVE lab column model and a model having radial symmetry, which may be useful in connection with SVE operations underneath paving, a concrete building floor, or some other impermeable barrier that overlies the domain to be treated. The column model, similar to that discussed in Chapter 2, leads to the following modeling equations:

$$
\frac{dC_i}{dt} = \frac{K_A A}{(\sigma K_A + 1)\Delta V} \cdot [v_i^I C_{i-1} - v_i^o C_i], \ i = 1,2,\ldots,n_x
\tag{31}
$$

where C_0, the concentration of VOC in the gas entering the column, is taken as zero. Here A is the cross-sectional area of the column, and

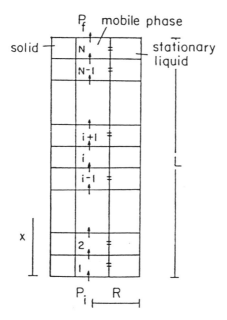

Figure 5.3 Notation and column partitioning for modeling a laboratory aeration column. P_i = inlet pressure, P_f = outlet pressure.

$\Delta V = A\Delta x$, where Δx is the length of a volume element (see Figure 5.3). The gas velocities are given by

$$v_i^I = v[(i - 1)\Delta x] \tag{32}$$

and

$$v_i^0 = v[i\Delta x] \tag{33}$$

where $v(x)$ is given by Equation (17) in Chapter 2.

The second model, for vapor stripping underneath an impermeable cap of radius r_m with a vertical well screened along its entire length h, is diagrammed in Figure 5.4. The system is radially symmetrical, so the steady-state gas pressure distribution is determined by

$$\frac{1}{r}\frac{\partial}{\partial r}\left[r\frac{\partial P^2}{\partial r}\right] = 0 \tag{34}$$

with boundary conditions

$$P(r_w) = P(\text{well}) = P_w \tag{35}$$

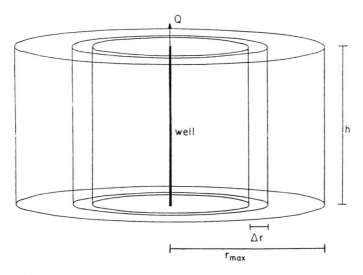

Figure 5.4 Notation and column partitioning for modeling a one-dimensional radial SVE system having an overlying cap of radius $r_{max} = r_m \gg$ the thickness h of the vadose zone. Here, the well is screened along its entire length.

where r_w is the radius of the well gravel packing, and

$$P(r_m) = P(\text{ambient}) = P_a \qquad (36)$$

where r_m is roughly the radius of the overlying impermeable cap.

The solution to Equation (34), which satisfies the two boundary conditions, is readily shown to be

$$P^2(r) = \frac{(P_a^2 - P_w^2) \log_e r + P_w^2 \log_e r_m - P_a^2 \log_e r_w}{\log_e (r_m/r_w)} \qquad (37)$$

The superficial gas velocity has only a radial component, which is given by

$$v = -\frac{K_D}{2P} \frac{dP^2}{dr} \qquad (38)$$

where

$$P(r) = \left[\frac{(P_a^2 - P_w^2) \log_e r + P_w^2 \log_e r_m - P_a^2 \log_e r_w}{\log_e (r_m/r_w)} \right]^{1/2} \qquad (39)$$

and

$$\frac{dP^2}{dr} = \frac{P_a^2 - P_w^2}{\log_e (r_m/r_w)} \cdot \frac{1}{r} \tag{40}$$

The total molar gas flow rate is obtained by integrating the molar gas flux over the surface of a cylinder containing the well:

$$Q = \int_0^h \int_0^{2\pi} \frac{K_D}{2RT} \cdot \frac{P_a^2 - P_w^2}{\log_e (r_m/r_w)} \cdot \frac{1}{r} \cdot r d\theta dz \tag{41}$$

which gives

$$Q = \frac{\pi K_D h}{RT} \cdot \frac{P_a^2 - P_w^2}{\log_e (r_m/r_w)} \tag{42}$$

Let the midpoint of each volume element be given by

$$r_i = (i + 1/2)\Delta r \tag{43}$$

The inner and outer surfaces of an annular volume element are then

$$A_i^I = 2\pi i \Delta r h \tag{44}$$

$$A_i^O = 2\pi (i + 1)\Delta r h \tag{45}$$

The volume of the element is

$$V_i = (2i + 1)\pi \Delta r h \tag{46}$$

The modeling equations are then

$$\frac{dC_i}{dt} = \frac{K_A}{(\sigma K_A + 1)\Delta V_i} \cdot [-v_i^O A_i^O C_{i+1} + v_i^I A_i^I C_i] \quad i = 0,1,\ldots,n_r \tag{47}$$

The boundary condition is $C_{nr+1} = 0$. VOC entering into the 0th volume element is assumed to be removed, so C_0 is kept at a value of zero.

MORE COMPLEX LOCAL EQUILIBRIUM SVE MODELS

In the preceding models it was assumed that the stationary VOC was present in a single form and that the adsorption isotherm was linear. Here, we relax this assumption and develop a procedure that allows one, within the framework of the local equilibrium approximation, to handle systems in which the stationary VOC is present as NAPL, in aqueous solution, adsorbed (arbitrary adsorption isotherm), and dissolved in a greasy or tarry phase. Recall from Chapter 4 that a mass balance on VOC for this case for any volume element gives

$$C = \omega C^w + \sigma C^v + g C^g + C^a + C^N \qquad (48)$$

where we have omitted subscripts specifying the volume element, and

C = total VOC concentration, kg/m³ of soil
C^w = VOC concentration in the aqueous phase, kg/m³ of aqueous phase
C^v = VOC concentration in the vapor phase, kg/m³ of vapor phase

$$= K_H C^w \qquad (49)$$

C^g = VOC concentration in the greasy phase, kg/m³ of greasy phase

$$= K_g C^w \qquad (50)$$

C^a = VOC concentration in the adsorbed phase, kg/m³ of soil

$$= f(C^w) \qquad (51)$$

C^N = VOC concentration as NAPL, kg/m³ of soil
σ = air-filled porosity, dimensionless
ω = water-filled porosity, dimensionless
g = volumetric grease/tar content, dimensionless

The equilibrium relationships allow us to write the right side of Equation (48) in terms of a single variable; we choose C^w, obtaining

$$C = (\omega + \sigma K_H + g K_g) C^w + f(C^w) \qquad (52)$$

from which we calculate (generally by numerical methods) a value for C^w. If this is less than the aqueous solubility of the VOC, C_{sat}, one then sets $C^N = 0$ and calculates C^v, C^g, and C^a from Equations (49)–(51). If the value of C^w turns out to exceed C_{sat}, one sets $C^w = C_{sat}$, calculates C^v, C^g, and C^a from Equations (49)–(51), and then solves Equation (48) for C^N, the concentration of nonaqueous phase liquid that is present. In either case, one obtains a value of $C^v = C^v_{ij}$, the VOC concentration in the advecting gaseous phase in that volume element. All other forms of VOC are assumed to be stationary.

A mass balance on total VOC in the ijth volume element includes only advection in the vapor phase on the right-hand side, giving

$$\frac{\partial C}{\partial t} = -\nabla \cdot \mathbf{v}C^v \tag{53}$$

as the partial differential equation to be satisfied. For numerical solution we approximate this as a set of ordinary differential equations as before. We use largely the same notation as in Equation (16) for the case of a single vertical well screened for a short length at the bottom; this gives

$$\frac{dC_{ij}}{dt} = \frac{1}{\Delta V_{ij}} \{A^I_{ij}[v^I_{ij}S(v^I)C^v_{i-1,j} + v^I_{ij}S(-v^I)C^v_{i,j}]$$

$$+ A^O_{ij}[-v^O_{ij}S(-v^O)C^v_{i+1,j} - v^O_{ij}S(v^O)C^v_{i,j}]$$

$$+ A^B_{ij}[v^B_{ij}S(v^B)C^v_{i,j-1} + v^B_{ij}S(-v^B)C^v_{i,j}]$$

$$+ A^T_{ij}[-v^T_{ij}S(-v^T)C^v_{i,j+1} - v^T_{ij}S(v^T)C^v_{i,j}]\} \tag{54}$$

A modeling run is initialized (system parameters, initial distribution of VOC, etc.), and the set of coefficients needed in Equation (54) are computed. The C^v_{ij} are calculated as described above, and Equation (54) is then used to calculate values of the C_{ij} at time Δt. This process is then repeated to carry the simulation forward in time until it is completed.

The progress of the cleanup is most readily followed by calculating the total mass of residual VOC as a function of time, and the off-gas VOC concentration is given by the value of C^v_{1q} in the volume element adjacent to that containing the well. Note that this model reduces to our previous case (linear adsorption isotherm) if a linear adsorption isotherm is assumed and there is no NAPL present.

The buried horizontal slotted pipe SVE configuration is handled in the same way as was the vertical well; we use essentially the same notation as in Equation (29). The result is

$$\frac{dC_{ij}}{dt} = \frac{1}{\Delta v} \{A_L[v^L_{ij}S(v^L)C^v_{i-1,j} + v^L_{ij}S(-v^L)C^v_{i,j}]$$

$$+ A^R[-v^R_{ij}S(-v^R)C^v_{i+1,j} - v^R_{ij}S(v^R)C^v_{i,j}]$$

$$+ A^B[v^B_{ij}S(v^B)C^v_{i,j-1} + v^B_{ij}S(-v^B)C^v_{i,j}]$$

$$+ A^T[-v^T_{ij}S(-v^T)C^v_{i,j+1} - v^T_{ij}S(v^T)C^v_{i,j}]\} \tag{55}$$

DIFFUSION/DESORPTION KINETIC LIMITATIONS

The models presented above all include the local equilibrium approximation—the assumption that diffusion and desorption processes are sufficiently faster than advective transport that the vapor phase at any point in the domain of interest is essentially in equilibrium with all other phases with respect to VOC transport. As noted in Chapter 1, there are considerable experimental data that support this position. However, Chapter 1 also cites substantial evidence that the local equilibrium assumption is not a good approximation. Presumably, sites of both types, in fact, occur.

We therefore explore two approaches to the development of SVE models, which take into account the rates of diffusion/desorption processes. The first is a lumped parameter approximation, which is a fairly generic aproach; the second examines the evaporation of VOC from droplets of NAPL, which have a surrounding layer of water through which the VOC must diffuse before it reaches the advecting gas.

A LUMPED PARAMETER APPROACH

In the lumped parameter approach to diffusion/desorption kinetics, it is assumed that the rate of VOC transport between the condensed phase(s) and the vapor phase at a point in the domain of interest is proportional to the difference between the actual VOC vapor concentration C^v at that point and the VOC concentration that would be in equilibrium with the immobile VOC present as NAPL, adsorbed VOC, dissolved VOC, etc. [see Wilson (1990)]. Here, we examine a somewhat more general case in which the equilibrium partitioning of VOC between the vapor phase and the stationary phase(s) can be described by Equation (56).

$$C^v(\text{equil}) = F[C^\alpha(\text{equil})] \tag{56}$$

Here C^α represents all immobile VOC at the point of interest.

The lumped parameter approximation gives

$$\left[\frac{C^\alpha}{dt}\right]_{mass\ transport} = -\lambda[C^v - C^v(\text{equil})] \tag{57}$$

$$= -\lambda[C^v - F(C^\alpha)] \tag{58}$$

for the rate of change of C^α with time due to mass transport only. Now

$$C = \sigma C^v + C^\alpha \tag{59}$$

Since C is only affected by advection, not by mass transport,

$$\left[\frac{\partial C}{\partial t}\right]_{mass\ transport} = 0 = \sigma\left[\frac{\partial C^v}{\partial t}\right]_{mass\ transport} + \left[\frac{\partial C^\alpha}{\partial t}\right]_{mass\ transport} \tag{60}$$

Equations (58) and (60) yield an equation for dC^a/dt:

$$\left[\frac{\partial C^a}{\partial t}\right]_{mass\ transport} = \frac{dC^a}{dt} = -\sigma\left[\frac{\partial C^v}{\partial t}\right]_{mass\ transport} \tag{61}$$

which yields finally

$$\frac{dC^a_{ij}}{dt} = \sigma\lambda[C^v_{ij} - F(C^a_{ij})] \tag{62}$$

where the subscripts specifying the volume element have been included.

The vapor phase VOC concentration C^v is governed by both advection of vapor phase VOC and mass transport of VOC from the condensed stationary phases. For a single vertical SVE well, combination of these gives

$$\frac{dC^v_{ij}}{dt} = \frac{1}{\Delta v_{ij}}\ \{A^I_{ij}[v^I_{ij}S(v^I)C^v_{i-1,j} + v^I_{ij}S(-v^I)C^v_{i,j}]$$

$$+ A^O_{ij}[-v^O_{ij}S(-v^O)C^v_{i+1,j} - v^O_{ij}S(v^O)C^v_{i,j}]$$

$$+ A^B_{ij}[v^B_{ij}S(v^B)C^v_{i,j-1} + v^B_{ij}S(-v^B)C^v_{i,j}]$$

$$+ A^T_{ij}[-v^T_{ij}S(-v^T)C^v_{i,j+1} - v^TS(v^T)C^v_{i,j}]\}$$

$$- \sigma\lambda[C^v_{ij} - F(C^a_{ij})] \tag{63}$$

The modeling equations are then Equations (62) and (63), which can be integrated forward in time as described previously for the simpler models. If a horizontal slotted pipe configuration is to be modeled, Equation (63) is replaced by

$$\frac{dC^v_{ij}}{dt} = \frac{1}{\Delta v}\ \{A^L[v^L_{ij}S(v^L)C^v_{i-1,j} + v^L_{ij}S(-v^L)C^v_{i,j}]$$

$$+ A^R[-v^R_{ij}S(-v^R)C^v_{i+1,j} - v^R_{ij}S(v^R)C^v_{i,j}]$$

$$+ A^B[v^B_{ij}S(v^B)C^v_{i,j-1} + v^B_{ij}S(-v^B)C^v_{i,j}]$$

$$+ A^T[-v^T_{ij}S(-v^T)C^v_{i,j+1} - v^T_{ij}S(v^T)C^v_{i,j}]\}$$

$$- \sigma\lambda[C^v_{ij} - F(C^a_{ij})] \tag{64}$$

Unfortunately, rarely do the data available for a site permit using an adsorption isotherm of any great sophistication, so one usually sets $F(C^a) = K_aC^a$ and hopes for the best. Also unfortunately, values of K_a in such linear isotherms have a marked tendency to decrease as the remediation proceeds and the less strongly bound VOC molecules are removed, leaving only the more strongly bound ones to be desorbed later in the remediation. One should not rely on K_a values that have been obtained from lab or pilot-scale measurements made only during the removal of the first 50% or so of the VOC.

THE STEADY-STATE APPROXIMATION

The inclusion of diffusion/desorption kinetics doubles the number of differential equations that must be integrated, so it can be expected to substantially increase the computer time required for a simulation over that needed with a local equilibrium model for the same system. Actually, this difficulty can be even more severe. The value of Δt, which can be used without leading to catastrophic instabilities, is generally determined by the advection terms and is relatively small. If the rate constant for diffusion/desorption is relatively small, one may have to run the simulation for quite a long time (both model time and computer time) with small values of Δt. This seriously interferes with the usefulness of the model if one wishes to make a large number of runs for design purposes, sensitivity analysis, or estimation of the cost-effectiveness of various options. Mathematically, the system of differential equations is stiff.

One can get around this difficulty by making use of the steady-state approximation, which is very commonly used in the analysis of mechanisms in chemical kinetics. [For a discussion of the steady-state approximation, see Levine (1988) or Laidler (1965).] Generally, the quantity of VOC actually present in the gas phase at any one time is a very small fraction of the total mass of VOC present. With a steady flow of gas to the well, the values of C^v throughout the domain will be monotonically decreasing, and the magnitude of this rate of change must be quite small, because C_v itself is quite small and the time period involved is substantial. The steady-state approximation consists of setting these very small derivatives equal to zero, i.e., assuming that the rate of depletion of gas phase VOC by advection is exactly matched by the rate of its replenishment by mass transport from the stationary phase(s). See Rodríguez-Maroto and Wilson (1991). For the case of a single vertical well, this converts the differential Equation (63) into an algebraic equation,

$$
\begin{aligned}
0 = \frac{1}{\Delta V_{ij}} \{ &A_{ij}^I [v_{ij}^I S(v^I) C_{i-1,j}^v + v_{ij}^I S(-v^I) C_{i,j}^v] \\
&+ A_{ij}^O [-v_{ij}^O S(-v^O) C_{i+1,j}^v - v_{ij}^O S(v^O) C_{i,j}^v] \\
&+ A_{ij}^B [v_{ij}^B S(v^B) C_{i,j-1}^v + v_{ij}^B S(-v^B) C_{i,j}^v] \\
&+ A_{ij}^T [-v_{ij}^T S(-v^T) C_{i,j+1}^v - v_{ij}^T S(v^T) C_{i,j}^v] \} \\
&- \sigma\lambda [C_{ij}^v - F(C_{ij}^a)]
\end{aligned} \tag{65}
$$

which, together with Equation (62), constitutes the model. A similar modification of Equation (64) yields a steady-state model for the horizontal slotted pipe configuration.

For one-dimensional models the equations analogous to Equation (65) are easily solved exactly by starting at the upwind end and working toward the well or the exit end of the column. For systems in which it is difficult to determine which volume elements are upwind from which, one must solve Equation (65) by solving for C_{ij}^v and then iterating the equation a couple of times, using as starting values the values of the C_{ij}^v from the immediately previous time point and the values of the C_{ij}^a calculated from Equation (62).

The steady-state models for kinetically limited SVE typically run about twenty times as fast as the exact models, and the results obtained by the two methods are virtually indistinguishable. Computer time requirements for the steady-state models of kinetically limited SVE are roughly comparable to those required for local equilibrium models of the same degree of complexity.

DIFFUSION-LIMITED SVE OF NAPL

We next turn to the vapor stripping of soil in which the VOC is present as NAPL, which must dissolve in a fairly thick surrounding boundary layer of water and diffuse through it before being released into the advecting soil gas. The situation is illustrated in Figure 5.5.

This mass transport model was explored in Chapter 4, and the following

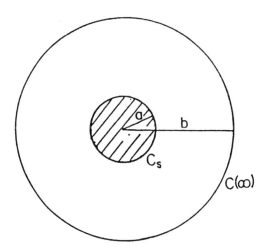

Figure 5.5 Model for diffusion of VOC from an NAPL droplet of radius $a(t)$ through a surrounding spherical aqueous layer of radius b to the advecting vapor phase.

result was obtained for the rate of solution of NAPL, with some minor changes in notation to be consistent with the notation used in this chapter.

$$\frac{dC_{ij}^N}{dt} = -\frac{3D(C_0^N)^{2/3}(C_{sat} - C_{ij}^w)b(C_{ij}^N)^{1/3}}{\varrho a_0^2[b - a_0(C_{ij}^N/C_0^N)^{1/3}]} \tag{66}$$

Here

C_{ij}^N = VOC NAPL concentration in the ijth volume element, kg/m³ of soil

D = VOC diffusivity in water, m²/sec

C_{sat} = saturation concentration of VOC in water, kg/m³

K_H = Henry's constant of VOC, dimensionless

C_{ij}^w = C_{ij}^v/K_H, VOC concentration in water at equilibrium with the vapor phase in the ijth volume element, kg/m³ of water

C_{ij}^v = vapor phase VOC concentration in the ijth volume element, kg/m³ of air

ϱ = density of NAPL, kg/m³

b = thickness of water boundary layer, m

a_0 = initial radius of NAPL droplet, m

σ = air-filled porosity, dimensionless

The differential equations for the C_{ij}^v include both advection terms and mass transport terms, as before; for a single vertical SVE well they are

$$\frac{dC_{ij}^v}{dt} = \frac{1}{\Delta V_{ij}} \{A_{ij}^I[v_{ij}^I S(v^I)C_{i-1,j}^v + v_{ij}^I S(-v^I)C_{i,j}^v]$$

$$+ A_{ij}^O[-v_{ij}^O S(-v^O)C_{i+1,j}^v - v_{ij}^O S(v^O)C_{i,j}^v]$$

$$+ A_{ij}^B[v_{ij}^B S(-v^B)C_{i,j-1}^v + v_{ij}^B S(-v^B)C_{i,j}^v]$$

$$+ A_{ij}^T[-v_{ij}^T S(-v^T)C_{i,j+1}^v - v_{ij}^T S(v^T)C_{i,j}^v]\}$$

$$- \frac{1}{\sigma} \cdot \frac{dC_{ij}^N}{dt} \tag{67}$$

As with the lumped parameter model, these differential equations are often stiff, so that excessive computer time is required for a simulation.

The steady-state approximation for Equation (67) gives

$$
0 = \frac{1}{\Delta V_{ij}} \{A_{ij}^I[v_{ij}^I S(v^I) C_{i-1,j}^v + v_{ij}^I S(-v^I) C_{i,j}^v]
$$

$$
+ A_{ij}^O[-v_{ij}^O S(-v^O) C_{i+1,j}^v - v_{ij}^O S(v^O) C_{i,j}^v]
$$

$$
+ A_{ij}^B[v_{ij}^B S(v^B) C_{i,j-1}^v + v_{ij}^B S(-v^B) C_{i,j}^v]
$$

$$
+ A_{ij}^T[-v_{ij}^T S(-v^T) C_{i,j+1}^v - v_{ij}^T S(v^T) C_{i,j}^v]\}
$$

$$
- \frac{1}{\sigma} \cdot \frac{dC_{ij}^N}{dt} \tag{68}
$$

and the model finally consists of the differential Equation (66) and the algebraic Equation (68), which are solved as described above.

SVE OF NAPL MIXTURES OBEYING RAOULT'S LAW

Many of the VOCs that occur at hazardous waste sites are mixtures such as gasoline and other petroleum-based fuels. If these are present as NAPL at a site, the vapor pressures of the various components will not be those of the pure compounds, but will be substantially reduced by dilution. This very markedly affects their behavior during vapor stripping operations. Johnson et al. (1989a, 1989b, 1990) have explored this in some detail in connection with gasoline spills. Here, we follow the analysis of Kayano and Wilson (1992), who developed models for single vertical wells and horizontal slotted pipes. These models are local equilibrium models and assume that the only nonmobile VOC present is NAPL.

The gas flow fields needed were developed in Chapter 3 and will be used without further discussion. Here, we first address the calculation of the mole numbers of the component VOCs in the gas phase and the liquid phase for a single volume element. Notation is as follows:

V = volume of the volume element, m³
P_i^0 = vapor pressure of component i in the pure state at the ambient soil temperature, atm
σ = air-filled porosity of the medium
n_i^g = number of moles of component i in the gas phase
n_i^l = number of moles of component i in the NAPL phase
n_i = total number of moles of component i in the volume element
X_i = mole fraction of component i in the liquid phase

P_i = vapor pressure of component i in the mixture in the volume element, atm

J = number of components in the mixture

Raoult's Law gives

$$P_i = P_i^0 X_i \tag{69}$$

Also

$$X_i = \frac{n_i^l}{\displaystyle\sum_{j=1}^{J} n_i^l} \tag{70}$$

Then the concentration of component i in the vapor phase (mol/m³) is given by

$$c_i^g = \frac{P_i^0}{RT} \cdot \frac{n_i^l}{\displaystyle\sum_{j}^{J} n_j^l} = \frac{n_i^g}{\sigma \Delta V} \tag{71}$$

By definition

$$n_i = n_i^g + n_i^l \tag{72}$$

From Equation (71) we have

$$n_i^g = \frac{\sigma \Delta V P_i^0}{RT} \cdot \frac{n_i^l}{\displaystyle\sum_{j} n_j^l} \tag{73}$$

Substitution of Equation (73) into Equation (72) yields

$$n_i = n_i^l \left[1 + \frac{\sigma \Delta V P_i^0}{RT} \cdot \frac{1}{\displaystyle\sum_{j} n_j^l} \right] \tag{74}$$

Define

$$a_i = \frac{\sigma \Delta V P_i^0}{RT} \tag{75}$$

$$u = \sum_{j} n_j^l \tag{76}$$

and rearrange Equation (74) to obtain

$$n_i^l = \frac{n_i^u}{a_i + u}$$ (77)

Sum Equation (77) over i and use Equation (76) to get

$$u = u \sum_i \frac{n_i}{a_i + u}$$ (78)

or

$$1 = \sum_i \frac{n_i}{a_i + u}$$ (79)

Equation (79) is then solved numerically for u. Our programs use a simple binary search procedure, starting with $u_{min} = 0$, $u_{max} = \Sigma_j n_j$, $\Delta u = u_{max}/2$. If no solution exists in this range, no liquid phase is present. Then $n_i^l = 0$ and $n_i^g = n_i$ for all i. If $0 < u < u_{max}$, then Equation (79) is used to obtain the n_i^l, after which the n_i^g are obtained from

$$n_i^g = a_i n_i^l / u$$ (80)

The molar concentration of component i in the vapor phase in this volume element is given by

$$c_i^g = n_i^g / \sigma \Delta V$$ (81)

and the concentrations C_i^g are given by

$$C_i^g = c_i^g \cdot (MW)_i$$ (82)

where

C_i^g = vapor phase concentration of ith component, kg/m^e
$(MW)_i$ = molecular weight of ith component, kg/mol

The development of a local equilibrium SVE model for Raoult's Law mixtures by means of a horizontal slotted pipe follows along the lines leading to Equation (29); one carries out a mass balance on the ijth volume element and obtains for component q

$$
\begin{aligned}
\frac{dC_{ij}^q}{dt} = \frac{1}{\Delta V} \{ & A^L [v_{ij}^L S(v^L) C_{i-1,j}^{qg} + v_{ij}^L S(-v^L) C_{i,j}] \\
& + A^R [-v_{ij}^R S(-v^R) C_{i+1,j}^{qg} - v_{ij}^R S(v^R) C_{i,j}^{qg}] \\
& + A^B [v_{ij}^B S(v^B) C_{i,j-1}^{qg} + v_{i,j}^B S(-v^B) C_{i,j}^{qg}] \\
& + A^T [-v_{ij}^T S(-v^T) C_{i,j+1}^{qg} - v_{ij}^T S(v^T) C_{i,j}^{qg}] \}
\end{aligned}
\tag{83}
$$

where

C_{ij}^q = total concentration of component q in the ijth volume element, kg/m³ of soil

C_{ij}^{qg} = concentration of component q in the gas phase in the ijth volume element, kg/m³ of gas

and the other notation is the same as in Equation (29).

REMOVAL OF FLOATING NAPL

The penetration of NAPL through the vadose zone to the water table is fairly common at hazardous waste sites, so one needs to explore the possibility of removing NAPL, which is floating on the water table, by SVE. If the NAPL is denser than water (DNAPL) and its pressure head is sufficient to overcome the capillary pressure associated with the DNAPL/water interface in the porous medium, the DNAPL penetrates into the aquifer, perhaps even pooling on the underlying aquitard. Removal of such material by SVE is out of the question, since the DNAPL VOC must dissolve in the water and then move by diffusion/dispersion through a water layer that may be several meters thick before the VOC reaches the capillary fringe of the water table, at which point it can move into the vapor phase. Rates of solution and diffusion in the aqueous phase are three orders of magnitude or more smaller than the corresponding processes in the vapor phase, so SVE is practically useless for the removal of any VOC below the water table.

The removal of NAPL floating on the water table is a different matter. We next examine the modeling of this process.

Diffusivities and dispersivities are important in modeling the vapor stripping of underlying NAPL. Scheidegger's (1974) equations for calculating longitudinal and transverse dispersivities seem to be well established; these are

$$D_{long} = D_{mol} + 1.75\delta v \tag{84}$$

$$D_{trans} = D_{mol} + 0.055\delta v \tag{85}$$

where

D_{mol} = molecular diffusivity in the porous medium, m²/sec
δ = grain size parameter, m
v = linear gas velocity, m/sec

Methods for estimating diffusivities in porous media have been developed, and a summary of these is given in Chapter 4. Jury and Valentine (1986) regard the model of Millington and Quirk (1961) as probably the most satisfactory. This relates the diffusivity of a volatile chemical in the soil to its diffusivity in air and is

$$D_{soil} = \frac{(\nu - \omega)^{10/3}}{\nu^2} \cdot D_{air} \tag{86}$$

where

ν = total soil porosity, dimensionless
ω = specific moisture content, dimensionless

Thibodeaux et al. (1988) used this expression in modeling chemical vapor losses from landfills.

We consider the evaporation of NAPL from a plane surface underlying the vadose zone into the vadose zone. The evaporated VOC is then swept away by a constant, uniform flow of soil gas, such as would be generated in the vicinity of a vapor stripping well. We wish to estimate the rate of removal of the NAPL. The geometry and boundary conditions are indicated in Figure 5.6. The equilibrium vapor concentration C_0 (kg/m³) is given in terms of the vapor pressure by

$$C_0 = [0.01603(MW)/T]P_0(T) \tag{87}$$

where

(MW) = molecular weight of the VOC, gm/mol

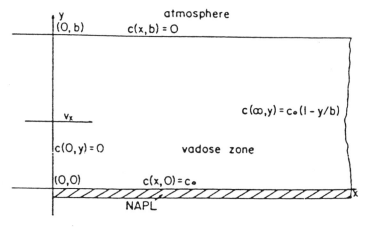

Figure 5.6 Model for SVE of an LNAPL pool underlying the vadose zone. Air flow is from left to right and the LNAPL pool starts at $x = 0$ and extends to the right.

T = temperature, deg K
$P_0(T)$ = equilibrium vapor pressure of VOC at temperature T, torr

Let

σ = soil air-filled porosity
b = thickness of the vadose zone, m
v_x = soil gas linear velocity, m/sec
D_x = longitudinal dispersivity, m²/sec
D_y = transverse dispersivity, m²/sec
C = gas phase VOC concentration, kg/m³

The steady-state advection-dispersion equation for the system is

$$D_x \frac{\partial^2 C}{\partial x^2} - v_x \frac{\partial C}{\partial x} + D_y \frac{\partial^2 C}{\partial y^2} = 0 \tag{88}$$

with boundary conditions (from Figure 5.6)

$$C(x,b) = 0 \tag{89}$$

$$C(0,y) = 0 \tag{90}$$

$$C(x,0) = C_0 \tag{91}$$

$$\lim_{x \to \infty} C(x,y) = C_0(b - y)/b \tag{92}$$

Equation (88) is solved by separation of variables; one assumes that

$$C(x,y) = C_0(b - y)/b + \sum_\lambda X_\lambda(x)Y_\lambda(y) \qquad (93)$$

This gives

$$0 = D_x \frac{X_\lambda''}{X_\lambda} - v_x \frac{X_\lambda'}{X_\lambda} + D_y \frac{Y_\lambda''}{Y_\lambda} \qquad (94)$$

which is split in the usual way to yield

$$Y_\lambda'' + (\lambda/D_y)Y_\lambda = 0 \qquad (95)$$

and

$$D_x X_\lambda'' - v_x X_\lambda' - \lambda X_\lambda = 0 \qquad (96)$$

The solution to Equation (95) is

$$Y_\lambda = A_\lambda \sin [(\lambda/D_y)^{1/2}y] + B_\lambda \cos [(\lambda/D_y)^{1/2}y] \qquad (97)$$

The boundary conditions require that the B_λ vanish and that

$$\sin [(\lambda/D_y)^{1/2}b] = 0 \qquad (98)$$

which yields

$$\lambda = \lambda_n = D_y \left(\frac{n\pi}{b}\right)^2 \qquad (99)$$

so

$$Y_\lambda = Y_n = A_n \sin (n\pi y/b) \qquad (100)$$

Equation (96) is solved by the usual exponential substitution, which yields the characteristic equation

$$D_x m^2 - v_x m - \lambda_n = 0 \qquad (101)$$

for which the roots $m^\pm$ are

$$m_n^\pm = \frac{v_x \pm [v_x^2 + 4D_x D_y(n\pi/b)^2]^{1/2}}{2D_x} \qquad (102)$$

The m_n^+ are all >0 and the m_n^- are all <0; to avoid exponentially increasing solutions we must therefore drop the solutions $\exp(m_n^+ x)$. The general solution to Equation (88) is therefore

$$C(x,y) = C_0(b - y)/b + \sum_{n=1}^{\infty} A_n \exp(-m_n x) \sin(n\pi y/b) \quad (103)$$

where

$$m_n = \frac{v_x}{2D_x}\left\{\left[1 + 4D_x D_y\left(\frac{n\pi}{v_x b}\right)^2\right]^{1/2} - 1\right\} \quad (104)$$

The A_n must be chosen so that the solution satisifies Equation (90), which yields

$$-C_0(b - y)/b = \sum_{n=1}^{\infty} A_n \sin(n\pi y/b) \quad (105a)$$

Multiplying Equation (105) by $\sin(m\pi y/b)$ and integrating from 0 to b yields

$$A_m = -2C_0/m \quad (105b)$$

so the VOC vapor concentration distribution above the pool of NAPL is given by

$$C(x,y) = C_0(b - y)/b - \frac{2C_0}{\pi}\sum_{n=1}^{\infty} \frac{1}{n} \exp(-m_n x) \sin(n\pi y/b) \quad (106)$$

We next calculate the total contaminant flux $F(x)$ through a vertical surface normal to the direction of gas flow, extending from $y = 0$ to $y = b$, and 1 m wide. This is given by

$$F(x) = \int_0^b v_x C(x,y) dy \quad (107)$$

Substitution of Equation (106) into Equation (107) and integration then yields

$$F(x) = \frac{v_x C_0 b}{2}\left[1 - \frac{8}{\pi^2}\sum_{n=1}^{\infty} \frac{1}{(2n + 1)^2} \exp(-m_{2n+1} x)\right] \quad (108)$$

where

$$m_{2n+1} = \frac{v_x}{2D_x}\left\{\left[1 + 4D_xD_y\left(\frac{(2n + 1)\pi}{v_xb}\right)^2\right]^{1/2} - 1\right\}$$ (109)

for the total flux of VOC being swept away at a point where the air stream has traversed a distance x over the underlying NAPL pool.

One can easily modify the model by requiring that there be an impermeable barrier at the top of the vadose zone. In this second case, Equation (89) is replaced by

$$\lim_{x \to \infty} C(x,y) = C_0$$ (110)

The trial form of the solution is taken to be

$$C(x,y) = C_0 + \sum_\lambda X_\lambda(x)Y_\lambda(y)$$ (111)

which eventually gives

$$C(x,y) = C_0 + \sum_{n=1}^{\infty} A_n \sin\left[\frac{(2n - 1)\pi y}{2b}\right] \exp(-M_nx)$$ (112)

where

$$M_n = \frac{v_x}{2D_x}\left\{\left[1 + 4D_xD_y\left(\frac{(2n + 1)}{2bv_x}\right)^2\right]^{1/2} - 1\right\}$$ (113)

The Fourier coefficients A_n are calculated from the requirement that

$$-C_0 = \sum_{n=1}^{\infty} A_n \sin\left[\frac{(2n - 1)\pi y}{2b}\right]$$ (114)

which yields

$$C(x,y) = C_0 - \frac{4C_0}{\pi}\sum_{n=1}^{\infty}\frac{1}{2n - 1} \sin\left[\frac{(2n - 1)\pi y}{2b}\right] \exp(-M_nx)$$ (115)

The flux is calculated as before; the result is

$$F(x) = v_xC_0b\left[1 - \frac{8}{\pi^2}\sum_{n=1}^{\infty}\frac{1}{(2n - 1)^2} \exp(-M_nx)\right]$$ (116)

where the M_n are defined by Equation (113).

EFFECTS OF VARIABLE AIRFLOW RATES IN
DIFFUSION-LIMITED SVE OPERATION

In this section a model for SVE is developed which includes the effects of mass transport kinetics of VOC between nonaqueous phase liquid (NAPL) droplets and the aqueous phase and between the aqueous and vapor phases. The model permits time-dependent gas flow rates in the vapor extraction well. It will be employed in the next chapter to demonstrate the effectiveness of certain types of pilot-scale SVE experiments in determining the rate of mass transport processes. It will also be used to explore several time-dependent airflow schedules for SVE well operation. The results will indicate that the use of suitably selected airflow schedules in SVE can result in greatly reduced volumes of air to be treated for VOC removal with relatively little increase in the time required to meet remediation standards.

One of the more significant of the site-specific aspects of SVE is the extent to which the kinetics of diffusion and/or desorption may limit the rate at which VOCs can be removed. If one has a site with a highly homogeneous sandy soil containing very little natural organic material and relatively little moisture, one may expect to find that diffusion/desorption rates present no problem and that a local equilibrium treatment of the process is quite adequate. On the other hand, if the porous medium has a highly heterogeneous permeability, if it contains significant amounts of clay or humic organic material, or if it contains substantial amounts of water, the kinetics of diffusion and/or desorption may prove to be serious bottlenecks in the removal of VOCs by SVE. DiGiulio et al. (1990) have discussed this problem in some detail and have described experiments that could be done during pilot studies to ascertain the extent to which these mass transport kinetics problems may slow down the remediation. Oma et al. (1990) and Gómez-Lahoz et al. (1993) have explored some aspects of the economic advantages to be obtained by pulsed operation of SVE systems within the framework of a one-dimensional model.

Here, we present a mathematical model for soil vapor extraction which includes two possible kinetic bottlenecks and which allows one to vary arbitrarily the airflow rate through the vacuum well. The kinetic bottlenecks included are (1) the rate of aqueous solution of droplets of nonaqueous phase liquid (NAPL) distributed within the porous medium and (2) the rate of mass transport of dissolved VOC into the moving gaseous phase. The pilot-scale experiments proposed by DiGiulio et al. (1990) are then simulated with the model and found to provide valuable information about the rates of these kinetically limited processes. Lastly, an air injection experiment for investigating the kinetics of diffusion/desorption will be analyzed. In Chapter 6 several gas flow operating schedules for an SVE well will be simulated with this model, with the objective of substantially reducing the total volume of soil gas that must be treated without substantially increasing the time required to achieve the target level of remediation.

We model here the operation of a single soil vapor extraction well screened at the bottom and drilled in a homogeneous, isotropic medium. The VOC contaminant is assumed to be initially present as NAPL, as dissolved VOC in the soil moisture, and as vapor in the soil gas. Mass transport of VOC between the NAPL phase and the aqueous phase is handled by means of a technique described earlier for modeling the solution of DNAPL droplets in groundwater in pump-and-treat operations (Kayano and Wilson, 1993) and in sparging (Roberts and Wilson, 1993a, 1993b). Mass transport of VOC between the aqueous phase and the moving vapor phase is handled by means of the lumped parameter method used previously in this chapter and employed earlier in SVE modeling (Wilson et al., 1992, for example).

We use the gas flow field for a single well in a medium of constant isotropic permeability as obtained in Chapter 3, deriving this from a potential function W.

Note that, while W and its derivatives are directly proportional to Q_a, P and the velocity components v_r and v_z are not. If we denote by primes ($'$) values of W, $\partial W/\partial r$, and $\partial W/\partial z$ calculated with the molar airflow rate $Q_a = 1$ mol/sec, we can then express the soil gas pressure and gas velocities at other values of Q_a by Equations (117)–(119).

$$P(r,z,Q_a) = [1 \text{ atm}^2 + Q_a W'(r,z)]^{1/2} \tag{117}$$

$$v_r(r,z,Q_a) = -\frac{KQ_a \partial W'/\partial r}{2[1 \text{ atm}^2 + Q_a W'(r,z)]^{1/2}} \tag{118}$$

$$v_z(r,z,Q_a) = -\frac{KQ_a \partial W'/\partial z}{2[1 \text{ atm}^2 + Q_a W'(r,z)]^{1/2}} \tag{119}$$

This permits us to evaluate W' and its derivatives at the necessary mesh points initially and then to use the much simpler Equations (118) and (119) to calculate the gas velocities as functions of $Q_a(t)$ during the course of the simulation.

The second phase of the calculation is to use the gas flow field generated above in carrying out mass balances on the three phases in which the VOC is present (NAPL, aqueous, gaseous) in the ijth ring-shaped volume element, illustrated in Figure 5.7. Let

$$r_i = (i + 1/2)\Delta r \tag{120}$$

$$z_j = (j + 1/2)\Delta z \tag{121}$$

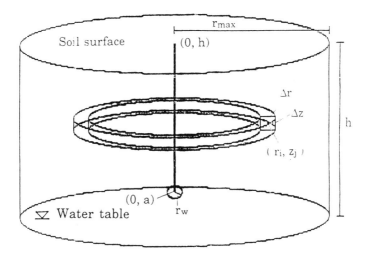

Figure 5.7 Geometry and representative volume element for SVE of soil by means of a single vertical well. Mass transport is from NAPL to aqueous VOC to VOC in the advecting vapor phase.

Then

$$V_{ij} = (2i + 1)\pi(\Delta r)^2 \Delta^2 \tag{122}$$

The inner surface of the volume element is given by

$$A_{ij}^I = 2\pi i \Delta r \Delta z \tag{123}$$

The outer surface is

$$A_{ij}^O = 2\pi(i+1)\Delta r \Delta z \tag{124}$$

The top (S_{ij}^T) and bottom (S_{ij}^B) surfaces are

$$A_{ij}^T = A_{ij}^B = (2i+1)\pi(\Delta r)^2 \tag{125}$$

Define

m_{ij} = total mass of VOC in ΔV_{ij}, kg
C_{ij}^N = concentration of NAPL in ΔV_{ij}, kg/m³
C_{ij}^w = concentration of aqueous VOC, kg/m³ of aqueous phase
C_{ij}^g = vapor concentration of VOC, kg/m³ of vapor phase

K_H = effective Henry's constant of VOC, C_{ij}^g/C_{ij}^w at equilibrium, dimensionless

ω = water-filled soil porosity, dimensionless

σ = air-filled soil porosity, dimensionless

The total mass of VOC in the ijth volume element is given by

$$m_{ij} = \Delta V_{ij}(C_{ij}^N + \omega C_{ij}^w + \sigma C_{ij}^g) \tag{126}$$

The soil gas superficial velocities at the four surfaces of the volume element are given by

$$v_{ij}^I = v_r[i\Delta r, (j + 1/2)\Delta z] \quad \text{(inner)} \tag{127}$$

$$v_{ij}^O = v_r[(i + 1)\Delta r, (j + 1/2)\Delta z] \quad \text{(outer)} \tag{128}$$

$$v_{ij}^B = v_z[(i + 1)\Delta r, (j\Delta z] \quad \text{(bottom)} \tag{129}$$

$$v_{ij}^T = v_z[(i + 1/2)\Delta r, (j + 1)\Delta z] \quad \text{(top)} \tag{130}$$

Define

$$S(v) = 1, v > 0$$

$$= 0, v \le 0 \tag{131}$$

Then a mass balance on total VOC in the ijth volume element yields

$$\frac{dm_{ij}}{dt} = A_{ij}^I v_{ij}^I \cdot [S(v^I) \cdot C_{i-1,j}^g + S(-v^I) \cdot C_{ij}^g]$$

$$- A_{ij}^O v_{ij}^O \cdot [S(-v^O) \cdot C_{i+1,j}^g + S(v^O) \cdot C_{ij}^g]$$

$$+ A_{ij}^B v_{ij}^B \cdot [S(v^B) \cdot C_{i,j-1}^g + S(-v^B) \cdot C_{ij}^g]$$

$$- A_{ij}^T v_{ij}^T \cdot [S(-v^T) \cdot C_{i,j+1}^g + S(v^T) \cdot C_{ij}^g] \tag{132}$$

For the rate of solution of NAPL, we use an expression developed previously for the rate of solution of DNAPL droplets in pump-and-treat and in sparging (Kayano and Wilson, 1993; Roberts and Wilson, 1993a, 1993b); this is described in Chapter 4 and is

$$\frac{dC_{ij}^N}{dt} = - \frac{3C_0^{N2/3}D(C_{sat} - C_{ij}^w)C_{ij}^{N1/3}}{\varrho a_0^2} \tag{133}$$

where

C_0 = initial NAPL concentration, kg/m³
D = VOC diffusivity in the aqueous phase in the porous medium, m²/sec
c_s = solubility of VOC in soil water, kg/m³ of aqueous phase
ϱ = density of NAPL, kg/m³
a_0 = initial NAPL droplet radius, m

From our assumed linear isotherm (i.e., an effective Henry's Law), we have

$$C_{ij}^g(\text{equilibrium}) = K_H C_{ij}^w \tag{134}$$

We use a lumped parameter approximation for mass transport (mt) of VOC between the aqueous and vapor phases,

$$\left[\frac{\partial C_{ij}^g}{\partial t}\right]_{mt} = \lambda[C_{ij}^g(\text{equil}) - C_{ij}^g]$$

$$= \lambda[K_H C_{ij}^w - C_{ij}^g] \tag{135}$$

Now

$$\omega \frac{dC_{ij}^w}{dt} = -\frac{dC_{ij}^N}{dt} - \sigma \left[\frac{\partial C_{ij}^g}{\partial t}\right]_{mt} \tag{136}$$

so

$$\frac{dC_{ij}^w}{dt} = -\frac{1}{\omega} \cdot \frac{dC_{ij}^N}{dt} - \frac{\lambda\sigma}{\omega} \cdot (K_H C_{ij}^w - C_{ij}^g) \tag{137}$$

Differentiating Equation (126) with respect to time yields

$$\frac{dm_{ij}}{dt} = \Delta V_{ij}\left[\frac{dC_{ij}^N}{dt} + \omega\frac{dC_{ij}^w}{dt} + \sigma\frac{dC_{ij}^g}{dt}\right] \tag{138}$$

Solving Equation (138) for dC_{ij}^g/dt and utilizing Equation (137) then yields

$$\frac{dC_{ij}^g}{dt} = \frac{1}{\sigma\Delta V_{ij}} \cdot \frac{dm_{ij}}{dt} + \lambda(K_H C_{ij}^w - C_{ij}^g) \tag{139}$$

The differential equations constituting the model are Equations (132), (133), (138), and (139). Equation (132) may be used merely to handle the advection terms; the m_{ij} can be calculated from Equation (126) if desired. The total mass of residual contaminant in the system is calculated from

$$M_{total} = \sum_{i=1}^{N_r} \sum_{j=1}^{N_z} m_{ij}(t) = \sum_{i=1}^{N_r} \sum_{j=1}^{N_z} \Delta V_{ij} [C_{ij}^N + \omega C_{ij}^w + \sigma C_{ij}^g] \quad (140)$$

The rate of removal of contaminant is readily calculated by

$$\left| \frac{dM_{total}}{dt} \right| = -\frac{M_{total}(t + \Delta t) - M_{total}(t)}{\Delta t} \quad (141)$$

If the gas flow rate in the system is not zero, the concentration of VOC in the exhausted soil gas is calculated by noting that the volumetric flow rate of the evacuated air (corrected to 1 atm) is given by

$$Q_{vol} = Q_a RT \quad (142)$$

where $R = 8.206 \times 10^{-5}$ m^3 atm/mol deg and T is the absolute temperature. Then the effluent VOC concentration is given by

$$C_{effl} = \frac{dM_{total}}{dt} / Q_{vol} \quad (143)$$

If the well is turned off, one can still follow the concentration of VOC in the vicinity of the well by examining the values of C_{ij}^g in the immediate vicinity of the well. In the present study, values of C_{ij}^g in the volume elements immediately above and immediately outside of the volume element containing the well were examined. These concentrations were found to be nearly identical. Following these values of the C_{ij}^g permits one to investigate the rate of "rebound" of the VOC concentration. This is useful in estimating the magnitude of the diffusion rate of the system during the course of the cleanup.

It is possible to develop a quite rough method for approximating the mass transport kinetics constant by comparing the equilibrium VOC gas phase concentration with the VOC gas phase concentration at steady state. We assume a well-stirred one-compartment model having a volume V. At steady-state operation the rate of removal of VOC from the system by advection is equal to the rate of release of VOC from condensed phases in the system; use of a lumped parameter approach then gives

$$Q \cdot C_{ss}^g = k' \cdot V \cdot (C_{equil}^g - C_{ss}^g) \quad (144)$$

which yields

$$k' = \frac{QC_{ss}^g}{V(C_{equil}^g - C_{ss}^g)} \qquad (145)$$

A reasonable value for V is that of a paraboloid of height h' and radius at the top of r', where r' is the effective radius of influence of the well (roughly equal to its depth) and h is the depth to which the well is drilled; this gives $V = (\Delta/2) \cdot h'r'^2$.

An alternative, perhaps better, approach to a rough estimate of the mass transport kinetics constant involves the use of a plug-flow model in the steady-state approximation. We assume a one-dimensional volumetric gas flow Q through a volume V. At any point in the volume, the rate of removal of VOC by advection is equal to the rate of its replenishment from the condensed phase(s); this gives

$$Q \frac{dC_{ss}^g}{dV} = k''[C_{equil}^g - C_{ss}^g] \qquad (146)$$

where k'' is the rate constant for mass transport. Separation of variables and integration then yields

$$\log_e \frac{C_{equil}^g}{C_{equil}^g - C_{ss}^g} = k''V/Q \qquad (147)$$

from which we obtain

$$k'' = (Q/V) \cdot \log_e \frac{C_{equil}^g}{C_{equil}^g - C_{ss}^g} \qquad (148)$$

We choose V to be the paraboloidal volume mentioned above.

In the limit in which mass transport between the aqueous and the vapor phases is rapid compared to that between the NAPL and the aqueous phases, one can calculate an expression for the rate constant for the rebound rate of the VOC vapor concentration after the well has been shut off. This is done as follows. Once the gas flow in the well is stopped, the aqueous, vapor, and NAPL concentrations at any point in the system are related by

$$\omega \frac{dC^w}{dt} + \sigma \frac{dC^g}{dt} = -\frac{dC^N}{dt} \qquad (149)$$

where the subscripts, not needed here, have been dropped. The limit mentioned above yields

$$C^w = C^g/K_H \tag{150a}$$

which, on substitution into Equation (149), yields

$$(\omega/K_H + \sigma)\frac{dC^g}{dt} = -\frac{dC^N}{dt} \tag{150b}$$

An expression for dC^N/dt is obtained from Equation (133); substitution of this into Equation (150) and rearrangement then gives

$$\frac{dC^g}{dt} = \frac{3C_0^{N2/3}DC^{N1/3}}{(\omega + K_H\sigma)\varrho a_0^2} \cdot (K_H C_{sat} - C^g) \tag{151}$$

or

$$\frac{dC^g}{dt} = \beta K_H C_{sat} - \beta C^g \tag{152}$$

Integration of this equation then yields

$$C^g(t) = K_H C_{sat} - [K_H C_{sat} - C^g(0)] \cdot \exp(-\beta t) \tag{153}$$

where

$$\beta = \frac{3C_0^{N2/3}DC^{N1/3}}{(\omega + K_H\sigma)\varrho a_0^2} \tag{154}$$

is the rate constant for the rebound of the soil gas VOC concentration toward equilibrium once the SVE well has been turned off. This gives us an equation relating the observable rebound rate constant to the solution kinetics parameters that govern the rate of remediation of the domain. Note that Equation (154) predicts that these rebound rate constants will decrease during the course of the cleanup, since $C^{N1/3}$ must decrease as the remediation proceeds.

This approach can be extended to the situation in which the rates of mass transfer between the NAPL and aqueous phases and the aqueous and gaseous phases are comparable, although one pays a price in terms of more complex formulas. We proceed as follows. After the well has been turned off so that there is no advection, we have

$$\omega \frac{dC^w}{dt} + \sigma \frac{dC^g}{dt} = -\frac{dC^N}{dt} \tag{155}$$

$$\frac{dC^g}{dt} = \lambda \left(K_H C^w - C^g \right) \tag{156}$$

and

$$\frac{dC^N}{dt} = -\frac{3DC_0^{N2/3}C^{N1/3}}{\varrho a_o^2} \cdot (C_{sat} - C^g) \tag{157}$$

Let us approximate that $C^{N1/3}$ remains constant during the rebound process, and define

$$\gamma = \frac{3DC_0^{N2/3}C^{N1/3}}{\varrho a_o^2} \tag{158}$$

$$G = C^g - K_H C_{sat} \tag{159}$$

and

$$H = C^w - C_{sat} \tag{160}$$

Then substitution of Equations (157)–(160) into Equations (155) and (156) and rearrangement yields

$$\frac{dH}{dt} = -\left[\frac{\gamma}{\omega} + \frac{\sigma\lambda K_H}{\omega}\right] \cdot H + \frac{\sigma\lambda}{\omega} \cdot G \tag{161}$$

and

$$\frac{dG}{dt} = \lambda K_H \cdot H - \lambda G \tag{162}$$

Let

$$A = \left(\frac{\gamma}{\omega} + \frac{\sigma\lambda}{\omega} K_H\right) \tag{163}$$

$$B = \sigma\lambda/\omega \tag{164}$$

$$C = \lambda K_H \tag{165}$$

$$D = \lambda \tag{166}$$

The time constants associated with Equations (161) and (162) are then the values of Λ satisfying the equation

$$\begin{vmatrix} -A + \Lambda & B \\ C & -D + \Lambda \end{vmatrix} = 0 \qquad (167)$$

or

$$\Lambda^2 - (A + D)\Lambda + (AD - BC) = 0 \qquad (168)$$

Of these, the smaller of the two will determine the length of time required for rebound of the VOC vapor concentration.

An alternative method for exploring the limits imposed by diffusion and desorption kinetics during pilot studies is to rapidly inject a slug of clean air into the vapor extraction well and then to follow the subsequent increase in VOC concentration in this newly injected air. To model this we use the same basic approach as described above. The equations that describe the changes in the distribution of VOC between phases (NAPL, aqueous-adsorbed, vapor) are

$$\frac{dC^N}{dt} = -\frac{3C_0^{N2/3}D(C_{sat} - C^w)C^{N1/3}}{\varrho a_0^2} \qquad (169)$$

for the NAPL concentration,

$$\frac{dC^w}{dt} = \frac{3C_0^{N2/3}D(C_{sat} - C^w)C^{N1/3}}{\omega\varrho a_0^2} - \frac{\sigma\lambda}{\omega}(K_H C^w - C^g) \qquad (170)$$

for the aqueous-adsorbed VOC concentration, and

$$\frac{dC^g}{dt} = \lambda(K_H C^w - C^g) \qquad (171)$$

for the vapor phase VOC concentration. Note that, since there is no gas flow during the test after the initial rapid injection of the slug of clean air, there are no advection terms in Equation (171), so subscripting of the concentrations is not necessary. Note also that the total concentration of VOC in the domain being tested remains constant during the rebound process, so that

$$C_{total} = C^N + \omega C^w + \sigma C^g \qquad (172)$$

The desired initial values of C^w and C^N are selected, C^g is initialized to zero, and Equations (169)–(171) are integrated forward in time to model the test. The concentration of VOC in the vapor phase, C^g, is then plotted as a function of time for various values of the model parameters to give insight into their effects on VOC vapor concentration rebound.

DISTRIBUTED DIFFUSION MODELS FOR THE SVE OF NAPL AND DISSOLVED VOC

A LABORATORY COLUMN MODEL

The handling of diffusion by means of the lumped parameter approach restricts one to a single time constant, as mentioned above. The simple lumped parameter method for including mass transport limitations in SVE models can give removal rates greatly reduced below those from models in which local equilibrium was assumed and has the advantage of rapid computation. This model, however, cannot yield with the same parameter set the very rapid initial VOC removal rates and the quite slow removal rates toward the end of the remediation which are observed experimentally. The lumped parameter approach to mass transport is evidently oversimplified.

In this section we develop a laboratory column SVE model that provides a detailed treatment of the diffusion of VOC through a stagnant aqueous boundary layer, includes the solution kinetics of NAPL droplets by the procedure described above, and permits time-dependent gas flow rates in the SVE column (Wilson, 1994). The objective here is to develop a model for SVE kinetics limitations, which yields initial rapid rates of VOC removal, followed by a transition region during which off-gas VOC concentrations decrease fairly rapidly, followed in turn by pronounced tailing of the off-gas VOC concentrations as cleanup approaches completion. This is a pattern that has been observed frequently at remediation sites, and its occurrence can cause some consternation and recriminations when short cleanup times have been predicted by local equilibrium models or by lumped parameter kinetics models where parameters have been selected on the basis of pilot-scale runs of short duration.

The laboratory SVE column is partitioned for mathematical analysis, as indicated in Figure 5.8. It is divided into n_x disk-shaped volume elements, each of thickness Δx. The water layer coating the soil particles in each of these volume elements is further divided into n_y slabs; the first is in contact with the advecting air, and the last is bordered by the solid soil surface. Let

h = height of column, cm
A = cross-sectional area of column, cm^2

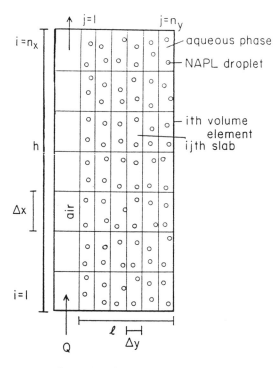

Figure 5.8 Geometry, notation, and mathematical partitioning of an SVE laboratory column, NAPL solution, and distributed aqueous diffusion model. [Reprinted from Wilson (1994) p. 583, by courtesy of Marcel Dekker, Inc.]

$\Delta x = h/n_x$

$A\Delta x$ = volume of a single volume element, cm³

Q = gas flow rate through column, mL/sec

ω = water-filled porosity of soil

σ = air-filled porosity of soil

l = average thickness of (stationary) soil water layer, cm

n_y = number of slabs into which the soil water layer is divided

y = l/n_y, thickness of one of the slabs into which the soil water layer is divided, cm

a_0 = average initial NAPL droplet diameter, cm

D = diffusivity of the VOC in soil water, cm²/sec

C_{sat} = solubility of VOC in water, gm/cm³

ϱ_{voc} = density of VOC in water, gm/cm³

ϱ_{soil} = density of bulk soil, gm/cm³

K_H = Henry's constant of VOC, dimensionless

m_{ij} = mass of MAPL in the jth slab of the ith volume element, gm

C_{ij}^w = dissolved VOC concentration in the jth slab of the ith volume element, gm/cm^3

C_i^g = VOC concentration in the gas phase in the ith volume element, gm/cm^3

It was shown in Chapter 4 that the rate of dissolution of an NAPL droplet into a spherical boundary layer surrounding it is given by Equation (173):

$$\frac{dm}{dt} = -4\pi D a_0 (C_{sat} - C_0)(m/m_a)^{1/3} \tag{173}$$

where m is the mass of the droplet at time t, m_a is the initial mass of the droplet, and C_0 is the VOC concentration at the periphery of the spherical boundary layer.

The initial amounts of VOC present in the vapor, aqueous, and NAPL phases are calculated as follows. Assume these initial concentrations are constant from volume element to volume element and that the aqueous and NAPL phase concentrations are constant from slab to slab within a volume element. Then

$$C_{tot} = \sigma C_0^g + \omega C_0^w + C_0^n \tag{174}$$

where C_0^g, C_0^w, and C_0^N, are the initial gaseous, aqueous, and NAPL concentrations, respectively. Assume that $C_0^N = 0$ and that the aqueous and gaseous phases are at equilibrium with each other with respect to VOC transport. Then, on using Henry's Law, it is easily shown that

$$C_0^w = \frac{C_{tot}}{\sigma K_H + \omega} \tag{175}$$

and

$$C_0^g = K_H C_0^w \tag{176}$$

If $C_0^w > C_{sat}$, however, set $C_0^w = C_{sat}$, $C_0^g = K_H C_{sat}$, and calculate C_0^N from

$$C_0^N = C_{tot} - (\sigma K_H + \omega) C_{sat} \tag{177}$$

The rate of change of NAPL mass is calculated as follows. The number of NAPL droplets in a volume element is given by n, where

$$n \cdot \frac{4\pi a_0^3 \varrho_{voc}}{3} = A\Delta x C_0^N$$

so

$$n = \frac{3A\Delta x C_0^N}{4\pi a_0^3 \varrho_{voc}} \qquad (178)$$

The number of NAPL droplets in a single slab in a volume element is then given by

$$n_s = n/n_y = \frac{3A\Delta x C_0^N}{4\pi a_0^3 \varrho_{voc} n_y} \qquad (179)$$

The initial NAPL mass in a single slab is

$$m_0 = A\Delta x C_0^N/n_y \qquad (180)$$

and the initial mass of a droplet is

$$m_d = 4\pi a_0^3 \varrho_{voc}/3 \qquad (181)$$

Finally, on using Equation (173), we find that the mass of NAPL in the jth slab of the ith volume element is governed by

$$\frac{dm_{ij}}{dt} = -\frac{3A\Delta x C_0^N D(C_{sat} - C_{ij}^w)(m_{ij}/m_0)^{1/3}}{a_0^2 \varrho_{voc} n_y} \qquad (182)$$

The changes in the aqueous VOC concentrations are calculated as follows. The volume of water in a single volume element is given by

$$V_w = \omega A\Delta x \qquad (183)$$

This water is assumed to be spread in a layer of thickness l. The areal extent of this volume of water is therefore given by

$$S_w = \omega A\Delta x/l \qquad (184)$$

which is also the area of the interface between any two adjacent slabs within the volume element into which the aqueous phase is partitioned and between which diffusion transport of VOC takes place.

A mass balance on the aqueous phase VOC in the jth slab of the ith volume element then yields after rearrangement

$$\frac{dC_{ij}^w}{dt} = \frac{n_y}{\omega A\Delta x}\left[\frac{S_w D}{\Delta y} \cdot (C_{i,j+1}^w - 2C_{i,j}^w + C_{i,j-1}^w) - \frac{dm_{ij}}{dt}\right] \qquad (185)$$

$$(j = 2, 3, \ldots, n_y - 1)$$

The first group of terms on the right-hand side of Equation (185) corresponds to diffusion transport of dissolved VOC from slab to slab; the last term represents mass transport to the aqueous phase from the dissolving NAPL droplets. For the slab adjacent to the solid medium we have

$$\frac{dC_{i,ny}^w}{dt} = \frac{n_y}{\omega A \Delta x}\left[\frac{S_w D}{\Delta y}\left(-C_{i,ny}^w + C_{i,ny-1}^w\right) - \frac{dm_{i,ny}}{dt}\right] \quad (186)$$

For the slab adjacent to the advecting gas phase, we assume that the aqueous VOC concentration at the air-water interface is given by Henry's Law, so

$$\frac{dC_{i1}^w}{dt} = \frac{n_y}{\omega A \Delta x}\left[\frac{S_w D}{\Delta y}\cdot\left(C_{i2}^w - C_1^w\right) + \frac{2S_w D}{\Delta y}\cdot\left(C_i^g/K_H - C_{i1}^w\right) - \frac{dm_{i1}}{dt}\right]$$

$$(187)$$

A mass balance on the gas phase VOC in the ith volume element yields

$$\frac{dC_i^g}{dt} = \frac{Q}{A \Delta x \sigma}\cdot\left(C_{i-1}^g - C_i^g\right) - \frac{2S_w D}{A \sigma \Delta x \Delta y}\cdot\left(C_i^g/K_H - C_{i1}^w\right) \quad (188)$$

The model then consists of Equations (182) and (185)–(188), together with the procedure for calculating the initial values of the vapor, aqueous, and NAPL phase VOC concentrations. The model parameters and concentrations are initialized and the differential equations are integrated forward in time in the usual way.

A FIELD-SCALE MODEL WITH DISTRIBUTED DIFFUSION AND NAPL

The Vanderbilt-Málaga group has developed several field-scale models for SVE by means of vertical wells or buried horizontal slotted pipes, which include solution of NAPL and distributed diffusion of VOC in soil water in low-permeability clay structures (Wilson et al., 1994a; Gómez-Lahoz et al., 1994a, 1994b; Rodríguez-Maroto et al., 1994; Wilson et al., 1994b). Here, what is probably the most physically realistic of these is presented. The configuration modeled is that of a single vertical well screened a short length near its bottom.

All these distributed diffusion models assume that VOC diffuses from water-saturated layers of finite thickness before it reaches the advecting soil gas and is removed. In one approach, the NAPL is present as droplets distributed throughout the water-saturated low-porosity layers (Gómez-Lahoz et al., 1994a, 1994b); in another, the NAPL is present as a film within the water-saturated lamellae (Rodríguez-Maroto et al., 1994). Both approaches could be made to yield rather similar results on suitable selection of the parameters in the models. The models easily produced the high initial VOC removal rates, the rapid declines in off-gas VOC concentration, and the lengthy plateaus and tailing observed experimentally.

The models seemed somewhat artificial, however. For the first model it was not clear how droplets of NAPL could migrate to or be formed in the interiors of the low-permeability domains. For the second, it was difficult to see how the postulated thin layer of NAPL was to be created deep within the low-permeability structures in the first place. This left these models of the diffusion process lacking an easily visualized physico-chemical basis. They had some meaning in terms of the least dimension of the low-permeability structures, and they produced reasonable results, but they also seemed somewhat contrived.

We avoid these conceptual difficulties if the model permits the NAPL to be present as droplets only in the mobile (air-filled) porosity and excludes it from the water-saturated low-permeability porous domains. We assume that VOC can migrate into these domains only by diffusion of dissolved VOC through the aqueous phase. We assume that, initially (at the time of the spill or sudden leak), the VOC is present only as vapor and NAPL, both in the air-filled porosity, and that subsequently the VOC diffuses into the water-saturated domains. In remediation, we therefore expect to see rapid removal of VOC initially as the NAPL droplets evaporate in the advecting gas stream, followed by a much slower rate of removal as VOC diffuses back out of the water-saturated domains.

The configuration of the single vertical SVE well is shown in Figure 5.7, along with much of the notation. The model for diffusion transport, together with notation, is shown in Figure 5.9. The development of an SVE model breaks down into three major parts: the calculation of the soil gas flow field in the vicinity of the vacuum well, the analysis of the equilibria and mass transport factors governing the release of the VOC being vapor stripped, and the combining of the two to form the model.

The Rate of Evaporation of Droplets of NAPL

We first look at evaporation of VOC from an NAPL droplet into the vapor phase (see Figure 5.10). The equation for steady-state diffusion from a spherical droplet is

$$\frac{1}{r^2} \cdot \frac{d}{dr}\left[r^2\frac{dC}{dr}\right] = 0 \qquad (189)$$

with boundary conditions

$$C(a) = C_{sat}^g \qquad (190)$$

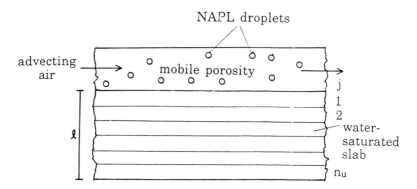

Figure 5.9 Model for diffusion transport of dissolved VOC into and from the aqueous phase in a low-permeability porous clay lens that is saturated with water. The slabs used to mathematically represent the lens are shown. [Reprinted from Wilson et al. (1994b) p. 1649, by courtesy of Marcel Dekker, Inc.]

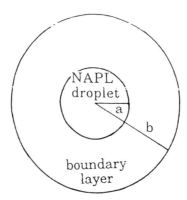

Figure 5.10 Evaporation of a single NAPL droplet. [Reprinted from Wilson et al. (1994b) p. 1649, by courtesy of Marcel Dekker, Inc.]

218

and

$$C(b) = C_0^g \qquad (191)$$

where C_{sat}^g is the saturation vapor concentration of the VOC and C_0^g is the VOC concentration at the outer surface ($r = b$) of the boundary layer surrounding the drop. Equation (189) integrates to

$$C(r) = c_1/r + c_2 \qquad (192)$$

Use of the boundary conditions then gives

$$C(r) = \frac{ab}{b - a}(C_{sat}^g - C_0^g)/r + c_2 \qquad (193)$$

from which

$$\frac{dC}{dr} = -\frac{ab}{b - a}(C_{sat}^g - C_0^g)/r^2 \qquad (194)$$

Fick's first law and Equation (194) then give for the mass m of the droplet

$$\frac{dm}{dt} = -\frac{4\pi D_g a(C_{sat}^g - C_0^g)}{1 - a/b} \qquad (195)$$

It is easily shown that

$$a = a_0(m/m_d)^{1/3} \qquad (196)$$

where m_d is the initial mass of the droplet, a_0 is its initial radius, and m and a are values at a later time t. So

$$\frac{dm}{dt} = -\frac{4\pi D_g a_0(C_{sat}^g - C_0^g)(m/m_d)^{1/3}}{1 - (a_0/b)(m/m_d)^{1/3}} \qquad (197)$$

A reasonable value for b, the boundary layer thickness around a droplet, is half the average distance between droplets. This is obtained as follows. Let the number of NAPL droplets in a volume element ΔV_{ij} be given by n. Then

$$n \cdot \frac{4\pi a_0^3 \varrho_{voc}}{3} = \Delta V_{ij} C_0^N \qquad (198)$$

and

$$n = \frac{3\Delta V_{ij} C_0^N}{4\pi a_0^3 \varrho_{voc}} \tag{199}$$

Let

σ = air-filled porosity
ω = water-filled porosity, assumed to be due only to the saturated clay lenses
ν_{clay} = total porosity (all water-filled) of the clay lenses

Then

$$\omega \Delta V_{ij} = \nu_{clay} f \Delta V_{ij}$$

where f is the fraction of the domain, which consists of clay lenses. So

$$f = \omega / \nu_{clay} \tag{200}$$

and $1 - f$ is the fraction of the domain that involves air-filled porosity. We therefore have n NAPL droplets distributed in a volume $(1 - f)\Delta V_{ij}$. The volume per droplet is therefore given by

$$(1 - f)\Delta V_{ij}/n = \frac{(1 - f)4\pi a_0^3 \varrho_{voc}}{3C_0^N} \tag{201}$$

The distance between droplets, $2b$, is just the cube root of this, and b, the thickness of the boundary layer, is then given by

$$b = a_0 \left[\frac{(\nu_{clay} - \omega)\pi \varrho_{voc}}{\nu_{clay} 6C_0^N}\right]^{1/3} \tag{202}$$

Rate of Change of NAPL Mass

Recall that the number of NAPL droplets in a volume element is given by Equation (199). The initial mass of a droplet, m_d, is

$$m_d = \frac{4\pi a_0^3 \varrho_{voc}}{3} \tag{203}$$

Finally, on using Equation (197), we find that the mass of NAPL in the ijth volume element is governed by

$$\frac{dm_{ij}}{dt} = -\frac{3\Delta V_{ij} C_0^N D_g (C_{sat}^g - C_{ij}^g)(m_{ij}/m_0)^{1/3}}{a_0^2 \varrho_{voc}[1 - (a_0/b)(m_{ij}/m_0)^{1/3}]} \tag{204}$$

Gas Flow Field

We assume a porous medium of constant, isotropic permeability and so may use the method of images from electrostatics (Smythe, 1953) for calculating the soil gas pressures near the SVE well, as described in detail in Chapter 3. Let

h = thickness of porous medium (depth to water table), m
r_{max} = radius of domain of interest, m
r_w = radius of gravel packing of the well, m
P_w = wellhead gas pressure (<1 atm), atm
P_a = ambient pressure, atm
$P(r,z)$ = soil gas pressure at the point (r,z), atm
K_D = Darcy's constant, m²/atm sec
a = distance of well above the water table, m
Q = molar gas flow rate to well, mol/sec
q = standard volumetric gas flow rate to well, m³/sec
v_r = r-component of superficial velocity, m/sec (m³/m² sec)
v_z = z-component of superficial velocity, m/sec (m³/m² sec)
R = gas constant, 8.206×10^{-5} m³ atm/mol deg
T = temperature, deg K

The analysis in Chapter 3 yields a velocity potential W such that

$$W(r,z) + P_a^2 = P^2(r,z) \tag{205}$$

W is given by

$$W = A \sum_{n=-\infty}^{\infty} \left[-\frac{1}{\{r^2 + [z - 4nh - a]^2\}^{1/2}} \right.$$

$$- \frac{1}{\{r^2 + [z - 4nh + a]^2\}^{1/2}}$$

$$+ \frac{1}{\{r^2 + [z - (4n - 2)h - a]^2\}^{1/2}}$$

$$\left. + \frac{1}{\{r^2 + [z - 4n - 2)h + a]^2\}^{1/2}} \right] \tag{206}$$

The constant A is given by

$$A = (P_w^2 - P_a^2)/S \tag{207}$$

where

$$S = \sum_{n=-\infty}^{\infty} \left[-\frac{1}{|r_w - 4nh|} - \frac{1}{|2a + r_w - 4nh|} \right.$$

$$\left. + \frac{1}{\{r_w - (4n - 2)h|} + \frac{1}{|2a + r_w - (4n - 2)h|} \right] \qquad (208)$$

The Darcy's constant for the medium is given in terms of the wellhead pressure and the volumetric flow rate by

$$K_D = \frac{qS}{2\pi(P_w^2 - P_a^2)} \qquad (209)$$

The superficial velocity of the gas is given by

$$v = -K_D \nabla P \qquad (210)$$

where the components of ∇W are $\partial W / \partial r$ and $\partial W / \partial z$, and the velocity components are v_r and v_z.

**Volume Elements and Surfaces of Volume Elements:
Advective Mass Balance**

See Figure 5.9. The volume of the annular volume element is given by

$$\Delta V_{ij} = (2i - 1)\pi(\Delta r)^2 \Delta z \qquad (211)$$

The surfaces of this volume element are as follows:

$$\text{Inner} \quad S_{ij}^I = 2(i - 1)\pi \Delta r \Delta z \qquad (212)$$

$$\text{Outer} \quad S_{ij}^O = 2i\pi \Delta r \Delta z \qquad (213)$$

$$\text{Top and Bottom} \quad S_{ij}^T = S_{ij}^B = (2i - 1)\pi(\Delta r)^2 \qquad (214)$$

The advective mass balance for VOC in this volume element is then

$$\Delta V_{ij} \left[\frac{dC_{ij}^g}{dt} \right]_{adv} = S_{ij}^I v_{ij}^I [S(v^I) C_{i-1,j}^g + S(-v^I) C_{ij}^g]$$

$$+ S_{ij}^O v_{ij}^O [-S(-v^O) C_{i+1,j}^g - S(v^O) C_{ij}^g]$$

$$+ S_{ij}^B v_{ij}^B [S(v^B) C_{i,j-1}^g + S(-v^B) C_{ij}^g]$$

$$+ S_{ij}^T v_{ij}^T [-S(-v^T) C_{i,j+1}^g - S(v^T) C_{ij}^g] \qquad (215)$$

where

$$v_{ij}^I = v_r[(i - 1)\Delta r, (j - 1/2)\Delta z] \tag{216}$$

$$v_{ij}^O = v_r[i\Delta r, (j - 1/2)\Delta z] \tag{217}$$

$$v_{ij}^B = v_z[(i - 1/2)\Delta r, (j - 1)\Delta z] \tag{218}$$

$$v_{ij}^T = v_z[(i - 1/2)\Delta r, j\Delta z] \tag{219}$$

and the function $S(v)$ is zero if $v \le 0$ and one if $v > 0$.

Initial Distribution of VOC among the Phases

This question is addressed as follows. The physical picture is that at time zero a spill occurs and that the NAPL rapidly flows through the vadose zone, leaving residual NAPL droplets and vapor phase VOC, but that there is insufficient time for diffusion of VOC into the water-saturated clay lenses to take place to any extent during this period. We may therefore set the initial aqueous VOC concentration $C_0^w = 0$. Let us assume that the initial VOC concentrations in the gas and NAPL phases are constant from volume element to volume element. Then

$$C_{tot} = \sigma C_0^g + C_0^N \tag{220}$$

where C_0^g and C_0^N are the initial gaseous and NAPL concentrations, respectively. If $C_{tot} < \sigma C_{sat}^g$, where C_{sat}^g is the saturation vapor concentration of the VOC, then

$$C_0^g = C_{tot}/\sigma \tag{221}$$

and

$$C_0^N = 0 \tag{222}$$

If $C_{tot} > \sigma C_{sat}^g$, then

$$C_0^g = C_{sat}^g \tag{223}$$

and

$$C_0^N = C_{tot} - \sigma C_{sat}^g \tag{224}$$

Change in Aqueous VOC Concentration

Let us assume that the clay lenses in which diffusion is taking place are of thickness $2l$, and that they contain the great bulk of the water in the soil (see Figure 5.10). Then the volume of water in a volume element can be written as

$$V_w = \omega \Delta V_{ij} = 2lA_{ij}\nu_{clay} \tag{225}$$

where

A_{ij} = total cross-sectional area of saturated clay lenses in the volume element, m²
ν_{clay} = porosity of the clay

Then

$$A_{ij} = \frac{\omega \Delta V_{ij}}{2l\nu_{clay}} \tag{226}$$

and the total area of lenses from which VOC may diffuse (counting top halves and bottom halves separately) is

$$2A = \frac{\omega \Delta V_{ij}}{l\nu_{clay}} \tag{227}$$

This is also the area of the interface between any two adjacent slabs within the volume element into which the aqueous phase is partitioned and between which diffusion transport of VOC may take place.

A mass balance on the aqueous phase VOC in the kth slab of the ijth volume element then yields

$$\frac{\omega \Delta V_{ij}}{n_u} \cdot \frac{dC^w_{ijk}}{dt} = \frac{\Delta V_{ij}\omega}{l\nu_{clay}} \cdot \frac{D}{\Delta u}(C^w_{ijk+1} - 2C^w_{ijk} + C^w_{ijk-1}) \tag{228}$$

or

$$\frac{dC^w_{ijk}}{dt} = \frac{D}{(\Delta u)^2 \nu_{clay}} \cdot (C^w_{ijk+1} - 2C^w_{ijk} + C^w_{ijk-1}) \tag{229}$$

For the innermost slab ($k = n_u$, on either side of the center plane of the lenses), we have

$$\frac{dC^w_{ijnu}}{dt} = \frac{D}{(\Delta u)^2 \nu_{clay}} \cdot (-C^w_{ijnu} + C^w_{ijnu-1}) \tag{230}$$

For the slab adjacent to the advecting gas phase we assume that the aqueous VOC concentration at the air-water interface is given by Henry's Law, so

$$\frac{dC_{ij1}^w}{dt} = \frac{D}{(\Delta u)^2 \, \nu_{clay}} \cdot [C_{ij2}^w - C_{ij1}^w + 2(C_{ij}^g/K_H - C_{ij1}^w)] \quad (231)$$

Completion of Gas Phase VOC Material Balance: The Model

We return to Equation (215) for the vapor phase advection terms, to which we must adjoin a term corresponding to mass transport of VOC by diffusion to or from the outermost aqueous slab and a term corresponding to vaporization of VOC from the NAPL droplets. The first term is given by

$$\sigma \Delta V_{ij} \left[\frac{dC_{ij}^g}{dt} \right]_{diff} = \frac{\omega \Delta V_{ij}}{l \nu_{clay}} \cdot \frac{D}{(\Delta u/2)} \cdot [C_{ij1}^w - C_{ij}^g/K_H] \quad (232)$$

or

$$\left[\frac{dC_{ij}^g}{dt} \right]_{diff} = \frac{\omega D}{\sigma l \nu_{clay}(\Delta u/2)} \cdot [C_{ij1}^w - C_{ij}^g/K_H] \quad (233)$$

The second term (corresponding to evaporation from NAPL droplets) is

$$\sigma \Delta V_{ij} \left[\frac{dC_{ij}^g}{dt} \right]_{evap} = -\frac{dm_{ij}}{dt} \quad (234)$$

where dm_{ij}/dt is given by Equation (197).

The complete equation is therefore given by

$$\frac{dC_{ij}^g}{dt} = \frac{S_{ij}^I v_{ij}^I}{\sigma \Delta V_{ij}} \cdot [S(v^I) C_{i-1,j}^g + S(-v^I) C_{ij}^g]$$

$$+ \frac{S_{ij}^O v_{ij}^O}{\sigma \Delta V_{ij}} \cdot [-S(-v^O) C_{i+1,j}^g - S(v^O) C_{ij}^g]$$

$$+ \frac{S_{ij}^B v_{ij}^B}{\sigma \Delta V_{ij}} \cdot [S(v^B) C_{i,j-1}^g + S(-v^B) C_{ij}^g]$$

$$+ \frac{S_{ij}^T v_{ij}^T}{\sigma \Delta V_{ij}} \cdot [-S(-v^T) C_{i,j+1}^g - S(v^T) C_{ij}^g]$$

$$+ \frac{D}{\sigma l \nu_{clay}(\Delta u/2)} \cdot (C_{ij1}^w - C_{ij}^g/K_H)$$

$$- (1/\sigma \Delta v_{ij}) \frac{dm_{ij}}{dt} \quad (235)$$

Equations (197), (229), (230), (231), and (235) then constitute the model.

The mass of residual VOC at any time during the course of a simulation is given by

$$M_{tot} = \sum_{i=1}^{n_r}\sum_{j=1}^{n_z}\left[\Delta V_{ij}\sigma C_{ij}^g + m_{ij} + \sum_{k=1}^{n_u}\frac{\omega\Delta V_{ij}}{n_u}\cdot C_{ijk}^w\right] \quad (236)$$

The effluent soil gas concentration is given by Equation (237).

$$C_{eff}^g = -\frac{M_{tot}(t + \Delta t) - M_{tot}(t)}{q(t)\Delta t} \quad (237)$$

An alternative approach to C_{eff}^g is to define it as follows. Let ΔV_{1J} be the volume element containing the well. Then

$$C_{eff}^g = \frac{S_{1J}^T|v_{1J}^T|C_{1,J+1}^g + S_{1J}^O|v_{1J}^O|C_{2J}^g + S_{1J}^B|v_{1J}^B|C_{1,J-1}^g}{S_{1J}^T|v_{1J}^T| + S_{1J}^O|v_{1J}^O| + S_{1J}^B|v_{1J}^B|} \quad (238)$$

CONCLUSIONS

The models for SVE developed in this chapter range from simple local equilibrium models with a linear adsorption isotherm for the VOC to models that include NAPL and the diffusion of VOC through water-saturated layers of the porous medium. They show a progressive increase in sophistication, complexity, and, unfortunately, computational demands as they become more realistic. All, however, can be run on currently available (1993) microcomputers. The techniques illustrated here should also be useful in the development of future SVE models, in which more elaborate adsorption isotherms for the VOCs are used and in which the features of various models presented here (Raoult's Law mixtures, NAPL, distributed diffusion, variable permeabilities, etc.) are combined.

In the next chapter, results from these models will be used to obtain clearer insight into the various factors affecting SVE, to see how some of the problems arising in SVE operations can be mitigated, and to develop an intuition for SVE system design.

REFERENCES

Cho, J. S., 1991, "Forced Air Ventilation for Remediation of Unsaturated Soils Contaminated by VOC," U.S. EPA Report No. EPA/600/2-91/016, July.

DiGiulio, D. C., J. S. Cho, R. R. Dupont, and M. W. Kemblowski, 1990, "Conduction Field Tests for Evaluation of Soil Vacuum Extraction Application," *Proc., 4th Natl. Outdoor Action Conf. on Aquifer Restoration, Ground Water Monitoring and Geophysical Methods,* May 14–17, Las Vegas, p. 587.

Gómez-Lahoz, C., J. M. Rodríguez-Maroto, and D. J. Wilson, 1994a, "Soil Cleanup by in situ Aeration. XVII. Field Scale Model with Distributed Diffusion," *Separ. Sci. Technol.,* 29:1251.

Gómez-Lahoz, C., J. M. Rodríguez-Maroto, D. J. Wilson, and K. Tamamushi, 1994b, "Soil Cleanup by in situ Aeration. XV. Effects of Variable Air Flow Rates in Diffusion-Limited Operation," *Separ. Sci. Technol.,* 29:943.

Gómez-Lahoz, C., R. A. García Delgado, F. García-Herruzo, J. M Rodríguez-Maroto, and D. J. Wilson, 1993, "Extracción a Vacio de Contaminantes Orgánicos del Suelo. Fenómenos de No-Equilibrio," *III Congreso do Ingeniería Ambiental, Proma '93,* Bilbao, Spain.

Hoag, G. E. and B. L. Cliff, 1985, "The Use of the Soil Venting Technique for the Remediation of Petroleum-Contaminated Soils," in *Soils Contaminated by Petroleum: Environmental and Public Health Effects,* E. J. Calabrese and P. T. Kostechi, eds., Wiley, New York.

Johnson, P. C., M. W. Kemblowski, and J. D. Colthart, 1989a, "Practical Screening Models for Soil Venting Applications," presented at the *Workshop on Soil Vacuum Extraction,* April 27–29, U.S. EPA Robert S. Kerr Environmental Research Laboratory, Ada, OK.

Johnson, P. C., M. W. Kemblowski, and J. D. Colthart, 1990, "Quantitative Analysis for the Cleanup of Hydrocarbon-Contaminated Soils by in situ Soil Venting," *Ground Water,* 28:413.

Johnson, P. C., M. W. Kemblowski, J. D. Colthart, D. L. Byers, and C. C. Stanley, 1989b, "A Practical Approach to the Design, Operation, and Monitoring of in-situ Soil Venting Systems," presented at the *Soil Vapor Extraction Technology Workshop,* June 28–29, U.S. EPA Risk Reduction Engineering Laboratory, Edison, NJ.

Jury, W. A. and R. L. Valentine, 1986, "Transport Mechanisms and Loss Pathways for Chemical in Soils," in *Vadose Zone Modeling of Organic Pollutants,* S. C. Hern and S. M. Melancon, eds., Lewis Publishers, Chelsea, MI, Chapter 2.

Kayano, S. and D. J. Wilson, 1992, "Soil Cleanup by in situ Aeration. X. Vapor Stripping of Mixtures of Volatile Organics Obeying Raoult's Law," *Separ. Sci. Technol.,* 27:1525.

Kayano, S. and D. J. Wilson, 1993, "Migration of Pollutants in Ground-Water. VI. Flushing of DNAPL Droplets/Ganglia," *Environ. Monitor. Assess.,* 25:193.

Laidler, K. J., 1965, *Chemical Kinetics,* 2nd ed., McGraw-Hill, Inc. New York, NY, p. 327.

Levine, I. N., 1988, *Physical Chemistry,* 3rd ed., McGraw-Hill, Inc., New York, NY, p. 532.

Marley, M. C., 1991, "Development and Application of a Three-Dimensional Air Flow Model in the Design of a Vapor Extraction System," presented at the *Symposium on Soil Venting,* April 29–May 1, RSKERL, Ada, OK.

Marley, M. C., S. D. Richter, B. L. Cliff, and P. E. Nangeroni, 1989, "Design of Soil Vapor Extraction Systems—A Scientific Approach," presented at the *Soil Vapor Extraction Technology Workshop,* June 28–29, U.S. EPA RREL, Edison, NJ.

Millington, R. J. and J. M. Quirk, 1961, "Permeability of Porous Solids," *Trans Faraday Soc.,* 57:1200.

Oma, K. H., D. J. Wilson, and R. D. Mutch, 1990, "In situ Vapor Stripping: The Importance of Nonequilibrium Effects in Predicting Cleanup Time and Cost," *Proc., Hazardous Materials Management Conferences and Exhibition/International,* June 5–7, Atlantic City, NJ.

Osejo, R. E. and D. J. Wilson, 1991, "Soil Cleanup by in situ Aeration. IX. Diffusion Constants of Volatile Organics and Removal of Underlying Liquid," *Separ. Sci. Technol.,* 26:1433.

Roberts, L. A. and D. J. Wilson, 1993a, "Groundwater Cleanup by in situ Sparging. III. Modeling of Dense Nonaqueous Phase Liquid Droplet Removal," *Separ. Sci. Technol.,* 28:1127.

Roberts, L. A. and D. J. Wilson, 1993b, "Soil Cleanup by in situ Aeration. XI. Cleanup Time Distributions for Statistically Equivalent Permeabilities," *Separ. Sci. Technol.,* 28:1539.

Rodríguez-Maroto, J. M. and D. J. Wilson, 1991, "Soil Cleanup by in situ Aeration. VII. High-Speed Modeling of Diffusion Kinetics," *Separ. Sci. Technol.,* 25:743.

Rodríguez-Maroto, J. M., C. Gómez-Lahoz, and D. J. Wilson, 1994,"Soil Cleanup by in situ Aeration. XVIII. Field Scale Models with Diffusion from Clay Structures," *Separ. Sci. Technol.,* 29:1367.

Scheidegger, A. E., 1974, *The Physics of Flow through Porous Media,* 3rd ed., University of Toronto Press, p. 306.

Smythe, W. R., 1953, *Static and Dynamic Electricity,* 2nd ed., McGraw-Hill, New York, NY.

Thibodeaux, L. J., K. T. Valsaraj, C. Springer, and G. Hildebrand, 1988, "Mathematical Models for Predicting Chemical Vapor Emissions from Landfills," *J. Hazardous Materials,* 19:101.

Wilson, D. J., 1990, "Soil Cleanup by in situ Aeration. V. Vapor Stripping from Fractured Bedrock," *Separ. Sci. Technol.,* 25:243.

Wilson, D. J., 1994, "Soil Cleanup by in situ Aeration. XIII. Effects of Solution Rates and Diffusion in Mass Transport-Limited Operation," *Separ. Sci. Technol.,* 29:579.

Wilson, D. J., A. N. Clarke, and J. H. Clarke, 1988, "Soil Cleanup by in situ Aeration. I. Mathematical Modeling," *Separ. Sci. Technol.,* 23:991.

Wilson, D. J., C. Gómez-Lahoz, and J. M. Rodríguez-Maroto, 1992, "Mathematical Modeling of SVE: Effects of Diffusion Kinetics and Variable Permeabilities," *Proc., Symposium on Soil Venting,* April 29–May 1, Houston, TX, U.S. EPA Report No. EPA/600/R-92/174, Sept.

Wilson, D. J., C. Gómez-Lahoz, and J. M. Rodríguez-Maroto, 1994a, "Soil Cleanup by in situ Aeration. XVI. Solution and Diffusion in Mass Transport-Limited Operation and Calculation of Darcy's Constants," *Separ. Sci. Technol.,* 29:1133.

Wilson, D. J., J. M. Rodríguez-Maroto, and C. Gómez-Lahoz, 1994b, "Soil Cleanup by in situ Aeration. XIX. Effects of Spill Age on Soil Vapor Extraction Remediation Rates," *Separ. Sci. Technol.,* 29:1645.

CHAPTER 6

Soil Vapor Extraction Modeling — Results

INTRODUCTION

IN this chapter we explore the results of calculations carried out with the mathematical models described in Chapter 5. The reader whose enthusiasm for differential equations and computer programming is limited will find that most of this chapter is much easier going than Chapter 5, and the material in Chapter 5 is not prerequisite to a physical understanding of the points to be discussed here. In this chapter we shall mostly be looking for trends, developing an intuition for SVE, and seeing what the models can tell us, rather than immersing ourselves in the theoretical and computational details of mathematical modeling.

With simple local equilibrium models it is possible to scale the results of calculations to ascertain the effects of changes in several of the model parameters. Cleanup times (90, 99, 99.9%) depend on these parameters, as indicated in Table 6.1. Some of these dependences are linked. Gas flow rate is proportional to pneumatic permeability, wellhead vacuum, and packed radius of a well, for example. Table 6.2 gives data showing the interdependence of well packing radius, permeability, and molar airflow rate. Also, while the 99% cleanup time is independent of contaminant concentration with this model, a domain having a very high contaminant concentration will require a higher percent cleanup. Note, also, that the conclusions summarized in Table 6.1 pertain only to the linear isotherm, local equilibrium model.

Usually, the major portion of the flow resistance to the soil gas moving to the well occurs in the near vicinity of the well, where pressure gradients are large and gas linear velocities are high. If the permeability of the soil is low, one can compensate, within limits, by increasing the radius of the gravel packing surrounding the screened section of the well. If one must screen a well in a domain of low permeability, one can increase the air flow by this means.

229

TABLE 6.1. Dependence of Cleanup Times on Local
Equilibrium Model Parameters.

Parameter	Dependence of Cleanup Time
Linear adsorption isotherm parameter	Inverse
Pneumatic permeability	Inverse
Gas flow rate	Inverse
Wellhead vacuum	Inverse
Contaminant concentration	Independent
Gravel packed radius of well	Inverse

Excessive distances between wells can have a very damaging impact on the performance of an SVE system, increasing cleanup times and total operating costs drastically. On the other hand, capital expenses are excessive if the wells are drilled closer together than is necessary. Figure 6.1 shows the effect of well spacing on the rate of cleanup. If multiple wells are placed on a square grid with a distance a between wells, the desired effective radius of a well's domain of influence is about $0.707a$. If the wells are on a hexagonal grid, each with six nearest neighbors, the desired effective radius of a well's domain of influence is about $0.577a$. If the distribution of the contaminant is fairly uniform, the wells are drilled through the contaminated zone, and the soil permeability is isotropic and constant, cleanup times start to increase rapidly as the effective radius of the domain of influence is increased above about 1 to 1.5 times the well depth, provided that the well is screened only near the bottom and that cleanup is advection-limited.

TABLE 6.2. Effects of Well Radius and Permeability
on Gas Flow Rate.[a]

Well Radius (m)	Permeability (m^2/atm sec)	Molar Airflow Rate (mol/sec)
0.12	0.6	0.9076
0.24	0.3	0.8905
0.24	0.6	1.7864
0.12	0.3	0.4543
0.36	0.2	0.8999
0.72	0.1	0.8760

[a]Well depth 15 m; overlying impermeable cap radius 15 m; radius of zone of influence 20 m; well is screened 3 m above the water table; $T = 25°C$; porosity 0.2; wellhead pressure 0.866 atm.

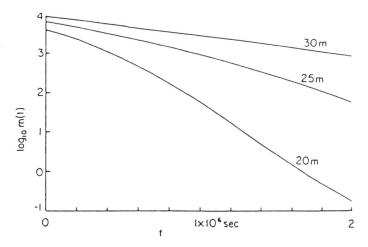

Figure 6.1 Plots of $\log_{10}$ total contaminant mass versus time, showing the effects of well spacing on the rate of removal. No cap was used in these runs. Depth of water table, 20 m; depth of well, 17 m; radius of zone of influence, 20, 25, and 30 m as indicated; screened radius of well, 0.12 m; molar gas flow rate, 9.1 mol/sec; gas-filled porosity, 0.2; specific moisture content, 0.2; effective Henry's constant, 0.01; wellhead pressure, 0.866 atm; permeability, 6.0 m²/atm sec; soil density, 1.6 gm/cm³; initial VOC concentration, 100 mg/kg. [Reprinted from Gannon, et al. (1989) p. 831, by courtesy of Marcel Dekker, Inc.]

Wells should also be drilled to at least the bottom of the zone of contamination if this lies in the vadose zone or down to the capillary fringe if the entire vadose zone is contaminated. The impact of well depth on cleanup time is shown for a simple local equilibrium model in Figure 6.2; effects should be qualitatively the same for more complex models, too. In these runs the wells were drilled to within 3, 6, and 9 m of the water table, which was at a depth of 15 m. These results also indicate that well efficiency is probably maximized if the well is screened only fairly near the bottom. Air that is drawn into the upper portion of a long screened section is being drawn through soil that lies near the axis of the domain of influence of the well. This region will be cleaned up most rapidly in any case, so this portion of the air flow only serves to increase the quantity of air that must be pumped and the amount of air and soil water that must be handled by the vapor treatment system. The screened well section must be long enough, however, to handle expected fluctuations in the height of the water table.

The effect of well depth on cleanup rate with a horizontal slotted pipe configuration was studied by Rodríguez-Maroto et al. (1991). The standard parameter set used is given in Table 6.3. Plots of $\log_{10}$ total residual contaminant mass versus time for runs made with the horizontal slotted pipe

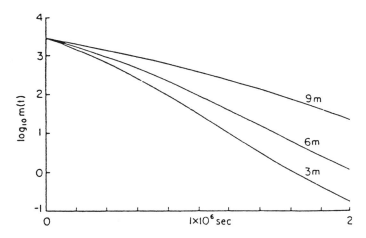

Figure 6.2 Plots of $\log_{10}$ total contaminant mass versus time, effect of well depth. Depth of water table, 15 m; radius of zone of influence of well, 20 m; height of well above water table, 3, 6, and 9 m. Other parameters as in Figure 6.1. [Reprinted from Gannon et al. (1989) p. 831, by courtesy of Marcel Dekker, Inc.]

at various distances above the water table are shown in Figure 6.3. As with the vertical wells, the impact of an excessively shallow well on the rate of cleanup is very damaging.

In the calculation of the gas flow field by numerical relaxation, it is easy to modify the boundary condition at the surface of the soil to model the presence of an impermeable circular cap; one simply replaces the requirement that $P = 1$ atm at the surface by the no-flow boundary condition $\partial P/\partial z = 0$ on that part of the boundary where the cap is to be located.

TABLE 6.3. **Standard Parameter Set for Simulations of SVE with a Horizontal Slotted Pipe.**

Parameter	Value
Domain length	13 m
Domain depth	8 m
Δx, Δy	1 m
Packed radius of well	0.1 m
Wellhead pressure	0.75 atm
Temperature	14°C
Soil porosity	0.375
Specific moisture content	0.2
K_x, K_y	0.100 m²/atm sec
Initial soil contaminant concentration	100 mg/kg
Soil density	1.7 gm/cm³
Effective Henry's constant	0.005

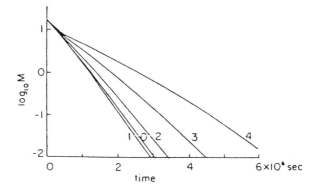

Figure 6.3 Log_{10} residual contaminant mass versus time; effect of height of horizontal slotted pipe above the water table. Numbers beside the curves indicate distance of the pipe above the water table. [Reprinted from Rodríguez-Maroto et al. (1991) p. 1054, by courtesy of Marcel Dekker, Inc.]

The presence of an impermeable cap influences the flow pattern of the soil gas in the vicinity of a vacuum well, tending to increase gas flow out near the periphery of the domain of influence.

For a system in which passive vent wells are not present and the vacuum well is one in an array of wells (so that $\partial P/\partial r = 0$ at the periphery of the domain of influence), the effects of impermeable caps of various sizes are shown in Figure 6.4. An overlying impermeable cap coaxial with the well

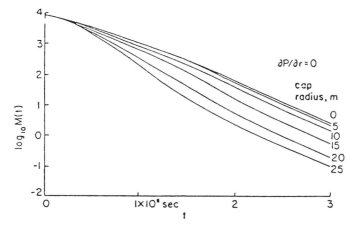

Figure 6.4 Log plots of total contaminant mass versus time; effect of overlying impermeable caps on SVE. Permeability, 0.1 m²/atm sec; radius of influence of well, 30 m; molar gas flow rate, 0.1744 mol/sec; (volumetric flow rate 0.00409 m³/sec). Other parameters as in Figure 6.1. A no-flow boundary condition is used at the periphery of the domain of influence. [Reprinted from Gannon et al. (1989) p. 831, by courtesy of Marcel Dekker, Inc.]

is seen to increase the rate of cleanup by as much as 50%; however, the cost and nuisance of the cap may make this a questionable bargain.

On the other hand, the results do indicate that SVE cleanups underneath parking lots, streets, building floors, etc., may actually be accelerated by the presence of the overlying barrier. When such a barrier is present, the vacuum well should be placed as near the center of the overlying barrier as possible. If multiple wells are used and are operated at the same vacuum, the gas pressure gradient in the region between the wells will be quite small, and cleanup of this domain of stagnation will be extremely slow. If a large gas flow is needed, it should be achieved in this situation either by the use of a single large-bore well with a long gravel packing of large radius or by the placement of passive wells between the vacuum wells to prevent the occurrence of zones of stagnation. Note that the previous stricture against long well screens does not apply to the situation where the well is drilled through a large overlying impermeable layer.

Passive wells may produce either positive or negative results. Figure 6.5 shows the results of SVE runs for systems identical to those illustrated in Figure 6.4 except that the vacuum wells in Figure 6.5 are surrounded by passive wells screened along their entire lengths. In the absence of an overlying cap, the passive wells actually appear to decrease the efficiency of this system, at least near the end of the cleanup, as seen in Figure 6.6. If a 25-m–radius cap is present, on the other hand, the presence of passive wells in this system results in some improvement in cleanup rate, as seen

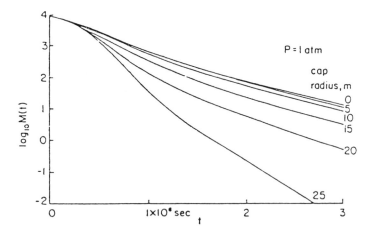

Figure 6.5 Log plots of total contaminant mass versus time; effect of impermeable cap radius on SVE. The system parameters are as in Figure 6.4, except that the boundary condition at the periphery of the domain of influence is $P = 1$ atm, corresponding to the presence of passive vent wells around the periphery. [Reprinted from Wilson (1990) p. 243, by courtesy of Marcel Dekker, Inc.]

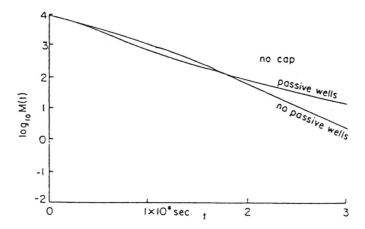

Figure 6.6 Comparison of SVE systems with and without passive vent wells. The system parameters are as in Figure 6.4. No impermeable cap is present in these runs. [Reprinted from Wilson (1990) p. 243, by courtesy of Marcel Dekker, Inc.]

in Figure 6.7. In general, if it is evident that the use of one or more passive wells will reduce or eliminate zones in which the soil gas pressure gradients are small, without resulting in excessive short-circuiting of gas from other regions, passive wells will be beneficial. Otherwise, they are likely to have relatively little effect.

The effects of overlying strip caps and passive wells, alone and in com-

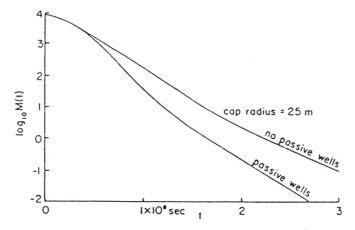

Figure 6.7 Comparison of SVE systems with and without passive vent wells. The system parameters are as in Figure 6.4. An impermeable cap of 25 m radius is present in these runs. [Reprinted from Wilson (1990) p. 243, by courtesy of Marcel Dekker, Inc.]

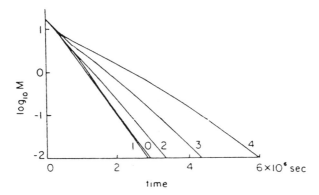

Figure 6.8 Plots of residual contaminant mass versus time, horizontal slotted pipe configuration; effect of an overlying impermeable strip cap of 5 m width. The standard parameter set is given in Table 6.3. The numbers indicate height of well above the water table. [Reprinted from Rodríguez-Maroto et al. (1991) p. 1051, by courtesy of Marcel Dekker, Inc.]

bination, on an array of horizontal slotted pipe SVE wells was explored by Rodríguez-Maroto et al. (1991). Figures 6.8–6.14 exhibit SVE runs that are identical to those shown in Figure 6.3, except that they include the effects of passive wells and/or strip caps of various sizes. In Figure 6.8 we see that an overlying strip cap of 5 m width causes very little increase in cleanup rate above that found for the runs in Figure 6.3. Increasing the cap width to 9 m yields the results seen in Figure 6.9; these are actually a little worse than those obtained with a 5-m cap (Figure 6.8). Evidently, for an array of fairly closely spaced horizontal slotted pipes, overlying caps would be a

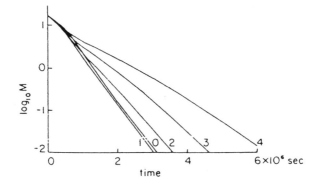

Figure 6.9 Plots of residual contaminant mass versus time, horizontal slotted pipe configuration; effect of an overlying impermeable strip cap of 9 m width. The standard parameter set is given in Table 6.3. [Reprinted from Rodríguez-Maroto et al. (1991) p. 1051, by courtesy of Marcel Dekker, Inc.]

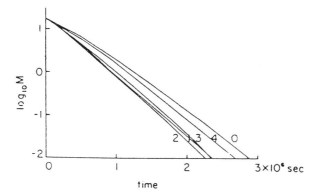

Figure 6.10 Plots of residual contaminant mass versus time, horizontal slotted pipe configuration; effect of passive horizontal vent pipes along the lower borders of the domain of influence. The numbers by the curves indicate the height of the vacuum well above the water table in meters. No cap is present. Parameters as in Table 6.3. [Reprinted from Rodríguez-Maroto et al. (1991) p. 1051, by courtesy of Marcel Dekker, Inc.]

waste of time and money. In the runs shown in Figure 6.10, no caps are present, and the lower borders of the domain of influence contain passive horizontal vent pipes. The cleanup rates of the best runs are slightly improved as compared to Figure 6.3 but probably not enough to warrant the extra expense of installing the passive wells.

Figures 6.11, 6.12, and 6.13 show the effects of combined strip caps of 5, 9, and 11 m width, respectively, with passive horizontal wells in the lower

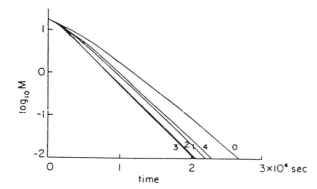

Figure 6.11 Plots of residual contaminant mass versus time, horizontal slotted pipe configuration. Passive horizontal vent pipes are located at the lower corners of the domain, and an overlying impermeable cap of 5 m width is present. Numbers by the curves indicate height of the vacuum pipe above the water table in meters. See Table 6.3. [Reprinted from Rodríguez-Maroto et al. (1991) p. 1051, by courtesy of Marcel Dekker, Inc.]

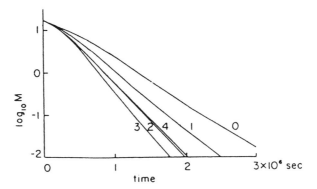

Figure 6.12 Plots of residual contaminant mass versus time, horizontal slotted pipe configuration. Passive horizontal vent pipes are located at the lower corners of the domain, and an overlying impermeable cap of 9 m width is present. Numbers by the curves indicate height of the vacuum pipe above the water table in meters. See Table 6.3. [Reprinted from Rodríguez-Maroto et al. (1991) p. 1051, by courtesy of Marcel Dekker, Inc.]

corners of the domain. Again, the results are not particularly impressive, and one would probably not wish to incur the extra expense for such minor improvements. One might wish to cap the entire domain of interest and rely only on passive wells for the air supply; this might be to avoid water infiltration or possibly to reduce exposure of the public to VOC vapors. The results of modeling such configurations with the horizontal well at various depths (Figure 6.14) indicate very clearly the advantages of placing

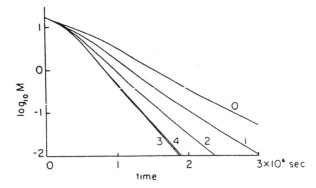

Figure 6.13 Plots of residual contaminant mass versus time, horizontal slotted pipe configuration. Passive horizontal vent pipes are located at the lower corners of the domain, and an overlying impermeable cap of 11 m width is present. Numbers by the curves indicate height of the vacuum pipe above the water table in meters. See Table 6.3. [Reprinted from Rodríguez-Maroto et al. (1991) p. 1051, by courtesy of Marcel Dekker, Inc.]

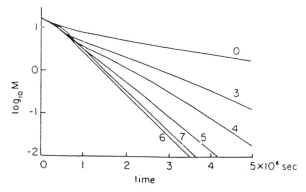

Figure 6.14 Plots of residual contaminant mass versus time, horizontal slotted pipe configuration. Passive horizontal vent pipes are located at the lower corners of the domain, and an overlying impermeable cap covering the entire domain, is present. Numbers by the curves indicate height of the vacuum pipe above the water table in meters. See Table 6.3. [Reprinted from Rodríguez-Maroto et al. (1991) p. 1051, by courtesy of Marcel Dekker, Inc.]

the well shallow in this situation, contrary to what one normally expects to be the optimum placement.

One situation in which passive wells may be quite useful is when one is attempting to isolate a well-defined region of the contaminated zone to study diffusion/desorption kinetics, as suggested by DiGiulio et al. (1990).

The presence of soil strata of markedly differing permeabilities or with permeabilities of high anisotropy (typically $K_{vertical} \ll K_{horizontal}$) can profoundly affect the soil gas pressure distribution around an SVE well and may drastically increase the effective range of influence of the well. Soil gas pressure measurements made during pilot studies in the vicinity of an SVE well in Toms River, New Jersey, showed that the influence of this vacuum well extended far beyond what one might normally expect. Well log data indicated that the top meter of soil was clay and till fill, underlain by mostly sand and gravel. Assignment of a permeability to the overlying layer which was 1/200 of the permeability of the underlying sand and gravel permitted the model to yield soil pressures in the vicinity of the well in good agreement with the experimentally observed values, as shown in Figure 6.15. The presence of such an overlying low-permeability stratum or of a highly anisotropic porous medium may permit one to space the vacuum extraction wells substantially farther apart than normal, with corresponding reduction in cost.

The effects of anisotropic permeabilities are shown for a horizontal slotted pipe configuration in Figures 6.16 and 6.17, which show SVE simulations identical to those plotted in Figure 6.3, except that in Figure 6.16 the vertical component of the permeability has been reduced by 50% to 0.05

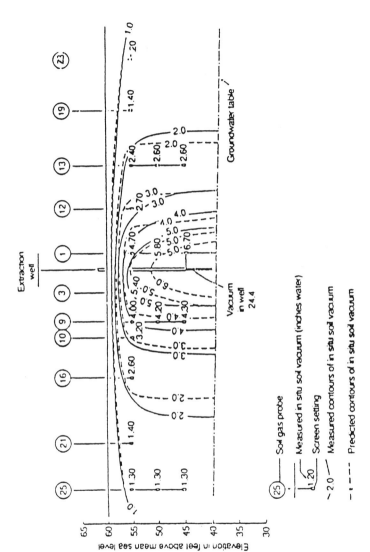

Figure 6.15 Impact of an overlying 1-m layer of low permeability clay on soil gas pressures in the vicinity of a vacuum extraction well. Toms River, NJ. Cross-sectional in situ soil vacuum contour maps (measured and modeled) (vacuum in inches of water). [Reproduced from Mutch et al. (1989), "In situ Vapor Stripping Research Project: A Progress Report," *Proc.*, *2nd Annual Hazardous Materials Conf./Central*, with permission from Advanstar Expositive.]

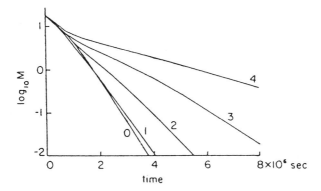

Figure 6.16 Residual contaminant mass versus time, horizontal slotted pipe configuration; effect of a 50% decrease in the vertical component of the pneumatic permeability ($K_y = 0.05$ m²/atm sec). Numbers beside the curves indicate distance of the pipe above the water table, m. Other parameters as in Table 6.3. [Reprinted from Rodríguez-Maroto et al. (1991) p. 1051, by courtesy of Marcel Dekker, Inc.]

m²/atm sec, and in Figure 6.17 the horizontal component of the permeability has been decreased to 0.05 m²/atm sec. Other parameters are given in Table 6.3.

Soil permeabilities show a lot of variation even over fairly short distances, particularly if structures such as clay or silt lenses are present. As seen in Chapter 3, these inhomogeneities can cause substantial changes in the streamlines of the soil gas in the vicinity of a horizontal slotted pipe

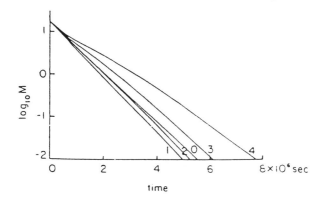

Figure 6.17 Residual contaminant mass versus time, horizontal slotted pipe configuration; effect of a 50% decrease in the horizontal component of the pneumatic permeability ($K_x = 0.05$ m²/atm sec). Numbers beside the curves indicate distance of the pipe above the water table, m. Other parameters as in Table 6.3. [Reprinted from Rodríguez-Maroto et al. (1991) p. 1051, by courtesy of Marcel Dekker, Inc.]

vacuum well, which in turn can markedly affect cleanup rates. Permeability functions can readily be constructed, which exhibit lens-like domains of low permeability; the function used in Chapter 3 for calculating streamlines and transit times is

$$K_x(x,y) = K_{x0} - \sum_{i=1}^{m} A_i \exp\left\{-\left[\left(\frac{x - x^i}{r_i}\right)^2 + \left(\frac{y - y_i}{s_i}\right)^2\right]^n\right\} \quad (1)$$

with a similar function for $K_y(x,y)$. Several SVE simulations were run with a single long clay lens located at different places in the domain of interest; plots of $\log_{10} M_{total}(t)$ for these runs are shown in Figure 6.18. In Run 1 no lens is present. In Run 5 the horizontal slotted pipe is screened right in the middle of the lens, which greatly decreases the gas flow rate through the well and has disastrous effects on the rate of cleanup by SVE. In Run 4 the lens is located far on the left side of the domain of interest. The streamlines for the gas flow field of this run are shown in Figure 3.12 in Chapter 3.

Runs 6, 7, and 8, shown in Figure 6.19, pertain to systems having low-permeability lenses near the sides of the domain of influence of the horizontal pipe, in high-high, low-low, and high-low configuration. The presence of these lenses has roughly doubled the cleanup times over that observed in Run 1, shown in Figure 6.18, in which there are no low-permeability lenses present. (Note that Figures 6.18 and 6.19 have different horizontal axis scales.)

Gómez-Lahoz et al. (1991) published the results from another set of nine

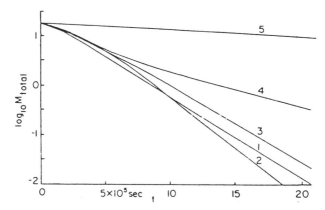

Figure 6.18 Residual contaminant mass versus time, horizontal slotted pipe configuration, Runs 1–5; effect of low-permeability lenses. Standard parameter set as in Table 6.4. [Reprinted from Gómez-Lahoz et al. (1991) p. 133, by courtesy of Marcel Dekker, Inc.]

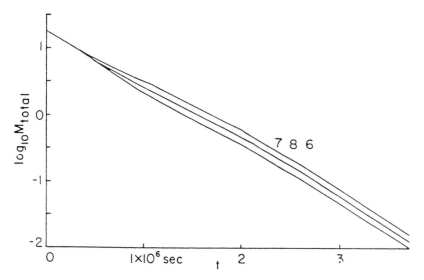

Figure 6.19 Residual contaminant mass versus time, horizontal slotted pipe configuration, Runs 6–8; effect of low-permeability lenses. Standard parameter set as in Table 6.4. [Reprinted from Gómez-Lahoz et al. (1991) p. 133, by courtesy of Marcel Dekker, Inc.]

runs made with model parameters given in Table 6.4 in exploring the effects of lenses of decreased permeability. The parameters describing the lenses are given in Table 6.5. [See Equation (113) in Chapter 3 for notation.] The locations of the lenses in the various runs are shown in Figure 6.20. Plots showing the progress of cleanup for Runs 9–12 are exhibited in Figure 6.21; those for Runs 9 and 13–17 are shown in Figure 6.22.

The 99.9% cleanup times for three of the systems containing lenses are

TABLE 6.4. Geometrical and Physical Parameters for the Horizontal Pipe SVE Model (Runs 9–17).

Depth of domain	12 m
Width of domain	25 m
Δx	1 m
Δy	0.6 m
Wellhead pressure	0.75 atm
Height of bottom of screened well section above the water table	1.2 m
Height of top of screened well section above the water table	3.0 m
Effective Henry's constant	0.001
Initial total contaminant mass	10 kg
Permeabilities K_{x0} and K_{y0}	0.1 m²/atm sec
Temperature	25°C

TABLE 6.5. Parameters Describing the Lenses
(Runs 9–17).

A_i	0.095 m²/atm sec
B_i	0.095 m₂/atm sec
r_i	6.0 m
s_i	1.5 m
n	1

actually shorter than the time required for 99.9% cleanup in Run 9, in which no lens is present. In Run 10 the lens close to the surface causes a decrease in the very rapid rate of cleanup of the soil near the well but also increases the rate of cleanup of the soil near the sides of the domain by directing gas flow into these regions. Since they are particularly slow to clean up, the result is an increase in the overall rate of VOC removal. In Run 14 the presence of the lens directs the flow of soil gas out to the lower left corner of the domain, increasing the rate of cleanup. In Run 16 lenses on both sides of the domain divert the air flow toward the lower left and right corners of the domain and results in a noticeable improvement in remediation rate. The two runs showing rather large increases in 99% cleanup times are 12 and 15, in which the lenses are very near the edge of the domain and therefore tend to block gas from passing through the lower corners of the domain.

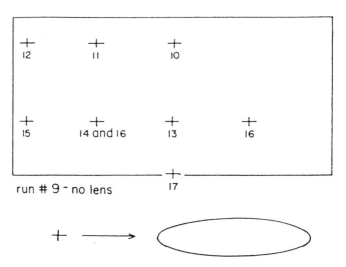

Figure 6.20 Positions of the centers of the low-permeability lenses for Runs 10–17. Run 9, the reference run, has no lenses. See Tables 6.4 and 6.5 for parameter values. The SVE configuration is that of a horizontal slotted pipe. [Reprinted, from Gómez-Lahoz et al. (1991) p. 133, by courtesy of Marcel Dekker, Inc.]

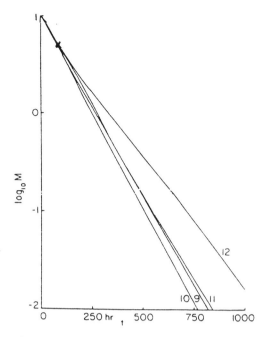

Figure 6.21 Plots of residual contaminant mass versus *t* for Runs 9–12, as indicated by the numbers by the plots. See Figure 6.20 and Tables 6.4 and 6.5 for parameter values. [Reprinted from Gómez-Lahoz et al. (1991) p. 133, by courtesy of Marcel Dekker, Inc.]

In Run 17 the low-permeability lens is located directly beneath the well. Although the time required for 50% cleanup in this run is comparable to those of the other runs, the time required for 99% cleanup is substantially longer than those of the other runs and is approximately double that of the reference run (Run 9, with no lenses).

The results obtained for these two sets of runs (1–8, 9–17) indicate that one can optimize the siting of SVE wells if one knows the location(s) of any low-permeability features from, say, well log information. These structures should be near the middle of the domain of influence of the well, and the well should be screened underneath them.

In Chapter 3 the effect of specific soil moisture content on permeability and on gas streamlines and transit times in the vicinity of an SVE well was discussed. High levels of soil moisture drastically reduce the pneumatic permeability of the soil, with equally drastic effect on the rate of VOC removal by SVE. Some representative runs for a rather dry soil are plotted in Figure 6.23 and show relatively short cleanup times. Figure 6.24, on the other hand, shows some runs made in which SVE from quite wet soils is being simulated. Note the differences in time scales in the two figures.

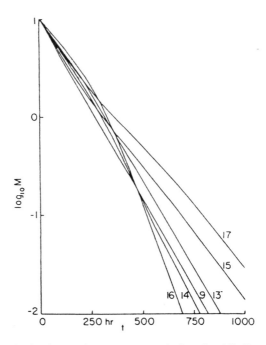

Figure 6.22 Plots of residual contaminant mass versus t for Runs 9 and 13–17, as indicated by the numbers by the plots. See Figure 6.20 and Tables 6.4 and 6.5 for parameter values. [Reprinted from Gómez-Lahoz et al. (1991) p. 133, by courtesy of Marcel Dekker, Inc.]

Remediations for these quite wet soils are intolerably slow because of the very low pneumatic permeabilities and the tendency of the streamlines to avoid the lower corners of the domain of interest, as seen in Figure 3.21 in Chapter 3.

CLEANUP TIME DISTRIBUTIONS FOR STATISTICALLY EQUIVALENT VARIABLE PERMEABILITIES

Roberts and Wilson (1993b) and Bolick and Wilson (1994) have explored another type of variation in pneumatic permeability. Generally, the data set one has to work with in SVE modeling includes a minimal number of permeability measurements to characterize the site. One is very fortunate if sufficient data are provided (1) to locate most of the low-permeability strata and lenses (if these types of heterogeneities are predominant) or (2) to calculate a mean value, a standard deviation, and a correlation length for the permeability if its variations are not associated with clearly defined structures.

We examine the second case here, with the objective of determining the extent of the uncertainty in a cleanup time that is calculated from data from which it is possible to estimate the mean, standard deviation, and correlation length of the permeability. First, we explore some possible ways for constructing and characterizing such families of permeability functions. Then, we use sets of these permeabilities in an SVE model to calculate cleanup times. These results then give us some idea of the uncertainty in the cleanup time that is associated with the uncertainty inherent in our probabilistic experimental information about the pneumatic permeability.

RANDOM PERMEABILITY FUNCTIONS

Here, we examine some methods for constructing permeability functions with random variations. First, we must consider the constraints intrinsic in the nature of the types of permeabilities we wish to examine. In an earlier section of this chapter and in Chapter 3, we considered permeabilities that are discontinuous at the boundaries between strata.

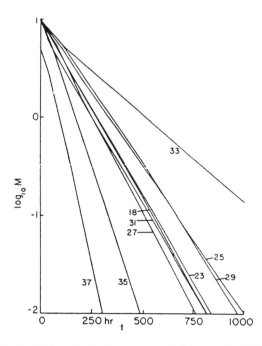

Figure 6.23 Plots of residual contaminant mass versus t (hr) for runs simulating SVE in rather dry soils ($\alpha = 10^{-4}$). Model parameters as in Table 6.4. [Reprinted from Gómez-Lahoz et al. (1991) p. 133, by courtesy of Marcel Dekker, Inc.]

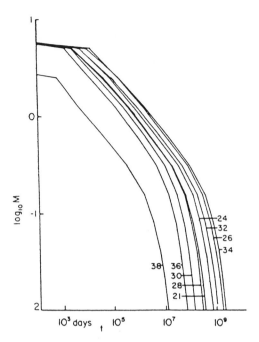

Figure 6.24 Plots of residual contaminant mass versus t (days, logarithmic scale) for runs simulating SVE in quite wet soils. Model parameters as in Table 6.4. [Reprinted from Gómez-Lahoz et al. (1991) p. 133, by courtesy of Marcel Dekker, Inc.]

Here, we shall require that the permeability be a continuous function of the space variables. It must also be nonnegative, and in order that mean values and standard deviations exist, it must have a finite upper bound in the domain of interest. We address a two-dimensional model, with one horizontal and one vertical coordinate. This should allow the exploration of the effects of variable permeabilities and simultaneously keep computational requirements within acceptable limits.

Let us first consider families of permeability functions of the form

$$K(x,y,\{\psi,\phi\}) = \sum_{m}^{M_x} \sum_{n}^{N_y} A_{mn} \sin (m\pi x/L_x + \psi_{mn}) \sin (n\pi y/L_y + \phi_{mn}) + \overline{K} \tag{2}$$

or, more compactly,

$$K(x,y,\{\psi,\phi\}) = f(x,y) + \overline{K} \tag{3}$$

where the family is determined by the set of coefficients A_{mn} and the value of $\overline{K}$ and a particular member of the family is determined by the choice of the set of random phase angles (ψ_{mn}, ϕ_{mn}). A set of these phase angles is selected randomly on the interval $(0,2\pi)$ to specify a particular permeability function in the family. Note that each phase angle, ψ_{mn} and ϕ_{mn}, is selected randomly and independently of the others. L_x and L_y are the dimensions of the domain of interest in the x- and y-directions, and M_x and N_y, the upper limits to the double summation, are determined by the minimum wavelength of variation one wishes to consider, usually determined in modeling computations by the dimensions of the volume elements used in the model.

We wish to explore the statistical properties of these families of permeability functions. This can either be done by carrying out the necessary averaging by first integrating over the space coordinates and then integrating over the phases, or by first integrating over the phases and then integrating (if necessary) over the space coordinates. The second approach is easier than the first. We let $\langle g \rangle$ be the average of $g(x,y,\{\psi,\phi\})$ over the uniformly distributed random phases $\{\psi,\phi\}$, where g is any function of x,y and the set $\{\psi,\phi\}$. Averaging a function over a single phase angle involves integrating that function with respect to the phase angle over the interval $(0,2\pi)$ and dividing by 2π. Averaging over the entire set of phase angles involves $2M_xN_y$ integrations and division by $(2\pi)^{2M_xN_y}$. Thus, the phase average of K is given by

$$\langle K(x,y,) \rangle =$$

$$\sum_{m}^{M_x} \sum_{n}^{N_y} A_{mn} \langle \sin (m\pi x/L_x + \psi_{mn}) \rangle \langle \sin (n\pi y/L_y + \phi_{mn}) \rangle + \overline{K} \quad (4)$$

The phase averages of the sine factors are zero, so the phase average of K is

$$\langle K(x,y) \rangle = \langle K \rangle = \overline{K} \quad (5)$$

The phase averaged mean square deviation of K is given by

$$\langle \sigma_K \rangle^2 = \langle K^2 \rangle - \langle K \rangle^2 \quad (6)$$

Now

$$K^2 = \sum_{m}^{M_x} \sum_{n}^{N_y} \sum_{m'}^{M_x} \sum_{n'}^{N_y} A_{mn} A_{m'n'} \sin (m\pi x/L_x + \psi_{mn})$$

$$\cdot \sin (m'\pi x/L_x + \psi_{m'n'}) \cdot \sin (n\pi y/L_y + \phi_{mn})$$

$$\cdot \sin (n'\pi y/L_y + \phi_{m'n'}) + 2f(x,y)\langle K \rangle + \langle K \rangle^2 \quad (7)$$

From the work above, the phase average of $f(x,y)$ vanishes, so we have

$$\langle K^2 \rangle - \langle K \rangle^2 = \sum_m \sum_{m'} \sum_n \sum_{n'} A_{mn} A_{m'n'}$$

$$\cdot \; \langle \sin (m\pi x/L_x + \psi_{mn}) \sin (m'\pi x/L_x + \psi_{m'n'})$$

$$\cdot \; \sin (n\pi y/L_y + \phi_{mn}) \sin (n'y/L_y + \phi_{m'n'}) \rangle \quad (8)$$

Terms in Equation (8) for which m is not equal to m' and/or for which n is not equal to n' average to zero. The remaining terms are all of the form

$$T_{mnmn} = \langle \sin^2 (m\pi x/L_x + \psi_{mn}) \cdot \sin^2 (n\pi y/L_y + \phi_{mn}) \rangle \quad (9)$$

which yield

$$T_{mnmn} = \langle \sin^2 (m\pi x/L_x + \psi_{mn}) \rangle \cdot \sin^2 (n\pi y/L_y + \phi_{mn}) \rangle \quad (10)$$

$$= (1/2)(1/2) = 1/4 \quad (11)$$

So

$$\langle K^2 \rangle - \langle K \rangle^2 = (1/4) \sum_m \sum_n A_{mn}^2 \quad (12)$$

Then

$$\langle \sigma_K \rangle = 1/2 \left[\sum_m \sum_n A_{mn}^2 \right]^{1/2} \quad (13)$$

gives the phase average standard deviation of the permeability. Note that the phase averages $\langle K \rangle$ and $\langle \sigma_K \rangle$ are independent of x and y, so we need not carry out averages over these variables. The technique of phase averaging is commonly used in classical statistical mechanics.

There is another piece of information about the permeability which is of considerable importance; this is its correlation length, which gives information about the following question: If one measures the permeability to be K_0 at a particular point (x_0, y_0) in the domain, how far from this point can one go [to some new point (x,y)] and still have K_0 to be at least a reasonable approximation to $K(x,y)$? We would generally expect that correlation lengths in the vertical direction will differ from correlation lengths in horizontal directions and that the correlation length in the horizontal plane will depend upon direction. In the following we address correlation lengths in the x-direction; extension to other directions is straightforward.

We explore this question by first calculating the autocorrelation function of the permeability in the x-direction, phase averaging, and then determining the value of the least positive root of this phase-averaged autocorrelation function. The procedure is as follows.

The autocorrelation function $C_x(r,y,\{\psi,\phi\})$ is defined as

$$C_x(r,y,\{\psi,\phi\}) = \frac{1}{L_x} \int_0^{L_x} [K(x,y) - \langle K \rangle] \cdot [K(x + r,y) - \langle K \rangle] dx$$

(14)

Use of Equations (3) and (5) permits us to write Equation (14) as

$$C_x(r,y,\{\psi,\phi\}) = \frac{1}{L_x} \int_0^{L_x} f(x,y) \cdot f(x + r,y) dx$$

(15)

Phase averaging Equation (15) then yields

$$\langle C_x(r,y,) \rangle = \frac{1}{L_x} \int_0^{L_x} \langle f(x,y) \cdot f(x + r,y) \rangle dx$$

(16)

or, by replacing f by its expression as a double trigonometric series,

$$\langle C_x(r,y) \rangle = \frac{1}{L_x} \int_0^{L_x} \sum_m \sum_n \sum_{m'} \sum_{n'} A_{mn} A_{m'n'}$$

$$\cdot \langle \sin (m\pi x/L_x + \psi_{mn}) \cdot \sin (m'\pi x/L_x$$

$$+ \psi_{m'n'} + m'\pi r/L_x) \cdot \sin (n\pi y/L_y + \phi_{mn})$$

$$\cdot \sin (n'\pi y/L_y + \phi_{m'n'}) \rangle dx$$

(17)

The terms in the sum vanish unless $m = m'$ and $n = n'$ as a result of the phase averaging, so

$$\langle C_x(r,y) \rangle = \frac{1}{L_x} \int_0^{L_x} \sum_m \sum_n A_{mn}^2 \langle \sin (m\pi x/L_x + \psi_{mn})$$

$$\cdot \sin (m\pi x/L_x + \psi_{mn} + m\pi r/L_x) \rangle$$

$$\cdot \langle \sin^2 (n\pi y/L_y + \phi_{mn}) \rangle dx$$

(18)

The phase average of the sine squared term gives a factor of 1/2, as before. The product of sines involving x is handled by means of a trigonometric identity, as follows:

$$\langle \sin(m\pi x/L_x + \psi_{mn}) \sin(m\pi x/L_x + \psi_{mn} + m\pi r/L_x)\rangle$$

$$= 1/2\langle\cos(m\pi r/L_x) + \cos(2m\pi x/L_x$$

$$+ 2\psi_{mn} + m\pi r/L_x)\rangle \tag{19}$$

The phase average of the term containing $2\psi_{mn}$ vanishes; the other yields $1/2\cos(m\pi r/L_x)$. Substitution of these results in Equation (18) and integration with respect to x then gives

$$\langle C_x(r,y)\rangle = 1/4\sum_m \sum_n A_{mn}^2 \cos(m\pi r/L_x) \tag{20}$$

We see that $C_x(0,y)$ is given by the mean square deviation, which is always positive. We define the phase-averaged correlation length l_c as the least positive root r_1 of Equation (20).

If we are fortunate, we will have three pieces of experimental information—the mean permeability, its standard deviation, and its correlation length. In our theoretical expressions we have $M_x N_y + 1$ constants—the A_{mn} and $\overline{K}$. Evidently, we must make some assumptions about the A_{mn} if we are to progress further; we can evaluate no more than three independent constants from our three pieces of experimental information. Therefore, let us assume that the A_{mn} are of the form

$$A_{mn} = \frac{1}{(m^2 + n^2)^\alpha} \tag{21}$$

where α is a parameter controlling the magnitude of the higher order (shorter wavelength) terms in the Fourier series. A large value of α yields rapidly vanishing higher order terms, resulting in a large correlation length $\langle l_c \rangle$. A small value of α results in coefficients of the higher order terms that are of significant size, resulting in a small correlation length.

The correlation length $\langle l_c \rangle$ can then be used with Equations (20) and (21) to calculate the value of α; one substitutes Equation (21) and Equation (20), sets the left-hand side of Equation (20) to zero, replaces r by $\langle l_c \rangle$ on the right-hand side of Equation (20), and solves the result [Equation (22)] for α numerically. This is done by a simple search routine

$$0 = \sum_m \sum_n \frac{\cos(m\pi\langle l_c\rangle/L_x)}{(m^2 + n^2)^{2\alpha}} \tag{22}$$

which determines for what values of α the right-hand side of Equation (22) changes sign and then progressively refines this interval.

The constant A is then obtained from Equations (13) and (21) in terms of $\langle \sigma_K \rangle$; the result is

$$A = \frac{2\langle \sigma_K \rangle}{\left[\sum_m \sum_n (m^2 + n^2)^{-2} \right]^{1/2}} \tag{23}$$

Lastly, we recall from Equation (5) that

$$\langle K \rangle = \overline{K} \tag{24}$$

Thus, we have the three model parameters determined in terms of the correlation length, the root mean square deviation of the permeability, and the mean value of the permeability.

We note that use of excessively large root mean square deviations $\langle \sigma_K \rangle$ can result in negative values of the permeability function at some points, so it must be avoided. Practically, this problem does not appear to arise if $\langle \sigma_K \rangle < 1/3\langle K \rangle$.

The random phase Fourier series appearing on the right-hand side of Equation (4) can also be used to generate more complex functions, which may be of use in defining randomly varying permeabilities with particular characteristics. Let

$$f(x,y,\{\psi,\phi\}) = \sum_m \sum_n A_{mn} \sin(m\pi x/L_x + \psi_{mn}) \sin(n\pi y/L_y + \phi_{mn}) \tag{25}$$

as before. Then families of functions f can be used to calculate new families of permeability functions by means of such equations as

$$K_1 = \frac{K_0}{1 + [f(x,y)]^2} \tag{26}$$

$$K_2 = \frac{K_0}{1 + |f(x,y)|} \tag{27}$$

$$K_3 = K_0 \exp[-f^2(x,y)] \tag{28}$$

$$K_4 = K_0 \exp[-|f(x,y)|] \tag{29}$$

and so forth.

One pays a price for this increase in flexibility, however, in that the only model parameter that can be readily approximated from the experimental data is K_0, which is an upper bound to the permeability. Correlation lengths, mean permeabilities, and mean square deviations must all be calculated numerically, so that fitting the model parameters A and α by means of the experimental quantities is of necessity a laborious numerical process. Still, there are some advantages. These functions all give permeability values that most certainly lie between zero and K_0, with no possibility of negative values. Their use allows one to obtain different types of distributions of permeability than are possible with the simple Fourier series.

Some of these functions may lead to permeabilities that are unrealistic; for instance, K_2 and K_4 show cusp maxima with discontinuous slopes at points where $f(x,y) = 0$. Such cusps are not observed in practice, so these two functions must be discarded. The behaviors of both K_1 and K_3 are reasonable over a range of parameter values, with smoothly varying maxima.

We next examine some sets of plots of K versus x for fixed y, A, and α. We also examine the mean values, standard deviations, and mean correlation lengths of some sets of these plots to determine the extent to which the procedure described above permits us to calculate permeability functions with specified mean values, standard deviations, and correlation lengths. One is by no means sure that the use of phase averages, as described above, to specify the model parameters will, in fact, lead to individual permeability functions having the desired properties. We then turn to an examination of the behaviors of the permeability functions K_1 and K_3 defined above. The objective is to get some feeling for the types of permeabilities that are best represented by these functions.

MODEL RESULTS, PERMEABILITIES

PROPERTIES OF THE PERMEABILITY FUNCTION $K(x,y)$ DEFINED BY EQUATIONS (4) AND (21)

Sets of twenty permeability functions $K(x,y)$ were generated and their average mean values (averages of twenty mean values), average standard deviations, and average correlation lengths were calculated, along with the standard deviations of each of the quantities being averaged. The default model parameters are given in Table 6.6. Twenty sets of twenty runs each were made, with values of $\langle l_c \rangle$ ranging from 5 to 14 m, with the results shown in Table 6.7.

TABLE 6.6. Default Model Parameters for
the Calculations of $K(x,y)$.

$$\langle K \rangle = 1 \text{ m}^2/\text{atm sec}$$
$$\langle \sigma_K \rangle = 0.3 \text{ m}^2/\text{atm sec}$$
$$L_x = 30 \text{ m}$$
$$L_y = 20 \text{ m}$$
$$M_x = 30$$
$$N_y = 20$$
$$y = 10 \text{ m}$$

Regression lines of the three calculated quantities against $\langle l_c \rangle$ were calculated from these data; they are as follows:

$$\bar{K} = (0.991 \pm .019) + (0.0007 \pm .0024)\langle l_c \rangle, \quad r^2 = 0.0048 \quad (30)$$

$$\bar{\sigma}_K = (0.346 \pm .011) - (0.0084 \pm .0013)\langle l_c \rangle, \quad r^2 = 0.682 \quad (31)$$

$$\bar{l}_c = (1.140 \pm .076)\langle l_c \rangle - (3.63 \pm .77), \quad r^2 = 0.913 \quad (32)$$

Equation (30) indicates little or no dependence of $\bar{K}$ on $\langle l_c \rangle$, as one

TABLE 6.7. Statistical Properties of Calculated Permeability Functions
$K(x,y)$, Twenty Runs per Set, Twenty Sets.

$\langle l_c \rangle$, m	$\bar{K}$, m²/atm sec	$\langle \sigma_K \rangle$	$\bar{l}_c$, m
5	0.9859 ± .0591	0.2966 ± .0525	3.64 ± 2.94
5	0.9889 ± .0658	0.2904 ± .0523	2.55 ± 2.06
6	1.0275 ± .0503	0.2973 ± .0495	2.58 ± 1.62
6	0.9805 ± .0696	0.2942 ± .0455	3.86 ± 2.69
7	1.0075 ± .0809	0.2763 ± .0315	5.65 ± 5.55
7	0.9917 ± .0830	0.2846 ± .0515	3.53 ± 1.31
8	1.0101 ± .0757	0.2757 ± .0428	5.32 ± 5.23
8	0.9815 ± .0823	0.2672 ± .0423	4.14 ± 3.00
9	0.9876 ± .1161	0.2879 ± .0533	5.33 ± 3.12
9	0.9887 ± .1013	0.2829 ± .0544	5.10 ± 2.72
10	1.0161 ± .1034	0.2702 ± .0725	8.58 ± 6.07
10	0.9822 ± .0954	0.2938 ± .0606	6.60 ± 2.21
11	1.0276 ± .0990	0.2688 ± .0676	8.36 ± 4.26
11	1.0108 ± .1189	0.2808 ± .0851	10.22 ± 5.96
12	1.0369 ± .1256	0.2366 ± .0610	10.63 ± 5.69
12	0.9282 ± .1197	0.2222 ± .0808	9.35 ± 6.76
13	1.0429 ± .1788	0.2461 ± .1008	10.59 ± 4.97
13	0.9442 ± .1672	0.2515 ± .1109	12.36 ± 7.69
14	1.0351 ± .1840	0.2158 ± .0869	13.70 ± 6.14
14	0.9825 ± .2178	0.1922 ± .0915	11.89 ± 5.19

would expect on the basis of the phase average of K, which is independent of $\langle l_c \rangle$. Equation (31) shows only a rather weak dependence of $\overline{\sigma_K}$ on $\langle l_c \rangle$; $\langle \sigma_K \rangle$ for all these sets of runs was held constant at 0.3 m²/atm sec, so one would expect that these root mean square variations would have approximately this value, as is seen.

However, the expected dependence of $\overline{l_c}$ on $\langle l_c \rangle$, $\overline{l_c} = \langle l_c \rangle$, is markedly different from Equation (32). A possible option is to solve Equation (32) for $\langle l_c \rangle$ in terms of $\overline{l_c}$ and use the desired value of the correlation length to calculate a value for $\langle l_c \rangle$ for use in generating the desired family of permeabilities. One difficulty with this option is suggested by the rather large standard deviations given in Table 6.7 for l_c; in some sets of runs with fixed $\langle l_c \rangle$, the calculated values of the correlation length varied by as much as a factor of three, falling both substantially below and substantially above $\langle l_c \rangle$ itself. See Figure 6.25 for a plot of average correlation lengths versus $\langle l_c \rangle$; this also illustrates the variability of the results. Thus, one has no guarantee that following this procedure will generate a family of permeability functions having correlation lengths that are close to some specified value. A crude but feasible solution is to simply generate permeability functions from the desired parameter set (mean permeability, standard deviation of the permeability, and correlation length $\langle l_c \rangle$), compute correlation lengths $\overline{l_c}$ for these, and discard those functions that give correlation lengths lying outside of the desired range.

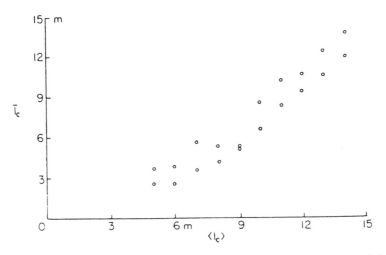

Figure 6.25 Plot of mean correlation length l_c (twenty runs) versus phase average correlation length $\langle l_c \rangle$. Parameters used in $K(x,y)$ as in Table 6.6. [Reprinted from Roberts and Wilson (1993) p. 1548, by courtesy of Marcel Dekker, Inc.]

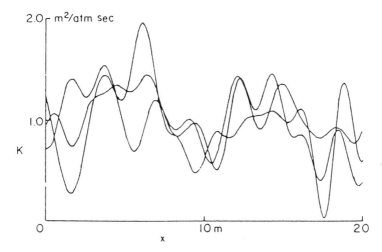

Figure 6.26 Representative plots of $K(x,y_0)$ vs x. $\langle K \rangle$ = 1 m²/atm sec, $\langle \sigma_K \rangle$ = 0.3, $\langle l_c \rangle$ = 3 m, L_x = 20 m, L_y = 10 m, M_x = 20, N_y = 10, α = 0.211, A = 0.1118 m²/atm sec, y_0 = 5 m. [Reprinted from Roberts and Wilson (1993) p. 1549, by courtesy of Marcel Dekker, Inc.]

Plots of some representative permeabilities $K(x,y)$ versus x at constant $y = y_0$ are shown for small and large values of α in Figures 6.26 and 6.27. Parameters not given in the captions are shown in Table 6.6. A plot of the frequency distribution of values of $K(x,y)$ for one of the sets of runs is shown in Figure 6.28. This type of distribution, a roughly bell-shaped curve centered approximately about the mean value of K, was observed for all of the sets of functions $K(x,y)$ of this type that we examined.

PROPERTIES OF THE PERMEABILITY FUNCTIONS $K_1(x,y)$ AND $K_3(x,y)$ DEFINED BY EQUATIONS (21), (26), AND (28)

The complexities of the functions K_1 and K_3 preclude the sort of detailed analysis that was possible in the previous section. One is limited to selecting values of the parameters K_0, A, and α, calculating a set of representative graphs of permeability versus x or y, and computing numerically such statistical properties of a single permeability function as its mean value and root mean square deviation along a certain direction or over some specified grid of points in space and its correlation length in the x- or y-direction for fixed values of y or x. These, however, should be more than sufficient to determine sets of permeability functions, which are reasonable representations for the unknown permeability function at a site for which permeability data are few and far between. These, when used in

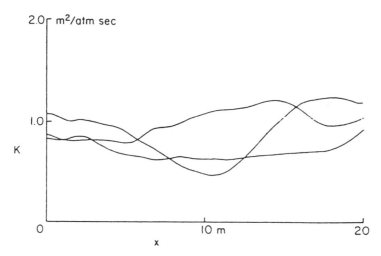

Figure 6.27 Representative plots of $K(x,y_0)$ vs x. $\langle K \rangle = 1$ m²/atm sec, $\langle \sigma_K \rangle = 0.3$, $\langle l_c \rangle = 9$ m, $L_x = 20$ m, $L_y = 10$ m, $M_x = 20$, $N_y = 10$, $\alpha = 1.151$, $A = 1.1128$ m²/atm sec, $y_0 = 5$ m. [Reprinted from Roberts and Wilson (1993) p. 1549, by courtesy of Marcel Dekker, Inc.]

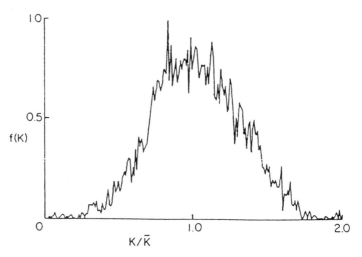

Figure 6.28 Frequency distribution of values of $K(x,y_0)$. $\langle K \rangle = 1$ m²/atm sec, $\langle \sigma_K \rangle = 0.3$, $\langle l_c \rangle = 6$ m, $L_x = 30$ m, $L_y = 20$ m, $M_x = 30$, $N_y = 20$, $\alpha = 0.296$, $A = 0.1211$ m²/atm sec. [Reprinted from Roberts and Wilson (1993) p. 1550, by courtesy of Marcel Dekker, Inc.]

TABLE 6.8. Default Model Parameters for the Calculations of $K_1(x,y)$.

$$K_{max} = 1 \text{ m}^2/\text{atm sec}$$
$$L_x = 20 \text{ m}$$
$$L_y = 10 \text{ m}$$
$$M_x = 20$$
$$N_y = 10$$
$$y = 5 \text{ m}$$

a vapor stripping model, will then permit us to get some semiquantitative idea of the uncertainty in cleanup time, which arises from our lack of detailed knowledge of the permeability throughout the contaminated domain.

The results for the two functions K_1 and K_3 were quite similar, so we focus principally on K_1. Default parameters for the sets of runs are given in Table 6.8. Twenty data sets of twenty runs each were generated. A statistical summary of the results is given in Table 6.9, and a plot of $\bar{l}_c$ versus

TABLE 6.9. Model Parameters and Statistical Results for the Calculations of $K_1(x,y)$, Twenty Runs per Set, Twenty Sets.

	A	$\overline{K}$, m²/atm sec	σ_K	$\overline{l}_c$, m
2.00	9.0	.69	.17	6.15
1.75	8.0	.61	.20	6.45
1.75	7.0	.61	.19	5.33
1.50	7.0	.55	.24	4.47
1.25	6.0	.45	.23	4.14
1.10	5.5	.52	.24	3.89
1.00	5.0	.51	.25	3.78
.90	4.5	.50	.27	3.97
.80	4.1	.46	.26	3.53
.75	3.7	.56	.27	4.59
.75	3.8	.47	.29	3.51
.75	3.9	.41	.28	3.53
.75	4.0	.43	.27	4.50
.70	3.0	.46	.28	3.23
.70	4.0	.39	.29	3.18
.65	2.5	.53	.29	4.31
.65	2.6	.48	.28	2.74
.60	2.5	.49	.30	3.10
.55	2.0	.54	.30	2.81
.50	1.5	.53	.28	2.26

α is shown in Figure 6.29. The least squares line fitted to the points plotted in Figure 6.29 is

$$\bar{l}_c = (2.088 \pm .274)\alpha + (1.948 \pm .297) \tag{33}$$

for which $r^2 = 0.740$. As before, we are faced with the fact that our prescription for generating families of permeability functions yields sets of permeability functions that show a good deal of variation in their correlation lengths. Evidently, if one wishes to specify the correlation length for a family of functions, one would have to choose a value of α on the basis of Equation (33) and then generate a large number of permeability functions, discarding those whose correlation lengths were outside the desired interval.

Figures 6.30 and 6.31 show some representative plots of permeability functions with small (0.39) and large (1.25) values of α, corresponding to short and long correlation lengths. Figure 6.32 shows the statistical distribution of values of a set of K_1 functions. Here, we see some marked differences between $K(x,y)$ and $K_1(x,y)$, as seen by comparing Figures 6.28 and 6.32. The distribution of values of K_1 tends to be bimodal, with a large number of values quite near the upper limit of the permeability and a broader and lower maximum in the low-permeability region. This is the sort of distribution one might expect if the site consisted of a fairly high-permeability sand in which there were heterogeneous regions of higher clay content.

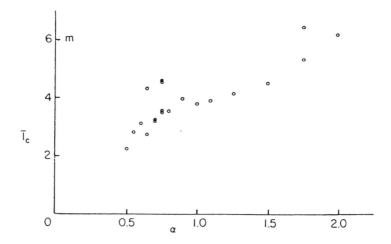

Figure 6.29 Plot of mean correlation length l_c (twenty runs) versus α. Parameters used in $K(x,y)$ as in Table 6.6. [Reprinted from Roberts and Wilson (1993) p. 1552, by courtesy of Marcel Dekker, Inc.]

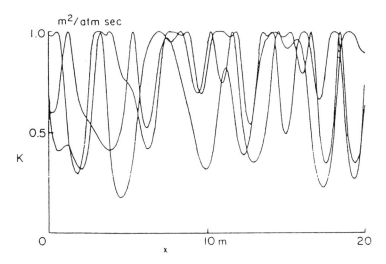

Figure 6.30 Representative plots of $K_1(x,y_0)$ vs x. $K_{max} = 1$ m²/atm sec, $\alpha = 0.3946$, $A = 0.4610$, $y_0 = 5$ m, $L_x = 20$ m, $L_y = 10$ m, $M_x = 20$, $N_y = 10$. [Reprinted from Roberts and Wilson (1993) p. 1552, by courtesy of Marcel Dekker, Inc.]

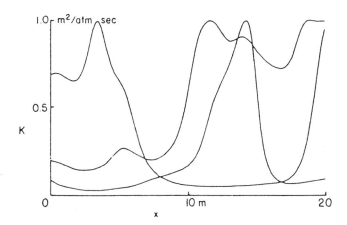

Figure 6.31 Representative plots of $K_1(x,y_0)$ vs x. $K_{max} = 1$ m²/atm sec, $\alpha = 1.25$, $A = 10$, $y_0 = 5$ m, $L_x = 20$ m, $L_y = 10$ m, $M_x = 20$, $N_y = 10$. [Reprinted from Roberts and Wilson (1993) p. 1553, by courtesy of Marcel Dekker, Inc.]

261

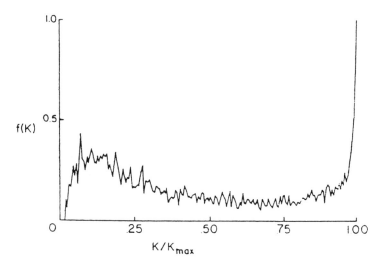

Figure 6.32 Frequency distribution of values of $K_1(x,y_0)$. $K_{max} = 1$ m²/atm sec, $A = 4$, $\alpha = 0.75$, $L_x = 20$ m, $L_y = 10$ m, $M_x = 20$, $N_y = 10$, $y_0 = 5$ m. [Reprinted from Roberts and Wilson (1993) p. 1553, by courtesy of Marcel Dekker, Inc.]

The characteristics of the permeability functions $K_3(x,y)$ defined by Equation (28) were explored; their behavior was sufficiently similar to that of the K_1 functions so that detailed investigation was unnecessary. Use of K_3 does permit one to generate a relatively large number of quite small permeability values and might be of interest in situations in which a relatively permeable medium was interspersed with a tight, low-permeability clay.

SOIL VAPOR EXTRACTION MODELING WITH RANDOM PERMEABILITIES

A local equilibrium linear isotherm soil vapor extraction model was used with the permeability functions discussed above to simulate SVE remediation in heterogeneous soils. A version of the model was discussed previously (Gómez-Lahoz et al., 1991); code for this version was written by Megehee (1993) for use in another connection. The permeability function is calculated in a subroutine in a program that calculates the soil gas velocity field by a numerical over-relaxation method and writes this information to a file. The file is then read by the program which actually simulates the SVE operation.

Default SVE model parameter values are indicated in Table 6.10. The configuration being modeled is that of a buried horizontal slotted pipe, so

TABLE 6.10. SVE Model Parameters, Fourier
Series Random Phase Runs.

Width of domain	13 m
Depth of domain	8 m
No. of horizontal divisions	13
No. of vertical divisions	8
Distance of well from left boundary	7 m
Distance of well from top boundary	7 m
Wellhead pressure	0.8959 atm

Cartesian coordinates (x,y) are used to represent horizontal and vertical distances at right angles to the direction of the pipe. Parameters used to calculate the various families of K_1 functions are given in Table 6.11; for each of the eight sets of parameters, a set of ten separate SVE simulations were run, each with its own K_1, defined by the parameter set and its own set of randomly selected phase angles. Cleanup time was defined as the time required to remove 99.9% of the initial mass of contaminant. Cleanup times were determined for all of the runs; these are listed in Table 6.12, along with the average and standard deviation for each set of cleanup times. The maximum and minimum permeability values in the set used in making each run were also determined.

One would expect a correlation between cleanup time and average permeability, and such a relationship is indicated by the data plotted in Figure 6.33. Cleanup time is plotted against the average permeability for each of the eight files. (The average permeability for a file is the average over the ten runs of the space-averaged permeability.) We see, as expected, a marked tendency for cleanup times to decrease with increasing average permeability, but it is also apparent that this is by no means the only sig-

TABLE 6.11. Parameters Used in $K_1(x,y)$.

File No.	Maximum Permeability, K_{max}, m²/atm sec		A
8	1.0	0.1	0.8
9	1.0	0.05	0.8
10	1.0	0.2	0.8
11	1.0	0.1	0.2
12	1.0	0.1	0.4
13	0.5	0.1	0.4
14	0.5	0.2	0.8
15	$K_h = 1.0, K_v = 0.333$ (anisotropic)	0.2	0.8

TABLE 6.12. **Results of SVE Simulations, Random Permeabilities.**

File No.	K_{ave}, m²/atm sec	σ_K	$t_{99.9}$, Days	$\bar{t}_{99.9}$
8	0.293	0.089	150, 180, 240, 140, 180, 180, 140, 140, 240, 150	174 ± 37
9	0.262	0.084	250, 180, 180, 200, 150, 160, 220, 150, 191.5, 249.5	193 ± 35
10	0.223	0.057	98.1, 98.5, 224.9, 765.5, 265.4, 78.8 183.1, 230.9, 100.1, 357.2	240 ± 195
11	0.690	0.064	57.8, 52.9, 52.8, 52.0, 56.4, 55.3, 53.6, 52.9, 58.0, 53.4	54.5 ± 2.1
12	0.484	0.093	82.2, 75.2, 75.6, 69.0, 70.8, 71.8, 67.8, 72.2, 82.5, 67.9	73.5 ± 5.1
13	0.248	0.024	122.2, 89.8, 104.1, 99.0, 106.5, 147.1, 85.1, 96.6, 100.0, 104.3	106 ± 17
14	0.091	0.013	275.3, 160.0, 678.4, 1,028.2, 520.1, 1,132.0, 1,650.0, 889.4, 193.1, 399.2	693 ± 457
15	0.226 (K_h)	0.059	369.7, 302.1, 117.0, 143.4, 241.6, 687.0, 825.3, 311.9, 531.7, 156.5	369 ± 228

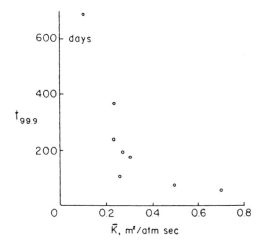

Figure 6.33 Plot of average cleanup time (days) vs average permeability for the eight files of SVE simulations using $K_1(x,y)$. See Tables 6.6 and 6.7 for parameters and results. [Reprinted from Roberts and Wilson (1993) p. 1557, by courtesy of Marcel Dekker, Inc.]

nificant factor. The correlation obviously does not take into account any spatial effects. The relationship between the permeability K_{ave} and the average cleanup time for each of the eight sets of runs suggested that we consider a dependence of the form

$$1/t_{99.9} = A\overline{K}_{ave} + B \tag{34}$$

where A and B are constants. A linear least squares fit of the eight points yielded

$$A = 0.0296 \pm 0.0017$$

$$B = 0.0017 \pm 0.0011$$

with $r^2 = 0.899$. A linear least squares fit of $1/t_{99.9}$ to $\overline{K}$ was then made for all eighty of the runs; the result was

$$1/t_{99.9} = (0.02560 \pm 0.00067)\overline{K} + (0.00020 \pm 0.00032) \tag{35}$$

with $r^2 = 0.857$.

A linear least squares fit of $1/t_{99.9}$ against K_{min} was made for all eighty of the runs. Here K_{min} is the minimum value of the permeability used in the SVE modeling calculations for a particular run. The result is

$$1/t_{99.9} = (0.1040 \pm 0.0084)K_{min} + (0.00476 \pm 0.00039) \tag{36}$$

with $r^2 = 0.717$. A linear least squares fit of $1/t_{99.9}$ against K_{min} and $\overline{K}$ was also made for all of the runs. This yielded

$$1/t_{99.9} = 0.028275K_{min} + 0.020244\overline{K} + 0.0009388 \qquad (37)$$

with $r^2 = 0.872$.

CONCLUSIONS REGARDING RANDOM PERMEABILITIES

Fourier series techniques involving the use of randomly selected phase angles have been used to generate families of pneumatic permeability functions having specified characteristics. These permeability functions have been used in a soil vapor extraction model to explore the effect of spatial variations in the permeability on SVE cleanup times.

One of the more interesting of the conclusions that can be drawn from the cleanup time results is that the effects of the heterogeneity introduced into the pneumatic permeability by functions of the $K_1(x,y)$ within a single set can be quite substantial indeed. The very large standard deviations of the cleanup times for file numbers 8, 9, 10, 14, and 15 give some idea of the uncertainties in cleanup times, which can be expected if there is a substantial amount of variation in the permeability. The relatively poor r^2 values for Equations (34)–(36) give a clear indication that neither the average permeability nor the minimum permeability is, by itself, an accurate predictor of cleanup time, although the cleanup time certainly has a tendency to depend upon one or the other of these quantities. This weak result is hardly surprising, since such correlations do not take into account the geometrical relationships between the well and domains of low permeability.

The results obtained with both low-permeability lenses and with random Fourier series permeabilities indicate that it would be imprudent to regard cleanup times estimated by mathematical modeling as being of high accuracy if well log data indicate that the permeability of the soil medium is highly variable. One must be careful in modeling not to encourage expectations regarding the accuracy of cleanup time estimates that are unrealistically high. The techniques presented in this chapter provide a means for making estimates of the uncertainty in the calculated cleanup times, which result from heterogeneity in the soil permeability. In some of our sets of runs, cleanup times varied by as much as a factor of five.

Finally, these results provide support for the fundamental rule of SVE, which is that you must be able to move air at a reasonable rate through any soil you propose to cleanup. Two practical implications of this are as follows: (1) Do not screen SVE wells in formations of low permeability, which will drastically reduce air flow. (2) Try to design wells in such a way as to maintain a substantial pressure gradient across contaminated domains which may be difficult to remediate. That is, don't screen wells over contaminated domains of low permeability, and don't attempt to remediate domains of low permeability that are toward the outer edge of the well's effective radius of influence.

VAPOR STRIPPING OF NAPL VOCs IN MIXTURES OBEYING RAOULT'S LAW

Results obtained with the SVE model described in Chapter 5 for handling mixtures of VOCs present as nonaqueous phase liquid and obeying Raoult's Law will be discussed. Gasoline is a good example of such a mixture. The parameters used to model runs made with a horizontal pipe configuration are given in Table 6.13, and parameters used for modeling operation of a single vertical well screened at the bottom are given in Table 6.14.

TABLE 6.13. Parameters Used in Modeling SVE of NAPL Mixtures Obeying Raoult's Law, Horizontal Slotted Pipe.

Depth to water table	6 m
Depth of well	5.5 m
Soil pneumatic permeability	0.1 m²/atm sec
Soil porosity	0.3
Soil density	1.7 gm/cm³
Domain width	12 m
Domain breadth	6 m
Δx	1 m
Δy	1 m
Radius of well gravel packing	0.1 m
Wellhead pressure	0.85 atm
Temperature	15°C
Components:	
Benzene	
Molecular weight	78 gm/mol
Vapor pressure	58.14 torr
Initial concentration	1,000 mg/kg of soil
Toluene	
Molecular weight	92 gm/mol
Vapor pressure	15.97 torr
Initial concentration	1,000 mg/kg of soil
p-Xylene	
Molecular weight	106 gm/mol
Vapor pressure	4.715 torr
Initial concentration	1,000 mg/kg of soil
Calculated gas flow rate	1.038 mol/sec
	0.02454 m³/sec

TABLE 6.14. Parameters Used in Modeling SVE of
NAPL Mixtures Obeying Raoult's Law,
Vertical Well Screened at the Bottom.

Depth to water table	6 m
Depth of well	5.5 m
Soil pneumatic permeability	0.1 m²/atm sec
Soil porosity	0.3
Soil density	1.7 gm/cm³
Domain radius	6 m
Δr	1 m
Δz	1 m
Effective radius of well gravel packing (assumed spherical)	0.3 m
Wellhead pressure	0.85 atm
Temperature	15°C
Components	
Benzene	See Table 6.13
Toluene	See Table 6.13
p-Xylene	See Table 6.13
Gas flow rate	0.692 mol/sec
	0.01635 m³/sec

These models were implemented in TurboBASIC and run on 80286 and 80386 SX microcomputers equipped with math coprocessors and running at 12 or 20 MHz. A typical run required between 1 and 2 hr of machine time. The vapor pressures of the three VOCs used (benzene, toluene, and *p*-xylene) were calculated at 15°C from the equation

$$\log_{10} P(T) = A - B/T$$

where the constants A and B were obtained by a least-squares fit to vapor pressure data from Montgomery and Welkom (1989) (see Table 2.2 in Chapter 2).

Plots of the residual masses of the three VOCs are shown in Figures 6.34 and 6.35 for a horizontal slotted pipe SVE well. The contaminated zone for the run plotted in Figure 6.34 is 8 m wide by 6 m broad by 3 m deep; it is 10 by 6 by 4 m for the run shown in Figure 6.35. We see most rapid removal of the most volatile component (benzene), followed by toluene, followed, in turn, by xylene. The rate of removal of xylene is seen to increase markedly during the course of the run, since its mole fraction increases as the more volatile components are removed. This increases its vapor pressure quite substantially, resulting in its accelerated removal toward the ends of the runs. The gas flow rate in these runs is 0.0245 m³/sec.

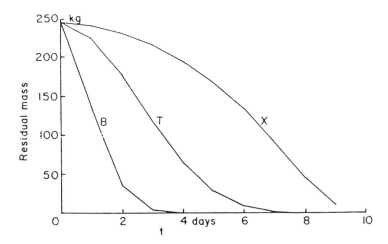

Figure 6.34 SVE of a Raoult's Law mixture with a horizontal slotted pipe. Plots of residual masses of benzene (B), toluene (T), and p-xylene (X) versus time. Model parameters are given in Table 6.13. The contaminated zone is of 8 m width, 6 m breadth, and extends to a depth of 3 m below the surface of the soil. [Reprinted from Kayano and Wilson (1992) p. 1525, by courtesy of Marcel Dekker, Inc.]

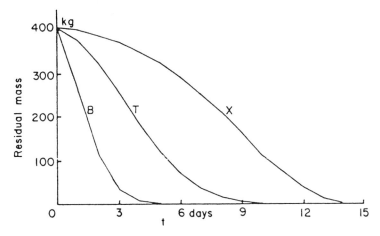

Figure 6.35 SVE of a Raoult's Law mixture with a horizontal slotted pipe. Plots of residual masses of benzene (B), toluene (T), and p-xylene (X) versus time. Model parameters are given in Table 6.13. The contaminated zone is of 10 m width, 6 m breadth, and extends to a depth of 4 m below the surface of the soil. [Reprinted from Kayano and Wilson (1992) p. 1525, by courtesy of Marcel Dekker, Inc.]

269

Figures 6.36 and 6.37 show runs modeling the removal of the same three VOCs, here by means of a single vertical well screened at the bottom. In Figure 6.36 the contaminated zone is 4 m in radius by 3 m in depth; in Figure 6.37 it is 6 by 4 m. The increased domain size requires roughly twice the time for cleanup as the smaller one. The gas flow rate through these wells is 0.0164 m³/sec. If one normalizes the cleanup times with respect to gas flow rate by taking the product of the cleanup time for, say, benzene times the gas flow rate, one concludes that the horizontal pipe configuration is slightly over 50% more efficient in its use of gas than the single vertical well configuration. This is based on a comparison of Figures 6.34 and 6.36, in which quite similar masses of VOCs are being removed. These results indicate a significant advantage for the use of horizontal buried slotted pipes in SVE, at least as long as the cost of putting these in is not excessive.

In Figures 6.38 and 6.39 a horizontal pipe and a vertical well, respectively, are used to vapor strip a domain that is contaminated throughout. As one would expect, cleanup times are somewhat longer than those found for the runs shown in Figures 6.35 and 6.37, but the increases are not large and there is not a significant increase in tailing. This is presumably due to the use of velocity fields calculated by the method of images, which do not have no-flow conditions at the peripheral boundaries of the domains, so that there are no volume elements in which the gas flow is essentially stagnant.

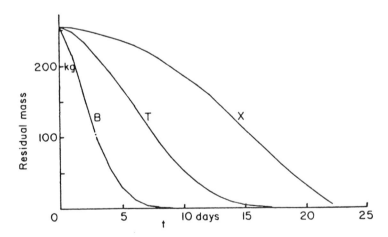

Figure 6.36 SVE of a Raoult's Law mixture with a single vertical well. Plots of residual masses of benzene, toluene, and p-xylene versus time. Model parameters are given in Table 6.14. The contaminated zone is of 4 m radius and extends to a depth of 3 m below the surface of the soil. [Reprinted from Kayano and Wilson (1992) p. 1525, by courtesy of Marcel Dekker, Inc.]

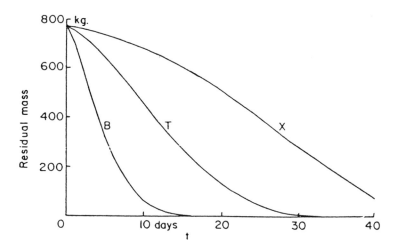

Figure 6.37 SVE of a Raoult's Law mixture with a single vertical well. Plots of residual masses of benzene, toluene, and *p*-xylene versus time. Model parameters are given in Table 6.14. The contaminated zone is of 6 m radius and extends to a depth of 4 m below the surface of the soil. [Reprinted from Kayano and Wilson (1992) p. 1525, by courtesy of Marcel Dekker, Inc.]

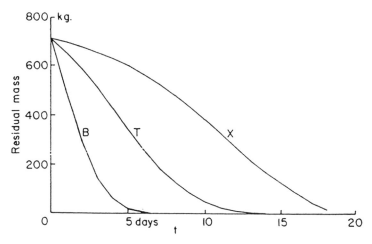

Figure 6.38 SVE of a Raoult's Law mixture with a horizontal pipe. Plots of residual masses of benzene, toluene, and *p*-xylene versus time. Model parameters are given in Table 6.13. The contaminated zone is of 12 m width, 6 m breadth, and extends to a depth of 6 m below the surface of the soil, to the water table. [Reprinted from Kayano and Wilson (1992) p. 1525, by courtesy of Marcel Dekker, Inc.]

271

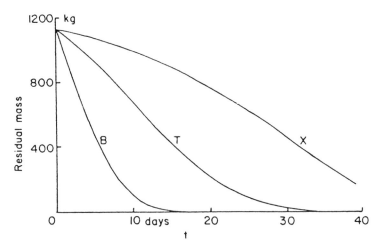

Figure 6.39 SVE of a Raoult's Law mixture with a single vertical well. Plots of residual masses of benzene, toluene, and *p*-xylene versus time. Model parameters are given in Table 6.14. The contaminated zone is of 6 m radius and extends to a depth of 6 m below the surface of the soil, to the water table. [Reprinted from Kayano and Wilson (1992) p. 1525, by courtesy of Marcel Dekker, Inc.]

Figure 6.40 plots a run with a horizontal pipe configuration and a contaminated domain of the same size and position as that shown in Figure 6.34; however, only xylene is present (at an initial concentration of 3,000 mg/kg of soil) in the run shown in Figure 6.40. Comparison of the results shown in Figures 6.34 and 6.40 support the thesis that single component SVE models can be used to obtain upper bounds for the cleanup times of mixtures if the component of lowest volatility is modeled.

The gas flow fields for the results reported above were obtained by the method of images. A second version of the model for horizontal slotted pipes makes use of velocity fields calculated by numerical solution of Laplace's equation by over-relaxation. The results are virtually identical to those reported above. This latter approach permits one to include passive vent wells, impermeable caps, and variable permeabilities but requires the separate calculation of the velocity field.

Plots of some representative runs made by this procedure are shown in Figures 6.41–6.43. For the run in Figure 6.41, only xylene NAPL is present, and we see the expected linear rate of removal of the xylene to nearly the end of the run. In Figure 6.42 we start with equal masses of benzene and toluene and see the expected inhibition of toluene removal by virtue of its dilution with benzene during the first twenty-five days or so of the run. The last run shows the removal of benzene, toluene, and xylene. Again, the inhibition of toluene removal (as well as that of xylene) is clear. Com-

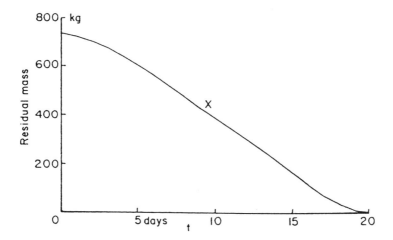

Figure 6.40 SVE of a single VOC NAPL (*p*-xylene) with a horizontal pipe; plot of residual mass of xylene versus time. Model parameters as in Table 6.13. The contaminated zone is of 8 m width, 6 m breadth, and extends to a depth of 3 m below the surface of the soil to the water table. The initial xylene concentration is 3,000 mg/kg of soil. [Reprinted from Kayano and Wilson (1992) p. 1525, by courtesy of Marcel Dekker, Inc.]

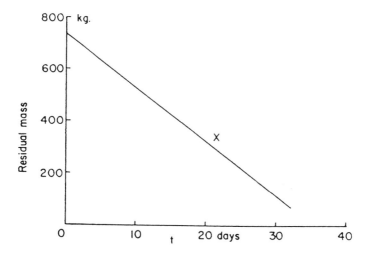

Figure 6.41 Modeling SVE with a horizontal pipe, *p*-xylene. The relaxation method was used to calculate the velocity field here. Model parameters as in Table 6.15. The contaminant is initially distributed uniformly throughout the entire domain. [Reprinted from Kayano and Wilson (1992) p. 1525, by courtesy of Marcel Dekker, Inc.]

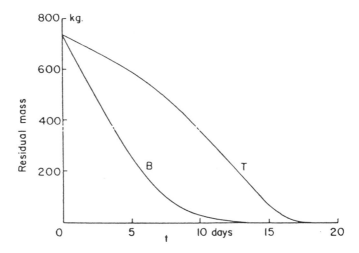

Figure 6.42 Modeling SVE with a horizontal pipe. The relaxation method is used for the velocity field. Model parameters as in Table 6.15. The contaminants (benzene and toluene) are initially distributed uniformly throughout the entire domain. [Reprinted from Kayano and Wilson (1992) p. 1525, by courtesy of Marcel Dekker, Inc.]

parison of Figure 6.41 with Figure 6.43 shows that one can obtain an upper bound for the removal of mixtures simply by assuming that the mixture consists only of the compound having the lowest vapor pressure (see Table 6.15).

DIFFUSION/DESORPTION KINETICS: THE LUMPED PARAMETER APPROXIMATION

If the soil being treated by SVE is homogeneous and reasonably porous, the local equilibrium approximation generally appears to be satisfactory [see, e.g., Hoag and Cliff (1985)]. However, Fall et al. (1989), Sterrett (1989), and others have found that soil gas VOC concentrations increase after the wells have been shut down for a period, which is strong evidence that diffusion/desorption kinetics are playing a role and that a nonequilibrium model should be considered. Well log data should be examined for indications of the presence of low-permeability domains, and pilot-scale studies should include tests to determine the extent to which diffusion/desorption kinetics may be a rate limiting factor in remediation by SVE. Intermittent operation of the well, with measurement of the rate of recovery of the soil gas VOC concentrations after the well has been turned off, gives information on this. The response of the system to such intermittent operation is also affected by the distribution of contaminants with re-

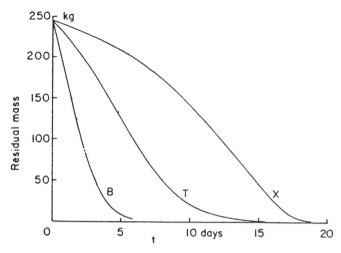

Figure 6.43 Modeling SVE with a horizontal pipe. The relaxation method is used for the velocity field. Model parameters as in Table 6.15. The contaminants (benzene, toluene and *p*-xylene) are initially distributed uniformly throughout the entire domain. [Reprinted from Kayano and Wilson (1992) p. 1525, by courtesy of Marcel Dekker, Inc.]

spect to the SVE well and the presence or absence of surrounding passive wells.

Bouchard (1989) and Bouchard et al. (1988) have presented batch and lab column data involving an aqueous phase; they noted that the effects of desorption and diffusion through a liquid layer in SVE were relatively sim-

TABLE 6.15. **Parameters Used for the Runs Shown in Figures 6.41–6.43 (horizontal slotted pipe configuration).**

Depth to water table	6 m
Depth of well	6 m
Soil pneumatic permeability	0.1 m²/atm sec
Soil porosity	0.3
Soil density	1.7 gm/cm³
Domain width	11 m
Domain breadth	6 m
Δx	1 m
Δy	1 m
Wellhead pressure	0.85 atm
Gravel-packed radius of well	0.1 m
Temperature	15°C
Molar gas flow rate	0.463 mol/sec
(Note that the location of the pipe at the very bottom of the vadose zone cuts off about half of its flow as compared to the earlier runs.)	
Properties of VOCs	See Table 6.13

ilar. In the lumped-parameter model discussed in Chapter 5, these two effects were combined and described as equivalent to diffusion from blocks of a porous medium. The diffusion time constant in lumped parameter models must be assigned on the basis of experimental data, since it includes desorption, as well as diffusion, effects. Diffusion through an occluding layer of soil water is shown in Figure 6.44; diffusion from low-

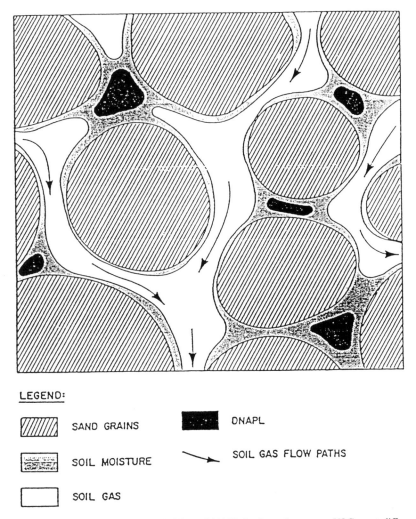

LEGEND:

SAND GRAINS		DNAPL	
SOIL MOISTURE		SOIL GAS FLOW PATHS	
SOIL GAS			

Figure 6.44 A mechanism for the holding of NAPL in the vadose zone. VOC must diffuse through a stationary layer of water before reaching the advecting soil gas. [Reprinted from Wilson and Clarke (1993) p. 171, by courtesy of Marcel Dekker, Inc.]

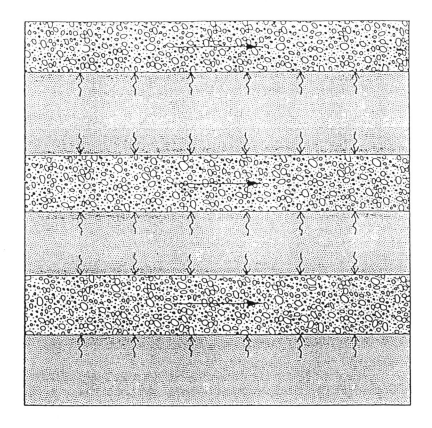

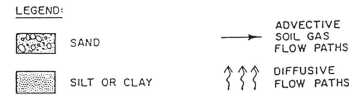

SAND

SILT OR CLAY

ADVECTIVE
SOIL GAS
FLOW PATHS

DIFFUSIVE
FLOW PATHS

Figure 6.45 Diffusion and advection of VOC in interbedded formations of low and high permeabilities. [Reprinted from Wilson and Clarke (1993) p. 171, by courtesy of Marcel Dekker, Inc.]

permeability silt or clay layers into high-permeability sand or gravel is illustrated in Figure 6.45.

Some lab column simulations showing the effects of the lumped parameter diffusion time constant are shown in Figure 6.46. The parameter set for these runs is given in Table 6.16. The curve labeled ∞ corresponds to the local equilibrium approximation. Departures from local equilibrium can

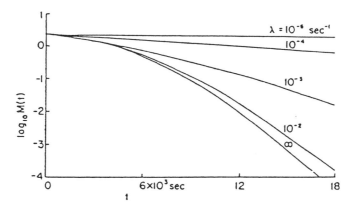

Figure 6.46 Plots of contaminant mass versus time, kinetics-limited SVE in a laboratory column. Effects of diffusion/desorption rate constant $\lambda = \infty$, 10^{-2}, 10^{-3}, 10^{-4} and 10^{-5} sec^{-1}. Other parameters as in Table 6.16. [Reprinted from Wilson (1990) p. 243, by courtesy of Marcel Dekker, Inc.]

be determined by making runs at different gas flow rates and plotting the residual contaminant mass against the volume of gas passed through the column. If mass transport kinetics are limiting, the larger the gas flow rate, the larger is the departure of the plot from the local equilibrium curve. An example of this is seen in Figure 6.47.

Laboratory column experiments should be satisfactory for investigating the effects of desorption kinetics and diffusion from low-permeability porous chunks of medium which are not broken up during the sampling process and packing of the column. To characterize the effects of low-permeability porous domains of larger size (or domains that are broken up during the preparation of lab column runs), one must carry out pilot-scale

TABLE 6.16. **Laboratory Column Parameter Set, Lumped Parameter Kinetics Model for SVE.**

Column length	50 cm
Column radius	10 cm
No. of volume elements into which the column is partitioned	10
Voids fraction associated with mobile gas	0.2
Voids fraction associated with immobile pore liquid	0.2
Gas flow rate	5 mL/sec
Effective Henry's constant	0.10
Soil density	1.6 gm/cm³
Initial contaminant concentration	100 mg/kg of soil

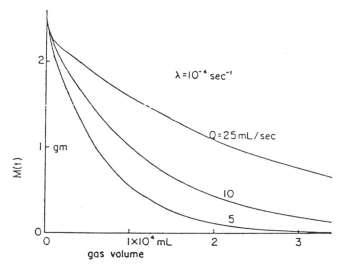

Figure 6.47 Plots of contaminant mass versus effluent gas volume, kinetics-limited SVE in a laboratory column. In these runs $\lambda = 10^{-4}$ sec^{-1}, and the gas flow rate Q through the column is as indicated. Other parameters as in Table 6.16. [Reprinted from Wilson (1990) p. 243, by courtesy of Marcel Dekker, Inc.]

field tests. Simulations for such runs are shown in Figure 6.48, and the parameters for these runs are given in Table 6.17.

Kinetic limitations have the potential in some situations to be quite severe, leading to much longer cleanup times than would be calculated by local equilibrium methods (Wilson, 1990; Oma et al., 1990; DiGiulio et al., 1990), so it is important that this point be investigated in any study of the feasibility of SVE for use at a site. The effects of kinetic limitations on SVE can be explored in a number of ways, as follows:

(1) The well may be operated for a time, shut down for a period, and then restarted. The VOC soil gas concentration will increase during the period of shutdown if kinetic effects are significant; the rate constant for the kinetic process is then obtained from the time dependence of this rebound process. DiGiulio et al. (1990) have discussed this approach in some detail.

(2) The well may be operated at several different flow rates and the soil gas VOC concentrations determined at each of these. If kinetic effects are significant, VOC concentrations will be lower when the gas flow rate is higher. The relationship between the VOC concentrations and the gas flow rate can be used to determine the rate constant for the kinetic processes.

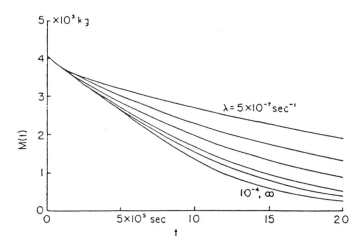

Figure 6.48 Plots of contaminant mass versus time, kinetics-limited SVE by a vertical extraction well. In these runs $\lambda = \infty$, 10^{-4} (plots superimposed), 10^{-5}, 5×10^{-6}, 2×10^{-6}, 10^{-6}, and 5×10^{-7} sec^{-1}; other parameters are given in Table 6.17. [Reprinted from Wilson (1990) p. 243, by courtesy of Marcel Dekker, Inc.]

(3) Mutch (1990) has suggested that clean air be injected through the well and into the surrounding domain. After an interval this gas is sampled and analyzed for VOCs. Several such experiments would map out the rate of equilibration of the injected air with the surrounding contaminated soil and yield the rate constant for the kinetic processes.

TABLE 6.17. Field-Scale Well Parameter Set for Lumped Parameter SVE Model.

Radius of domain of influence	30 m
Depth of water table	20 m
Depth of well	17.5 m
Radius of impermeable cap	25 m
Screened radius of well	0.12 m
Wellhead pressure	0.866 atm
Temperature	13°C
Voids fraction containing mobile gas	0.2
Voids fraction containing immobile pore liquid	0.2
Pneumatic permeability	1 m²/atm sec
Effective Henry's constant	0.1
Radius of zone of contamination	20 m
Initial contaminant concentration	100 mg/kg
Soil density	1.6 gm/cm³
Initial contaminant mass	4,018.2 kg
Molar gas flow rate	1.6246 mol/sec
Volumetric gas flow rate	0.03815 m³/sec

One should note, however, that the lumped parameter method used here replaces what is presumably a rather broad distribution of time constants with a single value. Tests should estimate a value representative of the long time constants (the small rate constants), since these are what will control the rate of removal of VOC along toward the end of the remediation. Mathematical analysis of diffusion problems of this sort leads to a spectral distribution of eigenvalues λ_n – the inverses of the time constants for the decay of the system toward equilibrium. The smallest eigenvalue yields the longest time constant, which ultimately controls the rate of cleanup.

DIFFUSION-ADVECTION FROM AN UNDERLYING FLOATING NAPL POOL

Osejo and Wilson's (1991) analysis of the evaporation of VOC from an underlying floating LNAPL pool was described in Chapter 5. Here, we discuss some of these results. Default parameters are given in Table 6.18. Preliminary measurements of the rate of loss of hexane by evaporation through tubes packed with fine sand gave a value for the molecular diffusivity of hexane in this medium of 2.6×10^{-6} m^2/sec. If one assumes a soil grain size δ of 0.1 cm and a soil gas velocity of 0.1 m/sec, Scheideg-ger's (1974) formulas yield the values given in the table for the longitudinal and transverse dispersivities.

The values of $C(x,y)$, the soil gas VOC concentration, obtained by the models were markedly dependent on the number of terms used in the Fourier series, exhibited oscillatory behavior, and were sometimes negative if large values of b (the vadose zone thickness) were used, particularly if the values of x, the horizontal distance along the NAPL pool in the direction of gas flow, were small. On the other hand, values of $C(x,y)$ were virtually independent of b as long as the values of x used were small enough that $C(x,y) \rightarrow 0$ for values of y significantly smaller than b. In most of our calculations, we used a value of b of 0.5 m, with values of $x < 20$ m. Under these conditions the first model, with boundary condition $C(x,b) = 0$, and the second model, with $\partial C(x,b)/\partial y = 0$, yield

TABLE 6.18. Model Parameters Used for Vapor Stripping Underlying NAPL.

Soil thickness b	0.5 m
Soil gas velocity v_x	0.1 m/sec
Longitudinal dispersivity D_x	1.78×10^{-4} m^2/sec
Transverse dispersivity D_y	8.1×10^{-6} m^2/sec
No. of terms in Fourier series	100
Concentration of saturated vapor C_0	0.666 kg/m^3
Porosity	0.2

results that are virtually identical and that are independent of the exact value of b selected (0.35, 0.5, 1.0 m). Using large values of b and small values of y necessitates the evaluation of a Fourier series near a discontinuity, convergence of which is slow or (at the discontinuity) may fail altogether. This problem is avoided by using artificially small values of b, which are still large enough that $C(x,b/2) = 0$ over the range of x of interest.

Plots of $C(x,y)/C_0$ versus y for $x = 1, 2, 5, 10$, and 20 m were calculated with the first model [$C(x,b) = 0$] and are shown in Figure 6.49. Since $C(20,y) \rightarrow 0$ as $y \rightarrow 0.20$ m, a value of 0.5 m for b is sufficient. From these plots one can estimate the flux of contaminant being carried in the moving gas stream at a distance x downwind of the edge of the NAPL pool. We do this as follows. Take the boundary layer thickness $l(x)$ as the value of y for which $C/C_0 = 1/2$, and approximate the curve $C(x,y)$ as linear in y. Then the flux per meter is given approximately by

$$F(x) = v_x \sigma l C_0 \tag{38}$$

For hexane at 25°C, $C_0 = 0.666$ kg/m³. From Table 6.18, $v_x = 0.1$ m/sec and $\sigma = 0.2$. From the plots in Figure 6.49, we see that $l(5 \text{ m}) = 1.87$ cm, $l(10 \text{ m}) = 2.68$ cm, and $l(20 \text{ m}) = 3.75$ cm. Substitution into Equation (38) then yields

$$F(5 \text{ m}) = 21.6 \text{ kg/m day}$$

$$F(10 \text{ m}) = 30.8 \text{ kg/m day}$$

$$F(20 \text{ m}) = 43.2 \text{ kg/m day}$$

One can also calculate the flux exactly from Equation (108) in Chapter 5. This was done using the parameters in Table 6.18, and the result is plotted in Figure 6.50. The exact calculation gives

$$F(5 \text{ m}) = 26.11 \text{ kg/m day}$$

$$F(10 \text{ m}) = 36.92 \text{ kg/m day}$$

$$F(20 \text{ m}) = 52.19 \text{ kg/m day}$$

indicating fairly good agreement between the estimate and the exact calculation.

The second model, for which $\partial C(x,b)/\partial y = 0$, was used to calculate plots of $C(x,y)$ and total flux $F(x)$ using the parameters given in Table 6.18.

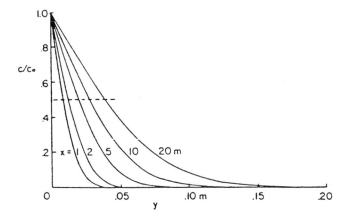

Figure 6.49 Concentration profiles in the vadose zone in the first 20 cm above the LNAPL pool at various distances x downwind from the edge of the pool. The boundary condition $C(x,b) = 0$ was used in this run. The parameters for the run are given in Table 6.18. Hexane at 25°C is being modeled in Figures 6.49–6.52. [Reprinted from Osejo and Wilson (1991) p. 1433, by courtesy of Marcel Dekker, Inc.]

Plots of $C(x,y)/C_0$ versus y for $x = 1, 2, 5, 10,$ and 20 m are given in Figure 6.51; a plot of $F(x)$ is given in Figure 6.52. These curves are indistinguishable from those of Figures 6.49 and 6.50, indicating the irrelevance of the boundary condition at the surface of the soil until solute has migrated up to that surface.

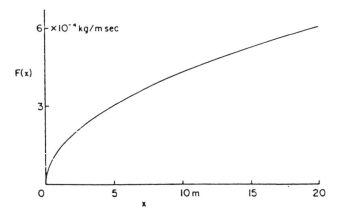

Figure 6.50 Flux $F(x)$ of VOC evaporated from an LNAPL pool of length x. Note: 10^{-4} kg/m sec = 8.64 kg/m day. See Table 6.18 for the run parameters. $C(x,b) = 0$. [Reprinted from Osejo and Wilson (1991) p. 1433, by courtesy of Marcel Dekker, Inc.]

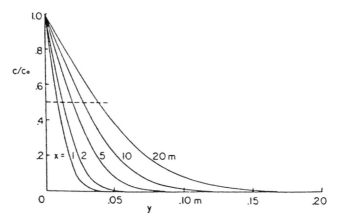

Figure 6.51 Concentration profiles in the vadose zone in the first 20 cm above the LNAPL pool at various distances x downwind from the edge of the pool. The boundary condition $\partial C(x,b)/\partial y = 0$ was used here. Run parameters are given in Table 6.18. [Reprinted from Osejo and Wilson (1991) p. 1433, by courtesy of Marcel Dekker, Inc.]

These fluxes are large enough to demonstrate the feasibility of vapor stripping underlying NAPL pools, provided that the vapor pressure of the VOC at the ambient temperature is high enough. One does not expect the dispersivities D_x and D_y to depend to any great extent on the identity of the NAPL being stripped. The only other term in the equations that depends

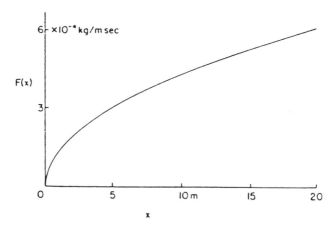

Figure 6.52 Flux $F(x)$ of VOC evaporated from a pool of width x. The boundary condition $\partial C(x,b)/\partial y = 0$ was used. Run parameters as in Table 6.18. [Reprinted from Osejo and Wilson (1991) p. 1433, by courtesy of Marcel Dekker, Inc.]

on the identity of the NAPL is C_0, the equilibrium vapor concentration, given by Equation (39):

$$C_0 = [0.01603 \ (MW)/T]P_0(T) \tag{39}$$

where

(MW) = VOC molecular weight, gm/mol
T = temperature, deg K
$P_0(T)$ = equilibrium vapor pressure at temperature T, torr

Table 6.19 lists these concentrations for twelve solvents over the temperature range 10 to 25°C; the lowest is that for p-xylene at 10°C, which is 0.0208 kg/m³. This generates a worst case for the soil vapor stripping of these NAPLs. For p-xylene we multiply the fluxes found for hexane by C_0(xylene, 10°C)/C_0(hexane, 25°C) = 0.0312 to get for p-xylene at 10°C

$$F(5 \ m) = 0.81 \ kg/m \ day$$

$$F(10 \ m) = 1.15 \ kg/m \ day$$

$$F(20 \ m) = 1.63 \ kg/m \ day$$

Even a solvent with a vapor pressure as low as that of xylene at 10°C can be evaporated at a reasonable rate from a pool of NAPL underlying the vadose zone. Vapor stripping of the other eleven solvents from a NAPL

TABLE 6.19. Equilibrium Concentrations for the Vapors of Some Common Organic Solvents.

Solvent	C_0, kg/m³			
	10°C	15°C	20°C	25°C
Benzene	.201	.253	.316	.391
Toluene	.0610	.0793	.102	.130
p-Xylene	.0208	.0276	.0364	.0475
n-Hexane	.351	.442	.550	.666
n-Heptane	.110	.141	.181	.230
n-Octane	.0361	.0479	.0630	.0815
C_2HCl_3 (TCE)	.256	.325	.409	.511
C_2Cl_4 (PCE)	.0737	.0960	.125	.161
CH_3CCl_3 (1,1,1-TCA)	.432	.544	.679	.839
CCl_4	.443	.559	.700	.869
C_6H_5Cl	.0309	.0408	.0533	.0690

pool at temperatures of 10°C or higher will be more rapid than that of
p-xylene. Fluxes for solvent M can be calculated by the formula

$$F_M(x,T) = [C_{0M}(T)/C_{0hexane}, 25°C] \times F_{hexane}(x, 25°C) \qquad (40)$$

EFFECTS OF VARIABLE AIRFLOW RATES IN DIFFUSION-LIMITED SVE OPERATION

In this section we explore a model for SVE developed in Chapter 5,
which includes the effects of mass transport kinetics of VOC between
nonaqueous phase liquid (NAPL) droplets and the aqueous phase and between the aqueous and vapor phases. The model permits time-dependent
gas flow rates in the vapor extraction well. The model is employed to demonstrate the effectiveness of certain types of pilot-scale SVE experiments
in determining the rate of mass transport processes. It is also used to explore several time-dependent airflow schedules for SVE well operation.
The results indicate that the use of suitable airflow schedules in SVE can
result in greatly reduced volumes of air to be treated for VOC removal with
relatively little increase in the time required to meet remediation standards.

As mentioned earlier in this chapter, one of the more significant site-
specific aspects of SVE is the extent to which the kinetics of diffusion
and/or desorption may limit the rate at which VOCs can be removed. If the
soil has a highly heterogeneous permeability, if it contains significant
amounts of clay or humic organic material, or if it contains substantial
amounts of water or grease/tar, the kinetics of diffusion and/or desorption
may be serious bottlenecks in the removal of VOCs by SVE. DiGiulio et
al. (1990) have discussed this and have described pilot-scale experiments
to ascertain the extent to which mass transport kinetics problems may slow
down the remediation. Oma et al. (1990) have looked at the impact of
nonequilibrium mass transport on SVE costs, and Gómez-Lahoz et al.
(1993) have explored some aspects of the economic advantages to be obtained by pulsed operation of SVE systems within the framework of a one-
dimensional model.

The mathematical model for soil vapor extraction for which we shall
present results here includes two possible kinetic bottlenecks and allows
one to vary arbitrarily the airflow rate through the vacuum well. The kinetic bottlenecks are (1) the rate of aqueous solution of droplets of nonaqueous phase liquid (NAPL) distributed within the porous medium and
(2) the rate of mass transport of dissolved VOC into the moving gaseous
phase. The pilot-scale experiments proposed by DiGiulio et al. (1990) are
then simulated with the model and found to provide valuable information

about the rates of these kinetically limited processes. Air injection experiments proposed by Mutch (1990) are also simulated. Lastly, several gas flow operating schedules for an SVE well are simulated with the objective of substantially reducing the total volume of soil gas, which must be treated without substantially increasing the time required to achieve the target level of remediation.

The model describes the operation of a single soil vapor extraction well screened at the bottom and drilled in a homogeneous, isotropic medium. The VOC contaminant is assumed to be initially present as NAPL, as dissolved VOC in the soil moisture, and as vapor in the soil gas. Mass transport of VOC between the NAPL phase and the aqueous phase is handled by means of a technique described earlier for modeling the solution of DNAPL droplets in groundwater in pump-and-treat operations (Kayano and Wilson, 1993) and in sparging (Roberts and Wilson, 1993a); see Chapter 4. Mass transport of VOC between the aqueous phase and the moving vapor phase is handled by means of a lumped parameter method used previously in SVE modeling (Wilson et al., 1992, for example; also Chapter 5).

VARIABLE AIRFLOW RATE RESULTS

A program implementing the variable airflow SVE model was written in TurboBASIC and run on MS-DOS microcomputers (80386-SX and 80386-DX microprocessors) equipped with math coprocessors and having clock speeds of 16 and 33 MHz, respectively. Run times ranged from about 15 min to as long as 4 hr.

There are three points of particular interest. The first is the extent to which one can gain useful information about kinetics limitations by examining the rebound of the soil gas VOC concentration in the vicinity of a well stripping an isolated domain after the well has been shut down, an experimental technique proposed by DiGiulio et al. (1990). Secondary points are the extent to which rate constants estimated from rebound rates agree with rate constants estimated from Equations (145) or (148) in Chapter 5 and with rate constants calculated from Equations (154) and (168) in Chapter 5. The second major point is the extent to which one can reduce the volume of water-saturated soil gas from which VOCs must be removed without seriously increasing the time required for remediation. The third is the extent to which one can gain information about kinetics limitations by examining the rebound of soil gas VOC concentration after injection of a slug of clean air into the domain of interest.

Default parameters for the runs to be described are given in Table 6.20. The VOC parameters were selected to correspond to those of trichloro-

TABLE 6.20. Default Parameters for the Variable Air Flow
SVE Modeling Runs Presented.

Radius of domain of interest	5 m
Thickness of vadose zone	5 m
Height of well above the bottom of the vadose zone	0.1 m
N_r, N_z	5, 5
Air-filled porosity	0.3
Water-filled porosity	0.1
Pneumatic permeability	1.0 m²/atm sec
Density of soil	1.7 gm/cm³
Identity of the VOC	Trichloroethylene (TCE)
Water solubility of VOC	1,100 mg/L
Effective Henry's constant of VOC, dimensionless	0.2
Density of NAPL VOC	1.46 gm/cm³
Diffusivity of VOC in the aqueous phase	2 × 10⁻¹⁰ m²/sec
Time constant λ for aqueous VOC/vapor transport	1 × 10⁻⁴ sec⁻¹
Initial NAPL concentration in the soil	2,000 mg/kg
Initial NAPL droplet diameter, $2a_0$	0.1 cm
Radius of zone of contamination	5 m
Depth of zone of contamination	5 m
Molar gas flow rate	1.0 mol/sec, 50.95 ft³/min
Ambient temperature	20°C
Initial total contaminant mass	1,404 kg
Δt	100 sec

ethylene (TCE). In Figure 6.53 we see the course of remediation as measured by plotting M_{total} against time. For this run the NAPL droplet diameter is 0.1 cm, so the rate of solution is relatively rapid, and the remediation is complete in about eighteen days. Examination of Equation (133) in Chapter 5 leads one to Equation (41) as an absolute lower bound for the 100% cleanup time for the case in which solution kinetics are limiting.

$$t(100\%) = \frac{\varrho a_0^2}{2Dc_s} \qquad (41)$$

Substitution of numerical values into Equation (41) gives a lower bound for the cleanup time of 9.6 days, so the system is not severely diffusion-limited.

In Figure 6.54 the soil gas effluent VOC concentrations (measured as $c^g_{(2.jwell)}$ are plotted against time (on the same time scale as used in Figure 6.53) for runs in which the gas flow was turned off after one, three, five, or fifteen days. In these runs the diameter of the NAPL droplets is 0.1 cm, so solution of the droplets is comparatively rapid. The rate of rebound of the soil gas VOC concentration toward its equilibrium value (.22 kg/m³) decreases somewhat with increasing duration of the evacuation phase, and

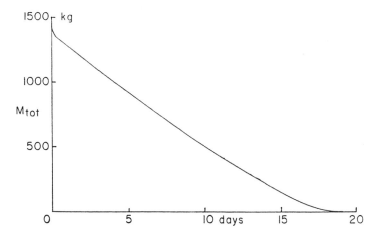

Figure 6.53 Plot of mass of residual contaminant versus time. $2a_0 = 0.1$ cm. Default parameters as in Table 6.20. [Reprinted from Gómez-Lahoz et al. (1994b) p. 956, by courtesy of Marcel Dekker, Inc.]

the rebound is fairly rapid. After fifteen days of treatment the rate of rebound is about 49% of the rebound rate after one day of treatment. The rate constants k for rebound were calculated by determining the half-life of the recovery to the equilibrium concentration (0.22 kg/m³) and then setting $k = 0.693/t_{1/2}$. This procedure gave results virtually identical to those obtained by least squares fits of the data and is much less laborious. Rate con-

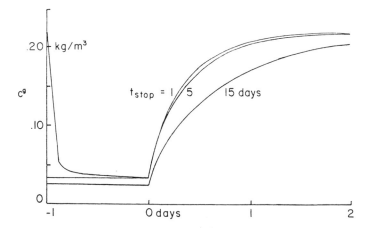

Figure 6.54 Plots of c^g_{effl} versus time since shutdown, $t_{stop} = 1$, 3, 5, and 15 days. $2a_0 = 0.1$ cm. Default parameters as in Table 6.20. [Reprinted from Gómez-Lahoz et al. (1994b) p. 957, by courtesy of Marcel Dekker, Inc.]

TABLE 6.21. Values of k, k', and k'' Calculated for
the Runs Shown in Figure 6.54.

Stop Time, Days	k, sec⁻¹	k', sec⁻¹	k'', sec⁻¹
1	3.54×10^{-5}	2.46×10^{-5}	2.25×10^{-5}
5	3.25×10^{-5}	2.34×10^{-5}	2.15×10^{-5}
15	1.72×10^{-5}	1.57×10^{-5}	1.48×10^{-5}

stants k' and k'' were estimated from Equations (145) and (148) in Chapter 5, making the assumption that the volume V has the dimensions given in Table 6.20 ($h' = 4.9$ m, $r' = h'$), giving it a value of 184.8 m³. Values of k, k', and k'' are given in Table 6.21. The rather drastically simplified single-compartment or plug flow calculations leading to Equations (145) and (148) in Chapter 5 both appear to estimate values for the rate constant for VOC concentration recovery, which are within roughly 30% of the values obtained from the recovery curves themselves.

In all the runs presented here the principal bottleneck is the rate of solution of NAPL from the droplets. Figure 6.55 shows plots of c^g_{effl} versus time for droplets of initial effective diameter $2a_0 = 0.1, 0.2,$ and 0.3 cm. In these three runs the gas flow was stopped after five days and the rebound of the soil gas VOC concentration was plotted for an additional seven days. Default parameters are given in Table 6.20. As expected, the increase in NAPL droplet diameter results in a severe decrease in the rate of recovery

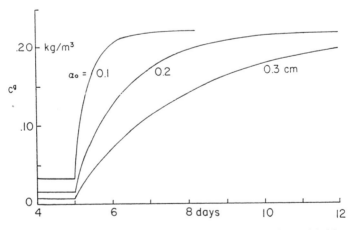

Figure 6.55 Plots of c^g_{effl} versus time since shutdown, $t_{stop} = 5$ days. $2a_0 = 0.1, 0.2,$ and 0.3 cm. Default parameters as in Table 6.20. [Reprinted from Gómez-Lahoz et al. (1994b) p. 958, by courtesy of Marcel Dekker, Inc.]

TABLE 6.22. Values of k, k', and k'' Calculated for
the Runs Shown in Figure 6.55.

$2a_0$, cm	k, sec^{-1}	k', sec^{-1}	k'', sec^{-1}
0.1	3.25×10^{-5}	2.34×10^{-5}	2.15×10^{-5}
0.2	8.81×10^{-6}	9.27×10^{-6}	8.95×10^{-6}
0.3	3.87×10^{-6}	4.68×10^{-6}	4.60×10^{-6}

of VOC concentration. Values of k, k', and k'' calculated for these runs are given in Table 6.22. Again, the values of k' and k'' are within roughly 30% of those of k.

In the runs to be considered next, the value of $2a_0$ used is 0.3 cm, so these runs are severely solution-kinetics limited. Other parameters are as in Table 6.20. Figure 6.56 shows a plot of M_{total} versus time for a run having a constant airflow rate of 1 mol/sec (51 ft^3/min). The time required for cleanup is slightly over ninety days. The lower bound estimate given by Equation (41) is 86.4 days, confirming the severe diffusion-kinetics limitation of the system.

Additional runs were made with these parameters in which the well was stopped, and c_{effl}^g was monitored to ascertain its recovery pattern at various times during the progress of the cleanup. These results are shown in Figure 6.57. The gas flow was stopped after five, twenty, and seventy days of SVE, as indicated on the figure. Here, one sees a marked decrease in

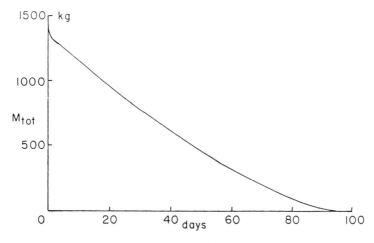

Figure 6.56 Plot of residual VOC mass versus time, $2a_0 = 0.3$. Default parameters as in Table 6.20. [Reprinted from Gómez-Lahoz et al. (1994b) p. 959, by courtesy of Marcel Dekker, Inc.]

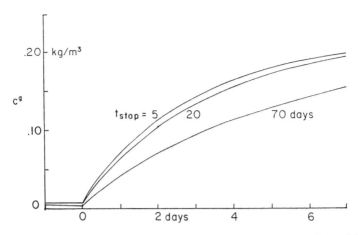

Figure 6.57 Plots of c^g_{zfft} versus time shutdown, t_{stop} = 5, 20, and 70 days. $2a_0$ = 0.3 cm. Default parameters as in Table 6.20. [Reprinted from Gómez-Lahoz et al. (1994b) p. 960, by courtesy of Marcel Dekker, Inc.]

rate of recovery toward the end of the cleanup. Evidently, one is not likely to obtain data indicating the full severity of mass transport limitations from runs carried out for a relatively short period of time. However, the effects do not appear to be extreme, as indicated by the relatively modest curvature seen in the plot of M_{total} in Figure 6.56 and suggested by the weak dependence of Equation (154) in Chapter 5 on C. The values of k, k', and k'' calculated from the runs shown in Figure 6.57 are given in Table 6.23. As before, we find that k' and k'' are roughly comparable to the rate constant for the recovery of the VOC concentration in the soil gas. It is evident that Equation (154) in Chapter 5 gives a very reasonable estimate for the value of the VOC rebound rate constant for this model.

Equation (168) in Chapter 5 was also used to calculate the rate constant associated with the soil gas VOC concentration rebound for this system. Table 6.24 shows the dependence of the smaller of the two roots of Equa-

TABLE 6.23. Values of k, k', and k'' Calculated for
the Runs Shown in Figure 6.57.

Stop Time, Days	k, sec^{-1}	k', sec^{-1}	k'', sec^{-1}	β, sec^{-1}
(0)	—	—		3.88×10^{-6}
5	3.87×10^{-6}	4.68×10^{-6}	4.60×10^{-6}	
20	3.52×10^{-6}	4.31×10^{-6}	4.24×10^{-6}	
70	2.02×10^{-6}	2.61×10^{-6}	2.58×10^{-6}	

TABLE 6.24. Dependence of Soil Gas VOC
Concentration Rebound Rate Constant on
the Progress of Cleanup. $2a_0 = 0.3$ cm. Other
Parameters as in Table 6.20.

C, kg/m³	Λ-, sec⁻¹
3.4	3.82×10^{-6}
3.0	3.67×10^{-6}
2.6	3.50×10^{-6}
2.2	3.31×10^{-6}
1.8	3.10×10^{-6}
1.4	2.86×10^{-6}
1.0	2.56×10^{-6}
0.8	2.37×10^{-6}
0.6	2.16×10^{-6}
0.4	1.89×10^{-6}
0.2	1.50×10^{-6}
0.15	1.36×10^{-6}
0.10	1.19×10^{-6}
0.05	0.947×10^{-6}
0.02	0.699×10^{-6}
0.01	0.555×10^{-6}

tion (168) in Chapter 5 on the extent to which the cleanup has progressed. The initial NAPL concentration was 2,000 mg/kg (3.4 kg/m³), and values of the rate constant are given down to an NAPL concentration of 0.01 kg/m³. In agreement with Equation (154) from Chapter 5 and with the numerical results, the rate constant is found to decrease during the progress of the cleanup, but the effect is not large until almost the very end.

The great impact of mass transport kinetics limitations on the remediation is evident when one calculates the volume of air required to move 1,404 kg of VOC (the initial mass present) if the gas is saturated ($c_{sat}^g = K_H \cdot c_s = 0.2 \cdot 1.1$ kg/m³ $= 0.22$ kg/m³); this is 6,383 m³. The volume of air actually used in the run shown in Figure 6.56 to achieve cleanup is about 191,000 m³, a volume about thirty times greater than this theoretical minimum. Evidently, in a run of this sort, one treats an enormous volume of gas that is highly dilute in VOC, an expensive proposition. We therefore turn to our second major objective, the exploration of means by which one might reduce the very large volumes of gas that are likely to be handled at sites that are kinetically controlled.

The runs shown in Figures 6.58, 6.59, and 6.60 test the feasibility of three approaches. Default parameters are as in Table 6.20, except that $2a_0 = 0.3$ cm in all these runs. Diffusion limitation is severe, as was the case in Figure 6.56. Recall that the 100% cleanup time [$t(100\%)$] for the

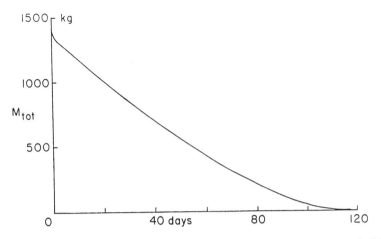

Figure 6.58 Plot of residual VOC mass versus time, $2a_0 = 0.3$. Default parameters as in Table 6.20. The airflow rate $Q(t)$, initially 1 mol/sec, is multiplied by 0.9 whenever c^g_{effl} is less than $0.2 \cdot c^g_{sat}$. [Reprinted from Gómez-Lahoz et al. (1994b) p. 962, by courtesy of Marcel Dekker, Inc.]

run plotted in Figure 6.56 is ninety-five days, the gas flow rate in that run is 51 ft³/min, and the total volume of gas used is 191,000 m³.

In Figure 6.58 the same parameters are used as in Figure 6.56, except that the concentration of the effluent soil gas, c^g_{effl}, was continuously monitored and whenever it got below $0.2 \cdot c^g_{sat}$, the current value of the molar

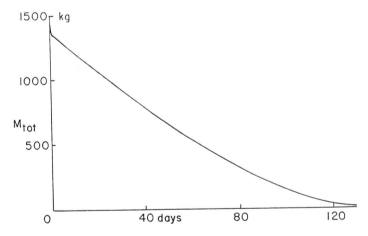

Figure 6.59 Plot of residual VOC mass versus time, $2a_0 = 0.3$. Default parameters as in Table 6.20. $Q(t) = 0.4$ mol/sec but is switched to 0 whenever c^g_{effl} is less than $0.2 \cdot c^g_{sat}$. [Reprinted from Gómez-Lahoz et al. (1994b) p. 962, by courtesy of Marcel Dekker, Inc.]

gas flow rate of the well was multiplied by 0.9. During the first day of the run, the molar gas flow rate dropped rapidly to about one-fifth of its initial value of 1.0 mol/sec (51 ft³/min), then slowly drifted down to about 4% of its initial value by the time cleanup was 98% complete. The time required for cleanup was 116 days, and the total volume of air drawn from the well was 29,600 m³, about 15.5% that required in the run pictured in Figure 6.56.

In Figure 6.59 we see the results of a run in which the molar airflow rate was 0.4 mol/sec (25 ft³/min), but for which the air flow was switched off completely when the value of c^g_{effl} went below $0.2 \cdot c^g_{sat}$. When c^g_{effl} recovered to a value above $0.4 \cdot c^g_{sat}$, the air flow was switched back on. This cycle was continued throughout the run. The cleanup time was 130 days, and the volume of gas drawn from the well was 22,100 m³, 11.6% that required in the run pictured in Figure 6.56.

Figure 6.60 shows results for a run in which the molar airflow rate was 0.1 mol/sec (5.1 ft³/min). This air flow was switched on and off using the same criteria as were used for the run shown in Figure 6.59, but the airflow rate was low enough that no switching occurred until the run was nearly complete. The cleanup time was 125 days, and the volume of gas drawn from the well was 21,400 m³, 11.2% that required in the run shown in Figure 6.56. It is apparent that very substantial reductions on the costs of soil gas pumping and treatment can be made with little increase in cleanup time by careful selection of pumping speeds.

We now turn to the results obtained modeling the clean air slug injection test. The computer program modeling this test was written in BASICA and

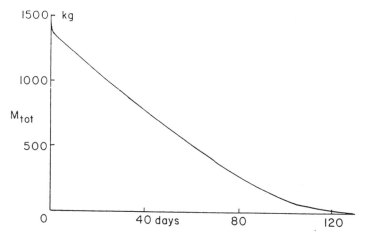

Figure 6.60 Plot of residual VOC mass versus time, $2a_0 = 0.3$. Default parameters as in Table 6.20. $Q(t) = 0.1$ mol/sec, but is switched to 0 whenever c^g_{effl} is less than $0.2 \cdot c^g_{sat}$. [Reprinted from Gómez-Lahoz et al. (1994b) p. 963, by courtesy of Marcel Dekker, Inc.]

TABLE 6.25. **Default Parameters Used in Modeling the VOC Vapor Concentration Rebound after Injection of a Slug of Clean Air.**

Initial NAPL concentration	2,000 mg/kg
Initial VOC concentration in the aqueous phase	1,100 mg/L
Soil density	1.7 gm/cm³
Density of NAPL	1.46 gm/cm³
Solubility of NAPL in water	1,100 mg/L
Diffusion constant of VOC in aqueous phase of porous medium	2×10^{-10} m²/sec
Effective diameter of NAPL droplets	0.2 cm
Effective Henry's constant of VOC, dimensionless	0.2
Rate constant λ for water-vapor VOC transport	1×10^{-4} sec⁻¹
Water-filled porosity	0.1
Air-filled porosity	0.3
Δt	112.5 sec
Duration of test	432,000 sec (5 days)

run with the BASICA interpreter; a run required just a few seconds. Default parameters for these runs are given in Table 6.25.

In Figure 6.61 plots of c^g/c^g_{sat} versus time are shown for various values of the effective NAPL droplet diameter. Here c^g_{sat} is the saturation vapor concentration of the VOC, given by K_{HC_s}. The initial very rapid rise in vapor phase VOC concentration is associated with mass transport from the aqueous phase to the vapor phase; the rate constant λ for this process in these runs was relatively large, 10^{-4} sec⁻¹. Long after this process has

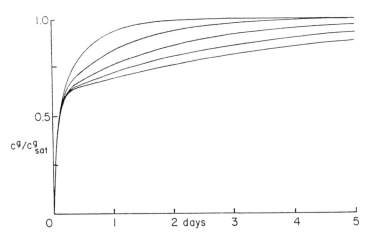

Figure 6.61 Plot of c^g/c^g_{sat} versus time, air injection slug test. Effect of NAPL effective droplet diameter. $2a_0 = 0.1, 0.15, 0.2, 0.25, 0.3$ cm from the top down. Other parameters as in Table 6.25. [Reprinted from Gómez-Lahoz et al. (1994b) p. 964, by courtesy of Marcel Dekker, Inc.]

come to a steady state, the mass transport of VOC from the NAPL phase through the aqueous phase to the vapor phase continues. The rate of mass transport of VOC from the NAPL phase decreases with decreasing surface to volume ratio of the NAPL, so the rate constant for solution of the larger NAPL droplets is much slower than it is for the smaller droplets.

The effect of the initial aqueous phase VOC concentration on the rate of rebound of the VOC vapor concentration is seen in Figure 6.62. The smaller the initial value of C^w, the smaller is the rapid initial rise in C^g and the more clearly is the much slower solution rate of the NAPL droplets displayed. These results suggest that one might wish to inject clean air for an extended period (perhaps several hours) in order to reduce C^w as much as possible before shutting off the air flow and following the kinetics of the VOC vapor concentration rebound. These results, like those in Figure 6.61, also indicate the importance of making sufficiently many measurements over a sufficiently long period of time so that the slow kinetic process(es) can be characterized. If one expects that "the" rate constant for mass transport is that associated with the very rapid initial rise of the rebound curve, one may be in for a very unpleasant surprise when the actual cleanup takes far longer than anticipated.

The effect of λ, the lumped parameter used to characterize the mass transport between the aqueous and vapor phases, is seen in Figure 6.63. For sufficiently small values of λ, the curves appear to be characterized by a single rate constant, and the "dog leg" seen for larger values of λ disappears. This relatively sharp break in the VOC vapor phase rebound curves

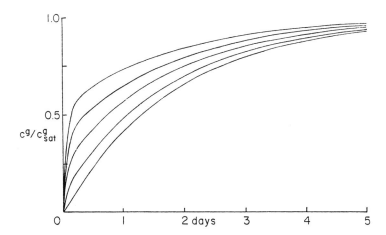

Figure 6.62 Plot of c_g/c_{sat}^g versus time, air injection slug test. Effect of initial aqueous phase VOC concentration. Initially $C^w = 50, 250, 500, 750, 1,000$ mg/L from the bottom up. Other parameters as in Table 6.25. [Reprinted from Gómez-Lahoz et al. (1994b) p. 964, by courtesy of Marcel Dekker, Inc.]

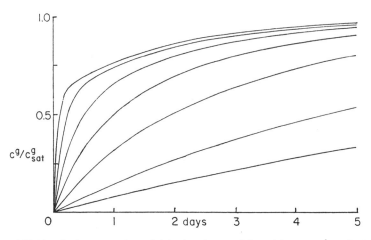

Figure 6.63 Plot of c^g/g_{sat}^g versus time, air injection slug test. Effect of the rate constant for VOC mass transport between the aqueous and vapor phases, λ. $\lambda = 1 \times 10^{-4}, 5 \times 10^{-5}, 2 \times 10^{-5},$ $1 \times 10^{-5}, 5 \times 10^{-6}, 2 \times 10^{-6}, 1 \times 10^{-6}$ sec^{-1}, from the top down. Other parameters as in Table 6.25. [Reprinted from Gómez-Lahoz et al. (1994b) p. 965, by courtesy of Marcel Dekker, Inc.]

is seen only when the rate of mass transport from the aqueous to the gaseous phase is markedly faster than the rate of mass transport from the NAPL to the aqueous phase.

The effect of the relative magnitudes of the air-filled porosity σ and the water-filled porosity ω is seen in Figure 6.64. Here, the rate constant for

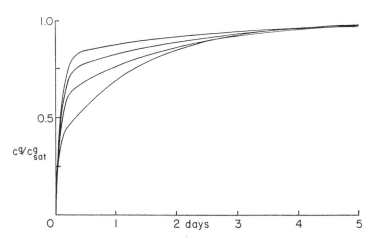

Figure 6.64 Plot of c^g/c_{sat}^g versus time, air injection slug test. Effect of increasing the water-filled porosity ω (and decreasing the air-filled porosity σ). $\omega + \sigma = 0.4$. Water-filled porosity = 0.2, 0.15, 0.1, 0.05 from the top down. Air-filled porosity = 0.2, 0.25, 0.3, 0.35 from the top down. Other parameters as in Table 6.25. [Reprinted from Gómez-Lahoz et al. (1994b) p. 966, by courtesy of Marcel Dekker, Inc.]

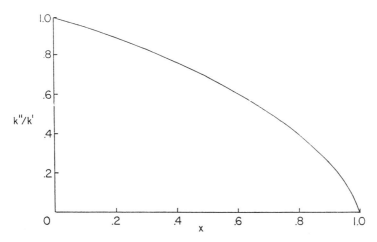

Figure 6.65 Plot of k''/k' versus c_{ss}^g/c_{equil}^g ($=x$). [Reprinted from Gómez-Lahoz et al. (1994b) p. 966, by courtesy of Marcel Dekker, Inc.]

mass transport of VOC between the aqueous and the vapor phases is relatively large (10^{-4} sec^{-1}), so the curves show the characteristic dogleg appearance. The larger the water-filled porosity (and the smaller the air-filled porosity), the higher is the initial rapid rebound of the curve and the poorer is the characterization of the NAPL-aqueous phase mass transfer process. Evidently, one would be well advised to try to carry out these tests on soils that were as well drained as possible and certainly not shortly after periods of very wet weather.

The relationship between the two methods for roughly estimating mass transfer kinetics rate constants [Equations (145) and (148) in Chapter 5] is indicated in Figure 6.65, in which the ratio k''/k' is plotted against c_{ss}^g/c_{equil}^g. This ratio is given by $-[(1-x)/x] \cdot \log_e(1-x)$, where $x = c_{ss}^g/c_{equil}^g$. We see that the values of k'' are always less than those of k', with k'' approaching k' as c_{ss}^g/c_{equil}^g approaches zero. We find no such simple inequality between the values of k' and k'', on the one hand, and k, on the other.

DISTRIBUTED DIFFUSION MODEL RESULTS

The distributed diffusion lab column model described in Chapter 5 was used to explore the effects of mass transport kinetics limitations in some detail. Parameters for the runs are given in Table 6.26 or in the captions to the figures. The VOC simulated was trichloroethylene.

In Figures 6.66–6.69 total residual VOC mass and effluent soil gas VOC concentration are plotted on a normalized basis—that is, these variables

TABLE 6.26. Default Parameters for Laboratory Column Simulations, Solution and Distributed Diffusion Model.

Column length	50 cm
Column diameter	10 cm
Soil air-filled porosity	0.2
Soil water-filled porosity	0.2
Soil density	1.7 gm/cm³
Water layer thickness	1.0 cm
VOC being simulated	Trichloroethylene
Henry's constant of VOC	0.2821
Aqueous solubility of VOC	1,100 mg/L
Density of NAPL VOC	1.46 gm/cm³
Diffusion constant of VOC in water-saturated porous medium	2×10^{-6} cm²/sec
Initial NAPL droplet diameter	0.1 cm
Airflow rate	0.1 mL/sec
Total VOC concentration in soil	2,000 mg/kg
Number of volume elements into which column is partitioned	10
Number of slabs into which each volume element is partitioned	10
Δt	450 sec
Duration of run	4,320,000 sec (50 days)

are divided by their values at the beginning of the run. The run durations are fifty days. The initial concentrations are: NAPL, 0.00312; aqueous, 0.00110 (saturated); vapor, 0.00031 gm/cm³ (saturated). The initial total VOC concentration in the soil is 2,000 mg/kg, so this soil is highly contaminated. For Figures 6.66–6.69 the initial NAPL droplet diameters are 0.01, 0.025, 0.05, and 0.1 cm, respectively.

The plots of effluent soil gas VOC concentration indicate initial saturation, followed quickly by a rapid fall-off through a transition region lasting only a few days, which, in turn, leads into a prolonged region of tailing before, rather abruptly, the soil gas VOC concentrations decrease to zero as remediation becomes complete. Despite the rather low gas flow rate (6 mL/min), the effluent VOC concentrations are far below saturation even when the bulk of the residual VOC in the column is present as NAPL. Attempts to fit linear or exponential curves to these effluent soil gas concentrations would be futile. Examination of the soil gas curves only in the initial phases of remediation would lead to wildly optimistic estimates of cleanup times. Examination of the rather flat regions between roughly days 20 and 45 might lead one to conclude that the tailing period would last for hundreds of days. In fact, as seen from the total mass curves, cleanup is proceeding in good order, with all cleanup times being roughly fifty days.

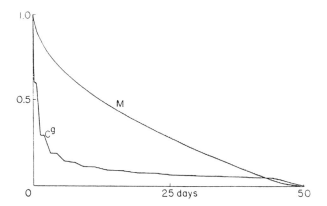

Figure 6.66 Plots of normalized residual mass of VOC and normalized effluent soil gas concentration versus time. In this run the initial NAPL droplet diameter is 0.01 cm. Other parameters are given in Table 6.26. Initial NAPL, aqueous, and vapor VOC concentrations are 0.00312, 0.00110, and 0.00031 gm/cm³. [Reprinted from Wilson (1994) p. 588, by courtesy of Marcel Dekker, Inc.]

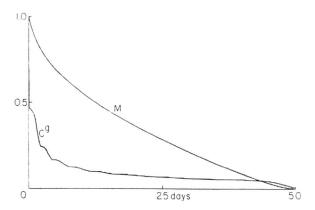

Figure 6.67 Plots of normalized residual mass of VOC and normalized effluent soil gas concentration versus time. Initial NAPL droplet diameter = 0.025 cm. Other parameters as in Table 6.26 and Figure 6.66. [Reprinted from Wilson (1994) p. 588, by courtesy of Marcel Dekker, Inc.]

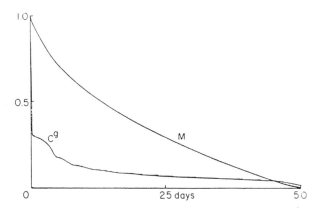

Figure 6.68 Plots of normalized residual mass of VOC and normalized effluent soil gas concentration versus time. Initial NAPL droplet diameter = 0.05 cm. Other parameters as in Table 6.26 and Figure 6.66. [Reprinted from Wilson (1994) p. 589, by courtesy of Marcel Dekker, Inc.]

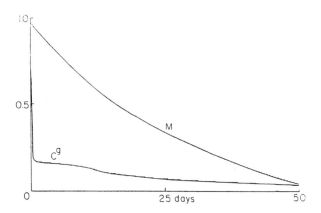

Figure 6.69 Plots of normalized residual mass of VOC and normalized effluent soil gas concentrations versus time. Initial NAPL droplet diameter = 0.10 cm. Other parameters as in Table 6.26 and Figure 6.66. [Reprinted from Wilson (1994) p. 589, by courtesy of Marcel Dekker, Inc.]

Systems with larger NAPL droplet sizes clean up more slowly than those with smaller droplets, since the total NAPL-water interfacial area is less if the droplets are large.

Figures 6.70–6.73 show the effect on cleanup rate of the thickness of the water layer in whch the VOC is dissolved. Here the initial VOC concentration is only 100 mg/kg, so no NAPL phase is present even initially. The gas flow rate in these runs is 1.2 mL/min. In these runs diffusion transport through the soil water is the only rate-limiting factor. It is evident that thick aqueous layers result in quite slow remediation.

If the system is in a diffusion or solution rate-limiting regime, increasing gas flow rate increases costs of blowers and off-gas treatment but does not result in any significant decrease in cleanup time. In Figures 6.74 and 6.75 the airflow rate Q is varied fivefold, but the impact on the time required for cleanup is quite small.

The effect of the initial total VOC concentration on the effluent soil gas concentration is shown in Figure 6.76. In these runs the water diffusion layer is 1 cm thick; the NAPL droplet diameter is 0.1 cm; the initial total VOC concentrations are 100, 250, 500, 1,000 and 2,000 mg/kg; and the gas flow rate is 1.2 mL/min. Cleanup times increase quite substantially with increasing initial VOC concentration. In all runs in which NAPL is present (the run with 100 mg/kg initial VOC has no NAPL present), there is extensive tailing after the initial drop-off in effluent soil gas VOC. However, cleanup is complete within fifty days for all runs except that with 2,000 mg/kg initial VOC, which required seventy-five days for cleanup.

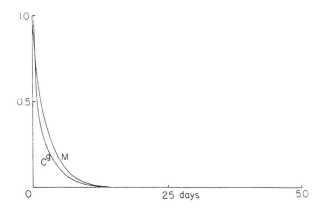

Figure 6.70 Plots of normalized residual mass of VOC and normalized effluent soil gas concentration. Initial total VOC concentration = 100 mg/kg; aqueous diffusion layer thickness 1 = 1.0 cm. No NAPL is present. Initial aqueous and vapor VOC concentrations are 0.0006633 and 0.000187 gm/cm³. Other parameters as in Table 6.26. [Reprinted from Wilson (1994) p. 592, by courtesy of Marcel Dekker, Inc.]

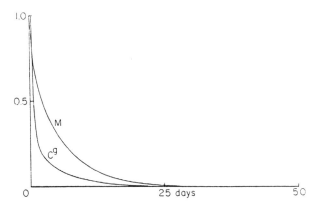

Figure 6.71 Plots of normalized residual mass of VOC and normalized effluent soil gas concentration. Initial total VOC concentration = 100 mg/kg; aqueous diffusion layer thickness 1 = 1.5 cm. No NAPL is present. Other parameters as in Table 6.26 and Figure 6.70. [Reprinted from Wilson (1994) p. 592, by courtesy of Marcel Dekker, Inc.]

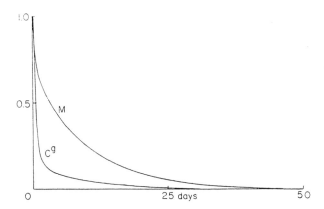

Figure 6.72 Plots of normalized residual mass of VOC and normalized effluent soil gas concentration. Initial total VOC concentration = 100 mg/kg; aqueous diffusion layer thickness 1 = 2.0 cm. No NAPL is present. Other parameters as in Table 6.26 and Figure 6.70. [Reprinted from Wilson (1994) p. 593, by courtesy of Marcel Dekker, Inc.]

304

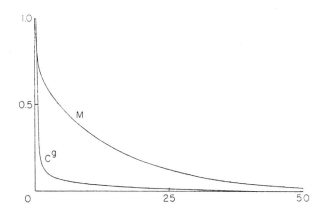

Figure 6.73 Plots of normalized residual mass of VOC and normalized effluent soil gas concentration. Initial total VOC concentration = 100 mg/kg; aqueous diffusion layer thickness 1 = 2.5 cm. No NAPL is present. Other parameters as in Table 6.26 and Figure 6.70. [Reprinted from Wilson (1994) p. 593, by courtesy of Marcel Dekker, Inc.]

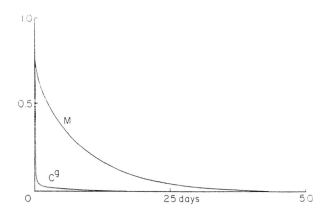

Figure 6.74 Plots of normalized residual mass of VOC and normalized effluent soil gas concentration. Initial total VOC concentration = 100 mg/kg; aqueous diffusion layer thickness l = 2.0 cm. No NAPL is present. The airflow rate is 0.1 mL/sec. Other parameters as in Table 6.26 and Figure 6.70. [Reprinted from Wilson (1994) p. 594, by courtesy of Marcel Dekker, Inc.]

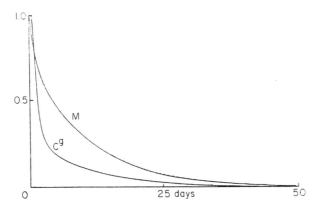

Figure 6.75 Plots of normalized residual mass of VOC and normalized effluent soil gas concentration. Initial total VOC concentration = 100 mg/kg; aqueous diffusion layer thickness l = 2.0 cm. No NAPL is present. The airflow rate is 0.01 mL/sec. Other parameters as in Table 6.26 and Figure 6.70. [Reprinted from Wilson (1994) p. 596, by courtesy of Marcel Dekker, Inc.]

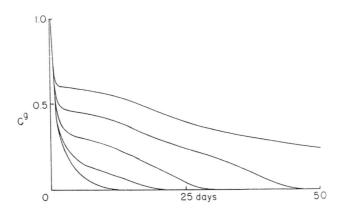

Figure 6.76 Plots of normalized effluent soil gas VOC concentration versus time. Effects of initial total VOC concentration. Initial total VOC concentrations are 100, 250, 500, 1,000 and 2,000 mg/kg. The aqueous diffusion layer thickness is 1.0 cm and the NAPL droplet size is 0.1 cm. (No NAPL is present in the run with initial VOC concentration of 100 mg/kg.) Gas flow rate = 0.02 mL/sec, Δt = 1,800 sec. Other parameters as in Table 6.26. [Reprinted from Wilson (1994) p. 596, by courtesy of Marcel Dekker, Inc.]

Figure 6.77 shows effluent soil gas VOC concentration plots for runs that were shut down after ten, twenty, thirty, forty, and fifty days. Cleanup is complete after about forty-five days if the run is not interrupted earlier. The effluent soil gas VOC concentration curves exhibit rebound after the gas flow is turned off; the vapor concentration rebounds to the saturation vapor pressure concentration if NAPL is still present. The rate of equilibration between the VOC in the vapor phase and the VOC in the condensed phase(s) decreases the longer the duration of the run before shutdown. This shows the error of using a single lumped parameter diffusion rate constant obtained from measurements made fairly near the beginning of a pilot-scale run. By the time forty days has elapsed, all the NAPL has dissolved; rebound in soil gas VOC concentration is therefore to a value determined by Henry's Law and the final aqueous VOC concentration, rather than to the saturation concentration. There is substantial rebound even after forty days of SVE, by which time only about 2% of the original quantity of VOC is still present in the column. Evidently, it would be difficult, if not impossible, to correlate the extent of cleanup with the effluent soil gas VOC concentration during operation of the well or with the final equilibrium value of the rebound soil gas VOC concentration unless cleanup was complete.

A number of field-scale SVE models have been developed that make use of the distributed diffusion approach (Wilson et al., 1994a; Gómez-Lahoz et al., 1994a, 1994b; Rodríguez-Maroto et al., 1994; Wilson et al., 1994b); all of these present results. We focus on the last of these papers since that model seems to be the most physically realistic.

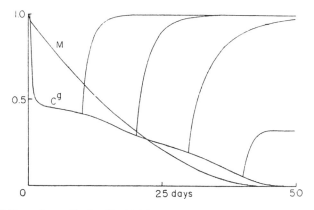

Figure 6.77 Plots of normalized residual VOC mass and normalized effluent soil gas VOC concentration; effect of shutting off the gas flow to the column after 5, 10, 20, 30, 40 and 50 days. Initial total VOC concentration = 1,000 mg/kg; water diffusion layer thickness = 1.0 cm; initial NAPL droplet diameter = 0.1 cm; gas flow rate = 0.02 mL/sec, t = 1,800 sec. Other parameters as in Table 6.26. [Reprinted from Wilson (1994) p. 597, by courtesy of Marcel Dekker, Inc.]

This model, of a single vertical SVE well, was implemented in TurboBASIC on an AlphaSystem 486-DX microcomputer running at 50 MHz. A simple Euler method was used for the integration. Typical runs required about 10–30 min, depending on how long an ageing period was simulated. The VOC that is simulated is trichloroethylene. Default parameters for the runs made with this model are given in Table 6.27.

Figure 6.78 shows the effect of ageing period (the period of time that elapses between the occurrence of the spill and the initiation of SVE) on SVE. If cleanup is initiated immediately, 10.75 days are required to achieve 99.5% removal of the VOC. If SVE is started one day after the spill, 19.34 days are required for 99.5% removal. For longer ageing periods, 99.5% removal is not achieved during the twenty-five–day period of SVE. For an ageing period of fifty days, twenty-five days of SVE leaves about 4.5% of the VOC in the soil; this run is shown in Figure 6.79. The run shows quite

TABLE 6.27. **Default Parameters Used with the Vertical Well Model.**

Radius of domain to be stripped	8 m
Depth of domain to be stripped	5 m
Depth of well	4.5 m
Volumetric gas flow rate of well	50 SCFM
	(0.0236 m³/sec)
Wellhead pressure	0.85 atm
Diameter of well gravel packing	30 m
Identity of VOC	Trichloroethylene, TCE
Aqueous solubility of VOC	1,100 mg/L
Henry's constant of VOC (dimensionless)	0.2821
Effective diffusion constant of VOC in water	
(diffusivity × tortuosity/ν_{clay})	2×10^{-10} m²/sec
Effective diffusion constant of VOC in air	
(diffusivity × tortuosity/ν_{soil})	2×10^{-8} m²/sec
Density of VOC	1.46 gm/cm³
Soil density	1.7 gm/cm³
Soil air-filled porosity σ	0.2
Soil water-filled porosity ω	0.2
Porosity of low-permeability clay lenses	0.4
Mean thickness 2l of porous clay lenses	5.0 cm
Inidial NAPL droplet diameter	0.1 cm
n_r	8
n_z	5
n_u	5
Total VOC concentration in the soil	1,000 mg/kg
Initial total mass of VOC	1,704 kg
Δt	100 sec
Duration of simulated SVE run	25 days (Figures 6.4–6.7)
	15 days (Figures 6.8–6.16)
Calculated Darcy's constant	0.1136 m²/atm sec

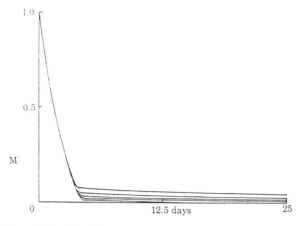

Figure 6.78 Plots of $M'[=M_{tot}(t)/M_{tot}(0)]$ versus time; effect of ageing. Ageing period = 0, 5, 10, 25 and 50 days, bottom to top. Duration of SVE = 25 days. Other parameters as in Table 6.27. [Reprinted from Wilson et al. (1994b) p. 1658, by courtesy of Marcel Dekker, Inc.]

rapid removal of VOC during the first four days (while the NAPL droplets are evaporating), followed by very slow removal as dissolved VOC diffuses out of the clay lenses (5 cm thick in this run). We note that the model parameters will be highly site-specific, so our results should be regarded as showing semiquantitative trends only. Still, the calculations make it very clear that one can expect to pay a heavy price for delays in SVE remediation if the soil contains low-permeability lenses.

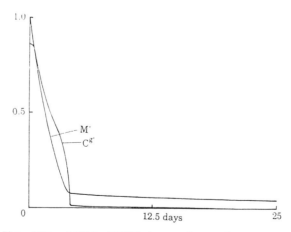

Figure 6.79 Plots of M' and $C^g{}'$ $[=C^g(t)/C^g_{sat}]$ versus time. Ageing period = 50 days. Other parameters as in Table 6.27. [Reprinted from Wilson et al. (1994b), p. 1660 by courtesy of Marcel Dekker, Inc.]

The effects of the thickness of the low-permeability lenses are shown in Figure 6.80. In these runs the same fraction of the medium is low-permeability clay (50%), and the lens thicknesses are 1.25, 2.5, and 5.0 cm; the ageing period is fifty days. In all cases the initial rapid evaporation of NAPL is followed by much slower removal rates as the dissolved VOC diffuses from the lenses. As expected, the thinner the lenses, the more rapid is this diffusion process. For a lens thickness of 1.25 cm, 99.5% cleanup required 14.45 days; systems having thicker lenses were not cleaned up to the 99.5% level during the twenty-five–day period of SVE. The crossing of the curves for the runs having lens thicknesses of 2.5 and 5.0 cm is not unexpected. Since the volume of clay is the same in these runs, the cross-sectional area of clay lens available for diffusion transport decreases with increasing lens thickness. During the ageing period, therefore, diffusion of VOC into the thicker lens occurs to a lesser extent than it occurs into the thinner lens, so the mass of dissolved VOC is smaller in the former case at the beginning of SVE. However, during SVE the rate of diffusion of dissolved VOC from the thicker lens is slower than it is from the thinner, resulting in the observed crossing of the curves plotting residual mass of VOC.

The effect of the size of the NAPL droplets is shown in Figure 6.81. In these runs a ten-day ageing period was followed by twenty-five days of SVE. As one would expect, the principal effect is on the rapid rate of removal of VOC during the period in which NAPL is evaporating; the larger the droplets, the slower the rate of evaporation. This is expected, since the

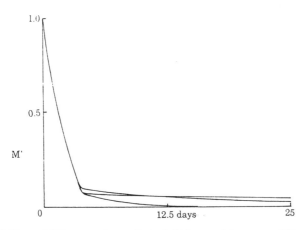

Figure 6.80 Plots of M' versus time; effect of thickness of low-permeability lenses, $2l$. $2l = 1.25$, 2.5 and 5.0 cm, bottom to top, right-hand side. Ageing period = 50 days. Other parameters as in Table 6.27. [Reprinted from Wilson et al. (1994b) p. 1660 by courtesy of Marcel Dekker, Inc.]

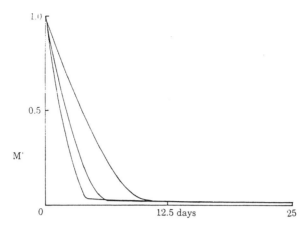

Figure 6.81 Plots of M' versus time; effect of NAPL droplet diameter, $2a_0$. Droplet diameter = 0.1, 0.3 and 0.5 cm, left to right. Ageing period = 10 days. Other parameters as in Table 6.27. [Reprinted from Wilson et al. (1994b) p. 1661 by courtesy of Marcel Dekker, Inc.]

total NAPL-air interface decreases with increasing droplet size for these runs in which the initial total mass of VOC is held constant. Even in the run for which the droplet diameter is 0.5 cm, an unrealistically large value, the VOC removal rate is sufficiently fast that one would not be concerned about it. Evidently, droplet size is not a significant parameter in determining SVE cleanup times. In all of the runs, we see prolonged tailing after the NAPL has been removed, and diffusion of dissolved VOC becomes the limiting factor in the cleanup.

In SVE it is common practice to rely rather heavily on soil gas analyses to follow the progress of the remediation, since these are quicker and cheaper than analyzing actual soil samples. As was seen in Figure 6.79, the effluent soil gas VOC concentrations become quite low after the NAPL has evaporated, even though substantial quantities of dissolved VOC remain in the soil. DiGiulio et al. (1990) has noted the importance of taking soil gas samples under static conditions, i.e., when the well has been turned off and the soil gas VOC has had sufficient time to come to equilibrium with the dissolved and adsorbed VOC in the soil and when we have examined this "rebound" phenomenon in the context of other models. Figures 6.82–6.85 show plots of total residual VOC mass and effluent soil gas VOC concentration for runs in which the sites have had ageing periods of one, ten, twenty-five, and fifty days (not shown in the plots). After ageing they were subjected to SVE for fifteen days, after which the well was turned off and the soil gas allowed to approach equilibrium with the soil for a period of ten days.

In all four runs the effluent soil gas VOC concentration becomes ex-

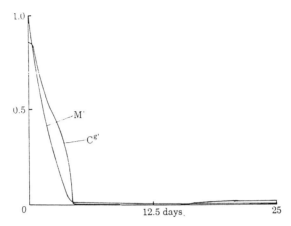

Figure 6.82 Plots of M' and $C^{g'}$ versus time; Figures 6.82–6.85 show the effects of ageing on the rebound of the soil gas VOC concentration. The ageing period here is 1 day, the duration of SVE is 15 days, and the equilibration period is 10 days. [Reprinted from Wilson et al. (1994b) p. 1663, by courtesy of Marcel Dekker, Inc.]

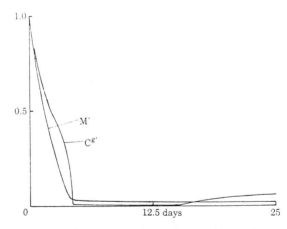

Figure 6.83 Plots of M' and $C^{g'}$ versus time. The ageing period here is 10 days, the duration of SVE is 15 days, and the equilibration period is 10 days. [Reprinted from Wilson et al. (1994b) p. 1663, by courtesy of Marcel Dekker, Inc.]

312

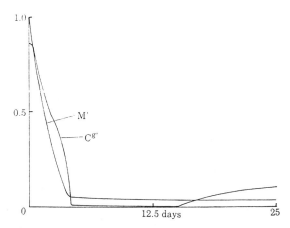

Figure 6.84 Plots of M' and $C^{g'}$ versus time. The ageing period here is 25 days, the duration of SVE is 15 days, and the equilibration period is 10 days. [Reprinted from Wilson et al. (1994b) p. 1664, by courtesy of Marcel Dekker, Inc.]

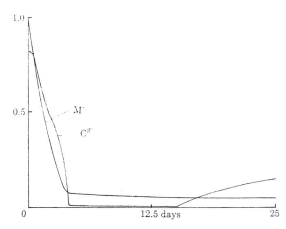

Figure 6.85 Plots of M' and $C^{g'}$ versus time. The ageing period here is 50 days, the duration of SVE is 15 days, and the equilibration period is 10 days. [Reprinted from Wilson et al. (1994b) p. 1664, by courtesy of Marcel Dekker, Inc.]

313

tremely small after the NAPL has been removed by evaporation. In all four runs, however, the plots of total mass residual VOC indicate that cleanup is not complete after fifteen days of SVE, and in all four runs a rebound of the soil gas VOC concentration is observed after the well is shut down. The extent of this rebound increases as the length of the ageing period increases, as expected, since the longer the ageing period, the larger is the quantity of VOC present in aqueous solution in the soil which is available for partitioning into the gas phase during equilibration.

Figures 6.86–6.90 simulate systems having different thicknesses of the low-permeability lenses; these are 1, 2, 3, 4, and 5 cm, respectively. In all cases the ageing period was twenty-five days (not shown), the period of SVE was fifteen days, and the equilibration period was ten days. The run shown in Figure 6.86 (lens thickness of 1 cm) shows no rebound because cleanup is complete after fifteen days of SVE. For the other four runs extensive rebound is observed, with the initial rate greatest for the lens of 2 cm thickness and least for the lens of 5 cm thickness. The ten-day equilibration period was not long enough to achieve equilibrium for these systems. The fraction of the original VOC remaining in the soil after treatment was 2.4% (2 cm), 3.6% (3 cm), 4.2% (4 cm), and 4.2% (5 cm). One expects this to go through a maximum for a finite ageing period, since the total cross-sectional area of the lenses available for diffusion transport decreases with increasing thickness of the lenses because the volume of clay is being held constant.

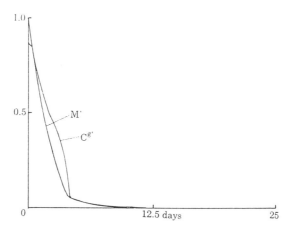

Figure 6.86 Plots of M' and $C^{g'}$ versus time; Figures 6.86–6.90 show the effect of the thickness of the clay layers on the rebound of the soil gas concentration. The ageing period in the runs shown in these figures is 25 days, the duration of SVE is 15 days, and the equilibration period is 10 days. The thickness of the clay layers is 1 cm. Other parameters as in Table 6.27. [Reprinted from Wilson et al. (1994b) p. 1665, by courtesy of Marcel Dekker, Inc.]

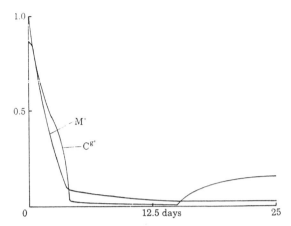

Figure 6.87 Plots of M' and $C^g{}'$ versus time. The thickness of the clay layers is 2 cm, the ageing period is 25 days, the duration of SVE is 15 days, and the equilibration period is 10 days. Other parameters as in Table 6.27. [Reprinted from Wilson et al. (1994b) p. 1666, by courtesy of Marcel Dekker, Inc.]

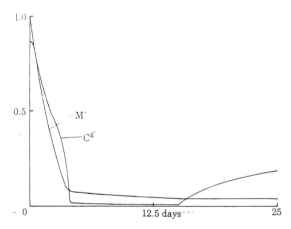

Figure 6.88 Plots of M' and $C^g{}'$ versus time. The thickness of the clay layers is 3 cm, the ageing period is 25 days, the duration of SVE is 15 days, and the equilibration period is 10 days. Other parameters as in Table 6.27. [Reprinted from Wilson et al. (1994b) p. 1666, by courtesy of Marcel Dekker, Inc.]

315

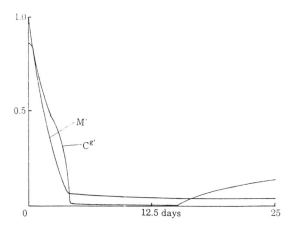

Figure 6.89 Plots of M' and $C^{g'}$ versus time. The thickness of the clay layers is 4 cm, the ageing period is 25 days, the duration of SVE is 15 days, and the equilibration period is 10 days. Other parameters as in Table 6.27. [Reprinted from Wilson et al. (1994b) p. 1667, by courtesy of Marcel Dekker, Inc.]

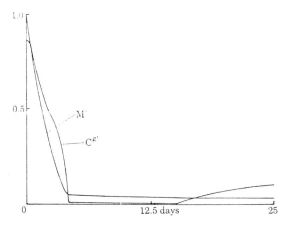

Figure 6.90 Plots of M' and $C^{g'}$ versus time. The thickness of the clay layers is 5 cm, the ageing period is 25 days, the duration of SVE is 15 days, and the equilibration period is 10 days. Other parameters as in Table 6.27. [Reprinted from Wilson et al. (1994b) p. 1667, by courtesy of Marcel Dekker, Inc.]

316

EFFECTS OF VOC CONCENTRATION DISTRIBUTION

Bolick and Wilson (1994) explored the effects of random variations in the contaminant VOC concentrations on the rate of cleanup. They found changes of cleanup times of the order of 2% or less from that resulting from an initial spatially constant VOC distribution within the same volume when the local initial VOC concentrations were randomly varied over ranges as large as 0 to 200 mg/kg with averages of approximately 100 mg/kg. These results suggest that the details of the contaminant concentration distribution are much less important in affecting cleanup times than the details of the soil permeability.

On the other hand, they found that the soil volume occupied by a given mass of VOC affected cleanup times profoundly. A given mass of VOC was distributed in each of the four regions indicated in Figure 6.91 and also throughout the entire domain. Progress of the cleanups of these contaminant distributions by identical SVE systems (horizontal slotted pipes) is shown in Figure 6.92. The cleanup times for the five cases were calculated to be 2.7, 10.2, 25.4, 47.0, and 123.7 days. The results indicate that very substantial savings could be made when spills occur if remedial intervention is made as rapidly as possible, giving the spilled material a minimum amount of time to spread and, as we have seen earlier, also reducing the time during which contaminant can soak or diffuse into porous domains of low gas permeability.

DIFFUSION/DESORPTION KINETICS MODELING CONCLUSIONS

Estimates of the extent to which mass transport (diffusion/solution/desorption) kinetics may be limiting in soil vapor extraction can be made

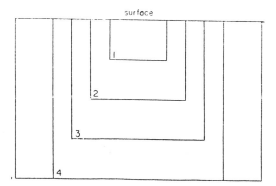

Figure 6.91 Schematic showing the distributions of contaminant (Regions 1, 2, 3, 4, and the entire domain) used in making the runs plotted in Figure 6.92. [Reprinted from Bolick and Wilson (1994) p. 701, by courtesy of Marcel Dekker, Inc.]

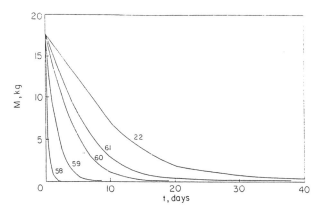

Figure 6.92 Plots of total contaminant mass versus time; effect of extent of spreading of the VOC. From the bottom up, the curves pertain to Regions 1, 2, 3, 4, and the entire domain. [Reprinted from Bolick and Wilson (1994) p. 701, by courtesy of Marcel Dekker, Inc.]

by examining the time dependence of soil gas VOC concentration for a period after the well has been shut down and also by comparing the steady-state soil gas VOC concentration during pumping of the well with the equilibrium soil gas VOC concentration reached after a well has been shut down. In certain cases the rate constants calculated by these means are related to diffusion parameters in the model by a simple algebraic equation. Diffusion/solution/desorption kinetics can also be studied by the rapid injection of a slug of clean air, after which one follows the time dependence of the soil gas VOC vapor concentration rebound. These techniques will yield excessively optimistic results, however, unless the well has been pumped extensively (and the bulk of the VOC removed) before they are carried out.

If a site is mass transport–limited, significant amounts of money can be wasted by pumping the SVE wells at excessive flow rates, in that a large volume of gas is being pumped, and also a large volume of water-saturated gas is being treated for VOC removal before discharge. Reductions in the volume of gas by a factor of one-fifth or less may be possible by working at a suitably selected constant low gas flow rate, by continuously adjusting the gas flow rate to maintain at least a certain minimum VOC concentration in the exhausted soil gas, or by cycling the air flow on and off to maintain at least a certain minimum VOC concentration in the exhausted soil gas.

Short pilot-scale tests in which only 5 to 20% of the VOC is removed from the domain of interest are not useful, as a general rule, for predicting the progress of the later stages of SVE remediation. Short tests can be relied on only if post-SVE rebound of the soil gas VOC concentration is clearly demonstrated not to occur, in which case the system can be described by a local equilibrium model.

The varied transit times of the gas flow paths to SVE wells, soil inhomogeneities, and the interaction between NAPL solution kinetics and VOC diffusion in the soil water lead to considerable complexity. It is therefore unlikely that any theoretical method will permit accurate estimation of the entire course of SVE remediation at a site from pilot-scale effluent soil gas VOC data, which do not extend over virtually the entire cleanup, i.e., to essentially 100% removal of the VOCs from the pilot test domain.

Proper treatment of the diffusion of VOC through an aqueous layer yields not one time constant, but a spectrum of time constants that vary over a rather wide range. Use of the lumped parameter approach for modeling diffusion/desorption kinetics is therefore fraught with some peril unless the lumped parameter diffusion rate constant is fitted against data taken near the end of the remediation of the pilot-scale domain.

It will be difficult to reliably correlate the extent of cleanup (the percent of the VOC that has been removed) with either effluent soil gas VOC concentration during SVE operation or with soil gas VOC concentration after the well has been shut down and rebound to equilibrium has occurred as long as NAPL is present. Fortunately, this preliminary phase of the cleanup should generally be of short duration. During the tailing phase of the cleanup, rebound (i.e., equilibrated) soil gas VOC concentrations should provide good correlations with total residual VOC if the partitioning of VOC between the soil and the vapor phase is governed by a linear isotherm.

The extent of the rebound of soil gas VOC concentrations under static conditions (after the well has been shut down) provides a useful method for assessing the completeness of SVE cleanups. Measurements of effluent soil gas VOC concentrations during well operation typically give results that do not reflect the true VOC content of the soil. Site closures should be based on measurements of soil gas VOC concentrations made after an extended period of equilibration after SVE has ceased.

Results from a model for the operation of an SVE well that includes the rates of NAPL droplet evaporation and diffusion of VOC through water-saturated clay lenses indicate that NAPL droplet evaporation is rapid and presents no problem in SVE cleanups. Diffusion of SVE through water-saturated layers, on the other hand, can result in a very prolonged terminal phase of the cleanup, during which effluent soil gas VOC concentrations are very low and SVE is quite inefficient. VOC removal rates during the period in which NAPL is evaporating (the first few days of operation) give no indication of the behavior to be expected of an SVE cleanup during the period in which dissolved VOC is diffusing into the advecting soil gas. Removal rates during this latter period may be very slow if diffusion kinetics limitations are unfavorable.

Runs with this field-scale model show that the ageing period (the interval between occurrence of a spill and the initiation of SVE) has a major impact on SVE cleanups; the longer the ageing period, the more severe is the tailing of the remediation.

In all distributed diffusion models studied, the thicker the low-permeability lenses, the more severe is the tailing of the remediation. Information on the extent to which this may be a problem can be obtained by examining the well logs for the site.

One can use the soil gas rebound data to adjust the gas flow rate during the terminal phases of the remediation. There is little point in using gas flow rates that yield effluent soil gas VOC concentrations an order of magnitude or more smaller than the rebound soil gas concentration after a period of static equilibration. An alternative approach is to use pulsed gas flow, with a down period (no gas flow) equal to the time required for the soil gas VOC concentration to rebound to roughly half of its maximum value and an up period (well in operation) long enough to reduce the effluent soil gas VOC concentration to a tenth or a twentieth of this maximum value. In either case one will be drastically reducing the volume of soil gas requiring treatment and will increase the cleanup time only marginally. This point was explored in more detail in connection with an earlier model (Gómez-Lahoz et al., 1994a, 1994b).

REFERENCES

Bolick, J. J. and D. J. Wilson, 1994, "Soil Cleanup by in situ Aeration. XIV. Effects of Random Permeability Variations on Soil Vapor Extraction Cleanup Times," *Separ. Sci. Technol.*, 29:701.

Bouchard, D. C., 1989, "The Role of Sorption in Contaminant Transport," *Workshop on Soil Vacuum Extraction*, April 27–28, U.S. EPA, Robert S. Kerr Environmental Research Laboratory, Ada, OK.

Bouchard, D. C., A. L. Wood, M. L. Campbell, P. Nkedi-Kissa, and P. C. S. Rao, 1988, "Sorption Nonequilibrium during Solute Transport," *J. Contaminant Hydrology*, 2:209.

Cho, J. S., 1991, "Forced Air Ventilation for Remediation of Unsaturated Soils Contaminated by VOC," U.S. EPA Report No. EPA/600/2-91/016, July.

Clarke, A. N., R. D. Mutch, Jr., P. D. Mutch, and D. J. Wilson, 1990, "In situ Vapor Stripping: Results of a Year-Long Pilot Study," *Hazardous Materials Control*, 3(6):25.

DiGiulio, D. C., J. S. Cho, R. R. Dupont, and M. W. Kemblowski, 1990, "Conducting Field Tests for Evaluation of Soil Vacuum Extraction," *Proc., 4th National Outdoor Action Conference on Aquifer Restoration, Ground Water Monitoring and Geophysical Methods*, Las Vegas, May 14–17, p. 587.

Fall, E. W., et al., 1988, "In situ Hydrocarbon Extraction: A Case Study," *Southwestern Ground Water Focus Conference*, March 23–25, Albuquerque, NM; see also *Hazardous Waste Consultant*, 1989 (Jan./Feb.):1-1.

Gannon, K., D. J. Wilson, A. N. Clarke, R. D. Mutch, Jr., and J. H. Clarke, 1989, "Soil Cleanup by in situ Aeration. II. Effects of Impermeable Caps, Soil Permeability, and Evaporative Cooling," *Separ. Sci. Technol.*, 26:831.

Gómez-Lahoz, C., J. M. Rodríguez-Maroto, and D. J. Wilson, 1991, "Soil Cleanup by in situ Aeration. VI. Effects of Variable Permeabilities," *Separ. Sci. Technol.*, 26:133.

Gómez-Lahoz, C., J. M. Rodríguez-Maroto, and D. J. Wilson, 1994a, "Soil Cleanup by in situ Aeration. XVII. Field Scale Scale Model with Distributed Diffusion," *Separ. Sci. Technol.*, 29:1251.

Gómez-Lahoz, C., J. M. Rodríguez-Maroto, D. J. Wilson, and K. Tamamushi, 1994b, "Soil Cleanup by in situ Aeration. XV. Effects of Variable Air Flow Rates in Diffusion-Limited Operation," *Separ. Sci. Technol.*, 29:943.

Gómez-Lahoz, C., R. A. García Delgado, F. García-Herruzo, J. M. Rodríguez-Maroto, and D. J. Wilson, 1993, "Extracción a Vacío de Contaminantes Orgánicos del Suelo. Fenómenos de No-Equilibrio," *III Congreso de Ingeniería Ambiental, Proma '93*, Bilbao, Spain.

Hoag, G. E. and B. L. Cliff, 1985, "The Use of the Soil Venting Technique for the Remediation of Petroleum-Contaminated Soils," in *Soils Contaminated by Petroleum: Environmental and Public Health Effects*, E. J. Calabrese and P. T. Kostechi, eds., Wiley, New York, NY.

Hutzler, N. J., B. E. Murphy, and J. S. Gierke, 1989, "Review of Soil Vapor Extraction System Technology," presented at the *Soil Vapor Extraction Technology Workshop*, June 28–29, U.S. EPA Office of Research and Development, Edison, NJ.

Johnson, P. C., M. W. Kemblowski, and J. D. Colthart, 1989, "Practical Screening Models for Soil Venting Applications," presented at the *Workshop on Soil Vacuum Extraction*, April 27–29, U.S. EPA Robert S. Kerr Environmental Research Laboratory (RSKERL), Ada, OK.

Johnson, P. C., M. W. Kemblowski, and J. D. Colthart, 1990, "Quantitative Analysis for the Cleanup of Hydrocarbon-Contaminated Soils by in situ Soil Venting," *Ground Water*, 28:413.

Johnson, P. C., M. W. Kemblowski, J. D. Colthart, D. L. Byers, and C. C. Stanley, 1989, "A Practical Approach to the Design, Operation, and Monitoring of in-situ Soil Venting Systems," presented at the *Soil Vapor Extraction Technology Workshop*, June 28–29, U.S. EPA Risk Reduction Engineering Laboratory (RREL), Edison, NJ.

Kayano, S. and D. J. Wilson, 1992, "Soil Cleanup by in situ Aeration. X. "Vapor Stripping of Volatile Organics Obeying Raoult's Law," *Separ. Sci. Technol.*, 27:1525.

Kayano, S. and D. J. Wilson, 1993, "Migration of Pollutants in Groundwater. VI. Flushing of DNAPL Droplets/Ganglia," *Environ. Monitor. Assess.*, 25:193.

Marley, M. C., 1991, "Development and Application of a Three-Dimensional Air Flow Model in the Design of a Vapor Extraction System," presented at the *Symposium on Soil Venting*, April 29–May 1, RSKERL, Ada, OK.

Marley, M. C., S. D. Richter, B. L. Cliff, and P. E. Nangeroni, 1989, "Design of Soil Vapor Extraction Systems—A Scientific Approach," presented at the *Soil Vapor Extraction Technology Workshop*, June 28–29, U.S. EPA RREL, Edison, NJ.

Megehee, M. M., 1990, unpublished work.

Megehee, M. M., 1993, unpublished work.

Montgomery, K. H. and L. M. Welkom, 1989, *Groundwater Chemicals Desk Reference*, Lewis Publishers, Ann Arbor, MI.

Mutch, R. D., Jr., 1990, personal communication.

Mutch, R. D., Jr., A. N. Clarke, and D. J. Wilson, 1989, "In situ Vapor Stripping Research Project: A Progress Report," in *Proc., 2nd Ann. Hazardous Materials Conf./Central*, March 14–16, Rosemont, IL, pp. 1–15.

Oma, K. H., D. J. Wilson, and R. D. Mutch, Jr., 1990, "In situ Vapor Stripping: The Importance of Nonequilibrium Effects in Predicting Cleanup Time and Cost," *Proc., Hazardous Materials Management Conferences and Exhibition/International*, June 5–7, Atlantic City, NJ.

Osejo, R. E. and D. J. Wilson, 1991, "Soil Cleanup by in situ Aeration. IX. Diffusion Constants of Volatile Organics and Removal of Underlying Liquid," *Separ. Sci. Technol.*, 26:1433.

Pedersen, T. A. and J. T. Curtis, 1991, "Soil Vapor Extraction Technology Reference Handbook," U.S. EPA Report No. EPA/540/2-91/003, Feb.

Roberts, L. A. and D. J. Wilson, 1993a "Groundwater Cleanup by in situ Sparging. III. Modeling of Dense Nonaqueous Phase Liquid Droplet Removal," *Separ. Sci. Technol.*, 28:1127.

Roberts, L. A. and D. J. Wilson, 1993b, "Soil Cleanup by in situ Aeration. XI. Cleanup Time Distributions for Statistically Equivalent Variable Permeabilities," *Separ. Sci. Technol.*, 28:1539.

Rodríguez-Maroto, J. M., C. Gómez-Lahoz, and D. J. Wilson, 1991, "Soil Cleanup by in situ Aeration. VIII. Effects of System Geometry on Vapor Extraction Efficiency," *Separ. Sci. Technol.*, 26:1051.

Rodríguez-Maroto, J. M., C. Gómez-Lahoz, and D. J. Wilson, 1994, "Soil Cleanup by in situ Aeration. XVIII. Field Scale Models and Diffusion from Clay Structures," *Separ. Sci.*, submitted.

Scheidegger, A. E., 1974, *The Physics of Flow through Porous Media*, 3rd ed., University of Toronto Press, Toronto, Canada, p. 306.

Sterett, R. J., 1989, "Analysis of in situ Soil Air Stripping Data," *Workshop on Soil Vapor Extraction*, April 27–28, U.S. EPA, RSKERL, Ada, OK.

Stumbar, J. P. and J. Rawe, 1991, "Guide for Conducting Treatability Studies under CERCLA: Soil Vapor Extraction Interim Guidance," U.S. EPA Report No. EPA/540/2-91/019A, Sept.

Wilson, D. J., 1990, "Soil Cleanup by in situ Aeration. V. Vapor Stripping from Fractured Bedrock," *Separ. Sci. Technol.*, 25:243.

Wilson, D. J., 1994, "Soil Cleanup by in situ Aeration. XIII. Effects of Solution Rates and Diffusion in Mass-Transport-Limited Operation," *Separ. Sci. Technol.*, 29:579.

Wilson, D. J. and A. N. Clarke, 1993, "Soil Vapor Stripping," in *Hazardous Waste Site Soil Remediation: Theory and Application of Innovative Technologies*, D. J. Wilson and A. N. Clarke, eds., Marcel Dekker, New York, NY, p. 171.

Wilson, D. J., A. N. Clarke, and J. H. Clarke, 1988, "Soil Cleanup by in situ Aeration. I. Mathematical Modeling," *Separ. Sci. Technol.*, 23:991.

Wilson, D. J., C. Gómez-Lahoz, and J. M. Rodríguez-Maroto, 1992, "Mathematical Modeling of SVE: Effects of Diffusion Kinetics and Variable Permeabilities," *Proc., Symp. on Soil Venting*, April 29–May 1, 1991, Houston, TX, U.S. EPA Report No. EPA/600/R-92/174, Sept.

Wilson, D. J., C. Gómez-Lahoz, and J. M. Rodríguez-Maroto, 1994a, "Soil Cleanup by in situ Aeration. XVI. Solution and Diffusion in Mass Transport-Limited Operation and Calculation of Darcy's Constants," *Separ. Sci. Technol.*, 29:1133.

Wilson, D. J., J. M. Rodríguez-Maroto, and C. Gómez-Lahoz, 1994b, "Soil Cleanup by in situ Aeration. XIX. Effects of Spill Age on Soil Vapor Extraction Rates," *Separ. Sci. Technol.*, 29:1645.

CHAPTER 7

Sparging: A New Technology for the Remediation of Aquifers Contaminated with Volatile Organic Compounds*

INTRODUCTION

W^E next turn to a problem of great concern to regulators and environmental engineers—the removal of dense nonaqueous phase liquid (DNAPL), which is distributed in an aquifer and trapped interstitially as blobs or droplets. One of the challenging problems in hazardous waste site remediation is the removal of DNAPLs from below the water table. These materials, mostly chlorinated hydrocarbon solvents, typically have low solubilities in water (of the order of 1,000 mg/L or less), and their diffusion constants in water are quite small, as are all diffusion constants in condensed phases. Kinetic limitations on the rates of solution of these compounds are therefore severe, resulting in long, drawn-out remediations if pump-and-treat methods are used. Related problems are the removal of light, nonaqueous phase liquid (LNAPL), which is either floating on the capillary fringe or is actually smeared in the upper portion of the aquifer during the course of drops and subsequent rises in the level of the water table. In situ air sparging, the technology to be addressed in this chapter and in Chapter 8, shows a good deal of promise with regard to the remediation of sites of the types described above. Before turning to the technique, however, we first examine the factors that cause problems with these sites, which will give us some insight as to why sparging is effective in remediating them.

An introduction to DNAPLs in groundwater has been provided by Feenstra and Cherry (1989). Schwille's (1988) elegant experimental work with chlorinated hydrocarbons demonstrated that these compounds quickly move down through most aquifers, leaving a substantial trail of DNAPL droplets/ganglia trapped interstitially in the porous aquifer medium. This DNAPL residue may amount to 5–50 L/m³, which may be the major part of the material that must be removed.

*This chapter was prepared by Richard A. Brown, Groundwater Technology, Inc.

Powers et al. (1991, 1992) have investigated the nonequilibrium factors involved with the solution of "blobs" of NAPL held interstitially in water-saturated porous media. They concluded that the low rates of mass transfer were due to (1) rate-limited mass transport between the nonaqueous and aqueous phases (the solution process itself), (2) the tendency of advecting aqueous phase to bypass contaminated regions of low aqueous permeability, and (3) nonuniform flow resulting from aquifer heterogeneities. These workers noted that mass transport of VOCs from trapped "blobs" in natural porous media is much slower than it is in highly regular media such as the glass beads used by Miller et al. (1990).

Mackay et al. (1991) have analyzed the equilibrium and kinetic processes involved in the dissolution of organics from a nonaqueous phase liquid in contact with groundwater and have developed equations for the concentrations of these components in groundwater as a function of the composition of the NAPL phase. Laboratory "generator column" data support the model, indicating that dissolution processes follow established physicochemical principles. The concentrations of compounds present in NAPL and dissolved in water can be estimated using a partition coefficient obtained from solubility data and information on the composition of the NAPL.

Conrad et al. (1992) [see also J. L. Wilson (1990)] have carried out a number of flow visualization experiments to examine the movement of organic liquids through the saturated zone, focusing principally on the behavior of the residual NAPL trapped by capillary forces. They confirmed that large amounts of NAPL are trapped as small blobs, indicating the importance of remediation techniques that address the problem of removing this distributed NAPL.

Wilson has published several models for the flushing of DNAPL droplets/ganglia by pump-and-treat operations (Kayano and Wilson, 1993; Wilson, 1993); these include the effects of mass transport kinetics and are simple enough to run on microcomputers. They bear on the problem of DNAPL removal from the zone of saturation by sparging, since one expects that the dissolution of the DNAPL ganglia in pump-and-treat and in sparging will be governed by the same basic mechanism. The modification of the parameters by the sparging technique, however, substantially accelerates the remediation of sites contaminated with DNAPL.

At the present time sparging suffers somewhat in comparison with soil vapor extraction, pump-and-treat operations, some bioremediation methods, and solidification/stabilization techniques. With these other techniques one can refer to a lengthy collection of successful case histories; there are a number of Superfund sites at which the technique is or has been used; and some sites at which these techniques have been used have achieved closure. Possibly, there may even be some bragging about the low cost with which rapid and complete cleanups were achieved. Admittedly, the pump-and-treat technology has some serious problems; it is turning out to be agonizingly slow, and with it very few sites have succeeded in achieving closure. But even with pump-and-treat, one has considerable past history of the technology in the field to give one an information base.

Unfortunately, at the present time this is not possible with sparging, as it is still rather new. It is perhaps at about the same place the soil vapor extraction was in 1988. At that point the SVE technology had little past history, and EPA was rather skeptical about its prospects. Now, six years later, SVE is no longer an untried weapon but has a number of successes to its credit, and most environmental engineers feel fairly comfortable with it. One might expect that six years in the future most environmental engineers will be feeling quite comfortable about the sparging technique, that it will be widely deployed at hazardous waste sites where VOCs in groundwater are involved, and that it will have a substantial number of successful site closures to its credit. If the reader is new to the sparging technology, this chapter will tell you a little about it, so that possible excessive enthusiasm for it may be tempered with caution, or, on the other hand, skepticism about it may be replaced by prudent enthusiasm.

Sparging is basically the injection of air into groundwater in situ for the purpose of removing contaminating VOCs. The technique is distinguished from air stripping above ground by trickle columns, by fine-bubble aeration, or by aeration in basins. The equilibrium and mass transport principles in sparging are basically the same as in these other water treatment techniques. Sparging is carried out at present in three configurations. These are (1) vacuum-vaporizer wells, (2) aeration curtains, and (3) simple sparging wells. A version of a vacuum-vaporizer well is shown in Figure 7.1. A diagram of an aeration curtain is given in Figure 7.2, and Figure 7.3 sketches the operation of a simple sparging well, the most common type of configuration in the United States. We shall be examining all of these configurations in more detail in Chapter 8 in connection with the mathematical modeling of sparging.

In all cases one desires to remove dissolved VOCs (and possibly NAPLs) from a contaminated aquifer. As is well known, soil vapor extraction, while very useful for the removal of VOCs above the water table, is ineffective below the water table. The equilibria and mass transport processes involved in sparging are as follows. First, if NAPL is present, it will generally be dissolved in the aqueous phase. This equilibrium is governed by the solubility of the VOC, and the mass transport is controlled by the aqueous-phase diffusion constant in the porous medium and by the boundary layer thickness around the NAPL blobs or droplets, which are trapped interstitially in the porous medium. In addition, the VOC will be distributed between the slowly moving aqueous phase and the much more mobile

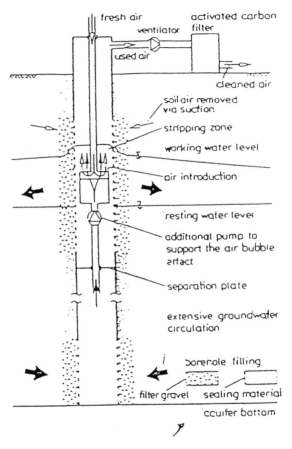

Figure 7.1 Vacuum-vaporizer-well with additional pump and separating plate. [From Herrling et al. (1992), EPA Document EPA/600/R-92/174.]

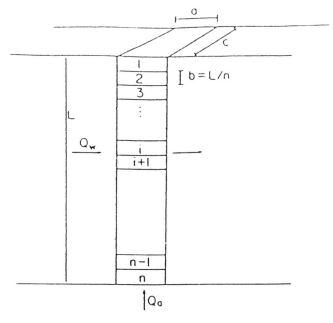

Figure 7.2 Aeration curtain schematic. Q_w = water flow rate (m³/sec) from the left; Q_a = air flow rate from the bottom; L = thickness of aquifer, m.

vapor phase, which is present in the aquifer as a result of the injection of air via the sparging well. This equilibrium is governed by Henry's Law, and the approach to equilibrium is governed by turbulence, diffusion, and dispersion, all of which are greatly affected by the air injected by the sparging well.

Sparging is a technique that is directed toward VOCs of low water solubility. As such, it is in direct competition with the pump-and-treat technique, by far the oldest and most widely used of the various in situ technologies. Pump-and-treat is also one of the most unsuccessful, in that many pump-and-treat operations have been started, but very few have been successfully closed if DNAPLs in the aquifer were involved. There are some excellent physical chemical reasons for this, which have been discussed in depth by Conrad et al. (1992), Feenstra and Cherry (undated, 1989), Mackay and Cherry (1989), Mackay et al. (1991), Miller et al. (1990), Mutch et al. (1993), Powers et al. (1991, 1992), Schwille (1988), and J. L. Wilson (1990) (among many others). Paramount is the fact that molecular diffusion constants of VOCs in the aqueous phase are extremely small, so diffusion transport of VOC out of nonmobile water or away from a blob of NAPL through a quiescent aqueous boundary layer of appreciable thick-

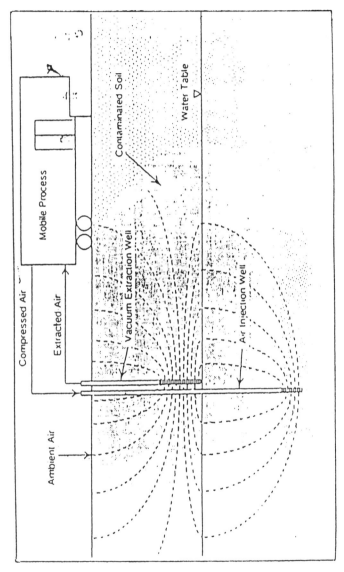

Figure 7.3 Sparging well with associated soil vapor extraction well.

330

ness to the slowly moving mobile groundwater is extremely slow. This bottleneck then limits the entire remediation process.

In sparging one is introducing a good deal of energy into the aquifer as one injects compressed air. As this air expands and rises, energy is dissipated as local turbulence, which reduces the thickness of the quiescent boundary layers around the NAPL blobs and, thereby, increases their rate of solution, which is inversely proportional to the square of the boundary layer thickness. See Figure 7.4 for an illustration of what is occurring. If air actually comes in contact with the NAPL blobs, so much the better for mass transport to the vapor phase. The principal potential advantage of sparging over pump-and-treat is improved mass transport between NAPL and the aqueous phase and between relatively quiescent aqueous phase and the freely flowing fluid (in this case air), which moves the contaminant from the aquifer to an agency for its capture or destruction. Angell (1992), Ardito and Billings (1990), and Brown (undated) have provided good overviews of the sparging technology. The principles and application of the technique have also been summarized by Marley and Billings (1992). Additional reviews are given in the references.

SPARGING WITH AIR INJECTION WELLS—A DESIGN PRIMER

We next take a somewhat more detailed look at sparging systems—what they are, what types of sites they are effective for, how they work, how they are designed, and what their limitations are. We deal here with simple air injection well systems, since it seems likely that these will continue to be the most widely used configuration in the United States. We distinguish between air sparging as used here and the sparging of air into wells in an in-well oxygenation system employed in the late 1970s and early 1980s to enhance in situ bioremediation. The two types of systems are compared in Figure 7.5. Sparging, in the bulk of this chapter, will be used to refer to the direct injection of air into the soil matrix.

SYSTEM DESCRIPTION

An air sparging system consists of four basic elements: an injector system, a compressor, a control system for fugitive emissions, and a monitoring system, as shown in Figure 7.6. What these are and how they function are explained next.

INJECTOR SYSTEM

The injector system is a conduit to carry air from the surface and inject it into the aquifer. A schematic diagram of a sparging injector (sparging

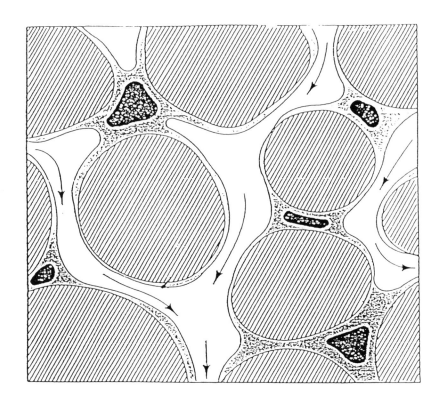

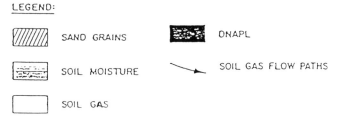
Figure 7.4 A mechanism for the holding of nonaqueous phase liquid. VOC must diffuse through a water boundary layer before being swept away.

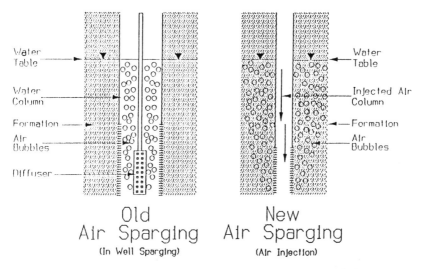

Figure 7.5 Differences between the old and new air sparging technologies.

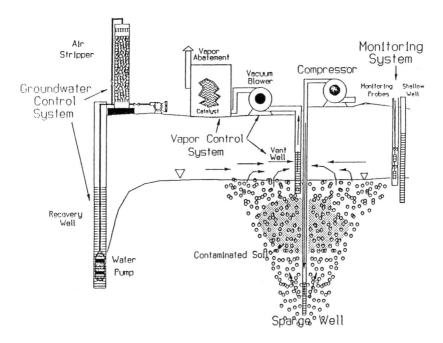

Figure 7.6 Elements of an air sparger system.

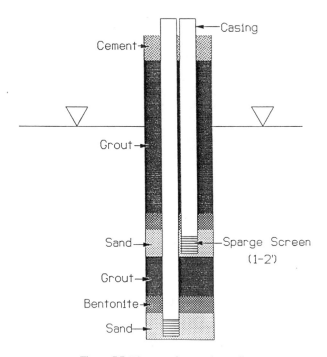

Figure 7.7 Diagram of a sparging well.

well) is given in Figure 7.7. The well consists of 1–2 ft of pervious section (well screen or porous pipe diffuser) connected to a pipe extending from the well screen to the surface. The well screen slot or hole size should be of a diameter similar to that of the average pore in the aquifer medium. This permits more efficient air injection by producing smaller air bubbles, which have less resistance to flow through the porous medium.

The sparge well is completed by placing a sand pack around the well screen. A 1- to 2-ft bentonite seal is placed above the sand pack. The well bore is then grouted to the top of the water table or to the bottom of a vent well or another sparge well. If there is no other well co-located in the well bore, it is grouted to the surface.

If the depth of the sparge well in the aquifer is greater than 20 ft, multiple sparge intervals may be installed as pictured in Figure 7.7. This nesting of sparge wells allows for greater control of the injected air.

COMPRESSOR

The second component of the air injector system is the air compressor. In sizing compressors for sparging systems, the two most important factors are the air pressure and the airflow rate that are needed. The air pressure is a function of the characteristics of the aquifer matrix and the depth of the sparge points below the water table. Air flow is a function of the number, location, and construction characteristics of the wells. Both the pressure and the flow rate for a sparge system have optimum ranges. Determining these will be discussed later. However, once the pressure and flow rate requirements have been determined, the compressor should be sized to supply 150% of the desired flow at 125% of the desired pressure.

EMISSIONS CONTROLS

Because air sparging involves the injection of air into a groundwater aquifer, it can mobilize contaminants in both the vapor state and in aqueous solution. Both vapor controls and/or groundwater controls may therefore be necessary. Vapor controls are necessary if there are vapor receptors within the area of impact of the air sparging operation or if there is concern as to the fate of the organic vapors released by sparging. Groundwater controls are necessary if there are potentially affected groundwater receptors or if there is need for active containment of the groundwater plume.

Organic vapors that are mobilized by air sparging are controlled by the application of a soil vapor extraction (SVE) system. The SVE system captures the released vapors by creating a negative pressure gradient toward the SVE well and away from any receptors, as shown in Figure 7.8. Once the vapors are captured by the SVE system, they can be treated as described earlier in this book.

Air sparging strips may dissolve VOCs from groundwater, provided that there is sufficient air flow. However, if the migration of dissolved contaminants is a concern, then a groundwater control system may be needed. The primary groundwater control system consists of a groundwater recovery well(s) and a groundwater treatment unit. A secondary groundwater control system would be barrier sparge wells, which ring the treatment area and strip the groundwater of VOCs at it migrates off the site.

Since air flow does not follow the hydraulic gradient and may flow radially and vertically in any direction from the sparge well, air sparging may "push" contaminants in any direction. Consequently, groundwater controls would be needed in any direction in which there is a potential receptor, not just in the downgradient direction (see Figure 7.9).

MONITORING

A number of parameters can be used to monitor the impact of an air sparging system. The best are dissolved oxygen (DO), water table eleva-

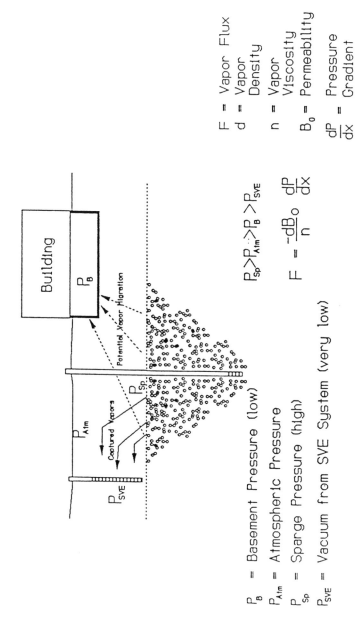

Figure 7.8 Vapor control by applying a preferential pressure gradient.

P_B = Basement Pressure (low)

P_{Atm} = Atmospheric Pressure

P_{Sp} = Sparge Pressure (high)

P_{SVE} = Vacuum from SVE System (very low)

$$P_{Sp} > P_{Atm} > P_B > P_{SVE}$$

$$F = \frac{-dB_0}{n}\frac{dP}{dx}$$

F = Vapor Flux

d = Vapor Density

n = Vapor Viscosity

B_0 = Permeability

$\dfrac{dP}{dx}$ = Pressure Gradient

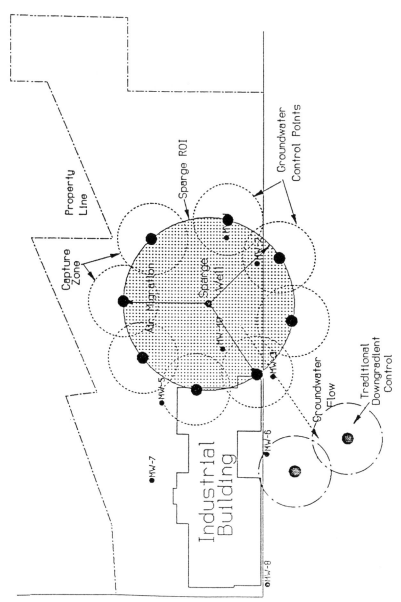

Figure 7.9 Groundwater control for air sparging compared with the traditional method.

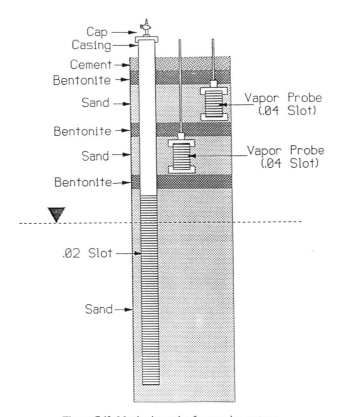

Figure 7.10 Monitoring point for sparging systems.

tion, soil gas pressure (or vacuum if an SVE system is also employed), and VOC concentration. The best monitoring configuration consists of a shallow monitoring well with vapor monitoring probes, as shown in Figure 7.10. The well can be used to monitor water table elevation, DO, VOC concentrations, and pressure at the water table. The probes, placed above the well in the vadose zone, are used to monitor pressure or vacuum and soil gas VOC concentration.

In addition to the above parameters, an active tracer such as helium can be used to monitor an air sparging system. The helium is injected into the air stream prior to the sparge well at a concentration of several percent. The concentrations of helium that appear in wells and vapor probes can then be used to calibrate the sparging system.

EFFECTS OF AIR SPARGING

There are four potential benefits that can be expected to result from air sparging. Three of these are primary benefits:

- the volatilization and removal of NAPL and adsorbed VOCs from the aquifer
- in situ stripping of dissolved VOCs from the groundwater
- enhanced biodegradation of adsorbed and dissolved organic compounds in the aquifer due to increased oxygenation

In addition to the treatment of NAPL, adsorbed, and dissolved organic compounds through biodegradation, solution, and volatilization and stripping, air sparging provides a fourth benefit: control of dissolved iron and manganese.

MODE OF OPERATION

VOCs can be removed from soils by direct volatilization. In unsaturated soils the VOCs adsorbed to the soil or present as NAPL are in equilibrium with their vapors. This equilibrium determines the vapor pressure of the VOC in the soil gas. If the VOC is present above the water table, it may be removed by inducing air flow by the application of a vacuum–SVE. Air passing through the soil removes the vapors from the soil pores, which causes more of the VOC present in stationary condensed phases to volatilize to re-establish the vapor equilibrium.

In the zone of saturation, the volatilization of VOCs is suppressed because there is no air-filled porosity. As a result, the direct vapor phase removal of VOCs is precluded. Air sparging addresses this problem. The injection of air into the saturated aquifer matrix displaces water from its pores, causing a transient air-filled porosity or partial saturation. NAPL and adsorbed VOCs exposed to this transient air-filled porosity evaporate into the airstream and are carried out of the saturated matrix (Figure 7.8), where they can be captured by a soil vapor extraction system.

There are considerable data that demonstrate the release of VOCs from below the water table during air sparging. These data are of two types: (1) measurement of VOC concentrations in the vadose zone soil gas during the injection of air and (2) changes in VOC concentrations in the SVE system off-gas resulting from the addition of an air sparge system. Both types of measurements show significant increases in VOC concentrations during sparging. Table 7.1 summarizes data taken for both chlorinated solvents and petroleum hydrocarbons at one site.

ENHANCEMENT OF BIODEGRADATION

Biodegradation in the saturated zone is limited by the availability of oxygen. When a biodegradable organic is present in groundwater or in soils below the water table, bacteria metabolize the organic, utilizing the available DO. The groundwater then often becomes anoxic or anaerobic,

TABLE 7.1. Increase in VOC Concentrations Observed during Air Sparging Pilot Tests.

Sparge Flow (scfm)	Vent Flow (scfm)	VOC Concentration, ppmv	
		Before Sparging	During Sparging
8	0	1,670[a]	29,000
10	100	330[a]	19,000
8	0	1,200[b]	4,500
25	0	890[a,b]	>10,000
15	55	100[b]	1,000

[a]Petroleum hydrocarbons.
[b]Chlorinated solvents.

as evidenced by the by-products of anaerobic metabolism such as the accumulation of dissolved iron (FeII) or the presence of partially dechlorinated organics. Additionally, DO measurements in areas in which there are biodegradable contaminants are often below 2 mg/L, as opposed to typical values of 5–6 mg/L for uncontaminated groundwater. Once depleted, DO concentrations are slow to recover, since the oxygen must diffuse/disperse into the aquifer from the surface of the water table. This is extremely slow because the diffusive path is long and diffusion constants in the liquid phase are extremely small.

Air sparging increases the availability of DO by greatly decreasing the diffusive pathlength of air into the water phase, as illustrated in Figure 7.11. Local turbulence is also increased, further enhancing oxygen transport. The net result is an enhanced supply of DO and a resulting significant enhancement of biodegradation.

The evidence supporting the impact of air sparging on biodegradation is both direct and indirect. The direct evidence is the increase in CO_2 concentrations observed during air sparging. Soil gas concentrations of CO_2 of 8–12% have been observed during air sparging of petroleum hydrocarbons at depths of 12–25 ft below the water table. The indirect evidence is the increase in DO levels measured in groundwater during air sparging operations. Within .5 to 2 hr after sparging is initiated, DO levels in aquifers contaminated with petroleum hydrocarbons often increase from 1–2 mg/L to saturation (>10 mg/L).

REMOVAL OF NAPL AND DISSOLVED VOCs

An air sparging operation acts as a crude air stripper, with the soil matrix acting as the packing. Air traveling through the soil contacts contaminated water (perhaps itself in contact with adsorbed VOCs and/or

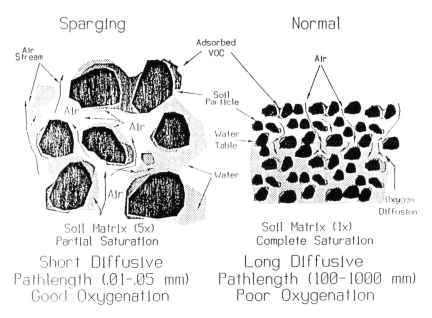

Figure 7.11 Volatilization of adsorbed VOCs and enhanced oxygenation during sparging.

NAPL) and strips the dissolved VOCs. One can roughly estimate the extent of removal of VOCs from contaminated groundwater by noting the strippability of the VOC as measured by its Henry's constant (and vapor pressure, if NAPL is present) and the ratio of air volume (airflow times duration of sparging) to water volume (total volume of the domain of influence times the porosity of the aquifer). Generally, the Henry's constant K_H must be $> 10^{-5}$ atm m³mol and the vapor pressure must be > 1 torr for a constituent to be strippable. Table 7.2 gives values of these parameters

TABLE 7.2. Henry's Constant and Vapor Pressure for Selected VOCs.

Compound	K_H, atm m³/mol[a]	Vapor Pressure, torr
Naphthalene	4.1 E-4 (10^{-4})	0.1
Benzene	5.6 E-3	28.0
Toluene	6.3 E-3	9.0
Xylene	5.7 E-3	3.0
Tetrachloroethylene	1.5 E-2	14.0
Trichloroethylene	9.9 E-3	28.0
Trans-1,2-Dichloroethylene	9.4 E-3	200
Acetone	6.8 E-6	76

[a] K_H (atm m³/mol) = K_H (dimensionless) · (8.206 E-5)T, where T = temperature, deg K.

TABLE 7.3. Change in Groundwater VOC Concentration (ppb) with Time.

Monitoring Well No.	Site A	
	Before Start-up	After 54 Days Sparge
MW-1S	2,108	4
MW-3S	2,161	2
MW-4S	4,328	444
MW-5S	6,940	357
	Site B	
	Before Start-up	After 2 Months Sparge
MW-3	10,230	4,530
MW-4S	173	67
MW-5	48	18
MW-7	473	18

for a few VOCs. The BTX compounds and most chlorinated solvents are readily strippable; soluble contaminants such as acetone are not.

Even if the contaminant is strippable, a sufficient volume of air must be used to remove it. A rough rule is that the air-to-water ratio should be between 10:1 and 20:1 to avoid unacceptable levels of residual contaminants after the operation is completed. This ratio is calculated by dividing (airflow rate × duration of operation) by (volume of water contained in a cone defined by the depth of the sparge point in the aquifer and the radius of influence of the well). This point is explored in more detail later in the chapter.

That air sparging strips dissolved VOCs is clearly evidenced in the rapid drop in groundwater VOC concentrations in downgradient monitoring wells that takes place in the course of a sparging operation. Table 7.3 shows the rapid drops in dissolved VOC concentrations that took place after only short periods of time at two sites.

EFFECT OF AIR SPARGING ON DISSOLVED METALS

Iron (Fe) and manganese (Mn) are mobilized by biological activity under anoxic conditions. They are both electron acceptors in their oxidized insoluble states (Fe^{+3}, Mn^{+4}) and are converted to the much more soluble reduced states (Fe^{+2}, Mn^{+2}) by the bacterial metabolism of organic compounds.

During sparging, these reduced species (Fe^{+2}, Mn^{+2}) are rapidly oxidized by oxygen to the original, much less soluble Fe^{+3} and Mn^{+4}, which are precipitated as $Fe(OH)_3$ and MnO_2. In addition to this oxidation of the soluble forms of the metals to the insoluble forms, sparging prevents further biologically mediated reduction of these metals by increasing the DO concentration of the groundwater and thus eliminating the anoxic conditions.

A commercial patented technology (Vyredox) used a form of air sparging to precipitate iron and manganese in an aquifer so that a water supply well is protected. The process used an array of aeration wells placed radially around the water recovery well.

LIMITATIONS TO AIR SPARGING

While air sparging appears to be a simple technology—the injection of air—it does have an inherent risk: the spread of the contaminants. The injected air can displace water in the aquifer matrix, and it can volatilize, entrain, and spread volatile contaminants. Both the displaced water and the injected airstream must be adequately controlled to prevent the spread of contaminants. Anything that interferes with this control limits the effectiveness of the technique. In this section we explore the limitations to the use of air sparging.

There are two primary types of limitations to the use of sparging—physical and operational. Physical limitations are any conditions that restrict the flow of air through the aquifer matrix: soil permeability, geological heterogeneity, and aquifer thickness. Operational limitations are conditions that affect the control of or the beneficial effects of air sparging. These limitations involve air pressure, airflow rate, and well depth.

AIR PERMEABILITY

The permeability of the aquifer matrix must be sufficient to allow the movement of air. This air movement must be both horizontal and vertical to allow the injected air to contact the domain of contamination and to leave the aquifer formation so that it can be captured. Thus, one must consider both the horizontal and vertical permeabilities. With respect to horizontal permeability, the matrix must be transmissive to air flow. In general, competent rock, soils having high silt or clay contents, and soils having hydraulic conductivities of $< 10^{-5}$ cm/sec are not amenable to air sparging; it is too difficult to inject air into such matrices. Weathered rock, however, may be sufficiently permeable to allow air sparging. With respect to vertical permeability, there should be no low-permeability barriers between the screen of the sparging well and the surface of the water table. Even if the permeability is generally sufficient to allow air injection, the presence of low-permeability barriers above the sparge interval can trap the injected air and channel it and any entrained contaminants away from the treatment area, thus spreading the contamination (see Figure 7.12). *A sparge point should never be set below a low-permeability zone.*

The best insurance for avoiding permeability problems is to adequately characterize the lithology of the site, both to determine the general soil conditions and to determine if there are any low-permeability barriers above the sparge interval. Ideally, each sparge well should be continuously cored from the surface of the water table to ensure that the sparge interval will not be set below a low-permeability barrier.

Any changes in permeability or in soil structure have the potential for trapping or channeling air flow. Air flows preferentially through areas of high permeability. If a high-permeability layer exists above the sparge interval, air flow can be channeled as illustrated in Figure 7.13.

These are two factors to consider with respect to heterogeneity. The first is the degree of heterogeneity. In general, the higher the degree of heterogeneity, the higher is the risk of channeling. Highly layered soils are not amenable to air sparging. The second factor is how the heterogeneity grades with depth. If layering exists, it may not impact the use of air sparging as long as the permeability decreases with increasing depth. If the permeability increases with depth, then the layering can trap and channel the airstream.

The first means of determining if heterogeneity is a problem is to characterize the lithology of the site. This, as with permeability, is best done

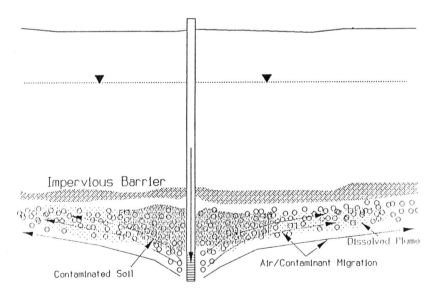

Figure 7.12 Inhibited vertical air flow due to the presence of an impervious barrier.

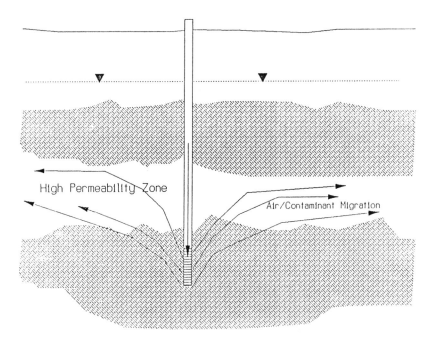

Figure 7.13 Inhibited vertical air flow due to channeling in a high-permeability zone.

by continuously coring all sparge points below the water table. The permeabilities of retained soil cores can be evaluated in a laboratory either by sieve analysis or by actual air permeability testing. Layering is determined by changes in permeability.

A second method of evaluating the potential impact of heterogeneity is to conduct a tracer test. A gas tracer such as helium is injected through the sparge well and monitored at various distances and directions from the sparge point. In homogeneous soils the distribution of the tracer will be radially symmetric. Any channeling will cause a distortion of this radially symmetric pattern. The distortions can locate preferential pathways or barriers.

EFFECT OF WELL DEPTH

Our concern here is not depth below grade, but sparge depth (depth below the water table). There are no known absolute limitations with respect to sparge depth. However, the greater the depth, the greater is the likelihood of lithological conditions (barriers or very permeable layers), which can trap or channel air. Experience to date has been at shallow to moderate depths (<30 ft). If air sparging is to be used at greater depths, the degree of care taken during design and installation should be substantially increased. If one is planning to sparge at depths below 30 ft, the following modifications in design and operating procedures should be used:

- All sparge wells should be continuously logged from the surface of the water table to the bottom.
- Sparge wells should be nested, with a sparge interval set every 12–15 ft.
- Tracer tests should be run at each sparge interval to check for channeling.
- Testing and operation of the sparge well should be carried out sequentially, starting at the shallowest sparge interval and working successively to greater depths.

OPERATING CONDITIONS AND THEIR LIMITATIONS

We next turn to the effects the operating conditions have on the performance of air sparging. The objective in sparging is to maximize the benefits (particularly the rapid and efficient elimination of contaminant VOCs from the site) while minimizing the potential detrimental effects that result from loss of control of the system. The primary operating parameters are well depth, air pressure, and airflow rate; another operating consideration is the extent and impact of water table mounding as a result of sparging.

Air pressure is what drives a sparging system. The minimum pressure is that which will overcome the hydraulic head and displace the water column in the sparge well, allowing air to enter the aquifer matrix. This minimum pressure is roughly 1 psi for every 2.3 ft of water column (1 atm for every 10.34 m of water column). This pressure, termed the "breakout pressure," may be somewhat larger than that calculated above from the hydrostatic head alone, due to the capillary pressure that may be necessary to force air into a saturated matrix having relatively fine pores.

There is also a maximum pressure in sparging. A common (and incorrect) assumption is that increasing the injection pressure simply increases the radius of influence, which obviously results in a greater effective domain of treatment of the well. The risk, however, is that when the pressure is too high, turbulent flow can occur, under which the injected air physically displaces the water, thereby rapidly spreading any contamination (see Figure 7.14). This is graphically illustrated when air rotary drilling, with pressures of > 200 psi, causes "geysers" in nearby uncapped monitoring wells. Under high pressure, control of the air system is lost; it is not always true that more is better.

The impact of the injection pressure can be seen from test data. Figure 7.15 depicts the relationship between horizontal response and air pressure

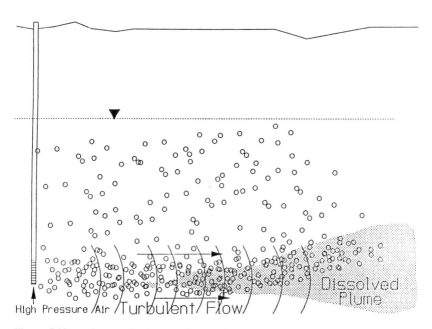

Figure 7.14 Accelerated dissolved VOC migration due to turbulent flow caused by over-pressurization of a sparging well.

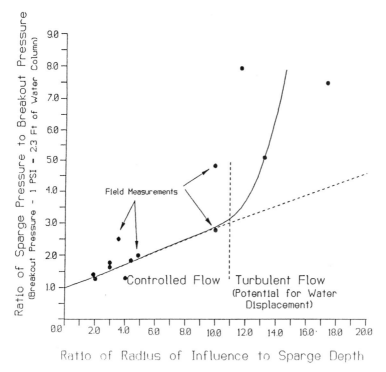

Figure 7.15 Impact of injection pressure on air sparging.

by plotting (radius of influence/sparge depth) versus (air pressure/breakout pressure). As the pressure is increased above the breakout pressure, air is driven laterally from the well. Initially, there is a linear relationship between the injection pressure and the horizontal response (radius of influence). However, once the injection pressure is more than about three times the breakout pressure, the response becomes nonlinear. This is the point at which large-scale physical movement of water and its displacement by air in the formation can occur. The design of an air sparging system, therefore, involves finding the optimum pressure range to provide the desired radius of influence without resulting in such "blow-out" of the well. The maximum operating pressure of a sparging well should never exceed three times the breakout pressure.

The airflow rate, driven by the injection pressure, provides several benefits, which depend on the magnitude of the airflow rate and the volume of air injected. At low flow rates the air oxygenates the water, enhancing biodegradation. At low to moderate flow rates, air sparging enhances the rate of partitioning of adsorbed and NAPL phase contaminants into groundwater by increasing the mixing of groundwater with the contaminated aquifer matrix, as shown schematically in Figure 7.16. This is the basis of an enhanced groundwater recovery process. This enhanced mass transport can be a benefit if a groundwater recovery system is operated concurrently with the air sparging system. However, if the dissolved

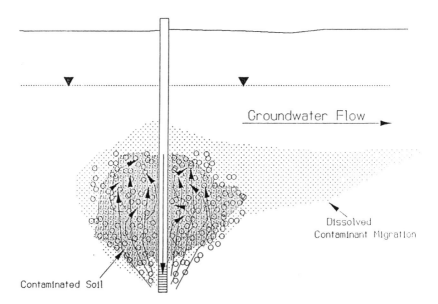

Figure 7.16 Dissolution of NAPL and adsorbed contaminant due to mixing at low airflow rates.

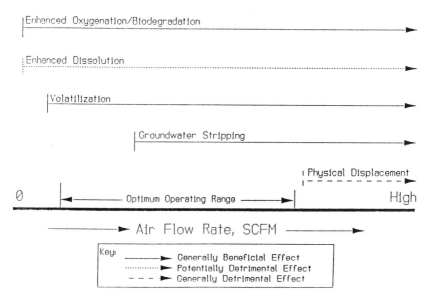

Figure 7.17 The effect of air sparging on VOC transport and treatment as a function of airflow rate.

plume is not captured, this increased rate of partitioning can increase downgradient contamination as the plume of dissolved contaminants is carried with the groundwater flow.

At moderate to high airflow rates, stripping of the dissolved contaminants is sufficiently rapid that treatment of the contaminated groundwater takes place before it is carried out of the zone of influence of the sparging system, reducing the risk of mobilized dissolved contaminants being carried off-site. At very high flow rates, however, there is a risk of physically displacing the groundwater by air and thereby spreading the contamination.

In summary, the effects of air sparging are dependent on the airflow rate. Whether sparging is beneficial or detrimental depends on how the airflow rate is designed and controlled. There is an optimal operating range, in which the air flow is sufficient to oxygenate groundwater, dissolve NAPL and adsorbed contaminants, and air strip dissolved VOCs, but it is not high enough to displace a significant quantity of groundwater from the domain of influence of the well. This is illustrated schematically in Figure 7.17.

Air sparging has a somewhat complex effect on the water table in the vicinity of a sparging well. When air is injected into a saturated matrix, the air displaces some of the water and creates a partially saturated domain. Since the air bubbles move both horizontally and vertically, this displacement can cause a mounding of the water table, as seen in Figure 7.18.

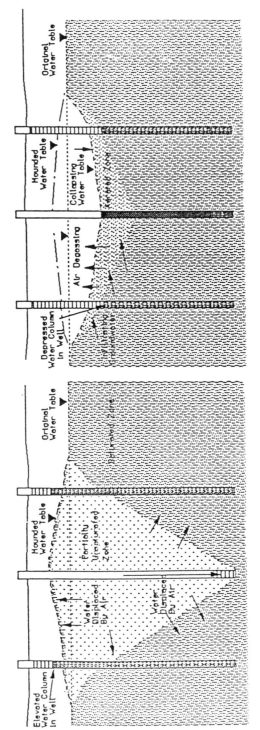

Figure 7.18 Effects of sparging on the water table.

350

When the air flow is terminated, there is a rapid collapse of the water table as the entrained air escapes and the partially saturated aquifer becomes, once again, fully saturated. Eventually, as water flows into the sparged area, the water table recovers to its original, presparging, level. The mounding and collapse of the water table are readily observed in monitoring wells in the area being sparged, as shown in Table 7.4.

The water levels in the monitoring wells may be affected by localized pressure effects—air pressure may support a water column in the well that exceeds the actual mounding. In addition to these monitoring well observations, on several occasions air sparging has been observed to actually drive water to the soil surface out of a formation with a shallow (<3 ft) water table.

One might well be concerned about the effect of water table mounding on groundwater flow and contaminant migration. We consider mounding from both short-term and long-term perspectives. On a short-term basis, water table mounding causes some local groundwater movement. On a long-term basis, it would appear that mounding results in static groundwater flow conditions on a large scale and has little or no impact on contaminant migration. Let us examine these points in more detail.

When air is injected under conditions of moderate pressure (and flow rate), water is displaced by the dispersed air bubbles. The flow of water out of the domain being sparged is in all directions, with the distribution determined by the resistance of the medium to flow, which depends upon permeability, saturated thickness, and layering. Because the air bubbles are dispersed throughout the aquifer matrix, the actual volume of water displaced is small—10–25% of the saturated volume based on a comparison of the water mounding observed with the sparging depth. Also, because the water and the air bubbles are mixed in the aquifer matrix, the air strips dissolved contaminants from the groundwater as it is displaced. We therefore expect the short-term impact of mounding to be negligible.

On a longer term basis, the sparged domain becomes static with respect to groundwater flow for two reasons. First, the hydraulic conductivity of the sparged domain decreases as the water is displaced by air, so that

TABLE 7.4. Water Table Mounding and Collapse.

Well No.	Distance from Sparging Well, ft	Water Table Depth, ft			
		Static	Sparging	5 min after	10 min after
MW-7	5	6.46	4.09	10.03	6.96
SE-19	19	6.42	6.20	6.93	6.54
S-26	29	6.71	6.55	6.96	6.77
NE-13	13	6.11	6.11	7.44	6.75

groundwater flow is actually diverted around the sparged domain. Second, the fluid density in the sparged domain is less than that of static groundwater because the fluid column is actually a mixture of air and water. Thus, the true piezometric surface is lower than what may be inferred from the apparent water table mounding.

Over a long time period, a properly operated air sparging operation will not cause or increase the spread of groundwater contamination for the following reasons. First, the groundwater flow through the area is, as discussed above, diverted around the domain that is being sparged. Second, the air flow through the contaminated groundwater strips out the dissolved VOCs. Third, the airstream also volatilizes and removes NAPL and adsorbed contaminants, thereby reducing the long-term sources of groundwater contamination.

However, under improper operating conditions, air sparging can spread contamination. As discussed above, at high air pressures and flow rates the air physically displaces much of the groundwater from the domain of influence. Rather than dispersed bubbles and a state of partial saturation in the zone of influence, such high flow rates create a single large air-filled domain from which most of the water has been displaced. Since the air does not mix with the groundwater, it does not strip the dissolved contaminants, and the displaced groundwater spreads contamination.

AIR SPARGING SYSTEM DESIGN

As with several other in situ technologies, air sparging system design is best handled by a phased approach. The first phase is the assessment of the applicability of air sparging to the site of interest. The second is the delineation, horizontally and vertically, of the domain of contamination requiring treatment. The third phase is the determination of the airflow properties of the site and the airflow requirements of the system. The fourth phase, system layout, is an overlay of the airflow properties (well location and spacing) with the area of contamination. The fifth and final phase, equipment specification, is the result of combining the number of wells required with the airflow and pressure requirements of each.

The design process for an air sparging system is a stepwise process, as illustrated in Figure 7.19. The two elements necessary before one can design an air sparging system are the contaminant distribution, which defines the area to be treated and the flow properties of the site. The flow properties are the radii of influence (ROI) of the sparge and SVE wells at the desired flows and pressures. These determine the number and location of the sparge and SVE wells. The flow rates and pressures determined during the ROI tests, together with the number of wells needed, then provide the basis for the equipment specifications.

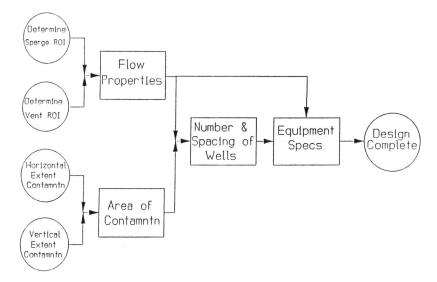

Figure 7.19 Design procedure for air sparging.

SYSTEM DESIGN, PHASE 1. APPLICABILITY OF THE TECHNIQUE

The applicability of sparging depends on the properties of the contaminant and on the geology and hydrogeology of the site. The contaminant must be volatile, strippable, or biodegradable, and the site geology and hydrogeology of the aquifer must be conducive to the flow of air, both vertically and horizontally. A flow chart of the screening procedure for sparging selection is given in Figure 7.20.

Contaminants Amenable to Air Sparging

Since air sparging will mobilize NAPL and adsorbed contaminants, stimulate biodegradation, and strip dissolved VOCs, contaminants must either be biodegradable or have sufficiently large vapor pressures and Henry's constants to permit them to be mobilized and stripped. Table 7.5 summarizes the limits of properties of contaminants for which air sparging is generally feasible. These are the properties of contaminants that are well suited to air sparging. Contaminants having properties with values outside these ranges may still be amenable to air sparging; however, the treatment time may be substantially increased and the effectiveness reduced. Table 7.6 gives the treatability of a number of common contaminants.

Geological and Hydrogeological Constraints

Air sparging is not applicable to all geological conditions. The site con-

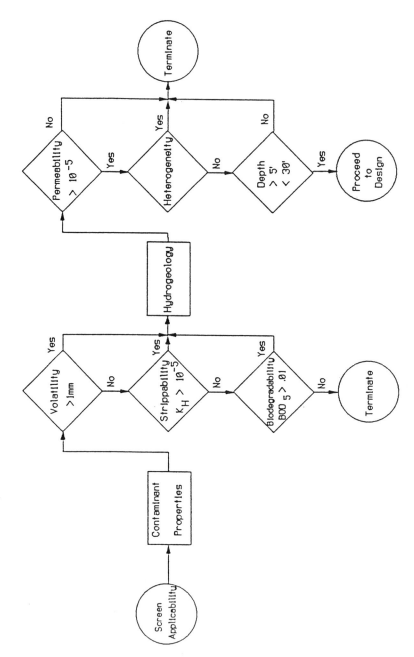

Figure 7.20 Screening procedure for determining the applicability of air sparging.

354

TABLE 7.5. Contaminant Properties; Limits for Air Sparging.

Property	Desired Range of Value
Vapor pressure	>1 torr (mm Hg)
Biodegradability	BOD_5 >0.01 mg/L, or
	$BOD_5/ThOD$ >0.001
Henry's constant	>10^{-5} atm m^3/mol
Solubility[a]	<20,000 mg/L

[a]If not biodegradable.

TABLE 7.6. Treatability of Various Contaminants.

Contaminant	Biodegradability	Volatility	Strippability	Solubility
High Treatability				
Benzene	high	moderate	high	low
Toluene	high	moderate	high	low
Xylenes	high	moderate	moderate	low
Gasoline	high	high	high	low
TCE	very low	high	moderate	moderate
1,1,1-TCA	very low	high	moderate	moderate
PCE	very low	high	moderate	moderate
Moderate Treatability				
Naphthalene	moderate	low	moderate	low
Acetone	high	high	very low	very high
Diesel	high	low	moderate	low
Low Treatability				
MTBE	low	moderate	low	very high
TBA	low	low	low	high
Anthracene	low	low	low	low
Benzopyrene	low	low	low	low

ditions must permit unimpeded flow of air both horizontally and vertically. The important geological characteristics are heterogeneity and permeability. Other important factors are depth of contamination below the water table and aquifer thickness. Heterogeneity is important because layering can cause barriers or channeling, which can lead to loss of control of the injected air and thereby spread contamination. Both vertical and horizontal components of the permeability are significant; a soil with a low ratio of horizontal to vertical permeability can be of lower general permeability than a soil with a high ratio, as there is less risk of channeling. A minimum saturated thickness of aquifer is necessary to prevent the injected air from "blowing out" the injection well.

To summarize, desirable site conditions for sparging are as follows:

- heterogeneity
  - no impervious layers above the sparge point
  - if layering is present, permeability increases toward grade (as one moves up)
- permeability
  - $> 10^{-5}$ cm/sec if horizontal:vertical ratio $< 3:1$
  - $> 10^{-4}$ cm/sec if horizontal:vertical ratio $> 5:1$
- physical
  - saturated thickness > 4 ft
  - sparge depth < 30 ft

Air sparging may be operated outside of these site conditions; however, its performance may be substantially impaired, and the risk of losing control and spreading contamination is substantially increased.

SYSTEM DESIGN, PHASE 2. DELINEATION OF CONTAMINANT DISTRIBUTION

An important part of the design process is the determination of the contaminant distribution, both its horizontal and vertical extents. The objective of this delination is to produce two "maps" of the contamination distribution, as illustrated in Figure 7.21. The first is a plan view showing the area to be treated and including both soil and groundwater contamination. This map is derived through soil gas surveying and well sampling. The second is a geological cross section upon which the contaminant levels are superimposed. This cross section provides a lithological profile that indicates the presence of any low-permeability barriers and/or high-permeability channels. It also shows areas of NAPL and adsorbed phase contamination. This cross section requires continuous coring and analysis (lab or field) of the cores for contaminant levels.

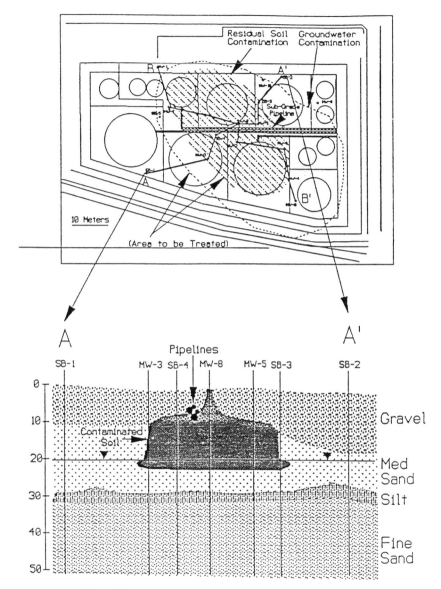

Figure 7.21 Site delineation data package; plan and profile views.

SYSTEM DESIGN, PHASE 3. FLOW PROPERTIES OF THE SITE

The design of an air sparging system requires understanding the horizontal and vertical air permeability of the domain to be treated. The best method is to run an on-site pilot test to determine the radii of influence for both the air sparging system and the SVE system.

The pilot test system consists of a combined or separate vent/sparge well(s) and a series of monitoring points placed radially around the vent/sparge well at varying distances. A monitoring point consists of a shallow monitoring well and several vapor probes separated by bentonite seals. See the schematics in Figure 7.22.

The pilot test is a series of short-term pump test (2–4 hr). The protocol is to first run a sparge test in which air is injected at the breakout pressure and then at 150%, 200%, and 250% of the breakout pressure, respectively. At each stage the pressure and flow are measured at the wellhead, and pressure, water table elevation, DO, and change in VOC concentration are measured at all probes. Carbon dioxide can also be monitored if the contaminant is biodegradable. The second test is a vent test of the SVE system. A vacuum is applied to the SVE well at 50%, 75%, and 100% of maximum flow. Wellhead vacuum and flow, and vacuum and VOC levels at each probe are measured at each stage. The final pump test is a combined sparge/SVE test. Two combinations of conditions are run: high vent, low sparge or low vent, high sparge. The purpose of the combined test is to prove that the SVE system is able to completely capture the

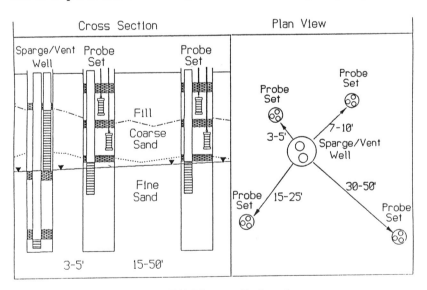

Figure 7.22 Pilot test cell schematic.

sparged air and thereby prevent fugitive vapor emissions. Additionally, tracer tests with helium can be run to confirm the ROIs, prove control, and check for barriers or channeling.

The data derived from these tests are then used to derive vent and sparge radii of influence (ROI), either through graphical reduction using simple linear regression or through the use of computer models. With the sparge test, the ROI should be calculated from all the parameters monitored — DO, pressure, water table elevation, and VOC concentration. Generally there is ±20% agreement in the values of the ROI derived from these different parameters. The ROIs then give the necessary spacing of the wells. The pressure (vacuum) and the flow times the number of wells then provide information for equipment specification. Flow/pressure values are derived from the pilot tests and from such considerations as stripping efficiency (the necessary air/water ratio to achieve the desired level of remediation).

These in-field pilot tests can be supplemented with laboratory tests. Sieve analyses may be run on retained cores to derive an estimate of permeability. Laboratory air permeability tests may also be run on retained cores or packed columns to determine airflow characteristics. However, until more data are available on sparge systems, an in-field pilot test should always be run; laboratory tests should be used only to supplement or confirm field results.

SYSTEM DESIGN, PHASES 4 AND 5. SYSTEM LAYOUT AND EQUIPMENT SPECIFICATION

While the contaminant distribution and the airflow properties are the critical factors in the design of an air sparging system, there are other factors that need to be considered in the overall design. Table 7.7 lists the elements of an air sparging system and the bases for their design.

MODES OF OPERATION

There is no standard way to operate an air sparging system. Different configurations and operating modes can be used to provide different short- and long-term results. There are four basic modes of operation, as illustrated in Figure 7.23; these are continuous, pulse, cyclical, and low intensity.

In the continuous mode of operation, air is injected continuously at a fairly constant rate through the sparge points. The length of time required is determined by treatment goals and system response. Individual wells may be shut down as treatment goals are achieved. The advantage of continuous operation is that it provides the most rapid reduction in gross

TABLE 7.7. **Sparging System Elements and the Bases for Their Design.**

System Component	Basis for Design
Sparge wells	
Number	site characterization, pilot test, treatment goals (time and target concentrations)
Depth	site characterization
Construction	general principles, site characterization (nesting)
Location	existing on-site construction, site characterization, regulatory requirements, receptors
Vent wells	
Number	site characterization, pilot test, receptors, treatment goals (time and target concentrations)
Depth	site characterization
Construction	general principles, site characterization (nesting)
Location	existing on-site construction, site characterization, regulatory requirements, receptors
Co-located wells	site characterization, pilot tests, on-site construction
Compressor	pilot tests, well design
Vacuum pump or blower	pilot tests, well design
Controls	
Noise	site location, regulatory requirements
Groundwater	receptors, regulatory requirements
Vapors	pilot tests, regulatory requirements, receptors
Moisture	pilot tests, site characterization
Monitoring systems	regulatory requirements, pilot tests, site characterization, treatment goals
Recovery wells	receptors, regulatory requirements, site characterization

levels of contamination and the best long-term control of groundwater contamination. Continuous operation, however, does not necessarily attain low-contaminant concentrations faster than other modes of operation, as diffusion mass transport kinetics may be the limiting factor. Continuous operation is best suited to highly permeable, homogeneous formations.

With pulse operation, the rate of air injection is varied with time. Pulse operation is best applied to low-permeability or heterogeneous formations where diffusion kinetics limitations reduce performance. The advantage of pulse operation is that the airflow pathways change, thus minimizing long-term channeling effects. Pulse operation can reduce equipment costs if the wells are operated in sequence. Additionally, if noise control is an issue, pulse operation can be used to minimize complaints. The disadvantage of pulse operation is that it does not provide long-term groundwater control and may require the use of groundwater recovery wells and treatment.

In cyclical operation, rings or banks of sparge wells are employed. Operation usually begins with the outermost or most downgradient ring/bank of sparge wells. The principle of operation is to strip the dissolved plume prior to treating the NAPL/adsorbed phase contaminants in the source area. This provides a "clean" barrier zone to any possible dissolved contaminant migration. Cyclical operation minimizes the need for groundwater control.

Low-intensity operation uses low air flow and wells having short radii of influence. It is used to maximize biodegradation and minimize stripping. Low-intensity operation is best used for biodegradable contaminants of low volatility and low solubility. It may also be used for polishing operations at the end of treatment to focus on small "hot spots." Low-intensity operation is an aeration or oxygenation process.

OUTSTANDING ISSUES CONCERNING AIR SPARGING

Air sparging is still a relatively new technology in the United States, although it has been in use for more than seven years in Europe. Conse-

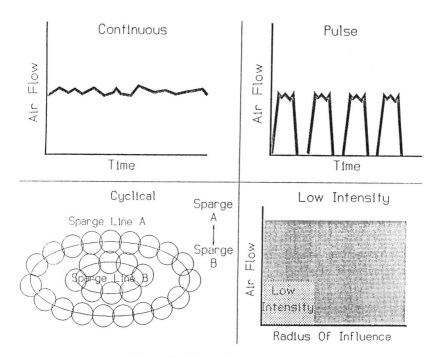

Figure 7.23 Modes of sparging operation.

quently, there is only limited information here on the application, design, operation, and effectiveness of air sparging. As a result, there are still some issues that need to be addressed. These are as follows:

(1) Liability/proof of control—Since air sparging is an injection process, it has the potential to move contaminants. Establishing and proving control is therefore important. Optimal systems for monitoring to prove control need to be developed, and historical data documenting control need to be accumulated and disseminated.

(2) Limits of operation—Because of the lack of operational data, there is limited information on the operational limits for air sparging—depth, permeability, heterogeneity. Two questions need to be addressed in this connection. What are the absolute limitations, and how may limitations are to be surmounted?

(3) Optimization of performance—Since this is a relatively new technology, there is limited information on how to optimize performance. Design and operating procedures need to be developed to maximize performance.

CONCLUSION

Air sparging is an important remedial technology, able to remove or treat contaminants in the zone of saturation through direct volatilization, biodegradation, or stripping. In the air sparging technique, air is injected under pressure below the water table, thereby creating a transient air-filled porosity. This enhances the biodegradation of such contaminants as petroleum hydrocarbons and rapidly strips such volatiles as gasoline, BETX compounds, and chlorinated solvents from the groundwater. Sparging is able to address the removal of NAPL and adsorbed contaminants.

Air sparging has two inherent dangers. First, the VOC-laden airstream can rapidly move through the vadose zone to low-pressure regions such as basements. This could cause vapor hazards. To prevent this from occurring, a sparge system is operated in conjunction with an SVE system. A second danger is that the injected air can mobilize groundwater contaminants rather than stripping them, causing their accelerated downgradient migration. This can occur if there are barriers to vertical air migration, which cause the air to be trapped and to spread laterally and excessively. It can also occur if excessive air pressures are used, resulting in physical displacement of the water column.

Because of these dangers, proper design is essential. This necessitates a field pilot test and careful site delineation. With proper design, the use of sparging can safely, rapidly, and efficiently remediate groundwater contamination. Proper design entails, first, understanding the distribution of contaminants across the site. Second, the flow dynamics of air must be determined in the vadose zone and in the zone of saturation, which requires a properly designed pilot study. Once the contaminant distribution and the airflow dynamics are known, the number, locations, and types of treatment wells can be specified. This, in turn, drives the equipment specifications.

Through a careful and phased design process that is supported by adequate testing, air sparging can be a very effective remedial system. If approach to the technology is oversimplified, however, and a faulty design results, sparging can be ineffective or even counterproductive.

A SCREENING MODEL FOR SPARGING

Sellers and Schreiber (1992) presented a simple analytical model for carrying out preliminary assessments of the feasibility of air sparging. We present here an extension of their approach, which permits one to include the interaction between equilibrium effects, controlled by Henry's Law, and mass transport effects to the sparging air bubbles, controlled by diffusion kinetics. The model presented here applies only to dissolved VOCs and leads to a simple analytical formula. The approach has been further extended to include the solution of NAPL as well, but this elaboration requires the numerical solution of the differential equations modeling the system (Wilson et al., 1994).

The model is a lumped parameter, one-compartment model for preliminary screening only. It assumes that the contaminant is present only as dissolved VOC and also that there is an even distribution of air bubbles throughout the domain of influence of the well or, equivalently, that there is complete mixing of the groundwater within the domain of influence. Evidently, the aquifer must have a fairly high permeability and a low chemical sorption capacity, and it must be sufficiently homogeneous that one can assume good mixing of the groundwater in the domain of influence of the well. There is no mixing of waters within and without the domain of influence, which is defined by the extent to which the bubbles spread laterally as they move up through the aquifer. The sparging air is treated as incompressible, and the model ignores removal of VOC by biodegradation.

Symbols are as follows:

t_t = bubble transit time across the aquifer, sec, best estimated by the user of tracers

a = bubble radius, m

$b - a$ = bubble boundary layer thickness, m, approximated as a
q = volumetric airflow rate, m³/sec
V = volume of the domain of influence, m³
ν = porosity of aquifer
K_H = Henry's constant of VOC, dimensionless
C_0 = initial average VOC concentration in the domain of influence, kg/m³

Reasonable estimates for the volume of influence are given by assuming either a conical or paraboloidal volume with height equal to the depth h of the screen in the water table and base radius r equal to the effective radius of the sparging well; these give $V = \pi r^2 h/3$ and $\pi r^2 h/2$, respectively.

The number of bubbles formed per second is then given by

$$n = 3q/(4\pi a^3) \tag{1}$$

We examine a single bubble as it moves across the aquifer. Let $m(t)$ = mass of VOC in the bubble at time t, kg; the bubble is formed and released from the sparger at $t = 0$.

We assume equilibrium with respect to VOC transport between the air in the bubble and the immediately adjacent portion of the water boundary layer of the bubble. The VOC concentration in the portion of the aqueous boundary layer next to the bubble is then given by

$$C(a) = \frac{3m}{4\pi a^3 K_H} \tag{2}$$

At steady state the VOC concentration in the boundary layer is given by

$$C(r) = A/r + B \tag{3}$$

where

$C(r)$ = VOC concentration a distance r from the center of the bubble

Then

$$C(a) = A/a + B \tag{4}$$

and

$$C_0 = A/b + B \tag{5}$$

from which we find

$$A = -[C_0 - C(a)] \cdot \frac{ab}{b - a} \tag{6}$$

and

$$\frac{dC}{dr} = \frac{ab[C_0 - C(a)]}{b - a} \cdot \frac{1}{r^2} \tag{7}$$

Use of Fick's First Law of Diffusion then gives (after some cancellation)

$$\frac{dm}{dt} = \frac{4\pi abD}{b - a}[C_0 - C(a)] \tag{8}$$

where D is the diffusivity of the VOC in water, m^2/sec.
  From Equation (2) we have

$$m = (4\pi a^3/3)K_H C(a) \tag{9}$$

which, with Equation (8), gives

$$\frac{dC(a)}{dt} = \frac{3bD}{a^2(b - a)K_H}[C_0 - C(a)] \tag{10}$$

Define

$$\alpha = \frac{3bD}{a^2(b - a)K_H} \tag{11}$$

Then the solution to Equation (10), which satisfies the initial condition that $C(a) = 0$ at $t = 0$, is

$$C(a) = C_0[1 - \exp(-\alpha t)] \tag{12}$$

  The concentration of VOC in the bubble as it escapes from the aquifer at time t_r is obtained from Equation (12); substitution of this result into Equation (9) yields

$$m(t_r) = \frac{4\pi a^3}{3} \cdot K_H C_0 \left\{1 - \exp\left[-\frac{3bDt_r}{a^2(b - a)K_H}\right]\right\} \tag{13}$$

This gives the mass of VOC removed by one bubble. A total of n bubbles are released per second, where n is given by Equation (1). Define

$$M(t) = \text{total mass of VOC in the domain of influence, kg}$$

Then

$$M(t) = \nu V C(t) \tag{14}$$

where

$C(t) = $ average VOC concentration in the domain of influence at time t, kg/m^3 of water

and

$$\frac{dM}{dt} = -nm(t_r) \tag{15}$$

Substitution of Equations (1), (13), and (14) into Equation (15) and rearranging gives

$$\frac{dC(t)}{dt} = -\frac{qK_H}{\nu V} \cdot \left\{ 1 - \exp\left[-\frac{3bDt_r}{a^2(b-a)K_H} \right] \right\} \cdot C \tag{16}$$

We define

$$\beta = \frac{qK_H}{V} \left\{ 1 - \exp\left[-\frac{3bDt_r}{a^2(b-a)K_H} \right] \right\} \tag{17}$$

Then integration of Equation (16) results in

$$C(t) = C_0 \cdot \exp(-\beta t) \tag{18}$$

The times required to remove 90, 99, and 99.9% of the contaminant are then given by

$$t_{90} = 2.303/\beta \tag{19}$$

$$t_{99} = 4.605/\beta \tag{20}$$

$$t_{99.9} = 6.908/\beta \tag{21}$$

In the limit of strict diffusion control, that is,

$$\frac{3bDt_t}{a^2(b - a)K_H} \ll 1 \qquad (22)$$

we have from Equation (17)

$$\beta = \frac{3qbDt_t}{\nu Va^2(b - a)} \qquad (23)$$

and find, somewhat surprisingly, that the removal rate of the VOC is independent of K_H under these conditions. If the process is equilibrium controlled, we have

$$\frac{3bDt_t}{a^2(b - a)K_H} \gg 1 \qquad (24)$$

and find that

$$\beta = qK_H/\nu V \qquad (25)$$

so that the removal rate is independent of D, b, t_t, and a, as one would expect for this limit.

Sellers and Schreiber (1992) describe methods for estimating the bubble residence time in the aquifer, t_t. These estimates are fraught with some considerable uncertainty, however, and it would be prudent to use estimates obtained from tracer measurements. Also, such simple one-compartment models are intended only for preliminary screening and should not be regarded as adequate for design purposes.

RECENT AND CURRENT ACTIVITY IN SPARGING

Bruno Herrling and his coworkers in Germany have been very active in the development and application of sparging techniques in Europe. In the United States they have collaborated with IEG Technologies, of Charlotte, North Carolina. They have specialized in the vacuum-vaporizer-well configuration, which permits simultaneous sparging below the water table and soil vapor extraction above, with a single rather complex well. One configuration is shown in Figure 7.8. In a recent EPA report they described use of this technique for the remediation of a site in the Rhine-Main area of Germany, which was contaminated with chlorinated hydrocarbons, principally tetrachloroethylene. This report also presents results of Herrling's elegant three-dimensional mathematical modeling of the flow patterns in the vicinities of these wells (Herrling et al., 1992).

In the United States Groundwater Technology, Inc., has been quite active in the field of sparging. Brown and Fraxedas (1992), in the EPA report just mentioned, has described the case history of a Delaware site that is the former location of a dry-cleaning establishment, at which both soil and groundwater contamination resulted from leaking underground storage tanks. Contaminants were tetrachloroethylene, trichloroethylene, dichloroethylene, and some petroleum hydrocarbons associated with heating oil. PCE was the major contaminant of concern. They give a fairly detailed description of the operation, including chlorinated hydrocarbon levels in groundwater initially, after 54 days of sparging, and after 125 days of sparging. An update on this project (Brown, 1992) gave the results summarized in Table 7.8, which certainly look promising, and he also provided some cost information.

BP Oil Co. and Engineering Science are using air sparging to clean up a gasoline release from an underground tank system at a gasoline station in Cleveland, Ohio (Chandler et al., 1992). This is an EPA demonstration site, so it should provide extensive field data on sparging. The operation includes both sparging and SVE wells, as well as a number of monitoring points. Effluent soil gas analyses indicate that VOCs are being removed, and BETX concentrations were reported as nondetectable.

A sparging well treatability study was conducted by Eckenfelder, Inc. at

TABLE 7.8. Removal of Chlorinated Hydrocarbons by Sparging at a Dry-Cleaning Site (from Brown and Fraxedas, 1992).

		Change in VOC Level with Time (VOC = PCE, TCE, DCE)			
			No. of Days of Sparging		3 Months after
Well	Before Start-Up	54[a]	147[b]	211[b]	Shutdown
MW-1S	2,108	3.5	4.9	BDL[c]	BDL
MW-1D	14	1.9	BDL	BDL	NS
MW-2S	41,000	290	897	704	1,260
MW-2D	BDL	BDL	1.5	BDL	NS
MW-3S	2,161	2.2	1.9	BDL	BDL
MW-3D	BDL	BDL	12	BDL	NS
MW-4S	4,328	444	240	250	87.8
MW-5S	6,940	357	124	115	249
MW-6S	166	5	BDL	BDL	BDL
MW-7S	134	31	5	23	34.6

[a]Results taken with system operating.
[b]Results taken after system shutdown for two weeks.
[c]Below detection level.

an industrial site in California. Sparging resulted in an average 49% decrease in the total VOC concentration in the groundwater at two test locations. The TCE present in the groundwater showed a 48% reduction (23 mg/L to 12 mg/L) at one location and a 25% reduction (10 mg/L to 7.5 mg/L) at the other after only 3 hr of sparging, which looked quite promising, although one would certainly expect a rebound in VOC concentrations after such a short period of sparging (Clarke et al., 1993).

P. C. Johnson et al. (1991) have described a combination SVE/ sparging/pump-and-treat operation for the remediation of a gasoline service station site in Costa Mesa, California. An estimated 4,000 kg of gasoline was present near the water table, and free product was floating on the groundwater. The system used four vapor extraction wells, two air sparging wells, and one groundwater pumping well. After the first year it was felt that vapor extraction well flow rates were too low, so dual vapor extraction/ groundwater recovery wells were installed. These increased the total airflow rate to 30–40 scfm; when the report was written, a total of 2,700 kg of product had been recovered.

Radecki et al. (1987) reported on the use of sparging and pre-aerated infiltration water in enhancing the biodegradation of hydrocarbons at a site in northern Michigan. The site is a shallow unconfined sand and sandy loam aquifer contaminated with a dissolved fraction from a gasoline spill in 1984. Twelve weeks of operation was reported to yield removal of more than 100 gallons of gasoline; the authors estimated that to remove that quantity of contaminant by pump-and-treat alone would have required years.

Horizontal wells have been installed and tested for in situ remediation of groundwater and soils at the DOE Savannah River site (Kaback et al., 1989a,b; Kaback and Looney, 1989). One deep horizontal well, installed below the water table, was used for air injection to strip VOCs from the groundwater, while a shallow horizontal well in the vadose zone recovered the vapor-phase VOCs. This concept is based on a directional drilling technology developed for the oil industry (Langseth, 1990).

Marley et al. (1990) have described the use of sparging at a gasoline spill site in Rhode Island; within two to three weeks the closure criteria were achieved, and at the time the article was written the site was on a quarterly groundwater monitoring program to ensure that the closure levels were being maintained. These authors also provide a description of the technique.

Leonard and Brown (1992) presented a discussion of the application of sparging at a New York site involving a plastics and rubber manufacturer. Toluene and several chlorinated solvents were present, and groundwater was affected. They discussed the impacts of the site geology and hydrogeology, contaminant properties, contaminant distribution, and regulatory aspects on the choice of technologies and described the technology screening and selection process, which determined that combined air sparging and soil vapor extraction system would be most effective for removing contaminant and that groundwater extraction was needed to provide plume control. Off-gas was to be treated with active carbon (off-site regeneration), and air stripping and active carbon polishing would be used to treat aqueous discharges. The pilot tests and their results were described in some detail, as was the system design; two sparge wells, seven SVE wells, and two recovery wells were installed.

R. L. Johnson and his coworkers (1992) have described experiments at the Oregon Graduate Institute in which hydrocarbon spills into a large sand bed are being used to study the efficiencies of SVE in combination with other techniques. Experiments to date have investigated stand-alone SVE, SVE in combination with air sparging below the water table, dewatering of the smear zone in which product is trapped as NAPL below the water table, and air injection into the dewatered smear zone. The quite shallow depth of the sparging wells in the system and the rather thin aquifer resulted in a radius of influence of the sparging wells of only about 2 m, so a substantial amount of the product floating on the water table did not receive direct air flow from the sparging wells. As a result, sparging for this system, while able to remove some product, left a significant mass of hydrocarbons in the system after the sparging phase of the study was complete.

Felten et al. (1992) described a case history in which a combined SVE–air sparging system was pilot tested and put into operation for the removal of petroleum hydrocarbons in groundwater and the vadose zone at a former gasoline station. The system was extensively monitored during its first six months of operation; VOC concentrations in groundwater, dissolved oxygen, effluent vapor VOC concentrations, subsurface vacuum/pressure, and counts of hydrocarbon-utilizing microorganisms were carried out. About 2,000 lb of product was removed.

Marley et al. (1992) have described an elegant three-dimensional model for use in the design of air sparging systems and have compared model predictions with field data from two sites and with analytical solutions for some special cases (homogeneous and layered domains). The Wisconsin site involves saturated and vadose zone soils contaminated with gasoline and VOCs from leaking underground storage tanks. The New Jersey site involves saturated and vadose zone contamination with VOCs, mainly toluene, with some chlorinated solvents. The authors regarded the agreement between the numerical model and both the analytical modeling and the field data as reasonably good but noted that the program makes rather heavy demands for computational resources.

The Vanderbilt-Eckenfelder-Málaga group has done extensive modeling work on in situ air sparging; since this is described in the next chapter, it will not be reviewed here. Space does not permit the review of a substantial number of other papers, which are listed (with titles) in the references to this chapter.

REFERENCES

Ahlfeld, D., A. Dahmani, M. Farrell, and W. Ji, 1992, "Detailed Field Measurements of an Air Sparging Pilot Test," *Ground Water* (submitted).

Angell, K. G., 1992, "In situ Remedial Methods: Air Sparging," *National Environmental Journal*, 2:20.

Ardito, C. P. and J. F. Billings, 1990, "Alternative Remediation Strategies: The Subsurface Volatilization and Ventilation System," *Proc., Conf. on Petroleum Hydrocarbons and Organic Chemicals in Groundwater*, Oct. 31–Nov. 2, p. 281.

Billings, J. F., A. I. Cooley, and G. K. Billings, 1994, "Microbial and Carbon Dioxide Aspects of Operating Air-Sparging Sites," in *Air Sparging for Site Remediation*, R. E. Hinchee, ed., Lewis Publishers, Chelsea, MI, p. 56.

Boersma, P., F. S. Petersen, P. Newman, and R. Huddleston, 1993, "Use of Groundwater Sparging to Effect Hydrocarbon Biodegradation," *Proc., Petroleum Hydrocarbons and Organic Chemicals in Ground Water: Prevention, Detection and Restoration*, Houston, TX, 17:557.

Bohler, U., J. Brauns, H. Hotzl, and M. Hahold, 1990, "Air Injection and Soil Air Extraction as a Combined Method for Cleaning Contaminated Sites – Observations from Test Sites in Sediments and Solid Rocks," in *Contaminated Soil '90*, F. Arendt, M. Hinsenveld, and W. J. van den Brink, eds., Kluwer Academic Publishers, the Netherlands.

Brown, R. A., Undated, "Air Sparging: A Primer for Application and Design," Groundwater Technology, Inc., 310 Horizon Center Dr., Trenton, NJ 08691.

Brown, R. A., 1992, "Air Sparging Project in Delaware," *Underground Tank Technology Update*, 6(3):8.

Brown, R. A. and R. Fraxedas, 1992, "Air Sparging – Extending Volatilization to Contaminated Aquifers," *Symposium on Soil Venting*, April 29–May 1, 1991, Houston, TX, U.S. EPA Report No. EPA/600/R-92/174, p. 249.

Brown, R. A. and F. Jasiulewicz, 1992, "Air Sparging Used to Cut Remediation Costs," *Pollution Engineering* (July):52.

Brown, R. A., C. Herman, and E. Henry, 1991a, "The Use of Aeration in Environmental Cleanups," *Proc., Haztech International Pittsburgh Waste Conference*, Pittsburgh, PA.

Brown, R. A., R. J. Hicks, and P. M. Hicks, 1994, "Use of Air Sparging for in situ Bioremediation," in *Air Sparging for Site Remediation*, R. E. Hinchee, ed., Lewis Publishers, Chelsea, MI, p. 38.

Brown, R. A., R. E. Payne, and P. F. Perlwitz, 1992, "Air Sparging Pilot Testing at a Site Contaminated with Gasoline," *Proc. Petroleum Hydrocarbons and Organic Chemicals in Ground Water: Prevention, Detection and Restoration*, Houston, TX, 14:429.

Brown, R. A., E. Henry, C. Hermann, and W. Leonard, 1991b, "The Use of Aeration in Environmental Cleanups," *Proc., Conf. on Petroleum Hydrocarbons and Organic Chemicals in Groundwater*, Nov. 20–22, p. 265.

Burchfield, S. D. and D. J. Wilson, 1993, "Groundwater Cleanup by in situ Sparging. IV. Removal of DNAPL by Sparging Pipes," *Separ. Sci. Technol.*, 28:1529.

Chandler, P., L. Beabes, and R. Banary, 1992, "Air Sparging Project in Cleveland, Ohio," *Underground Tank Technology Update*, 6(3):6.

Clarke, A. N., R. D. Norris, and D. J. Wilson, 1993, "Saturated Zone Remediation of VOCs through Sparging," in *Hazardous Waste Site Soil Remediation: Theory and Application of Innovative Technologies*, D. J. Wilson and A. N. Clarke, eds., Marcel Dekker, New York, NY.

Conrad, S. H., J. L. Wilson, W. R. Mason, and W. J. Peplinski, 1992, "Visualization of Residual Organic Liquid Trapped in Aquifers," *Water Resources Research*, 28:467.

Fairbanks, P. E., L. Pennington, and J. H. Rabaideau, 1993, "Air Sparging in Reduced Permeability Sediments," *Proc., Petroleum Hydrocarbons and Organic Chemicals in Ground Water: Prevention, Detection and Restoration*, Houston, TX, 17:561.

Feenstra, S. and J. A. Cherry, Undated, "Dense Organic Solvents in Ground Water: An Introduction," in *Dense Chlorinated Solvents in Ground Water*, Institute for Ground Water Research, University of Waterloo, Waterloo, Ont., Progress Report No. 0863985.

Feenstra, S. and J. A. Cherry, 1989, "Subsurface Contamination by Dense Non-Aqueous Phase (DNAPL) Liquid Chemicals," *Proc., International Groundwater Symposium on Hydrogeology of Cold and Temperate Climates and Hydrogeology of Mineralized Zones*, May 1-5, 1988, Halifax, Nova Scotia, p. 61.

Feenstra, S., D. M. Mackay, and J. A. Cherry, 1991, "Method for Assessing Residual NAPL Based on Organic Chemical Concentrations in Soil Samples," *Ground Water Monitoring Review*, 11:128.

Felten, D. W., M. C. Leahy, L. J. Bealer, and B. A. Kline, 1992, "Case Study: Remediation Using Air Sparging and Soil Vapor Extraction," *Proc., Conf. on Petroleum Hydrocarbons and Organic Chemicals in Ground Water: Prevention, Detection and Restoration*, Nov. 4-6, Eastern Regional Ground Water Issues, Houston, TX, 14:395.

Gómez-Lahoz, C., J. M. Rodríguez-Maroto, and D. J. Wilson, 1994, "Groundwater Cleanup by in situ Sparging. VII. VOC Concentration Rebound Caused by Diffusion after Shutdown," *Separ. Sci. Technol.*, 29:1509.

Gudemann, H. and D. Hiller, 1988, "In situ Remediation of VOC Contaminated Soil and Groundwater by Vapor Extraction and Groundwater Aeration," *Proc., Third Annual Haztech International Conference*, Cleveland, OH.

Gvirtzman, H. and S. M. Gorelick, 1992, "The Concept of in-situ Vapor Stripping for Removing VOCs from Groundwater," in *Transport in Porous Media*, Kluwer Academic Publishers, the Netherlands.

Herrling, B. and W. Buermann, 1990, "A New Method for in-situ Remediation of Volatile Contaminants in Groundwater – Numerical Simulation of the Flow Regime," in *Computational Methods in Subsurface Hydrology*, G. Gambolati, A. Rinaldo, C. A. Brebbia, W. G. Gray, and G. F. Pinder, eds., Springer, Berlin, p. 299.

Herrling, B., W. Buermann, and J. Stamm, 1991, "In situ Remediation of Volatile Contaminants in Groundwater by a New System of 'Vacuum-Vaporizer-Wells,' " in *Subsurface Contamination by Immiscible Fluids*, K. U. Weyer, ed., A. A. Balkema, Rotterdam.

Herrling, B., J. Stamm, B. J. Alesi, and P. Brinnel, 1992, "Vacuum-Vaporizer-Wells for in situ Remediation of Volatile and Strippable Contaminants in the Unsaturated and Saturated Zone," *Symposium on Soil Venting,* April 29–May 1, 1991, Houston, TX, U.S. EPA Report No. EPA/600/R-92/174.

Hinchee, R. E., 1994, "Air Sparging State of the Art," in *Air Sparging for Site Remediation,* R. E. Hinchee, ed., Lewis Publishers, Chelsea, MI, p. 56.

Howe, G. B., M. E. Mullins, and T. N. Rogers, 1986, *Evaluation and Prediction of Henry's Law Constants and Aqueous Solubilities for Solvents and Hydrocarbon Fuel Components. Vol. I.* Technical Discussion, USAFESE Report No. ESL-86-66, U.S. Air Force Engineering and Services Center, Tyndall AFB, FL.

Ji, W., A. Dahmani, D. Ahlfeld, J. Lin, and E. Hill, 1993, "Laboratory Study of Air Sparging: Air Flow Visualization," *Ground Water Monitoring Review* (Fall):127.

Johnson, R. L., 1994, "Enhancing Biodegradation with in situ Air Sparging: A Conceptual Model," in *Air Sparging for Site Remediation,* R. E. Hinchee, ed., Lewis Publishers, Chelsea, MI, p. 14.

Johnson, R. L., W. Bagby, M. Perrott, and C. Chen, 1992, "Experimental Examination of Integrated Soil Vapor Extraction Techniques," *Proc., Conf. on Petroleum Hydrocarbons and Organic Chemicals in Ground Water: Prevention, Detection and Restoration,* Nov. 4–6, Eastern Regional Ground Water Issues, Houston, TX, p. 441.

Johnson, R. L., P. C. Johnson, D. B. McWhorter, R. E. Hinchee, and I. Goodman, 1993, "An Overview of in situ Air Sparging," *Ground Water Monitoring Review* (Fall):127.

Johnson, P. C., C. C. Stanley, D. L. Byers, D. A. Benson, and M. A. Acton, 1991,"Soil Venting at a California Site: Field Data Reconciled with Theory," in *Hydrocarbon Contaminated Soils and Groundwater: Analysis, Fate, Environmental and Public Health Effects, and Remediation, Vol. 1,* Lewis Publishers, Chelsea, MI, p. 253.

Kaback, D. S. and B. B. Looney, 1989, "Status of in situ Air Stripping Tests and Proposed Modifications: Horizontal Wells AMH-1 and AMH-2 Savannah River Site," WSRC-RP-89-0544, Westinghouse Savannah River Co., Savannah River Laboratory, Aiken, SC.

Kaback, D. S., B. B. Looney, C. A. Eddy, and T. C. Hazen, 1989a, "Innovative Ground Water and Soil Remediation: In situ Air Stripping Using Horizontal Wells," presented at the *Fifth Outdoor Action Conference on Aquifer Restoration, Ground Water Monitoring and Geophysical Methods,* May 13–16, Las Vegas, NV.

Kaback, D. S., B. B. Looney, J. C. Corey, L. M. Write, and J. L. Steele, 1989b, "Horizontal Wells for in situ Remediation of Groundwater and Soils," *Proc., NWWA 3rd Outdoor Action Conf.,* Orlando, FL.

Kampbell, D., 1993, "U.S. EPA Air Sparging Demonstration at Traverse City, Michigan," in *Environmental Restoration Symposium,* sponsored by the U.S. Air Force Center for Environmental Excellence, Brooks AFB, TX.

Kayano, S. and D. J. Wilson, 1993, "Migration of Pollutants in Groundwater. VI. Flushing of DNAPL Droplets/Ganglia," *Environmental Monitoring and Assessment,* 25:193.

Langseth, D. E., 1990, "Hydraulic Performance of Horizontal Wells," *Superfund '90, Proc., 11th National Conf. of the Hazardous Materials Control Research Institute,* Washington, D.C., p. 398.

Leonard, W. C. and R. A. Brown, 1992, "Air Sparging: An Optimal Solution," *Proc., Conf. on Petroleum Hydrocarbons and Organic Chemicals in Ground Water: Pre-*

vention, Detection and Restoration, Nov. 4–6, Eastern Regional Ground Water Issues, Houston, TX, p. 349.

Loden, M. E., 1992, "Technology Assessment of Soil Vapor Extraction and Air Sparging," U.S. EPA Risk Reduction Engineering Laboratory, report no. EPA 600/R-92/173.

Lundegard, P. D. and G. Anderson, 1993, "Numerical Simulation of Air Sparging Performance," Proc., Petroleum Hydrocarbons and Organic Chemicals in Ground Water: Prevention, Detection and Restoration, Houston, TX, 17:461.

Mackay, D. M. and J. A. Cherry, 1989, "Ground Water Contamination: Limitations of Pump and Treat Remediation," Environ. Sci. Technol., 12:630.

Mackay, D. M., W. Y. Shiu, A. Maijanen, and S. Feenstra, 1991, "Dissolution of Non-Aqueous Phase Liquids in Groundwater," J. Contaminant Hydrology, 8:23.

Marley, M. C., 1991, "Air Sparging in Conjunction with Vapor Extraction for Source Removal at VOC Spill Sites," Proc., 5th National Outdoor Action Conf. on Aquifer Restoration, Groundwater Monitoring and Geophysical Methods, p. 89.

Marley, M. C. and R. Billings, 1992, "Principles of Air Sparging," Underground Tank Technology Update, 6(3):1.

Marley, M. C., D. J. Hazebrouck, and M. T. Walsh, 1991, "Air Sparging in Conjunction with Vapor Extraction for Source Removal at VOC Spill Sites," paper presented at 6th Annual Conference on Hydrocarbon Contaminated Soils, University of Massachusetts, Amherst, MA.

Marley, M. C., D. J. Hazebrouck, and M. T. Walsh, 1992a, "The Application of in-situ Air Sparging as an Innovative Soils and Groundwater Remediation Technology," Groundwater Monitoring Review, 12:137.

Marley, M. C., F. Li, and S. Magee, 1992b, "The Application of a 3-D Model in the Design of Air Sparging Systems," Proc., Conf. on Petroleum Hydrocarbons and Organic Chemicals in Ground Water: Prevention, Detection and Restoration, Nov. 4–6, Eastern Regional Ground Water Issues, Houston, TX, p. 377.

Marley, M. C., M. T. Walsh, and P. E. Nangeroni, 1990, "A Case Study on the Application of Air Sparging as a Complementary Technology to Vapor Extraction at a Gasoline Spill Site in Rhode Island," Hydrocarbon Contaminated Soil Conference, University of Massachusetts, Amherst, MA.

Middleton, A. C. and D. H. Hiller, 1990, "In situ Aeration of Groundwater, a Technology Overview," Conf. on Prevention and Treatment of Soil and Groundwater Contamination in the Petroleum Refining and Distribution Industry, Oct. 16–17, Montreal, Que.

Miller, C. T., M. M. Poirier-McNeill, and A. S. Mayer, 1990, "Dissolution of Trapped Nonaqueous Phase Liquids: Mass Transfer Characteristics," Water Resources Res., 26:2783.

Millington, R. J. and J. M. Quirk, 1961, "Permeability of Porous Solids," Trans. Faraday Soc., 57:1200.

Montgomery, J. H. and L. M. Welkom, 1990, Groundwater Chemicals Desk Reference, Vols. 1 and 2, Lewis Publishers, Ann Arbor, MI.

Mutch, R. D., J. I. Scott, and D. J. Wilson, 1993, "Cleanup of Fractured Rock Aquifers: Implications of Matrix Diffusion," Environmental Monitoring and Assessment, 24:45.

Neaville, C. C., P. C. Johnson, and R. L. Johnson, 1994, "An Evaluation of in-situ Air-Sparging Technology at a Gasoline UST Site, Hayden Island, Oregon," in Proc.,

1994 Conference on the Environmental Health of Soils, March 30–April 1, Long Beach, CA.

Norris, R. D. and D. J. Wilson, 1993, "Air Sparging of VOCs from Aquifers: Design and Modeling," *Proc. 6th Ann. Environmental Management and Technology Conf./Central (HAZMAT/CENTRAL 93),* March 9–11, Rosemont, IL, p. 417.

Ostendorf, D. W., E. E. Moyer, and E. S. Hinlein, 1993, "Petroleum Hydrocarbon Sparging from Intact Core Sleeve Samples," *Proc., Petroleum Hydrocarbons and Organic Chemicals in Ground Water: Prevention, Detection and Restoration,* Houston, TX, 17:415.

Pankow, J. F., R. L. Johnson, and J. A. Cherry, 1993, "Air Sparging in Gate Wells in Cutoff Walls and Trenches for Control of Plumes of Volatile Organic Compounds (VOCs)," *Ground Water,* 31:654.

Powers, S. E., L. M. Abriola, and W. J. Weber, 1992, "An Experimental Investigation of Nonaqueous Phase Liquid Dissolution in Saturated Subsurface Systems," *Water Resources Res.,* 28:2691.

Powers, S. E., C. O. Louriero, L. M. Abriola, and W. J. Weber, 1991, "Theoretical Study of the Significance of Nonequilibrium Dissoltuion of Nonaqueous Phase Liquids in Subsurface Systems," *Water Resources Res.,* 27:463.

Radecki, E. A., C. Matson, and M. R. Brenoel, 1987, "Enhanced Natural Degradation of a Shallow Hydrocarbon Contaminated Aquifer," *Proc., NWWA FOCUS Conf. on Midwestern Ground Water Issues,* National Water Well Assoc., Dublin, OH, p. 485.

Roberts, L. A. and D. J. Wilson, 1993, "Groundwater Cleanup by in situ Sparging. III. Modeling of Dense Nonaqueous Phase Liquid Droplet Removal," *Separ. Sci. Technol.,* 28:1127.

Roberts, L. A., S. Kayano, and D. J. Wilson, 1992, "Modeling of DNAPL Removal from Aquifers by Sparging and Pump-and-Treat Operations," *Proc., Special Symposium: Emerging Technologies for Hazardous Waste Management,* Sept. 21–23, Amer. Chem. Soc., Atlanta, GA, p. 364.

Schwille, F., 1988, *Dense Chlorinated Solvents in Porous and Fractured Media: Model Experiments,* Lewis Publishers, Chelsea, MI.

Sellers, K. L. and R. P. Schreiber, 1992, "Air Sparging Model for Predicting Groundwater Cleanup Rate," *Proc., Conf. on Petroleum Hydrocarbons and Organic Chemicals in Ground Water: Prevention, Detection and Restoration,* Nov. 4–6, Eastern Regional Ground Water Issues, Houston, TX, p. 365.

Underground Tank Technology Update, 1992, "Special Feature on Air Sparging; Principles of Air Sparging, Air Sparging Project in Cleveland, OH, Air Sparging Project in Delaware, Air Sparging and SVE Radius of Influence, Air Sparging/ SVE References," *Underground Tank Technology Update,* 6(2):1.

Wehrle, K., 1990, "In-situ Cleaning of CHC Contaminated Sites: Model-Scale Experiments Using the Air Injection (in situ Stripping) Method of Granular Soils," in *Contaminated Soil '90,* F. Arendt et al., eds., Kluwer Academic Publishers, the Netherlands, p. 1061.

Wilson, D. J., 1990, "Soil Cleanup by in-situ Aeration. V. Vapor Stripping from Fractured Bedrock," *Separ. Sci. Technol.,* 25:243.

Wilson, D. J., 1992, "Groundwater Cleanup by in situ Sparging. II. Modeling of Dissolved VOC Removal," *Separ. Sci. Technol.,* 27:1675.

Wilson, D. J., 1993, "Advances in the Modeling of Several Innovative Technologies,"

Proc., 6th Ann. Environmental Management and Technology Conf./Central (HAZMAT/CENTRAL 93), March 9–11, Rosemont, IL, p. 135.

Wilson, D. J., 1994, "Groundwater Cleanup by in situ Sparging. V. Mass Transport-Limited DNAPL and VOC Removal," *Separ. Sci. Technol.*, 29:71.

Wilson, D. J., C. Gómez-Lahoz, and J. M. Rodríguez-Maroto, 1994, "Groundwater Cleanup by in situ Sparging. VIII. Effect of Air Channelling on Dissolved VOC Removal Efficiency," *Separ. Sci. Technol.*, 29:2389.

Wilson, D. J., J. M. Rodríguez-Maroto, and C. Gómez-Lahoz, 1994, "Groundwater Cleanup by in situ Sparging. VI. A Solution/Distributed Diffusion Model for Nonaqueous Phase Liquid Removal," *Separ. Sci. Technol.*, 29:1401.

Wilson, D. J., S. Kayano, R. D. Mutch, and A. N. Clarke, 1992, "Groundwater Cleanup by in situ Sparging. I. Mathematical Modeling," *Separ. Sci. Technol.*, 27:1023.

Wilson, J. L., 1990, "Laboratory Investigation of Residual Liquid Organics from Spills, Leaks, and the Disposal of Hazardous Wastes in Groundwater," U.S. EPA Report No. EPA/600/6-90/004, April. Available from NTIS as PB90-235797.

CHAPTER 8

Sparging: Theory and Mathematical Models

INTRODUCTION

THERE are several reasons why one develops and uses mathematical models of a remediation technology. Simple models may be helpful in doing a preliminary feasibility assessment of a technique at a site using rough guesses as to parameters; this may also be helpful in designing and interpreting lab-scale experiments. For the design and interpretation of pilot-scale tests in the field, one will probably want a more elaborate model, which will require more computation but which will also be more realistic than the initial simple modeling. Such elaborate models are also useful in estimating cleanup times and costs, in optimizing the design of the full-field–scale remediation, and in carrying out sensitivity analyses to get some idea of how accurate you can expect the results of the computations to be. Lastly, as the actual cleanup progresses, one can compare theoretical and actual results to see if any unpleasant surprises are turning up that will require some modification of the remediation plan. Modeling is generally rather cheap and fast, so one can easily explore many options.

One word of warning about the modeling of remediation techniques at hazardous waste sites – Remedial Investigations/Feasibility Studies rarely, if ever, provide data in sufficient detail and of sufficient accuracy to permit highly precise mathematical modeling. The results of a mathematical modeling exercise are no better than the completeness and accuracy of the data set being used to describe the site. One does not want to discredit mathematical modeling, but the reader should be reminded that no amount of fancy partial differential equations and high-powered number crunching can make up for an inadequate data set, especially if some major feature of the site has been overlooked.

In addition to the applications mentioned above, one can also use mathematical modeling as a means to develop understanding and intuition about sparging – to develop some common sense about the technique by examin-

377

ing results of some experiments with sparging models. For this we shall use models that our group at Vanderbilt, the University of Málaga, and Eckenfelder, Inc., developed for use on microcomputers. We shall describe the physical models being used, any physical or chemical approximations that one needs to know about, the mathematical and computational details (which the casual reader may well wish to skip), and the results of the calculations.

THEORY OF SPARGING AND MODEL DEVELOPMENT

In the following sections we shall address several problems and three types of well configurations that arise in connection with sparging. First, we shall address the question of the airflow patterns that can be expected in the vicinity of an air injection well used for sparging. This is followed by the analysis of a particularly simple type of sparging configuration, the aeration curtain. We then turn to vacuum-vaporizer-wells of the sort commonly used in Germany and examine their operation for the removal of dissolved VOCs and, then, a somewhat more complex problem, the removal of DNAPL droplets dispersed in an aquifer. With these wells one has a circulation of water induced in the vicinity of each well, which transports the contaminant to the well, so it is necessary to estimate the local water flow pattern that is present when the well is in operation. Lastly, we turn to the modeling of simple air injection wells for sparging. In such calculations both water and air advection must be modeled; we include, in addition, the effects of diffusion kinetics in the solution of DNAPL droplets.

GAS FLOW PATTERNS: THE EFFECTIVE RADIUS OF A SPARGING WELL

One of the matters of interest in connection with simple sparging wells is the distribution pattern of the injected air as it percolates up through the zone of saturation to the water table. Calculation of this distribution from fundamental principles is an exceedingly difficult problem, in that it involves simultaneous flow of two fluid phases in a porous medium of heterogeneous permeability. We shall explore how one can use vadose zone soil gas pressures measured a short distance above the water table to determine the gas distribution radius of a sparging well at various flow rates, an approach suggested by Brown (see Brown, undated; Wilson et al., 1992).

Groundwater sparging can be carried out by means of a simple sparging well that discharges air into the aquifer down near the bottom. One wishes to know the size of the resulting aeration cone and how this depends on gas flow rate in order to design systems using sparging wells. One expects that soil gas pressure measurements in the vadose zone in the vicinity of a

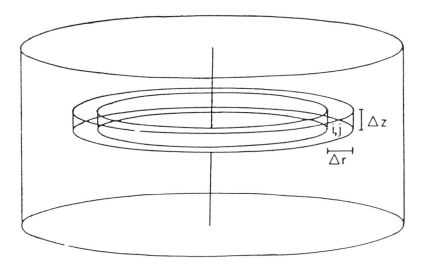

Figure 8.1 Mathematical partitioning of the domain of interest in the vadose zone for vadose zone gas pressure calculations. [Reprinted from Wilson et al. (1992) p. 1032, by courtesy of Marcel Dekker, Inc.]

sparging well would give information on these points, and this is, in fact, the case. In this section we explore the soil gas pressure distribution in a cylindrical domain in the vadose zone centered about a single sparging well discharging air well below the water table. Refer to Figure 7.3 for a possible design.

The equation governing the pressure of an ideal gas in a porous medium under steady-state conditions is well approximated by

$$0 = \nabla \cdot K \nabla (P^2) \tag{1}$$

This will be solved by approximating it using a finite difference representation and then solving the resulting linear equations (in P^2) by a relaxation method. The cylindrically symmetrical geometry and notation are indicated in Figure 8.1. The finite difference equation at an interior annular volume element (i, j) is

$$0 = K_r(P^2_{i-1,j} - P^2_{i,j}) \cdot 2\pi(i - 1)\Delta z$$

$$+ K_r(P^2_{i+1,j} - P^2_{i,j}) \cdot 2\pi i \Delta z$$

$$+ Kz(P^2_{i,j-1} - 2P^2_{i,j} + P^2_{i,j+1}) \cdot (2i - 1)\pi(\Delta r)^2/\Delta z \tag{2}$$

$$i = 2,3,\ldots,I_r; \qquad j = 2,3,\ldots,J_z$$

Along the axis of the cylindrical domain the equation is

$$0 = K_r(P^2_{i+1,j} - P^2_{i,j}) \cdot 2\pi i \Delta z$$

$$+ K_z(P^2_{i,j-1} - 2P^2_{i,j} + P^2_{i,j+1}) \cdot (2i - 1)\pi(\Delta r)^2/\Delta z \qquad (3)$$

$$i = 1; \qquad j = 2,3,\ldots,J_z - 1$$

If we assume a no-flow boundary condition at the cylindrical periphery of the domain, the equation is

$$0 = K_r(P^2_{i-1,j} - P^2_{i,j}) \cdot 2\pi(i - 1)\Delta z$$

$$+ K_z(P^2_{i,j-1} - 2P^2_{i,j} + P^2_{i,j+1}) \cdot (2i - 1)\pi(\Delta r)^2/\Delta z \qquad (4)$$

$$i = I_r; \qquad j = 2,3,\ldots,J_z - 1$$

This could be replaced by a boundary condition that the pressure be one atmosphere if desired. The development of the difference equations at the boundary is similar to that described above, except that gas flow through the cylindrical periphery is allowed, and it is assumed that the gas pressure at this surface is fixed at one atmosphere. In doing calculations for single wells, the radius of the cylindrical domain should be taken sufficiently large that the choice of boundary condition is immaterial. We shall explore this point further when we deal with the results. If multiple wells in a hexagonal grid are to be used (each well with six nearest neighbors), the radius of the cylindrical domain should be half the distance between adjacent wells and the no-flow peripheral boundary condition ($\partial P/\partial r = 0$) should be used.

At the bottom of the domain, we include a flux term $Q(r)$ to describe the incoming gas from the sparging well; this gives

$$0 = K_r(P^2_{i-1,j} - P^2_{i,j}) \cdot 2\pi(i - 1)\Delta z$$

$$+ K_r(P^2_{i+1,j} - P^2_{i,j}) \cdot 2\pi i \Delta z$$

$$+ K_z(P^2_{i,j+1} - P^2_{i,j}) \cdot (2i - 1)\pi(\Delta r)^2/\Delta z$$

$$+ Q[(i - .5)\Delta r] \cdot (2i - 1)\pi(\Delta r)^2 \qquad (5)$$

$$j = 1; \qquad i = 2,3,\ldots,I_r - 1$$

At the top of the domain the pressure is one atmosphere, and the difference equation becomes

$$0 = K_r(P_{i-1,j}^2 - P_{i,j}^2) \cdot 2\pi(i-1)\Delta z$$

$$+ K_r(P_{i+1,j}^2 - P_{i,j}^2) \cdot 2\pi i \Delta z$$

$$+ K_z[(P_{i,j-1}^2 - P_{i,j}^2) + 2(1 - P_{i,j}^2)] \cdot (2i - 1)\pi(\Delta r)^2/\Delta z \qquad (6)$$

$$j = J_z; \qquad i = 2,3,\ldots,I_r$$

In similar fashion one writes four difference equations for the four volume elements $(1,1)$, $(1,J_z)$, $(I_r,1)$, and (I_r,J_z).

Each of these equations is then solved for $P_{i,j}^2$ in terms of the surrounding pressures. One then assumes an expression for the gas flux term at the bottom of the domain, $Q(r)$, assigns a value of unity to all the $P_{s,t}^2$, and iterates the system until convergence has occurred. This is monitored during the course of the calculation by accumulating S, the sum of the squares of the differences between the $P_{s,t}^2$ over all mesh points for successive iterations.

$$S = \sum_{i=1}^{I_r} \sum_{j=1}^{J_z} (P_{i,j}^{2\prime} - P_{i,j}^{2\prime\prime})^2 \qquad (7)$$

where $P_{i,j}^{2\prime}$ and $P_{i,j}^{2\prime\prime}$ are the kth and the $(k + 1)$th iterated values for $P_{i,j}^2$.

One could speed up convergence by use of an over-relaxation method, but the calculations are sufficiently fast that it is not necessary to use this elaboration.

The choice of the function $Q(r)$, which specifies the flux coming into the vadose zone domain through its bottom surface, is somewhat uncertain. Initially, we explored three possibilities, as follows. The first choice was

$$Q(r) = \frac{Q_{tot}}{4\pi\sigma_r^2} \cdot \exp(-r^2/4\sigma_r^2) \qquad (8)$$

where Q_{tot} is the total gas flow rate and σ_r gives a measure of the effective radius of the distribution. The second choice was

$$Q(r) = \frac{3Q_{tot}}{a^3\pi} \cdot (a - r), r < a$$

$$= 0, r > a \qquad (9)$$

The third choice was

$$Q(r) = Q_{tot}/\pi a^2, r < a$$

$$= 0, r > a \qquad (10)$$

One can generalize Equation (9) for $Q(r)$ to explore a broader range of distribution functions by replacing the first power dependence on r by an nth power dependence, as follows:

$$Q(r) = \frac{Q_{tot} \cdot 2 \cdot (n + 2)}{a^2 \cdot n} \cdot [1 - (r/a)^n], r < a$$

$$= 0, r > a \qquad (11)$$

The mean radius of $Q(r)$ is given by

$$\bar{r} = \int_0^a Q(r)r^2 dr/Q_{tot} \qquad (12)$$

For Equation (11) this yields

$$\bar{r} = \frac{2(n + 2)}{3(n + 3)} \cdot a \qquad (13)$$

which in the limit as n approaches infinity applies to Equation (10) as well. For Equation (8), the Gaussian distribution, one obtains

$$\bar{r} = \sqrt{\pi}\sigma_r$$

$$= 1.772\sigma_r \qquad (14)$$

Let's now look at some of the results, to see how one can use vadose zone soil gas pressures to get information about the distribution of the sparging air as it reaches the water table. Figure 8.2 shows plots of soil gas pressure in cm of water versus distance from the domain axis at a height of 0.5 m above the water table. The Gaussian flux distribution was assumed, and the model parameters are given in Table 8.1. The gas flow rate from the well was held constant at 0.1 m³/sec for all the runs. It is apparent that the soil gas pressure distribution in the vadose zone near the water table depends very strongly on the effective radius parameter of the gas flux distribution at the bottom of the vadose zone.

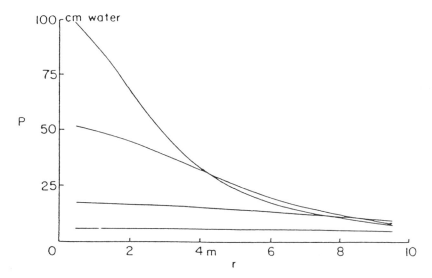

Figure 8.2 Effect of σ_r, effective radius parameter, on soil gas pressure distribution 0.5 m above the water table. $Q(r)$ is Gaussian [Equation (2)]. From the top down at $r = 1$ m, $\sigma_r = 1, 2, 5$, and 10 m. Other parameters as in Table 8.1. [Reprinted from Wilson et al. (1992) p. 1035, by courtesy of Marcel Dekker, Inc.]

Figure 8.3 shows plots of the soil gas pressure (cm of water) at a height of 0.5 m above the water table versus the distance from the center of the domain for the second, linear, distribution function, given by Equation (9). The model parameters are as in Table 8.1, except that the values for the effective radius parameter σ_r were replaced by values of the maximum radius parameter a in Equation (9); the values of a used were 2, 4, 10, and 20 m.

The results are qualitatively quite similar in appearance to those of Figure 8.2, so this technique cannot be used to discriminate between different functional forms of $Q(r)$ with any sensitivity. On the other hand, in both Figures 8.1 and 8.2 the radial dependence of the soil gas pressure depends strongly on the effective radius of the soil gas flux distribution func-

TABLE 8.1. Model Parameters for the Gaussian Gas
Flux Calculations.

Gas flowrate Q_a	0.1 m³/sec
Effective radius parameter σ_r	1, 2, 5, 10 m
Radius of domain of interest	30 m
Horizontal pneumatic permeability K_r	0.05 m²/atm sec
Vertical pneumatic permeability K_z	0.05 m²/atm sec

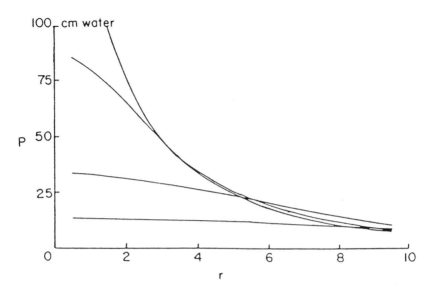

Figure 8.3 Effect of a, maximum radius parameter, on soil gas pressure distribution 0.5 m above the water table. $Q(r)$ is a linear function of r [$n = 1$ in Equation (3)]. From the top down at $r = 1$ m, $a = 12, 4, 10$, and 20 m. Other parameters as in Table 8.1. [Reprinted from Wilson et al. (1992) p. 1036, by courtesy of Marcel Dekker, Inc.]

tion $Q(r)$. Evidently, the measurement of soil gas pressures by means of piezometer wells screened near the bottom of the vadose zone is an effective method for estimating the radius of the gas flow distribution at the water table resulting from a sparging well screened in the zone of saturation.

Figure 8.4 shows plots of soil gas pressure (cm of water) at a height of 0.5 m above the water table versus the distance from the center of the domain for the second, constant, distribution function, given by Equation (10). The model parameters are as in Table 8.1, except that the values for the effective radius parameter σ_r were replaced by values of the maximum radius parameter a in Equation (10); the values of a used were 2, 4, 10, and 20 m. As before, the soil gas pressure distribution plots depend very markedly on the radius parameter of $Q(r)$, but it would be difficult to distinguish between the different functional forms of $Q(r)$ by means of soil gas pressure measurements.

The effects of the boundary condition used at the cylindrical periphery of the domain are shown in Figures 8.5 and 8.6. The gas flux $Q(r)$ is assumed to be Gaussian and the parameters used are those given in Table 8.1. In Figure 8.5 the effective radius parameter σ_r used is 5 m. In Figure 8.6 this parameter is 10 m; this is a least favorable case, in that it shows

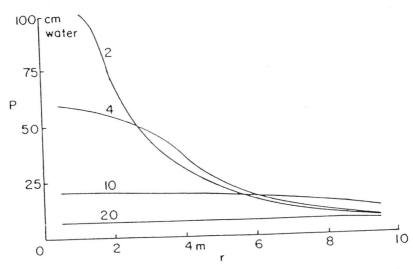

Figure 8.4 Effect of a, maximum radius parameter, on soil gas pressure distribution 0.5 m above the water table. $Q(r)$ is a constant for $r < a$ [Equation (10)]. From the top down at $r = 1$ m, $a = 2$, 4, 10, and 20 m. Other parameters as in Table 8.1. [Reprinted from Wilson et al. (1992) p. 1037, by courtesy of Marcel Dekker, Inc.]

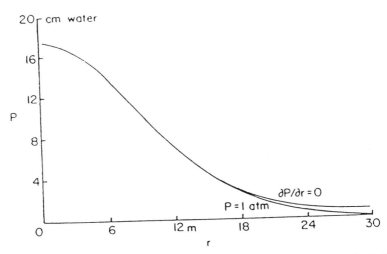

Figure 8.5 Effect of boundary condition at the domain periphery on soil gas pressure distribution 0.5 m above the water table. $Q(r)$ is Gaussian [Equation (8)] with $\sigma_r = 5$ m; other parameters as in Table 8.1. The boundary condition at $r = 30$ m for the upper curve is $\partial P/\partial r = 0$; for the lower curve it is $P = 1$ atm. [Reprinted from Wilson et al. (1992) p. 1038, by courtesy of Marcel Dekker, Inc.]

385

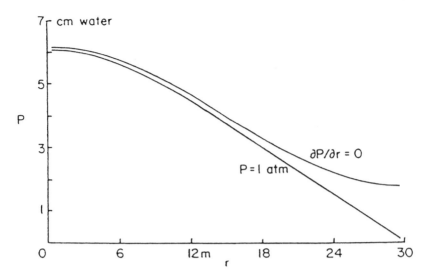

Figure 8.6 Effect of boundary condition at domain periphery on soil gas pressure distribution 0.5 m above the water table. $Q(r)$ is Gaussian [Equation (8)] with $\sigma_r = 10$ m; other parameters as in Table 8.1. The boundary condition at $r = 30$ m for the upper curve is $\partial P/\partial r = 0$; for the lower curve it is $P = 1$ atm. [Reprinted from Wilson et al. (1992) p. 1038, by courtesy of Marcel Dekker, Inc.]

quite marked differences in the soil gas pressures near the periphery of the domain, say from 20 to 30 m from the sparging well. In Figure 8.5, in which σ_r is 5 m, the two different boundary conditions ($P = 1$ atm and $\partial P/\partial r = 0$) yield quite similar results out almost to the periphery of the domain, at which the soil gas excess pressure from the well has fallen almost to zero. Evidently, the effect of the peripheral boundary condition can be made unimportant by selecting a domain radius that is substantially larger (a factor of five or six) than the effective radius parameter σ_r.

The effect of the functional form of $Q(r)$ is explored in Figure 8.7. The mean radius of $Q(r)$, given by Equation (12), was held constant at 5 m, and corresponding values of a or σ_r were calculated for the values of n given in Table 8.2 and for the Gaussian distribution. Runs were then made using these values of a or σ_r and the other parameters given in Table 8.1. Figure 8.7 shows the soil gas pressures 0.5 m above the water table as functions of r for these runs; pressure (cm of water) in excess of 1 atm is plotted. Note that the pressure scale does not start at 0. For values of r between 10 m and 30 m, the curves are virtually superimposed. We see that rather marked changes in the form of the gas distribution function $Q(r)$ result in relatively minor changes in the soil gas pressures in the vadose zone even

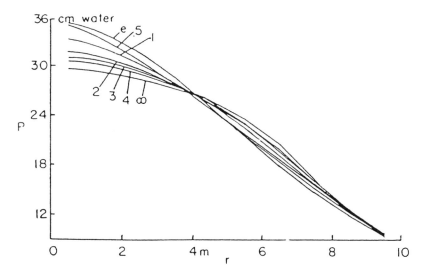

Figure 8.7 Effect of the functional form of $Q(r)$ on the soil gas pressure distribution 0.5 m above the water table. $Q(r)$ was calculated by Equation (8) (curve e) or by Equation (11) (all other plots); values of a or σ_r are from Table 8.3 and all other parameters are from Table 8.1. [Reprinted from Wilson et al. (1992) p. 1039, by courtesy of Marcel Dekker, Inc.]

quite close to the water table. While soil gas pressure measurements are a good probe for determining the effective radius of the spatial distribution of the flow of the injected gas, they are not sensitive to the details of this distribution.

In some situations one has a highly variable permeability, or the permeability is anisotropic, or the vadose zone has domains of low-permeability clay. It is an easy matter to include such modifications in the permeability into the calculation of the vadose zone gas pressures near a sparging well if this is desired; some possibilities are indicated in Chapter 3.

TABLE 8.2. Values of a (and σ_r) Corresponding to
$\bar{r} = 5$ m.

n	a (σ_r)
0.5	10.500 m
1	10.000
2	9.375
3	9.000
4	8.750
∞	7.500
(Gaussian)	2.82095 (σ_r)

In situ Sparging with an Aeration Curtain

In some situations in which an aquifer is rather close to the surface, an aeration curtain configuration for sparging may be suitable. Here, one places a trench at right angles to the direction of flow of the contaminant plume so as to intercept it. The trench contains a horizontal slotted aeration pipe at the bottom and is packed with crushed rock. (A diagram is given in Figure 8.8.) As the water flows across the aeration curtain, it is stripped of VOC. If necessary, an overlying horizontal slotted pipe recovers the injected air for treatment before discharge. We use Henry's Law and the local equilibrium approximation.

One carries out a mass balance on each of the volume elements indicated, with the groundwater flowing in from the left and the air flowing up from below. One then calculates the average VOC concentration coming out of the right side of the curtain.

The geometry of the aeration curtain system with a single curtain is indicated in Figure 8.8. The symbols are defined as follows:

L = thickness of aquifer, m
a = width of aeration curtain, m
c = length of aeration curtain, m

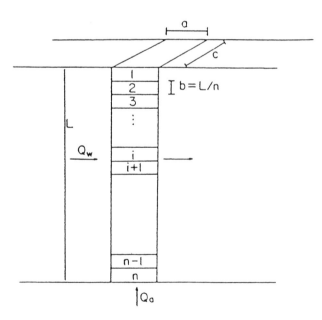

Figure 8.8 The aeration curtain. Geometry and notation. [Reprinted from Wilson et al. (1992) p. 1026, by courtesy of Marcel Dekker, Inc.]

n = number of equivalent theoretical transfer units into which the aeration curtain is partitioned for analysis

b = thickness of an equivalent theoretical transfer unit, m

Q_w = total water flow rate through the curtain, m³/sec

Q_a = total airflow rate through the curtain, m³/sec

C_0^w = incoming aqueous contaminant concentration, kg/m³

C_i^w = aqueous VOC concentration in compartment i, kg/m³

C_i^g = vapor phase VOC concentration in compartment i, kg/m³

K_H = Henry's constant of the contaminant, dimensionless

ν = voids fraction in the crushed rock/gravel curtain, dimensionless

The hydrostatic pressure at the bottom of the ith cell (transfer unit) is given by

$$P_i = (1 + ib/10.336) \text{ atm} \tag{15}$$

so the volumetric airflow rate at the bottom of the jth cell is given by

$$Q_j^a = \frac{Q_a}{1 + jb/10.336} \tag{16}$$

We neglect the volume of air compared to the volume of water in a cell. Let m_i = mass of contaminant in the ith cell, kg. Then

$$\frac{dm_i}{dt} = \frac{Q_w}{n} \cdot (C_0^w - C_i^w) + Q_i^a C_{i+1}^g - Q_{i-1}^a C_i^g \tag{17}$$

Making the steady-state assumption for the system then permits us to set Equation (17) equal to zero.

Henry's Law is assumed to apply to the VOC in the aquifer, which, along with the local equilibrium assumption, yields

$$C_j^g = K_H C_j^w, \qquad j = 1,2,\ldots,n \tag{18}$$

Substitution of Equations (16) and (18) into Equation (17) then gives

$$0 = \frac{Q_w}{n} \cdot (C_0^w - C_i^w) + \frac{Q_a K_H C_{i+1}^w}{1 + ib/10.336} - \frac{Q_a K_H C_i^w}{1 + (i - 1)b/10.336} \tag{19}$$

Solving Equation (19) for C_i^w then yields

$$C_i^w = \left| \frac{Q_w C_0^w}{n} + \frac{Q_a K_H C_{i+1}^w}{1 + ib/10.336} \right| \bigg/ \left| \frac{Q_w}{n} + \frac{Q_a K_H}{1 + (i - 1)b/10.336} \right| \tag{20}$$

and

$$C_n^w = \frac{Q_w C_0^w}{n} \left| \frac{Q_w}{n} + \frac{Q_a K_H}{1 + (n - 1)b/10.336} \right| \qquad (21)$$

The mean contaminant concentration in the water that has passed through the curtain is given by

$$\overline{C}_{out}^w = \sum_{i=1}^{n} C_i^w/n \qquad (22)$$

and the percent removal of VOC from the passing groundwater by the curtain is

$$R = 100(1 - \overline{C}_{out}^w/C_0^w) \qquad (23)$$

We next explore the dependence of R, the percent removal of VOC by the aeration curtain, on the parameters of the model. The default parameters used in the calculations are given in Table 8.3.

The effect of airflow rate is shown in Table 8.4. The expected increase in percent VOC removal with increasing airflow rate is observed, but, unlike the situation with countercurrent aerators, there is no abrupt transition. If the percent VOC removal is not adequate at a flow rate that is near the feasible maximum, a marginal increase in flow rate is not going to solve the problem.

The impact of the flow rate of the groundwater in the aquifer is given in Table 8.5. The expected decrease of percent VOC removal with increasing groundwater flow rate is seen, and again we find that there is no fairly abrupt transition to an overloaded condition such as is found with counter-

TABLE 8.3. Default Parameters Used in the Aeration Curtain Model.

Depth of aquifer	5 m
Thickness of aeration curtain	0.5 m
Length of aeration curtain	100 m
Number of theoretical transfer units	10
Voids fraction of the crushed rock curtain	0.4
Contaminant	Trichloroethylene
Henry's constant (15 °C)	0.2821
Contaminant concentration in groundwater incident on the aeration curtain	100 mg/L
Total water flow across the aeration curtain	0.1 m³/sec
Total airflow rate generating the curtain	2 m³/sec

TABLE 8.4. Effect of Airflow Rate on Percent VOC Removal, Aeration Curtain.

Airflow Rate (m³/sec)	Percent VOC Removal
0.5	67.02
1.0	80.96
1.5	86.64
2.0	89.71
3.0	92.96
5.0	95.68

current separations. The geometry of the aeration curtain is such that this is a cross-current separation method.

The number of theoretical transfer units used to represent the column has only a minor effect on the percent VOC removal (see Table 8.6). Since this is a cross-current separation, this is expected. The result also indicates that there is little to be gained from efforts to reduce longitudinal dispersion in the air in the curtain. The model assumes complete mixing across the thickness of the curtain for both vapor and aqueous phases; unless the curtain is uneconomically thick, this is a reasonable assumption.

One expects the percent VOC removal to be independent of the VOC concentration of the groundwater entering the curtain since the modeling equations are linear in the concentrations. Runs made with values of C_o^w of 10, 100, and 1,000 mg/L showed that this was the case.

One might expect that there would be no dependence of percent VOC removal on aquifer thickness so long as the air and groundwater flow rates and the height of a theoretical transfer unit are held constant. As seen in Table 8.7, this is not exactly true. The airflow rate, Q_a, is at a pressure of 1 atm, while the gas in any particular cell in the curtain is subject to addi-

TABLE 8.5. Effect of Groundwater Flow Rate on Percent VOC Removal, Aeration Curtain.

Groundwater Flow Rate (m³/sec)	Percent VOC Removal
0.02	97.80
0.025	97.80
0.05	94.65
0.10	89.71
0.15	85.16
0.20	80.96
0.50	61.43

TABLE 8.6. Effect of the Number of Theoretical Transfer Units n on Percent VOC Removal, Aeration Curtain.

n	Percent VOC Removal
20	90.05
10	89.71
5	89.06
3	88.25

tional pressure from the hydrostatic head of the curtain. Therefore, the flow rate of gas passing through the lower portions of the curtain is correspondingly reduced; the greater the depth, the greater is the reduction. This, in turn, results in a slight decrease in the amount of VOC that can be removed from the thicker aquifers.

The effect of the Henry's constant value on the percent VOC removal is of particular interest, since VOCs of interest show a rather wide range of Henry's constant values. This point is explored in Table 8.8. The Henry's constant for TCE (trichloroethylene) at 15°C was adapted from Howe et al. (1986). In the aquifer environment we expect that use of these Henry's constants is justified, unlike the situation in soil vapor extraction, where absorption effects may make the effective Henry's constants much smaller than the Henry's constants of the compounds in water.

An inspection of the Henry's constants compiled by Howe et al. shows that, while some VOCs have K_H values that are larger than 1.0 and many have K_Hs that are larger than 0.1, there are a number that have Henry's constants at 15°C, which are substantially less than 0.1. A few of these are 1,2-dichlorobenzene (0.060), 1,2-dichloroethane (0.055), 1,1,2-trichloroethane (0.027), tetralin (0.044), ethylene dibromide (0.020), 1,2-dichloropropane (0.053), dibromochloromethane (0.019), 1,2,4-trichlorobenzene (0.044), methylethyl ketone (0.016), and methylisobutyl ketone (0.016). The results of Table 8.8 raise some doubt as to the suitability of the aeration curtain technique for such compounds.

It was mentioned above that this is a cross-current technique, so the aer-

TABLE 8.7. Effect of Aquifer Thickness on Percent VOC Removal, Aeration Curtain.

Aquifer Thickness, m	Percent VOC Removal
3	89.73
5	89.71
10	88.87

TABLE 8.8. Effect of Henry's Constant K_H on Percent VOC Removal, Aeration Curtain.

K_H (dimensionless)	Percent VOC Removal
0.30	90.28
0.2821 (TCE)	89.71
0.20	85.95
0.10	74.73
0.05	58.23

ation curtain is, in essence, a single-stage technique. One might therefore expect to improve its performance substantially by using a system having several aeration curtains in series, and, indeed, this is the case. A simple analysis yields Equation (24) for R_n, the percent VOC removal resulting from n aeration curtains in series.

$$R_n = 100 \cdot [1 - (1 - R/100)^n] \qquad (24)$$

Table 8.9 shows the effect of using multiple aeration curtain systems. The useful range of the aeration curtain technique can be extended significantly by employing multiple curtains. These need to be separated only enough to ensure that there is no movement of sparging air from one curtain to another and could even be in the same trench.

VACUUM-VAPORIZER-WELL–TYPE CONFIGURATIONS FOR DISSOLVED VOC REMOVAL

Here we shall first consider the expected water flow pattern in the vicinity of a simple vacuum-vaporizer-well–type sparging configuration. Then we shall use the result in a model for the sparging of dissolved VOC from an aquifer [see Wilson (1992) for details].

TABLE 8.9. Percent VOC Removals from Multiple Aeration Curtain Systems.

	$R (= R_1)$			
	90	75	50	25
n		R_n		
2	99.00	93.75	75.00	43.75
3	99.90	98.44	87.50	57.81
4	99.99	99.61	93.75	68.36

Groundwater Flow Patterns in the Vicinity of a Sparging Well

In this section we explore the flow patterns that may be expected when an aquifer with little or no natural flow is sparged with a single well.

Consider the situation illustrated in Figure 8.9, where there is a sink $-Q$ at the bottom of the aquifer and a source Q at the top of the aquifer, representing the intake of water from the bottom of the aquifer into the bottom of the sparging zone and the discharge of water from the top of the sparging zone back into the top of the aquifer, respectively. We let h be the thickness of the aquifer. The method of images, from electrostatics, allows us to construct the velocity potential for the system from the array of sources and sinks shown in Figure 8.10, where the hatched region is the actual physical domain of interest (the aquifer).

A solution to Laplace's equation corresponding to this distribution of sources and sinks is the following:

$$W = \sum_{n=-\infty}^{\infty} A\left[\frac{2Q}{\sqrt{r^2 + [z - (2n + 1)h]^2}} - \frac{2Q}{\sqrt{r^2 + [z - 2nh]^2}}\right] \quad (25)$$

The constant A is determined by the requirement that

$$2Q = \int_0^{2\pi} \int_0^{\pi} A \cdot \frac{2Q}{\varrho^2} \cdot \varrho^2 \sin\theta d\theta d\phi \quad (26)$$

which yields $A = 1/4\pi$. Then

$$W = \sum_{n=-\infty}^{\infty} \frac{Q}{2\pi}\left[\frac{1}{\{r^2 + [z - (2n + 1)h]^2\}^{1/2}}\right.$$

$$\left. - \frac{1}{\{r^2 + [z - 2nh]^2\}^{1/2}}\right] \quad (27)$$

Now the linear velocity of the fluid is given by

$$\mathbf{v} = -\frac{1}{\nu}\nabla W \quad (28)$$

This gives, for the velocity components v_r and v_z, the following results:

$$v_r = \frac{Q}{2\pi\nu} \sum_{n=-\infty}^{\infty} \left[\frac{r}{\{r^2 + [z - (2n + 1)h]^2\}^{3/2}} - \frac{r}{\{r^2 + [z - 2nh]^2\}^{3/2}}\right]$$

$$(29)$$

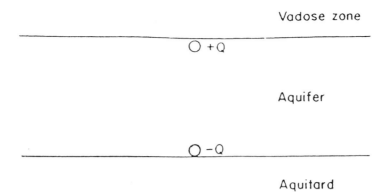

Vadose zone

○ +Q

Aquifer

○ -Q

Aquitard

Figure 8.9 Locations of source and sink representing the water intake and discharge of a vacuum-vaporizer-well configuration. [Reprinted from Wilson et al. (1992) p. 1682, by courtesy of Marcel Dekker, Inc.]

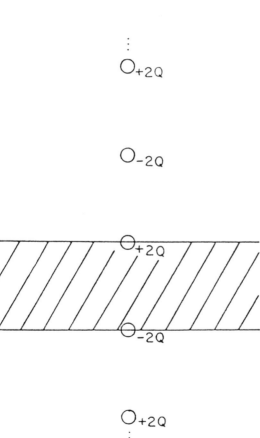

$\vdots$

$\bigcirc_{+2Q}$

$\bigcirc_{-2Q}$

$\bigcirc_{+2Q}$

$\bigcirc_{-2Q}$

$\bigcirc_{+2Q}$

$\vdots$

Figure 8.10 Distribution of sources and sinks for generating the water flow field in the vicinity of a sparging well by the method of images. The hatched region corresponds to the domain of physical interest. [Reprinted from Wilson et al. (1992) p. 1683, by courtesy of Marcel Dekker, Inc.]

and

$$v_z = \frac{Q}{2\pi v} \sum_{n=-\infty}^{\infty} \left[\frac{z - (2n + 1)h}{\{r^2 + [z - (2n + 1)h]^2\}^{3/2}} - \frac{z - 2nh}{\{r^2 + [z - 2nh]^2\}^{3/2}} \right]$$

(30)

One then calculates the flow paths and transit times by integrating the parametric equations

$$\frac{dr}{dt} = v_r(r,z)$$

(31)

and

$$\frac{dz}{dt} = v_z(r,z)$$

(32)

Streamlines (flow paths) for such a system are shown in Figure 8.11. The parameters used in calculating these, the transit time of each of the streamlines shown, and the volumes enclosed by the figures of revolution generated by rotating the streamlines around the axis between the sink and the source are shown in Table 8.10. A dimensionless time $\tau = h^3 t/Q$ and a dimensionless distance $x' = x/h$ can be used to calculate from Figure 8.11 and Table 8.10 the transit times for geometrically similar streamlines with

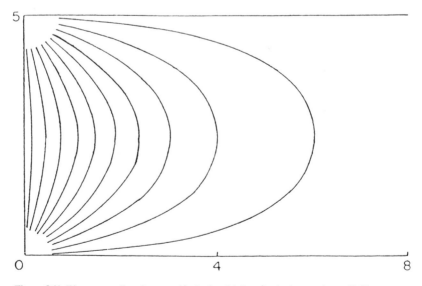

Figure 8.11 Water streamlines in an aquifer in the vicinity of a single sparging well. The parameters used and the transit times along each of the streamlines are given in Table 8.10. [Reprinted from Wilson et al. (1992) p. 1684, by courtesy of Marcel Dekker, Inc.]

TABLE 8.10. Streamlines and Transit Times around a Sparging Well
(see Figure 8.11).

i	Transit Time (sec)	R_{max} of Streamline (m)	Volume Contained within Surface Generated by Rotating Streamline (m³)
1	1,490	0.147	0.156
2	1,580	0.445	1.441
3	1,770	0.755	4.198
4	2,110	1.086	8.867
5	2,680	1.452	16.254
6	3,690	1.872	27.995
7	5,610	2.379	47.354
8	9,780	3.036	81.812
9	21,930	4.003	154.294
10	104,650	6.003	394.710

Thickness of aquifer = 5 m.
Aquifer porosity = 0.3.
Induced water flow rate Q_w = 0.01 m³/sec.
Radius of domain = 8 m.

different values of the aquifer thickness h and the flow rate Q. We see from Table 8.10 that a volume around the sparging well of approximately $1.2h^3$ is flushed out fairly rapidly (within about 6 hr with these parameters) but that the flushing time increases very markedly up to 29 hr if the volume of influence of the well is increased to a value of $3.2h^3$. Evidently, the effective radius of influence of a sparging well of this type in an aquifer having an isotropic permeability is somewhere around $0.8h$.

An n-Compartment Sparging Well Model

Here we use the velocity field calculated in the previous section to construct an n-compartment model for the operation of a sparging well operating in a stagnant or nearly stagnant aquifer. The set-up is shown in Figure 8.12.

The flow of fluid between the ith and the $(i + 1)$th annular volume elements in the top half of the domain is given by

$$Q_i^{out} = \int_0^{2\pi} \int_{h/2}^h \nu v_r \cdot r_i d\theta dz \tag{33}$$

$$= Q r_i^2 \int_{h/2}^h \left[\sum_{n=-\infty}^{\infty} \{r_i^2 + [z - (2n + 1)h]^2\}^{-3/2} \right.$$

$$\left. - \{r_i^2 + [z - 2nh]^2\}^{-3/2} \right] dz \tag{34}$$

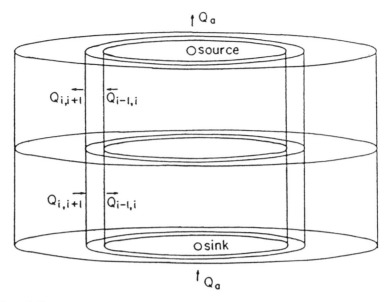

Figure 8.12 An n-compartment sparging well model, showing the partitioning of the domain in the vicinity of the sparging well into annular volume elements. [Reprinted from Wilson et al. (1992) p. 1686, by courtesy of Marcel Dekker, Inc.]

Integration and insertion of the limits of integration then yield

$$Q_i^{out} = Q \sum_{n=-\infty}^{\infty} \left[\frac{-1}{\{1 + [r_i/2nh]\}^{1/2}} + \frac{1}{\{1 + [r_i/(2n + 1/2)]^2\}^{1/2}} \right.$$

$$\left. + \frac{1}{\{1 + [r_i/(2n - 1)h]^2\}^{1/2}} - \frac{1}{\{1 + [r_i/(2n - 1/2)h]^2\}^{1/2}} \right]$$

$$(35)$$

as the flow rate outward from the ith to the $(i + 1)$th annular volume element in the top half of the domain. Henceforth, we denote this as $Q_{i,i+1}$. The individual series here do not converge; it is necessary to pair positive and negative terms in order to get a convergent series.

In the bottom half of the domain, the flow rates are the same in magnitude, but in the opposite direction (from the periphery toward the well). We use these flow rates to get an estimate of the movement of dissolved VOC between the volume elements by advection. This (together with the assumption that the central volume element is being sparged with air at a flow rate Q_a, that this volume element is well-mixed, and that Henry's Law applies) yields the following model equations for the sparging system:

$$V_i \frac{dC_i^w}{dt} = Q_{i-1,i}(C_{i-1}^w - C_i^w) + Q_{i,i+1}(C_{i+1}^w - C_i^w), \ i = 2,3,\ldots \quad (36)$$

and

$$V_1 \frac{dC_1^w}{dt} = Q_{1,2}(C_2^w - C_1^w) - Q_a K_H C_1^w \quad (37)$$

Here V_j is the volume of the jth annular volume element,

$$V_j = [j^2 - (j-1)^2]\pi(\Delta r)^2 \cdot h \quad (38)$$

Note that this model is for handling the removal of a dissolved VOC. The model is not applicable if nonaqueous phase liquid (NAPL) is present, since in that case the rate-limiting step is almost certain to be the rate of dissolution of the nonaqueous phase liquid into the aqueous phase, which is not handled in the model. One would expect, however, that the flow velocities calculated above would be helpful in the development of a model for NAPL removal, since mass transfer from the NAPL phase is surely affected by the velocity of the water streaming past it.

Results, Sparging of Dissolved VOC

The n-compartment model was implemented in TurboBASIC and run on microcomputers using MS-DOS, equipped with math coprocessors, and running at clock speeds between 12 and 33 MHz. Run times ranged from fifteen minutes to about an hour, depending on the run parameters.

Default parameters for all of the runs shown in Figures 8.13–8.15 are given in Table 8.11. Values of parameters having values other than the default values are given in the captions. Figure 8.13 illustrates the effect of varying the water flow rate Q_w and shows the expected increase in VOC removal rate with increasing Q_w. In these runs the airflow rate Q_a was held constant.

The effect of varying the airflow rate Q_a is shown in Figure 8.14. Here Q_w was held constant. The expected increase of removal rate with increasing airflow rate is observed. Air and water flow rates Q_a and Q_w are varied together in the runs shown in Figure 8.15; Q_a was taken equal to Q_w in these three runs. The VOC removal rate appears to be proportional to Q_a for these runs. In fact, one would expect Q_w to be linked to Q_a in some fashion such as this.

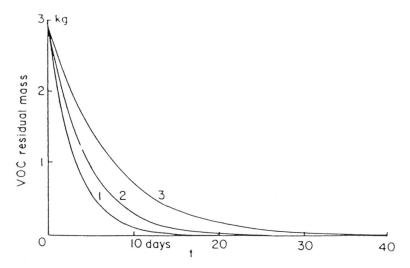

Figure 8.13 Dissolved VOC sparging simulation using the *n*-compartment model. Effect of airflow rate on removal rate of VOC. See Table 8.11 for default run parameters. Q_a = 0.01, 0.005, and 0.0025 m³/sec for Runs 1, 2, and 3, respectively. [Reprinted from Wilson et al. (1992) p. 1687, by courtesy of Marcel Dekker, Inc.]

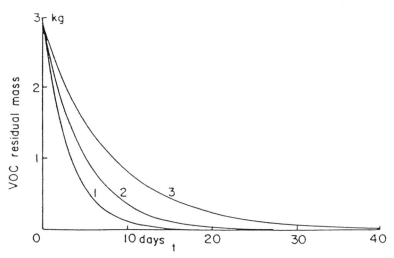

Figure 8.14 Dissolved VOC sparging simulation using the *n*-compartment model. Effect of water flow rate on VOC removal rate. See Table 8.11 for default run parameters. Q_w = 0.01, 0.005, and 0.0025 m³/sec for Runs 1, 2, and 3, respectively. [Reprinted from Wilson et al. (1992) p. 1688, by courtesy of Marcel Dekker, Inc.]

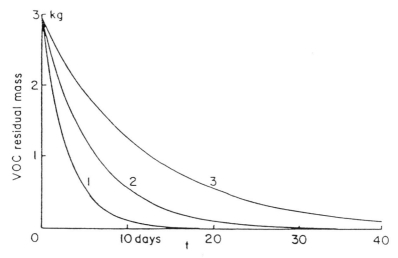

Figure 8.15 Dissolved VOC sparging simulation, n-compartment model. Effect of linked water and airflow rates on VOC removal rate. See Table 8.11 for default run parameters. In these runs $Q_a = Q_w = 0.01$ (Run 1), 0.005 (Run 2), and 0.0025 m³/sec (Run 3). [Reprinted from Wilson et al. (1992) p. 1688, by courtesy of Marcel Dekker, Inc.]

The effect of varying the Henry's constant of the VOC is shown in Figure 8.16. Removal rate increases with increasing Henry's constant but is not proportional to K_H with the parameter sets used here.

The manner in which Q_w depends on Q_a is likely to be highly site-specific, depending particularly on the permeability of the aquifer. One can reasonably expect that Q_w increases monotonically with increasing Q_a, but the precise functional form of this dependence is likely to be site-specific. Use of this model is therefore dependent upon the availability of

TABLE 8.11. Default Parameters for the Runs Plotted in
Figures 8.13–8.16.[a]

Domain radius	5 m
Aquifer thickness	5 m
Radius of zone of contamination	5 m
Number of annular compartments	20
Airflow rate Q_a	0.01 m³/sec
Water circulation rate Q_w	0.01 m³/sec
Henry's constant K_H (dimensionless)	0.1
Porosity of aquifer medium	0.3
Initial VOC concentration in groundwater	25 mg/L

[a]Departures from these values are indicated in the captions of the figures.

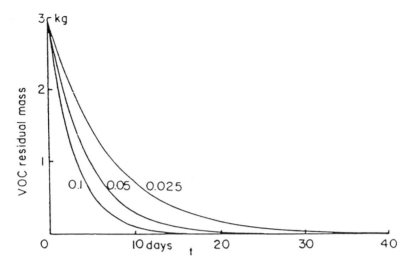

Figure 8.16 Dissolved VOC sparging simulation, n-compartment model. Effect of Henry's constant. Default run parameters from Table 8.11. Henry's constants are 0.1, 0.05, and 0.025 (dimensionless) as indicated. [Reprinted from Wilson et al. (1992) p. 1689, by courtesy of Marcel Dekker, Inc.]

pilot test data giving information on the dependence of the induced water flow rate Q_w on the air sparging rate Q_a.

Some words of warning before we continue: in the real world, removal rates even of dissolved VOCs are not likely to be as rapid as this model predicts. There are a number of reasons for this. First, aquifers often contain clay lenses and other domains of low permeability through which water will move only slowly, so these will be flushed out slowly. Second, the model assumes that all the groundwater in the domain of influence is mobile water. There may, however, be a significant amount of water that is in dead-end pores and, therefore, is immobile. VOC in that water must diffuse out into the mobile water before it can be swept away by advection [see Mutch et al. (1993), for example]. In short, if the model indicates that removal by sparging will be unacceptably slow, one should reject sparging. If the model indicates that removal by sparging will be at a reasonably rapid rate, one can start planning a pilot test, but should not be surprised if the results of the pilot test indicate that cleanup will be slower than this simple model predicts. Incidentally, it is an easy matter to patch diffusion transport onto this sparging model, provided that one has a reasonably fast microcomputer. Taking into account the effects of an inhomogeneous permeability is more difficult in practice, because one must use a three-dimensional model instead of a two-dimensional one, and this requires a substantially bigger and faster computer.

VACUUM-VAPORIZER-WELL CONFIGURATION FOR THE REMOVAL OF DISTRIBUTED DNAPL

In the following we combine the advective flow field developed earlier for sparging dissolved VOCs with the mass transport kinetics developed for the flushing of DNAPLs present as trapped droplets (Kayano and Wilson, 1993) to construct a model for the sparging of DNAPL droplets from the zone of saturation by means of vacuum-vaporizer-wells.

The scheme used for this DNAPL sparging model and the notation are shown in Figures 8.17 and 8.18. Air is injected near the bottom of the aquifer and is assumed to rise rather close to the pipe; it may be confined within a larger pipe coaxial with the air injection pipe and perhaps packed with crushed rock to provide a longer transit time for the rising air bubbles. In any case, it is assumed that the rising air induces a flow of water into the axis of the system at the bottom of the aquifer and that this water is air-stripped and then discharged at the top of the aquifer. We therefore require a water flow field of the same type that was used in the dissolved VOC model. We assume that the aquifer is of constant and isotropic permeability. We shall be carrying out the usual type of mass balance on the annular-shaped volume elements into which the domain of interest is partitioned for analysis.

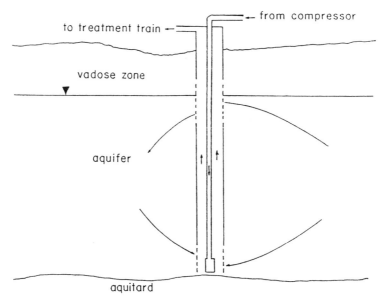

Figure 8.17 A simple vacuum-vaporizer-well configuration for the removal of DNAPL. [Reprinted from Herring et al. (1991).]

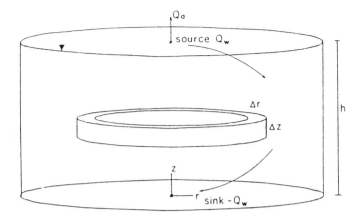

Figure 8.18 Notation and geometry for the vacuum-vaporizer-well DNAPL sparging model. [Reprinted from Roberts and Wilson (1993) p. 1129, by courtesy of Marcel Dekker, Inc.]

As in our previous work, we require a water flow field that provides a sink of magnitude $-Q_w$ at $(0,0)$ and a source of magnitude Q_w at $(0,h)$. We also require that there be no flow through either the top or the bottom of the aquifer. We further assume that the aquifer is of constant and isotropic permeability, so the system is axially symmetric. It was shown earlier in this chapter that the method of images can be used to construct a solution to Laplace's equation, which serves as a velocity potential function for this system; the resulting superficial velocities are given by

$$v_r = \frac{Q_w}{2\pi} \sum_{n=-\infty}^{\infty} \left[\frac{r}{\{r^2 + [z - (2n + 1)h]^2\}^{3/2}} \right.$$

$$\left. - \frac{r}{\{r^2 + [z - 2nh]^2\}^{2/3}} \right] \tag{39}$$

and

$$v_z = \frac{Q_w}{2\pi} \sum_{n=-\infty}^{\infty} \left[\frac{z - (2n + 1)h}{\{r^2 + [z - (2n + 1)h]^2\}^{3/2}} \right.$$

$$\left. - \frac{z - 2nh}{\{r^2 + [z - 2nh]^2\}^{2/3}} \right] \tag{40}$$

We define

$$r_i = (i - 1/2)\Delta r \tag{41}$$

$$z_j = (j - 1/2)\Delta z \tag{42}$$

and

$$v_{ij}^L = v_r[(i - 1)\Delta r, (j - 1/2)\Delta z] \qquad (43)$$

$$v_{ij}^R = v_r[i\Delta r, (j - 1/2)\Delta z] \qquad (44)$$

$$v_{ij}^B = v_z[(i - 1/2)\Delta r, (j - 1)\Delta z] \qquad (45)$$

$$v_{ij}^T = v_z[(i - 1/2)\Delta r, j\Delta z] \qquad (46)$$

Also, the *ij*th annular volume element is given by

$$V_{ij} = \pi(2i - 1)(\Delta r)^2\Delta z \qquad (47)$$

Let

C_{ij}^w = concentration of dissolved VOC in V_{ij}, kg/m³ of water
C_{ij}^N = concentration of DNAPL in V_{ij}, kg/m³ of medium
ν = porosity of the aquifer, dimensionless

and define

$$S(x) = 0, x < 0$$

$$= 1, 0 < x \qquad (48)$$

We then carry out a material balance on dissolved VOC in the *ij*th volume element; this yields

$$\frac{dC_{ij}^w}{dt} = \frac{2\pi(i - 1)\Delta r\Delta z}{\nu\Delta V_{ij}} \cdot v_{ij}^L \cdot [S(v^L)C_{i-1,j}^w + S(-v^L)C_{ij}^w]$$

$$- \frac{2\pi i\Delta r\Delta z}{\nu\Delta V_{ij}} \cdot v_{ij}^R \cdot [S(-v^R)C_{i+1,j}^w + S(v^R)C_{ij}^w]$$

$$+ \frac{\pi(2i - 1)(\Delta r)^2}{\nu\Delta V_{ij}} \cdot v_{ij}^B \cdot [S(v^B)C_{i,j-1}^w + S(-v^B)C_{ij}^w]$$

$$- \frac{\pi(2i - 1)(\Delta r)^2}{\nu\Delta V_{ij}} \cdot v_{ij}^T \cdot [S(-v^T)C_{i,j+1}^w + S(v^T)C_{ij}^w]$$

$$+ \left[\frac{\partial C_{ij}^w}{\partial t}\right]_{\text{from DNAPL}} \qquad (49)$$

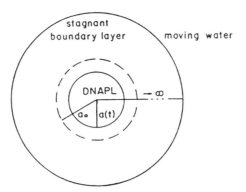

Figure 8.19 Model for solution and diffusion from a DNAPL droplet. [Reprinted from Roberts and Wilson (1993) p. 1131, by courtesy of Marcel Dekker, Inc.]

Figure 8.19 shows the model to be used for DNAPL droplet dissolution. The DNAPL slowly dissolves, the dissolved VOC diffuses across a thick quiescent aqueous boundary layer, and then it reaches the advecting water, which sweeps it away [see Wilson (1994), Kayano and Wilson (1993), or Chapter 4]. The result of the analysis of the model for droplet dissolution is

$$\frac{dC_{ij}^N}{dt} = -\frac{3C_0^{N2/3}D(C_{sat} - C_{ij}^w)}{\varrho a_0^2} C_{ij}^{N1/3} \tag{50}$$

where

ϱ = DNAPL density, kg/m³
C_0^N = initial DNAPL concentration, kg/m³ of aquifer
D = VOC diffusivity in water-saturated porous medium, m²/sec
C_{sat} = VOC solubility in water, kg/m³
a_0 = initial DNAPL droplet radius, m
C_{ij}^N = DNAPL concentration in the ijth volume element, kg/m³ of aquifer medium
C_{ij}^w = dissolved VOC concentration in the ijth volume element, kg/m³ of aqueous phase

Now, solution of DNAPL is a conservative process within the ijth volume element, which gives

$$\Delta V_{ij}\nu \left[\frac{\partial C_{ij}^w}{\partial t}\right]_{from\ DNAPL} + \Delta V_{ij} \cdot \frac{dC_{ij}^N}{dt} = 0 \tag{51}$$

From Equations (50) and (51) we obtain

$$\left[\frac{\partial C_{ij}^w}{\partial t}\right]_{\text{from DNAPL}} = \frac{3C_0^{N2/3}D(C_{sat} - C_{ij}^w)}{a_0^2 \varrho \nu} \cdot C_{ij}^{N1/3} \qquad (52)$$

Substitution of Equation (52) into Equation (49), plus Equation (51) then provides the modeling equations governing the C_{ij}^N and C_{ij}^w for most of the volume elements.

We next examine the effect of the central aeration pipe. It will be represented as a single, well-stirred tank aerator, with the dissolved VOC obeying Henry's Law and with local equilibrium between the aqueous and the vapor phases. We let

K_H = Henry's constant for the VOC, dimensionless
C_1^w = dissolved VOC concentration in the aerator, kg/m³
Q_a = airflow rate through the central aeration pipe, m³/sec

In steady state the flux of VOC into the aerator at the bottom must equal the sum of the two fluxes out of the aerator at the top, which gives

$$Q_w C_{11}^w = (Q_w + Q_a K_H) C_1^w \qquad (53)$$

so

$$C_1^w = \frac{C_{11}^w}{1 + Q_a K_H / Q_w} \qquad (54)$$

Then in the 1,1th volume element (at the bottom of the pipe) we must modify Equation (49) by subtracting a term on the right-hand side as indicated:

$$\frac{dC_{11}^w}{dt} = -\frac{Q_w C_{11}^w}{\Delta V_{11}} + \text{advective and dissolution terms} \qquad (55)$$

to take into account the movement of VOC from the 1,1th volume element into the aeration region. In similar fashion, in the 1,Jth element (at the top of the pipe) we must modify Equation (49) by adding a term on the right-hand side:

$$\frac{dC_{ij}^w}{dt} = \frac{Q^w C_{11}^w}{\Delta V_{ij}(1 + Q_a K_H / Q_w)} + \text{advective and dissolution terms} \qquad (56)$$

which accounts for the movement of VOC from the aeration region into the 1,*J*th volume element.

In addition to the parameters given above, it is necessary to assign the radius of the domain that is contaminated. Then one simply integrates Equations (49), (52), (55), and (56) forward in time.

Sparging with a Horizontal Slotted Pipe

We next develop a model for sparging by means of a horizontal slotted pipe buried in an aeration curtain configuration, as indicated in Figure 8.20. This approach might be useful when the contamination is rather shallow but extends over a fairly wide area. Again, we assume an isotropic, constant permeability for the aquifer. A velocity potential function that satisfies boundary conditions of no normal flow at the top and bottom of the aquifer ($y = h$ and $y = 0$, respectively) and has a source at $(0,h)$ and a sink at $(0,0)$ is easily constructed by the method of images; this is

$$
W = \frac{Q_w}{2\pi l\nu} \sum_{n=-\infty}^{\infty} \{ -\log [x^2 + (y - 2nh)^2]
$$

$$
+ \log [x^2 + (y - (2n + 1)h)^2] \} + \nu_0 x \qquad (57)
$$

This represents flow in the aquifer from a source Q_w at $(0,h)$ to a sink $-Q_w$ at $(0,0)$ superimposed on a uniform natural flow in the x-direction having a velocity ν_0 m/sec.

The potential function W yields the following expressions for the x- and y-components of the linear fluid velocity:

$$
\nu_x = \nu_0 + \frac{Q_w}{\pi l\nu} \sum_{n=-\infty}^{\infty} \left[\frac{x}{x^2 + [y - (2n + 1)h]^2} - \frac{x}{x^2 + [y - 2nh]^2} \right] \qquad (58)
$$

and

$$
\nu_y = \frac{Q_w}{\pi l\nu} \sum_{n=-\infty}^{\infty} \left[\frac{y - (2n + 1)h}{x^2 + [y - (2n + 1)h]^2} - \frac{y - 2nh}{x^2 + [y - 2nh]^2} \right] \qquad (59)
$$

The domain of interest is partitioned into rectangular prism volume elements such that the size of the ijth volume is given by

$$
\Delta V_{ij} = \Delta V = \Delta x \Delta y l \qquad (60)
$$

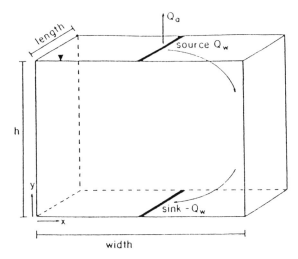

Figure 8.20 Sparging curtain geometry and notation. [Reprinted from Roberts and Wilson (1993) p. 1133, by courtesy of Marcel Dekker, Inc.]

Also,

$$x_i = (i - i_{well} - 1/2)\Delta x \qquad (61)$$

$$y_j = (j - 1/2)\Delta y \qquad (62)$$

Here i_{well} is the x-index of the volume elements containing the source and sink.

We use Equation (50) to model the rate of solution of the DNAPL droplets, as before. The equation describing advection of the dissolved VOC is

$$\frac{dC_{ij}^w}{dt} = \frac{v_{ij}^L}{\Delta x} \cdot [S(v^L)C_{i-1,j}^w + S(-v^L)C_{ij}^w]$$

$$- \frac{v_{ij}^R}{\Delta x} \cdot [S(-v^R)C_{i+1,j}^w + S(v^R)C_{ij}^w]$$

$$+ \frac{v_{ij}^B}{\Delta y} \cdot [S(v^B)C_{i,j-1}^w + S(-v^B)C_{ij}^w]$$

$$- \frac{v_{ij}^T}{\Delta y} \cdot [S(-v^T)C_{i,j+1}^w + S(v^T)C_{ij}^w]$$

$$+ \text{ a solution term and a source (sink) term if needed} \qquad (63)$$

Here

$$v_{ij}^L = v_x[(i - 1)\Delta x, (j - 1/2)\Delta y] \tag{64}$$

$$v_{ij}^R = v_x[i\Delta x, (j - 1/2)\Delta y] \tag{65}$$

$$v_{ij}^B = v_y[(i - 1/2)\Delta x, (j - 1)\Delta y] \tag{66}$$

$$v_{ij}^T = v_y[(i - 1/2)\Delta x, j\Delta y] \tag{67}$$

The solution term is given by Equation (52). The sink term at the bottom of the sparging unit is given by

$$\left[\frac{dC_{iwell,1}^w}{dt}\right]_{sink} = -\frac{Q_w C_{iwell,1}^w}{\Delta V \nu} \tag{68}$$

The source term at the top of the aquifer is given by

$$\left[\frac{dC_{iwell,J}^w}{dt}\right]_{source} = \frac{Q_w C_{iwell,J}^w}{\Delta V(1 + Q_a K_H/Q_w)\nu} \tag{69}$$

As in the earlier model, we have represented the sparging unit itself as a single-stage aerator with equilibrium between the vapor and liquid phases.

Diffusion-Controlled Sparging

We can obtain a formula useful in the limit as diffusion becomes the controlling factor in the solution of DNAPL droplets by setting $C_{ij}^w = 0$ in Equation (13), corresponding to the assumption that advective transport is sufficiently rapid to remove dissolved VOC as fast as it is formed. The resulting equation is readily integrated, yielding

$$C^N(t) = C_0^N\left[1 - \frac{2DC_{sat}}{\varrho a_0^2} \cdot t\right]^{3/2} \tag{70}$$

for $0 < t < (\varrho a_0^2/2DC_{sat}) = t_c$, the time required for complete dissolution of the DNAPL.

Results

Computer programs were written in TurboBASIC, implementing the models on microcomputers using 80286 or 80386 microprocessors. The results described below were obtained with machines having math coprocessors, using MS-DOS, and running at 12, 16, and 33 MHz. Running times for the runs presented below ranged from 45 min to 4 hr on a 16-MHz 80386 SX machine.

We first examine the results obtained for a single sparging well. Default values of the model parameters are given in Table 8.12; departures from these values are indicated in the captions. Figure 8.21 shows the effect on VOC removal of the initial effective diameter a_0 of the DNAPL droplets being dissolved. If one is in the diffusion limited regime, the occurrence of a_0^2 in the denominator of Equation (50) leads us to expect rather drastic increases in remediation times with increasing values of a_0, as is, in fact, observed in Figure 8.21. Evidently, the effective radius of the DNAPL blobs dispersed in the aquifer medium is an important parameter in the modeling.

One expects that the airflow rate in the sparging well (Q_a) and the associated induced water flow rate (Q_w) are linked, with increasing values of Q_a resulting in increased values of Q_w. The functional nature of this dependence is not known and surely depends on the nature of the aquifer medium. In Figure 8.22 it is assumed that Q_w is proportional to Q_a, and the dependence of VOC removal on the linked flow rates is shown. As the system parameters are such that diffusion limitations are important, cleanup times are not inversely proportional to the flow rates. For this system little would be gained by increasing the airflow rate above 0.01 m³/sec.

Another feature observed in Figure 8.22, as well as in Figure 8.21, is marked tailing. This occurs after all of the DNAPL has been dissolved and is due to the very slow movement of dissolved VOC out around the periphery of the domain of interest. Operation of a single sparging well results

TABLE 8.12. Default Parameters for DNAPL Sparging Simulations.

Thickness of aquifer	5 m
Radius of domain of interest	15 m
Porosity of aquifer medium	0.3
Airflow rate through sparging well	0.01 m³/sec
Induced water flow rate	0.0025 m³/sec
Aquifer medium density	1.7 gm/cm³
Aqueous solubility of DNAPL	1,100 mg/L
Henry's constant of DNAPL	0.2
Density of DNAPL	1.46 gm/cm³
Diffusivity of DNAPL in porous medium	2×10^{-10} m²/sec
Initial DNAPL concentration in aquifer	2,000 mg/kg
Radius of DNAPL-contaminated zone	5 m
Initial DNAPL droplet diameter $2a_0$	0.2 cm

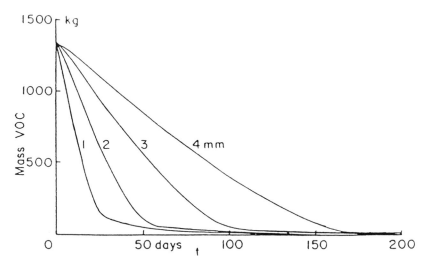

Figure 8.21 DNAPL removal. Plots of total residual VOC versus time for a single sparging well; effect of droplet size. Droplet diameters are as indicated. Other parameters are as in Table 8.12. [Reprinted from Roberts and Wilson (1993) p. 1136, by courtesy of Marcel Dekker, Inc.]

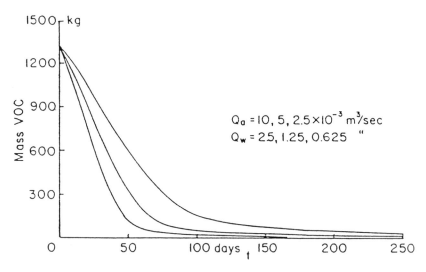

Figure 8.22 DNAPL removal. Plots of total residual VOC versus time for a single sparging well; effect of linked Q_w and Q_a. From left to right, values of (Q_a, Q_w) are (0.01, 0.0025), (0.005, 0.00125), and (0.0025, 0.000625) m³/sec. Other parameters as in Table 8.12. [Reprinted from Roberts and Wilson (1993) p. 1137, by courtesy of Marcel Dekker, Inc.]

412

in the slow movement of dissolved VOC *away* from the well in the upper half of the aquifer; the streamlines turn around at the median plane, and the VOC returns to the well in the lower half of the aquifer. The removal of dissolved VOC, which moves along streamlines that extend far out from the axis of the well, is quite slow, leading to quite long cleanup times if very high levels of removal (99 + %) are sought. These results suggest the use of arrays of sparging wells in which the wells around the periphery of the domain of contamination are sited in uncontaminated aquifer for the purpose of establishing no-flow boundary conditions for the wells operating in the domain of contamination. Alternatives might be the placement of impermeable barriers to restrict the outward flow of water around a sparging well and/or placement of groundwater recovery wells to intercept this water.

The effect of the radius of the contaminated zone is shown in Figure 8.23. Since diffusion kinetics are important, cleanup times are not strongly dependent upon the size of the DNAPL-contaminated domain, with an increase in contaminated zone radius from 3 to 5 m, resulting in a very minor increase in the time required to remove the DNAPL. Again, we see tailing toward the end of the run, which is associated with the movement of dissolved VOC out away from the well in the upper half of the aquifer.

The results shown in Figures 8.24–8.29 pertain to the horizontal slotted pipe sparging configuration. Default values of the parameters used in these

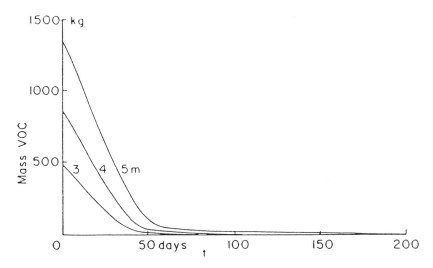

Figure 8.23 DNAPL removal. Plots of total residual VOC versus time for a single sparging well; effect of radius of contaminated zone. Contaminated zone radii are 5, 4, and 3 m as indicated. Other parameters as in Table 8.12. [Reprinted from Roberts and Wilson (1993) p. 1138, by courtesy of Marcel Dekker, Inc.]

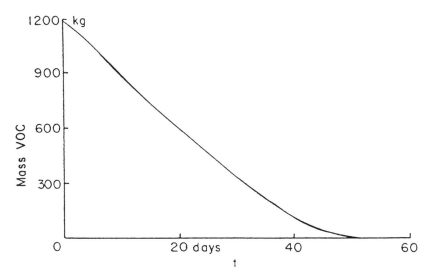

Figure 8.24 DNAPL removal. Plots of total residual VOC versus time for a single horizontal sparging pipe; effect of length of domain. Domain lengths (at right angles to the curtain) are 15, 17, and 19 m; other parameters as in Table 8.13. [Reprinted from Roberts and Wilson (1993) p. 1138, by courtesy of Marcel Dekker, Inc.]

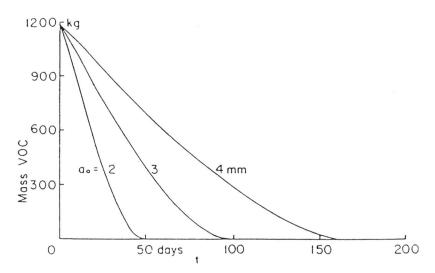

Figure 8.25 DNAPL removal. Plots of total residual VOC versus time for a single horizontal sparging pipe; effect of initial DNAPL droplet size. Droplet diameters are 2, 3, and 4 mm. Other parameters as in Table 8.13. [Reprinted from Roberts and Wilson (1993) p. 1139, by courtesy of Marcel Dekker, Inc.]

414

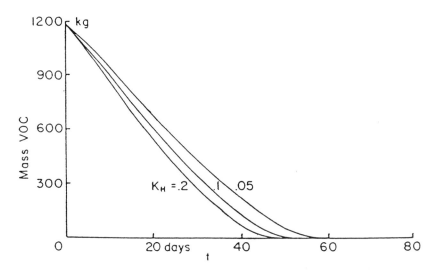

Figure 8.26 DNAPL removal. Plots of total residual VOC versus time for a single horizontal sparging pipe; effect of VOC Henry's constant. From left to right, K_H (dimensionless) = 0.2, 0.1, and 0.05. Other parameters as in Table 8.13. [Reprinted from Roberts and Wilson (1993) p. 1140, by courtesy of Marcel Dekker, Inc.]

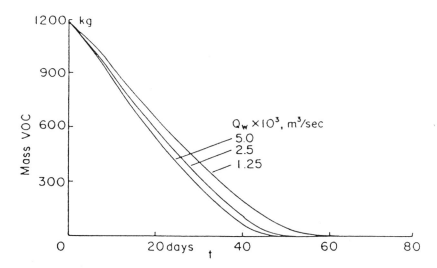

Figure 8.27 DNAPL removal. Plots of total residual VOC versus time for a single horizontal sparging pipe; effect of induced water flow rate Q_w. From left to right, Q_w = 0.005, 0.0025, and 0.00125 m³/sec. Other parameters as in Table 8.13. [Reprinted from Roberts and Wilson (1993) p. 1141, by courtesy of Marcel Dekker, Inc.]

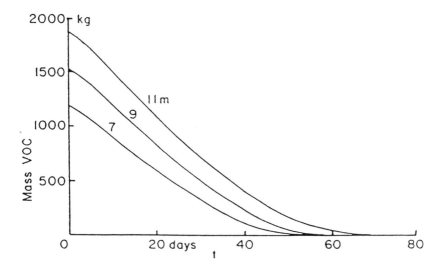

Figure 8.28 DNAPL removal. Plots of total residual VOC versus time for a single horizontal sparging pipe; effect of width of DNAPL-contaminated domain. From bottom to top, the width of the contaminated domain is 7, 9, and 11 m. Other parameters as in Table 8.13. [Reprinted from Roberts and Wilson (1993) p. 1141, by courtesy of Marcel Dekker, Inc.]

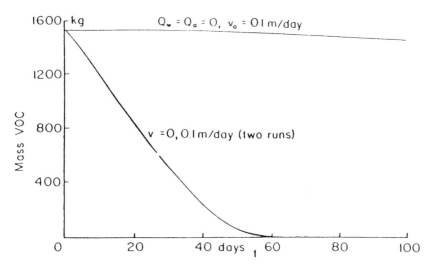

Figure 8.29 DNAPL removal. Plots of total residual VOC versus time for a single horizontal sparging pipe; effect of natural groundwater movement. The top curve describes a run in which no sparging is occurring. The two runs in which sparging is taking place (one with, one without natural groundwater movement) are essentially superimposed. Groundwater velocity is 0.1 m/day, domain lengths are 19 m, and the contaminated zone is 9 m wide and centered on the sparging curtain. The other parameters are as in Table 8.13. [Reprinted from Roberts and Wilson (1993) p. 1142, by courtesy of Marcel Dekker, Inc.]

416

runs are given in Table 8.13, with departures indicated in the captions. One matter of interest was the effect of the extent of the domain modeled on either side of the sparging curtain. This should be such that the results obtained are independent of the length of the domain. Figure 8.24 shows three runs, identical except that the domain lengths are 15, 17, and 19 m. On the scale of the figure, these runs are indistinguishable. Similar results were obtained with the vertical sparging well model.

The effect of the initial effective DNAPL droplet radius a_0 on DNAPL removal is shown in Figure 8.25. As was the case with the vertical sparging well model (see Figure 8.21), this is a crucial parameter in determining the cleanup time of a site. In the limit of diffusion-controlled sparging, cleanup time is proportional to a_0^2 [see Equation (50)]. The results plotted in Figure 8.25 show substantially less tailing with the horizontal pipe configuration than was found with a single vertical sparging well.

The effect of the Henry's constant of the VOC on DNAPL removal is seen in Figure 8.26. With the other parameters used in the runs modeled here, an increase in K_H by a factor of 4 (from 0.05 to 0.2, dimensionless) results in a quite modest decrease in cleanup time. Evidently, the removal of dissolved VOC from the water in the immediate vicinity of the well by aeration is not a major bottleneck under these conditions.

The transport of VOC dissolved from the stationary DNAPL droplets to the sparging curtain depends upon the water flow rate induced by the sparging curtain, Q_w. One therefore expects that the rate of VOC removal increases with increasing Q_w, as is seen to be the case in Figure 8.27. Here Q_a is being held constant. Since the principal bottleneck in the process is

TABLE 8.13. Default Parameters for DNAPL Sparging
with a Horizontal Slotted Pipe.

Thickness of aquifer	5 m
Length of domain of interest	15 m
Length of curtain	10 m
Porosity of aquifer medium	0.3
Airflow rate of sparging curtain	0.01 m³/sec
Induced water flow rate	0.0025 m³/sec
Natural groundwater flow velocity normal to curtain	0 m/day
Aquifer medium density	1.7 gm/cm³
Aqueous solubility of DNAPL	1,100 mg/L
Henry's constant of DNAPL	0.2
Density of DNAPL	1.46 gm/cm³
Diffusivity of DNAPL in porous medium	2×10^{-10} m²/sec
Initial DNAPL concentration in the aquifer	2,000 mg/kg
x-Coordinate of left boundary of contaminated zone	−3.5 m
x-Coordinate of right boundary of contaminated zone	3.5 m
Initial DNAPL droplet diameter $2a_0$	0.2 cm

diffusion from the DNAPL droplets under the conditions modeled, cleanup rates increase only relatively slightly with increasing Q_w.

The effect of the extent of the contaminated zone on either side of the sparging curtain on the VOC removal rate is shown in Figure 8.28. The time required for cleanup increases with increasing size of the domain of contamination, but the effect is not large under the conditions modeled here, where diffusion of VOC away from the DNAPL droplets is the major rate-limiting step. In all three runs shown here the circulation of the water and the aeration of the water at the curtain are sufficient to make diffusion kinetics the controlling factor.

In Figure 8.29 we see the results of superimposing a uniform natural flow of water upon the flow field resulting from the sparging curtain. The upper curve shows the natural solution and migration of the VOC from the domain of interest when there is no sparging; the lower curve depicts a run with the sparging curtain in operation. The distribution of VOC in the domain of interest during both runs demonstrated loss of VOC by migration from the zone of interest; evidently, sparging operations may require careful monitoring to avoid movement of VOC off of the site during the course of the sparging. The mixing caused by sparging results in more rapid solution of DNAPL than would otherwise take place, so that water recovery wells are often needed to prevent any of this VOC from escaping off-site.

The models presented for sparging units of the vacuum-vaporizer-well-type provide a simple, two-parameter method for including diffusion kinetics in the sparging of residual DNAPL droplets from an aquifer. Estimation of diffusion constants in porous media by the method of Millington and Quirk (1961) provides a means for getting this parameter from porosities and diffusion constants of the VOCs in free water. Estimation of the initial effective radius of the DNAPL droplets is more difficult but crucial to the success of a modeling effort. This will probably require some field measurements, perhaps a small pilot-scale experiment in which clean water is injected into the aquifer and allowed to stand for an extended period, during which samples are taken at various times to determine the way in which the dissolved VOC concentration increases with time.

The models predict that increases in air and water flow rates beyond the point where diffusion kinetics become limiting is a waste of effort in the removal of NAPL by sparging with vacuum-vaporizer-wells. The models allow one to estimate the distance to which a well may be effective in removing DNAPL at a reasonable rate and also permit one to estimate the effect of VOC Henry's constants on removal rates.

A major item of unfinished business is the relationship between the airflow rate through the sparging well or curtain and the water circulation that this airflow induces. Lab- or pilot-scale studies with dissolved tracers should provide some insight into this. The dependence of Q_w on Q_a is probably rather site-specific. A preliminary lab-scale study using a water-saturated sand bed and a sparging well of the type shown in Figure 8.17 suggests that the water circulation rate is roughly proportional to the airflow rate over a significant range.

Extension of these models to aquifers of varying or anisotropic permeability could be easily accomplished by using relaxation methods to calculate the velocity potential of the water. A full three-dimensional model could easily be coded but would be difficult to run on currently available microcomputers.

SIMPLE SPARGING WELLS FOR DISTRIBUTED DNAPL REMOVAL

The models discussed so far pertain to configurations in which the injected air is localized within a pipe or curtain at the center of the domain of influence of the well. A cheaper and probably more effective configuration is that in which air is injected directly into the aquifer and then spreads out as it rises up to the surface of the water table, as illustrated in Chapter 7, Figure 7.3, and as described by Brown and Fraxedas (1992). This generates more local turbulence in the aquifer, so it should speed solution of the DNAPL.

The precise solution of the problem of the flow of mixed phases in a porous medium is quite difficult and beyond the capability of microcomputers at present. We therefore construct our model by postulating a physically reasonable function for the molar flux in the vertical direction of the injected air, calculate from this the horizontal component of the molar gas flux, and assume for the circulating water the same flow field that was used in the previous models. This is probably the best that can be done on presently available microcomputers [see Burchfield and Wilson (1993); Gómez-Lahoz et al. (1994)].

Gas Flow with a Single Vertical Air Injection Pipe

The configuration of the sparging system to be examined is shown in Figure 7.3, Chapter 7, and Figure 8.30. Here, air is injected at the bottom of the aquifer from a single pipe. It is assumed that the rate of natural groundwater movement is sufficiently slow that it can be neglected on the time scale of interest, so that the problem has axial symmetry. We use cylindrical coordinates r and z. Symbols are defined as follows:

h = thickness of aquifer, m
Q_a = molar airflow rate through the sparging well
a_0 = maximum distance from the well at the top of the aquifer at which rising air is observed, m

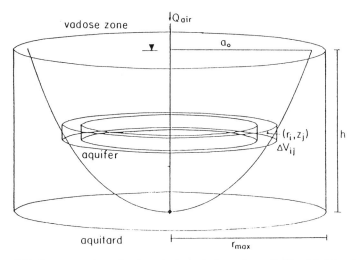

Figure 8.30 Geometry and notation for a single vertical sparging well. [Reprinted from Burchfield and Wilson (1993) p. 2531, by courtesy of Marcel Dekker, Inc.]

Let us assume a molar gas flux rate in the z-direction of

$$q_z(r,z) = A(a^2 - r^2), \text{ mol/m}^2 \text{ sec}, r < a$$

$$= 0, r > a \tag{71}$$

Here, we take

$$a = a(z) = a_0(z/h)^{1/2} \tag{72}$$

Then

$$Q_a = \int_0^a 2\pi A(a^2 - r^2)r\,dr$$

$$= \pi A a^4/2 \tag{73}$$

which yields

$$A = \frac{2Q_a}{\pi a_0^4} \cdot \frac{h^2}{z^2} \tag{74}$$

So

$$q_z(r,z) = \frac{2Q_a}{\pi a_0^4} \cdot \frac{h^2}{z^2} \cdot [a_0^2(z/h) - r^2] \tag{75}$$

Now the molar gas flux in steady-state flow is conservative, so

$$\nabla \cdot q = 0 = \frac{1}{r} \cdot \frac{\partial}{\partial r} (rq_r) + \frac{\partial q_z}{\partial z} \tag{76}$$

This can be used with Equation (75) to obtain the radial component of the molar gas flux, as follows. Differentiating $q_z(r,z)$ yields

$$\frac{\partial q_z}{\partial z} = \frac{2Q_a}{\pi a_0^4} \cdot \left[-\frac{2h^2}{z^3} [a_0^2(z/h) - r^2] + \frac{h^2}{z^2} \cdot \frac{a_0^2}{h} \right]$$

$$= \frac{2Q_a}{\pi a_0^4} \cdot \left[\frac{2r^2h^2}{z^3} - \frac{a_0^2h}{z^3} \right] \tag{77}$$

From Equations (76) and (77) we have

$$\frac{1}{r} \frac{\partial}{\partial r} (rq_r) = -\frac{2Q_a}{\pi a_0^4} \left[\frac{2r^2h^2}{z^3} - \frac{a_0^2h}{z^2} \right] \tag{78}$$

so

$$\frac{\partial}{\partial r} (rq_r) = -\frac{2Q_a}{\pi a_0^4} \left[\frac{2h^2r^3}{z^3} - \frac{a_0^2hr}{z^2} \right] \tag{79}$$

Integrating from $r' = 0$ to $r' = r$ then gives

$$rq_r = -\frac{2Q_a}{\pi a_0^4} \left[\frac{h^2r^4}{2z^3} - \frac{a_0^2hr^2}{2z^2} \right] \tag{80}$$

so

$$q_r = \frac{Q_a}{\pi a_0^4} \frac{h^2r}{z^3} [a_0^2(z/h) - r^2] \tag{81}$$

Recall

$$q_z(r,z) = \frac{2Q_a}{\pi a_0^4} \cdot \frac{h^2}{z^2} \cdot [a_0^2(z/h) - r^2] \tag{75}$$

Now what is needed for the model is the volumetric flux of gas, rather than the molar flux. We assume that the gas obeys the ideal gas law, so the components of the volumetric flux are given in terms of the components of the molar flux by

$$S_r = \frac{RT}{P} \cdot q_r \tag{82}$$

and

$$S_z = \frac{RT}{P} \cdot q_z \tag{83}$$

Let us take the total pressure as ambient plus hydrostatic, to an adequate approximation, so

$$P(z) = P_a + \sigma'(h - z) \tag{84}$$

where $\sigma' = 1$ atm/10.336 m. Then

$$S_r = \frac{RT}{P_a + \sigma'(h - z)} \cdot \frac{Q_a}{\pi a_0^4} \cdot \frac{h^2 r}{z^3} [a_0^2(z/h) - r^2] \tag{85}$$

and

$$S_z = \frac{RT}{P_a + \sigma'(h - z)} \cdot \frac{2Q_a}{\pi a_0^4} \cdot \frac{h^2}{z^2} [a_0^2(z/h) - r^2] \tag{86}$$

with S_r and $S_z = 0$ if $r^2 > a_0^2(z/h)$

Gas Flow with a Single Horizontal Slotted Pipe

An alternative sparging well configuration involves use of a horizontal slotted pipe for the air injection. A diagram of the system is given in Figure 8.31. Let

l = length of horizontal slotted pipe, m
h = thickness of aquifer, m
Q_a = molar air flow rate, moles/sec

We assume a molar gas flux in the z-direction of

$$q_z(x,z) = A(a^2 - x^2), \quad |x| < a$$

$$= 0, \quad |x| > a \tag{87}$$

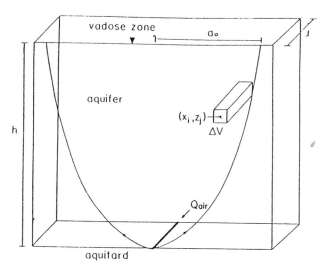

Figure 8.31 Geometry and notation for a single horizontal sparging pipe. [Reprinted from Burchfield and Wilson (1993) p. 2539, by courtesy of Marcel Dekker, Inc.]

Then

$$Q_a = 2l \int_0^a A(a^2 - x^2)dx \tag{88}$$

$$= 4lAa^3/3 \tag{89}$$

So

$$A = 3Q_a/4la^3 \tag{90}$$

We choose

$$a = a_0(z/h)^{1/2} \tag{91}$$

as before, so

$$A = \frac{3Q_a}{4la_0^3} \cdot (h/z)^{3/2} \tag{92}$$

Then

$$q_z(x,z) = \frac{3Q_a}{4la_0^3} \cdot (h/z)^{3/2} \cdot [a_0^2(z/h) - x^2] \tag{93}$$

Now $\nabla \cdot q = 0$ for steady flow, which gives

$$\frac{\partial q_x}{\partial x} = -\frac{\partial q_z}{\partial z} \tag{94}$$

So

$$\frac{\partial q_x}{\partial x} = -\frac{\partial}{\partial z}\left[\frac{3Q_a}{4la_0^3}(h/z)^{3/2} \cdot [a_0^2(z/h) - x^2]\right] \tag{95}$$

$$= -\frac{3Q_a h^{3/2}}{4a_0^3 l}\left[-\frac{a_0^2}{2hz^{3/2}} + \frac{3x^2}{2z^{5/2}}\right] \tag{96}$$

Integrate Equation (96) with respect to x' from 0 to x and note that $q_x(0,z) = 0$ to get

$$q_x(x,z) = \frac{3Q_a h^{3/2}}{8a_0^3 l} \cdot \frac{x}{z^{5/2}} \cdot [a_0^2(z/h) - x^2] \tag{97}$$

Recall

$$q_z(x,z) = \frac{3Q_a}{4la_0^3} \cdot (h/z)^{3/2} \cdot [a_0^2(z/h) - x^2] \tag{93}$$

As before, the volumetric gas flux components are given by

$$S_x = (RT/P)q_x \tag{98}$$

and

$$S_z = (RT/P)q_z \tag{99}$$

Also, as before,

$$P(z) = P_a + \sigma'(h - z) \tag{84}$$

Then

$$S_x = \frac{RT}{P_a + \sigma'(h - z)} \cdot \frac{3Q_a h^{3/2} x}{8a_0^3 lz^{5/2}} \cdot [a_0^2(z/h) - x^2] \tag{100}$$

and

$$S_z = \frac{RT}{P_a + \sigma'(h - z)} \cdot \frac{3Q_a h^{3/2}}{4a_0^3 l z^{3/2}} \cdot [a_0^2(z/h) - x^2] \tag{101}$$

if $a_0^2(z/h) > x^2$. Also, $S_z = S_x = 0$ if $a_0^2(z/h) < x^2$.

Sparging Model, Single Vertical Air Injection Pipe

We return to the single vertical sparging pipe and consider the mass balances for the volume elements. The volume of the ijth volume element is given by

$$V_{ij} = (2i - 1)\pi(\Delta r)^2 \Delta z \tag{102}$$

The areas of the top and bottom surfaces of the volume element are given by $(2i - 1)\pi(\Delta r)^2$. The area of the inner surface is $2(i - 1)\pi\Delta r\Delta z$, and the area of the outer surface is $2i\pi\Delta r\Delta z$.

Let

C_{ij}^N = DNAPL concentration in the ijth volume element, kg/m^3
C_{ij}^w = dissolved VOC concentration in the ijth volume element, kg/m^3 of aqueous phase
C_{ij}^g = gaseous VOC concentration in the ijth compartment, kg/m^3
ν = total porosity of medium
ω = water-filled porosity of medium
K_H = VOC Henry's constant, dimensionless
m_{ij} = total mass of VOC in ijth volume element

Then, advective transport is described by

$$\frac{dm_{ij}}{dt} = S_{ij}^B(2i - 1)\pi(\Delta r)^2 C_{i,j-1}^g + S_{ij}^I 2(i - 1)\pi\Delta r\Delta z C_{i-1,j}^g$$

$$- S_{ij}^T(2i - 1)\pi(\Delta r)^2 C_{ij}^g - S_{ij}^O 2i\pi\Delta r\Delta z C_{ij}^g \tag{103}$$

and

$$C_{ij}^g = K_H C_{ij}^w \tag{104}$$

on making the assumption that water advection can be ignored and that Henry's Law applies. In Equation (103) S_{ij}^B is the volumetric gas flux into the volume element through its bottom surface, S_{ij}^I is the flux in through the inner surface, S_{ij}^T is the flux through the top surface, and S_{ij}^O is the flux through the outer surface. These are defined as follows:

$$S_{ij}^B = S_z[(i - 1/2)\Delta r, (j - 1)\Delta z] \qquad (105)$$

$$S_{ij}^I = S_r[(i - 1)\Delta r, (j - 1/2)\Delta z] \qquad (106)$$

$$S_{ij}^O = S_r[i\Delta r, (j - 1/2)\Delta z] \qquad (107)$$

$$S_{ij}^T = S_z[(i - 1/2)\Delta r, j\Delta z] \qquad (108)$$

where S_r and S_z are defined in Equations (85) and (86). Calculating the total mass of VOC in the ijth volume element yields

$$m_{ij} = \Delta V_{ij}[C_{ij}^N + \omega C_{ij}^w + (\nu - \omega)C_{ij}^s] \qquad (109)$$

For the kinetics of solution of the DNAPL droplets, we take the expression used earlier, which was derived for use in modeling the solution of DNAPL droplets by pump-and-treat operations (Kayano and Wilson, 1993). With this, it is assumed that the droplets are spherical in shape and that solution must take place through a stationary aqueous boundary layer of thickness large compared to the radius of the droplets. The result is

$$\frac{dC_{ij}^N}{dt} = -\frac{3C_0^{N2/3}D(C_{sat} - C_{ij}^w)C_{ij}^{N1/3}}{\varrho a_0^2} \qquad (110)$$

where

D = diffusivity of VOC in the water-saturated porous medium, m²/sec
ϱ = DNAPL density, kg/m³
a_0 = initial DNAPL droplet radius, m
C_{sat} = saturation concentration of VOC in water, kg/m³

Now from the solution/diffusion process only, we have

$$\left(\frac{\partial m_{ij}}{\partial t}\right)_{diff} = 0 = \Delta V_{ij}\left[\left(\frac{\partial C_{ij}^N}{\partial t}\right)_{diff}\right.$$

$$\left. + [w + (\nu - \omega)K_H]\left(\frac{\partial C_{ij}^w}{\partial t}\right)_{diff}\right] \qquad (111)$$

from which we obtain

$$\left(\frac{\partial C_{ij}^w}{\partial t}\right)_{diff} = -\frac{1}{[\omega + (\nu - \omega)K_H]} \cdot \frac{dC_{ij}^N}{dt} \qquad (112)$$

Let the mass of VOC in the aqueous and vapor phases in ΔV_{ij} be μ_{ij}. Then

$$\mu_{ij} = \Delta V_{ij}[\omega + (\nu - \omega)K_H]C_{ij}^w \tag{113}$$

Now

$$\left(\frac{\partial \mu_{ij}}{\partial t}\right)_{diff} = -\Delta V_{ij}\frac{dC_{ij}^N}{dt} \tag{114}$$

and

$$\left(\frac{\partial \mu_{ij}}{\partial t}\right)_{advect}$$

is given by the right side of Equation (103). Then

$$\left(\frac{d\mu_{ij}}{dt}\right)_{total} = -\Delta V_{ij}\frac{dC_{ij}^N}{dt} + S_{ij}^B(2i - 1)\pi(\Delta r)^2C_{i,j-1}^q$$

$$+ S_{ij}^q 2(i - 1)\pi\Delta r\Delta z C_{i-1,j}^q$$

$$- S_{ij}^T(2i - 1)\pi(\Delta r)^2C_{ij}^q$$

$$- S_{ij}^O 2i\pi\Delta r\Delta z C_{ij}^q \tag{115}$$

Equation (113) yields

$$\frac{dC_{ij}^w}{dt} = \frac{1}{\Delta V_{ij}[\omega + (\nu - \omega)K_H]} \cdot \frac{d\mu_{ij}}{dt} \tag{116}$$

This, with Equation (115), then yields

$$\frac{dC_{ij}^w}{dt} = -\frac{1}{\omega + (\nu - \omega)K_H} \cdot \frac{dC_{ij}^N}{dt} + \frac{1}{\Delta V_{ij}[\omega + (\nu - \omega)K_H]}$$

$$\cdot \{S_{ij}^B(2i - 1)\pi(\Delta r)^2C_{i,j-1}^q$$

$$+ S_{ij}^q 2(i - 1)\pi\Delta r\Delta z C_{i-1,j}^q$$

$$- S_{ij}^O 2i\pi\Delta r\Delta z C_{ij}^q$$

$$- S_{ij}^T(2i - 1)\pi(\Delta r)^2C_{ij}^q\} \tag{117}$$

Now we must take account of the fact that the gas is expanding as it rises, so that in the z-direction we must include a dilution factor for the C_{ij}^g. We assume that gas enters at the bottom of a volume element, rises (with expansion) to the middle, equilibrates with the liquid in this volume element at that point, and then rises (with expansion) out through the top of the volume element. The dilution factors can readily be calculated; for gas rising from a height z_1 to a height z_2, we have

$$\frac{C^g(z_2)}{C^g(z_1)} = \frac{P_a + \sigma'(h - z_2)}{P_a + \sigma'(h - z_1)} \tag{118}$$

Including this dilution effect requires the following corrections. From the bottom of the i,jth volume element to the center of the i,jth volume element,

$$C^g[(i - 1/2)\Delta r, (j - 1/2)\Delta z] = C^g[(i - 1/2)\Delta r, (j - 1)\Delta z]$$
$$\cdot \{P_a + \sigma'[h - (j - 1/2)\Delta z]\}$$
$$/ \{P_a + \sigma'[h - (j - 1)\Delta z]\} \tag{119}$$

From the middle of the $(i, j - 1)$th volume element to the bottom of the i,jth volume element,

$$C^g[(i - 1/2)\Delta r, (j - 1)\Delta z] = C^g[(i - 1/2)\Delta r, (j - 3/2)\Delta z]$$
$$\cdot \{P_a + \sigma'[h - (j - 1)\Delta z]\}$$
$$/ \{P_a + \sigma'[h - (j - 3/3)\Delta z]\} \tag{120}$$

From the middle of the i,jth volume element to the top of the i,jth volume element,

$$C^g[(i - 1/2)\Delta r, j\Delta z] = C^g[(i - 1/2)\Delta r, (j - 1)\Delta z]$$
$$\cdot \{P_a + \sigma'[h - j\Delta z]\}$$
$$/ \{P_a + \sigma'[h - (j - 1/2)\Delta z]\} \tag{121}$$

Let us define

$$P(k\Delta z) = P_a + \sigma'(h - k\Delta z) \tag{122}$$

Then we return to Equation (117) and rescale the C_{ij}^s's to take this dilution by gas expansion into account. This yields

$$\frac{dC_{ij}^w}{dt} = -\frac{1}{\omega + (\omega - \nu)K_H} \cdot \frac{dC_{ij}^N}{dt} + \frac{1}{\Delta V_{ij}[\omega + (\omega - \nu)K_H]}$$

$$\cdot \left[S_{ij}^B (2i - 1)\pi(\Delta r)^2 \cdot \frac{P[(j-1)\Delta z]}{P[(j-3/2)\Delta z]} \cdot C_{i,j-1}^s \right.$$

$$+ S_{ij}^I 2(i - 1)\pi \Delta r \Delta z C_{i-1,j}^s - S_{ij}^O 2i\pi \Delta r \Delta z C_{ij}^s$$

$$\left. - S_{ij}^T (2i - 1)\pi(\Delta r)^2 \cdot \frac{P[j\Delta z]}{P[(j-1/2)\Delta z]} \cdot C_{ij}^s \right] \tag{123}$$

The model then consists of Equations (103), (110), and (123). The differential equations are integrated forward in time, Equation (103) is used to calculate the gas phase VOC concentrations at each step, and the residual mass of contaminant in the domain of interest is calculated from the equation

$$M_{total} = \sum_{i,\ j} \sum \Delta V_{ij} \{ C_{ij}^N + [\omega + (\nu - \omega)K_H]C_{ij}^w \} \tag{124}$$

Sparging Model, Horizontal Slotted Air Injection Pipe

We next turn to the modeling of sparging with a horizontal air injection pipe and carry out the calculation of the mass balances for the volume elements. Because of the symmetry of the problem, we need examine only the right half of the domain of interest. Let

$$x_i = (i - 1/2)\Delta x \tag{125}$$

$$z_j = (j - 1/2)\Delta z \tag{126}$$

The volume elements are of size

$$\Delta V = \Delta x \Delta z l \tag{127}$$

The inner and outer surfaces of a volume element are $l\Delta z$, and the upper and lower surfaces are $l\Delta x$. Other notation is as in the preceding section (see Figure 8.31). Again, we have

$$C_{ij}^s = K_H C_{ij}^w \tag{104}$$

and

$$\frac{dC_{ij}^N}{dt} = - \frac{3C_0^{N2/3}D(C_{sat} - C_{ij}^w)C_{ij}^{N1/3}}{\varrho a_0^2} \tag{110}$$

Equation (117) is replaced by

$$\frac{dC_{ij}^w}{dt} = - \frac{1}{\omega + (\nu - \omega)K_H} \cdot \frac{dC_{ij}^N}{dt} + \frac{1}{\Delta V[\omega + (\nu - \omega)K_H]}$$

$$\cdot \{S_{ij}^B l \Delta x C_{i,j-1}^g + S_{ij}^L l \Delta z C_{i-1,j}^g - S_{ij}^R l \Delta z C_{ij}^g - S_{ij}^T l \Delta x C_{ij}^g\} \tag{128}$$

Lastly, we need to correct the VOC concentrations at the top and bottom of the volume element for dilution effects as the rising gas expands. This is done in exactly the same way as was used to obtain Equation (123); one obtains

$$\frac{dC_{ij}^w}{dt} = - \frac{1}{\omega + (\nu - \omega)K_H} \cdot \frac{dC_{ij}^N}{dt} + \frac{1}{V[\omega + (\nu - \omega)K_H]}$$

$$\cdot \left[S_{ij}^B l \Delta x \cdot \frac{P[(j-1)\Delta z]}{P[(j-3/2)\Delta z]} \cdot C_{i,j-1}^g + S_{ij}^L l \Delta z C_{i-1,j}^g - S_{ij}^R l \Delta z C_{ij}^g \right.$$

$$- S_{ij}^T l \Delta x \cdot \frac{P[j\Delta z]}{P[(j-1/2)\Delta z]} \cdot C_{ij}^g \tag{129}$$

The gas fluxes are calculated from Equations (100) and (101) as follows:

$$S_{ij}^B = S_z[(i - 1/2)\Delta x, (j - 1)\Delta z] \tag{130}$$

$$S_{ij}^L = S_x[(i - 1)\Delta x, (j - 1/2)\Delta z] \tag{131}$$

$$S_{ij}^R = S_x[i\Delta x, (j - 1/2)\Delta z] \tag{132}$$

$$S_{ij}^T = S_z[(i - 1/2)\Delta x, j\Delta z] \tag{133}$$

To simulate a run, the model parameters are read in, the C_{ij}^N and C_{ij}^w are initialized, and Equations (110) and (129) are integrated forward in time. The C_{ij}^g are calculated from Equation (103). The total residual mass of VOC is given by

$$M_{total} = \Delta V \sum_{i,\ j} \sum \{C_{ij}^N + [\omega + (\nu - \omega)K_H]C_{ij}^w\} \tag{134}$$

Results: DNAPL Sparging with Air Injection Pipes

Programs implementing these two sparging models were written in TurboBASIC and run on microcomputers using 80386 SX (16 MHz) and 80386 DX (33 MHz) microprocessors and math coprocessors. Typical runs required only a few minutes. Figures 8.32 through 8.38 pertain to sparging with a single vertical pipe; Figures 8.39–8.41 pertain to sparging with a buried horizontal slotted pipe. Default parameters for Figures 8.32 through 8.38 are given in Table 8.14 and default parameters for Figures 8.39–8.41 in Table 8.15. The DNAPL characteristics were chosen to represent trichloroethylene unless otherwise specified.

Figure 8.32 shows plots of total mass of residual contaminant versus time for various sizes of the DNAPL droplets. Evidently, the model is capable of representing severely diffusion-limited solution of the DNAPL droplets if one selects droplet sizes that are relatively large. With the operating parameters used in these calculations, the sparging is diffusion-limited, so removal rates decrease like the reciprocal of the square of the initial droplet radius, a_0. This is as expected from Equation (110).

The effect of varying the Henry's constant of the DNAPL on the rate of DNAPL removal is shown in Figure 8.33. Since the sparging is diffusion-limited, a marked decrease in Henry's constant (from 0.2 to 0.0025, dimensionless) results in only a slight decrease in removal rate. One expects that the effect of a decrease in Henry's constant would be larger if the gas flow rate through the sparging well were reduced to the point where the

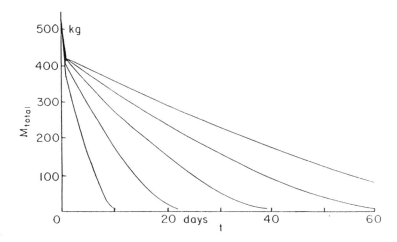

Figure 8.32 DNAPL sparging. Plots of residual contaminant mass versus time; effect of DNAPL droplet size. From bottom to top, initial DNAPL droplet diameter = 0.10, 0.15, 0.20, 0.25, and 0.30 cm. Initial contaminant concentration = 1,000 mg/kg. Other parameters as in Table 8.14. [Reprinted from Burchfield and Wilson (1993) p. 2542, by courtesy of Marcel Dekker, Inc.]

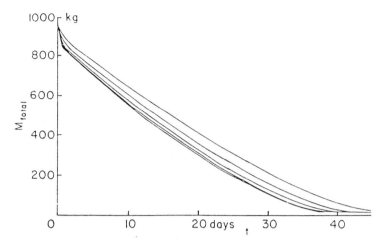

Figure 8.33 DNAPL sparging. Plots of residual contaminant mass versus time; effect of VOC Henry's constant. From bottom to top, K_H = 0.2, 0.1, 0.05, 0.025, and 0.0125 (dimensionless). Other parameters as in Table 8.14. [Reprinted from Burchfield and Wilson (1993) p. 2542, by courtesy of Marcel Dekker, Inc.]

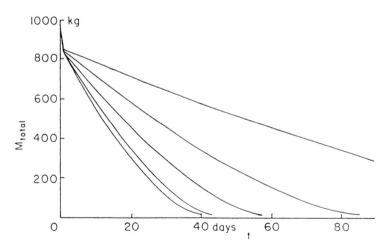

Figure 8.34 DNAPL sparging. Plots of residual contaminant mass versus time; effect of DNAPL aqueous solubility. DNAPL solubility = 1,100 (TCE solubility), 1,000, 750, 500, and 250 mg/L, from bottom to top. Other parameters as in Table 8.14. [Reprinted from Burchfield and Wilson (1993) p. 2543, by courtesy of Marcel Dekker, Inc.]

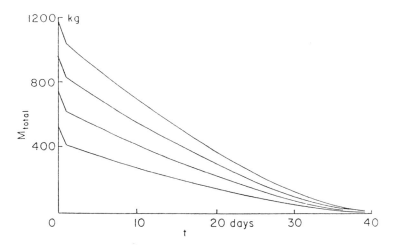

Figure 8.35 DNAPL sparging. Plots of residual contaminant mass versus time; effect of initial DNAPL concentration in the aquifer. C_0 = 1,000, 1,500, 2,000, and 2,500 mg/kg, from bottom to top. Other parameters as in Table 8.14. [Reprinted from Burchfield and Wilson (1993) p. 2544, by courtesy of Marcel Dekker, Inc.]

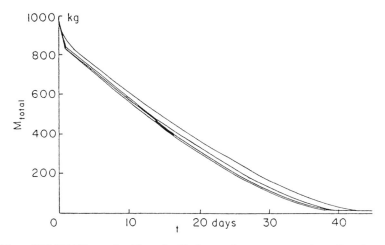

Figure 8.36 DNAPL sparging. Plots of residual contaminant mass versus time; effect of sparging well airflow rate. Q_a = 0.98, 0.49, 0.245, and 0.1225 moles/sec, from bottom to top. Other parameters as in Table 8.14. [Reprinted from Burchfield and Wilson (1993) p. 2544, by courtesy of Marcel Dekker, Inc.]

433

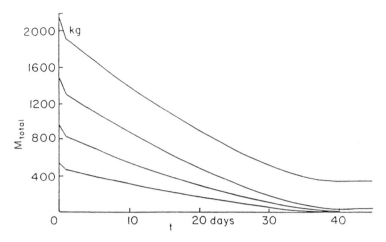

Figure 8.37 DNAPL sparging. Plots of residual contaminant mass versus time; effect of radius of zone of contamination. Radius zone of contamination = 3, 4, 5, and 6 m, from bottom to top. Other parameters as in Table 8.14. [Reprinted from Burchfield and Wilson (1993) p. 2545, by courtesy of Marcel Dekker, Inc.]

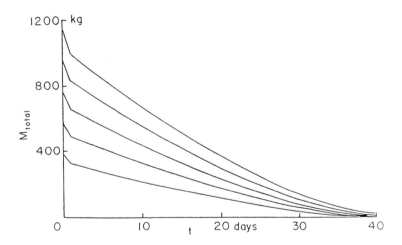

Figure 8.38 DNAPL sparging. Plots of residual contaminant mass versus time; effect of depth of zone of contamination below the surface. The zone of contamination extends from the surface of the aquifer to depths of 2, 3, 4, 5, and 6 m, from bottom to top. Other parameters as in Table 8.14. [Reprinted from Burchfield and Wilson (1993) p. 2546, by courtesy of Marcel Dekker, Inc.]

434

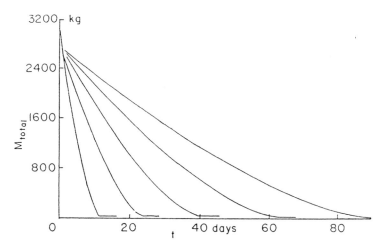

Figure 8.39 DNAPL sparging, horizontal pipe configuration. Plots of residual contaminant mass versus time; effect of DNAPL droplet size. Droplet diameters are 0.10, 0.15, 0.20, 0.25, and 0.30 cm from bottom to top. Other parameters as in Table 8.15. [Reprinted from Burchfield and Wilson (1993) p. 2547, by courtesy of Marcel Dekker, Inc.]

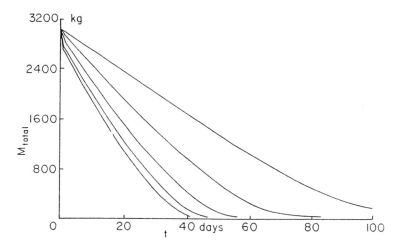

Figure 8.40 DNAPL sparging, horizontal pipe configuration. Plots of residual contaminant mass versus time; effect of airflow rate. Q_a = 0.98, 0.49, 0.245, 0.1225, and 0.06125 mole/sec, bottom to top. Other parameters as in Table 8.15. [Reprinted from Burchfield and Wilson (1993) p. 2548, by courtesy of Marcel Dekker, Inc.]

435

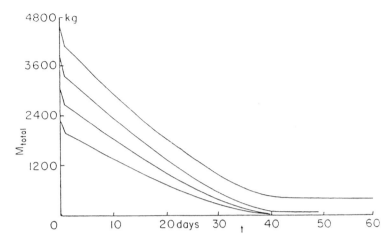

Figure 8.41 DNAPL sparging, horizontal pipe configuration. Plots of residual contaminant mass versus time; effect of width of zone of contamination. Full width of zone of contamination = 6, 8, 10, and 12 m. Other parameters as in Table 8.15. [Reprinted from Burchfield and Wilson (1993) p. 2548, by courtesy of Marcel Dekker, Inc.]

process is no longer so strongly diffusion-controlled; this was, in fact, found to be the case.

The effect of varying the aqueous solubility of the DNAPL (while holding the Henry's constant unchanged) is shown in Figure 8.34. Increases in solubility C_{sat} result in very marked increases in removal rates, as one

TABLE 8.14. Default Parameters for DNAPL Sparging with a Single Vertical Well.

Thickness of aquifer	7 m
Radius of influence of air injection well at top of aquifer	7 m
N_z	7
N_r	7
Molar airflow rate of sparging well	1.96 mol/sec
Ambient temperature	20°C
VOC Henry's constant	0.35
VOC solubility in water	1,100 mg/L
Soil density	1.7 gm/cm³
Total porosity of medium	0.4
Water-filled porosity of medium	0.36
Diffusion constant of VOC in porous medium	2×10^{-10} m²/sec
Density of VOC	1.46 gm/cm³
Initial DNAPL concentration	2,000 mg/kg
Initial diameter of trapped DNAPL droplets	0.2 cm
Depth to which contaminant extends in aquifer	5 m
Radius to which contaminant extends about well	4 m
Δt	900 sec

TABLE 8.15. Default Parameters for DNAPL Sparging with a
Single Horizontal Slotted Pipe.

Thickness of aquifer	7 m
Full width of influence of air injection well at top of aquifer	14 m
N_z, N_x	7, 7
Depth to which contaminant extends in aquifer	5 m
Distance to which contaminant extends on either side of buried horizontal pipe	4 m

Other parameters as in Table 8.14.

would expect from Equation (110). In interpreting these results, one should note that, as the solubility of the DNAPL is increased at constant K_H, the equilibrium vapor pressure of the DNAPL is increased proportionately. If, on the other hand, one holds the equilibrium vapor pressure constant while increasing the DNAPL solubility, K_H decreases proportional to $1/C_{sat}$, which results in decreasing vapor phase VOC concentrations according to Equation (103). This tends to decrease removal rates. In the diffusion-limited regime the two effects virtually cancel each other out. At lower gas flow rates (where diffusion is no longer so limiting) removal rate decreases with increasing DNAPL solubility at constant DNAPL vapor pressure.

Figure 8.35 exhibits the effect of initial DNAPL concentration on removal rate. In these runs the DNAPL droplet size has been held constant, so the number of droplets per m³ is proportional to the DNAPL concentration. We therefore find, as expected, that the rate of DNAPL removal is directly proportional to the initial DNAPL concentration in the aquifer.

Decreases in the airflow rate of the sparging well result in decreases in the removal rate, as seen in Figure 8.36. For the parameter sets used in these runs, diffusion kinetics are the dominant limiting factor in the removal; therefore, the results of decreasing the airflow rate are minor until one gets down to flow rates of 0.1225 moles/sec or so. Excessively high flow rates result in little increase in removal rate if one is approaching the diffusion-limited regime. Such high airflow rates merely result in excessive energy costs for compressing air and also excessive treatment costs for removing VOCs from a high-volume, low-concentration exhaust gas stream if this must be done.

In this model the domain of the aquifer that is actually being aerated is assumed to be a paraboloid of revolution [see Equations (75) and (81)]. As long as the zone of contamination lies entirely within this paraboloid of influence, one expects that complete remediation will occur. If, however, any of the contaminated region lies outside of this paraboloid, that portion of the domain will never be cleaned up, since these models neglect the circulation of water in the vicinity of the well. This is shown in Figure 8.37, which shows plots of residual contaminant mass versus time for zones of contamination of several different radii. The two lowest curves correspond to zones of contamination lying wholly within the paraboloid of influence. The third has a small portion lying outside of the paraboloid, and the fourth has a substantial portion lying outside. For these last two plots, at large times the residual contaminant mass curves approach a positive limiting value. Since in these runs solution/diffusion of the DNAPL droplets is rate-limiting, all the plots reach their final values of residual DNAPL at about the same time.

The effect of the depth to which the contamination extends is shown in Figure 8.38. In four of the five runs, the domain of contamination lies wholly within the paraboloid of influence of the well. In the fifth (top) run there is a small residual mass of DNAPL that is not removed since the cylinder of contaminated aquifer extends beyond the paraboloid of influence of the sparging well near the bottom of the aquifer.

Figures 8.39–8.41 pertain to sparging with a buried horizontal slotted pipe. Model default parameters are given in Table 8.15. In Figure 8.39 we see the effect of increasing DNAPL droplet radius (i.e., progressively decreasing the diffusion/solution rate). The plots are similar to those seen for the vertical pipe model, which are given in Figure 8.32. As before, removal rates are essentially proportional to a_0^2, since the systems are in the diffusion-limited regime.

The effect of variation in the sparging well airflow rate is seen in Figure 8.40. At the lowest airflow rate (0.06125 mole/sec, about 3 scfm) the rate of advective removal is becoming the major bottleneck in the remediation. Evidently, for this system airflow rates much larger than about 6 scfm will result in only modest increases in remediation rate.

Figure 8.41 exhibits the effect of the width of the zone of contamination (assumed to be a rectangular paralellopiped) on the course of the remediation. The zone of influence of the horizontal slotted pipe sparging well is assumed to be a parabolic cylinder [see Equations (93) and (97)]. As before, if the zone of contamination lies entirely within the domain of influence of the well, remediation is complete, as with the lower two plots. If, however, any of the zone of contamination lies outside of the domain of influence, contaminant in that portion of the zone will not be removed, as is seen for the upper two plots.

The effects of varying the DNAPL solubility while holding its vapor pressure constant are shown in Figures 8.42 and 8.43. In Figure 8.42 the sparging gas flow rate is rather large (0.98 mol/sec), and for this situation (where diffusion is definitely rate limiting) the rate of removal is essentially proportional to the aqueous solubility of the DNAPL. The situation is rather different if the sparging gas flow rate is only 0.0615 mol/sec, as is the case in Figure 8.43. There, removal rates decrease with decreasing sol-

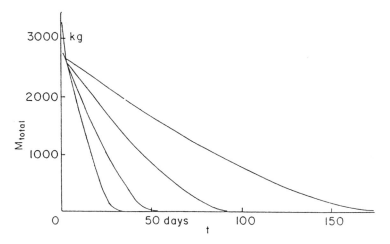

Figure 8.42 DNAPL sparging, horizontal pipe configuration. Plots of residual contaminant mass versus time; effect of varying DNAPL solubility at constant DNAPL vapor pressure, high sparging rate regime. $Q_a = 0.98$ mol/sec. From left to right, $(K_H, C_{sat}) = (0.10, 2,000$ mg/L), $(0.20, 1,000$ mg/L), $(0.40, 500$ mg/L), $(0.80, 250$ mg/L). Other parameters as in Table 8.15. [Reprinted from Burchfield and Wilson (1993) p. 2549, by courtesy of Marcel Dekker, Inc.]

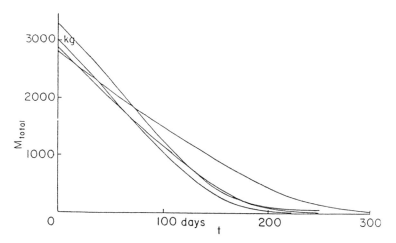

Figure 8.43 DNAPL sparging, horizontal pipe configuration. Plots of residual contaminant mass versus time; effect of varying DNAPL solubility at constant DNAPL vapor pressure, low sparging rate regime. $Q_a = 0.06125$ mol/sec. From left to right, $(K_H, C_{sat}) = (0.10, 2,000$ mg/L), $(0.20, 1,000$ mg/L), $(0.40, 500$ mg/L), and $(0.80, 250$ mg/L). Other parameters as in Table 8.15. [Reprinted from Burchfield and Wilson (1993) p. 2549, by courtesy of Marcel Dekker, Inc.]

ubility, but the effect is very much smaller. Recall that in Figures 8.42 and 8.43 the vapor pressure of the DNAPL is being held constant, so that at low sparging gas flow rates the concentration of VOC in the advecting gas is determined essentially by the equilibrium vapor pressure of the DNAPL, rather than by the rate of diffusion from the DNAPL droplets. The observed differences between Figures 8.42 and 8.43 are therefore as one would expect.

Conclusions: Models for Simple Sparging Well without Water Circulation

The models presented here enable one to get some insight into the factors affecting the sparging of DNAPL droplets/ganglia/blobs from contaminated aquifers. The model for solution/diffusion permits one to represent a virtually infinite range of rates for this process. The number of parameters required to use the models is relatively modest, and most of these are readily obtained. Computationally, the models are sufficiently simple that they can be run on readily available microcomputers.

On the other hand, the models also focus attention on aspects of sparging that require further study. Some of these are as follows:

(1) We have postulated the form of the flow field of the sparge gas, which is an approach leaving something to be desired in terms of rigor, even if the postulated flow field seems "reasonable." Solution of this problem in terms that permit microcomputer modeling may be rather difficult. In particular, we have little insight into the relationship between airflow rate, thickness of aquifer, and the radius of the circle at the top of the aquifer through which air is moving, a_0. Intuitively, one expects that a_0 increases with increasing airflow rate Q_a, but we do not know the functional dependence.

(2) We have assumed local equilibrium between the dissolved VOC and the VOC in the vapor phase. This approximation could be replaced without much difficulty by some kind of lumped parameter approach as experimental results become available, indicating that the kinetics of VOC mass transfer between the liquid and vapor phase is a significant factor; this is explored in the next section.

(3) We have asumed that the flow field of the sparging air induces no bulk movement of the water in the aquifer. Some preliminary bench-scale experiments were carried out in which dye is put into the pool of liquid at the top of a sand bed in which air sparging is taking place and the attenuation of the dye is then followed with time as sparging continues. These indicated that the rate of mixing of the liquid in the sand bed with the overlying liquid is fairly slow, suggesting that this assumption may be reasonable. One would expect, however, that a rig-

orous theoretical treatment will be difficult and probably beyond the scope of computer models suitable for use on microcomputers. The effect of water advection will be studied in the next section.

SPARGING WITH A HORIZONTAL PIPE: INCLUSION OF WATER-VAPOR MASS TRANSPORT KINETICS AND WATER CIRCULATION

In this section a model is presented for the removal of DNAPL droplets and dissolved VOC in an aquifer by sparging with a horizontal slotted pipe. VOC diffusion transport from DNAPL droplets to aqueous phase takes place through a thick stagnant water layer in the porous medium to mobile water. Transport of VOC from the aqueous phase into the gas phase is modeled by means of a lumped parameter approach. The air-induced circulation of water near the air injection pipe is modeled by the method of images. The effects of the model parameters on the rate of VOC removal are explored, both for cases in which DNAPL is present and in which contaminant is present only as VOC dissolved in the aqueous phase.

The models for sparging of DNAPLs presented in the previous section included the assumptions that the kinetics of solution of the DNAPL blobs could be a severely rate-limiting step but that the rate of mass transport of VOC from the aqueous phase to the moving gas stream could be adequately described by a local equilibrium approximation. In the modeling of the operation of simple sparging wells, it was further assumed that the movement of air through the aquifer did not bring about significant circulation of water.

In the model discussed in this section, both of these approximations are eliminated. The kinetics of mass transfer between the aqueous phase and the vapor phase is handled by means of a lumped parameter method used previously (Wilson, 1990) and described in Chapter 4. The air-induced circulation of water is represented by introducing the flow field resulting from a source at the top of the aquifer and a corresponding sink at the bottom; discharge to the sink is shunted to the source so that the flow field is conservative. Here, we deal with sparging by means of a horizontal slotted pipe of length that is long compared to its breadth of influence, which permits the use of a two-coordinate Cartesian system if end effects are ignored.

Modeling the Airflow-Induced Water Flow Field

We assume that the water circulation induced by the injected air can be described adequately by the flow field generated by a water source Q_w at the top of the aquifer (at $x = 0$, $z = h$), and a water sink $-Q_w$ at the bottom of the aquifer (at $x = 0$, $z = 0$). We assume no-normal-flow boundary conditions at the top and bottom of the aquifer. Then, one can generate the velocity potential by the method of images from electrostatics [this chapter; also Wilson (1992)]. This, in turn, yields the following expressions for the water superficial velocities v_x and v_z:

$$v_x = \frac{Q_w}{\pi l} \sum_{n=-\infty}^{\infty} \left[\frac{x}{x^2 + [z - (2n + 1)h]^2} - \frac{x}{x^2 + [z - 2nh]^2} \right] \qquad (135)$$

$$v_z = \frac{Q_w}{\pi l} \sum_{n=-\infty}^{\infty} \left[\frac{z - (2n + 1)h}{x^2 + [z - (2n + 1)h]^2} - \frac{z - 2nh}{x^2 + [z - 2nh]^2} \right] \qquad (136)$$

Here Q_w, the magnitude of the source and the sink generating the water flow field, is a function of the aquifer thickness h, the permeability of the aquifer medium, and Q_a, the airflow rate through the sparging well. Presumably, Q_w will be determined on a site-specific basis.

Solution of DNAPL Droplets: Equations for the C_{ij}^N

We use the method for modeling the solution of DNAPL droplets, which was employed in the earlier DNAPL sparging models. The physical picture for the process is that of spherical DNAPL droplet from which VOC is dissolving and diffusing through a thick stagnant water layer in the porous medium to the moving aqueous phase (which is in contact with the vapor phase). Analysis of this picture of the DNAPL solution process then gives

$$\frac{dC_{ij}^N}{dt} = - \frac{3 C_0^{N2/3} D (C_{sat} - C_{ij}^w) C_{ij}^{N1/3}}{\varrho a_0^2} \qquad (50)$$

where

C_{ij}^N = concentration of DNAPL in the ijth volume element of the system, kg/m³

C_0^N = initial concentration of DNAPL in the contaminated portions of the system, kg/m³

C_{ij}^w = aqueous concentration of VOC in the ijth volume element, kg/m³ of aqueous phase

C_{sat} = saturation concentration of VOC in water, kg/m³

D = diffusivity of VOC in the water-saturated porous medium, m²/sec

ϱ = DNAPL density, kg/m³

a_0 = initial DNAPL droplet radius, m

Material Balance for the Aqueous Phase: Equations for the C_{ij}^w

On carrying out a material balance on the VOC in the aqueous phase in the ijth volume element, we obtain the following set of equations:

$$\omega\Delta x\Delta zl\frac{dC_{ij}^w}{dt} = -\Delta x\Delta zl\frac{dC_{ij}^N}{dt} - \Delta x\Delta zl(\nu - \omega)\lambda(K_HC_{ij}^w - C_{ij}^g)$$

$$+ \Delta zlv_x[(i - 1)\Delta x, (j - 1/2)\Delta z]T(v_x)C_{i-1,j}^w$$

$$- \Delta zlv_x[i\Delta x, (j - 1/2)\Delta z]T(-v_x)C_{i+1,j}^w$$

$$+ \Delta xlv_z[(i - 1/2)\Delta x, (j - 1)\Delta z]T(v_z)C_{i,j-1}^w$$

$$- \Delta xlv_z[(i - 1/2)\Delta x, j\Delta z]T(v_z)C_{i,j+1}^w$$

$$+ \{\Delta zlv_x[(i - 1)\Delta x, (j - 1/2)\Delta z]T(-v_x)$$

$$- \Delta zlv_x[i\Delta x, (j - 1/2)\Delta z]T(v_x)$$

$$+ \Delta xlv_z[(i - 1/2)\Delta x, (j - 1)\Delta z]T(-v_z)$$

$$- \Delta xlv_z[(i - 1/2)\Delta x, j\Delta z]T(v_z)\} \cdot C_{ij}^w \tag{137}$$

In Equation (137) the first term on the right-hand side represents mass transport of VOC into the aqueous phase from DNAPL droplets. The second term represents a lumped parameter approximation for mass transport of VOC between the aqueous phase and the vapor phase. The remaining eight terms represent advective transport of aqueous VOC between the volume element of interest and its nearest neighbors. Here

ν = total porosity of the aquifer medium
ω = water-filled porosity of the medium
K_H = Henry's constant for the VOC, dimensionless
λ = lumped parameter rate constant for mass transport of VOC between the aqueous and vapor phases, sec^{-1}
C_{ij}^g = vapor phase VOC concentration in the ijth volume element, kg/m^3

The function $T(v)$ is a switching function, equal to zero if its argument is negative, and equal to one if its argument is positive. In Equation (137) the arguments of T in the various terms are the velocities that precede T in each term.

To close the loop on the water circulation field and prevent unphysical accumulation of contaminant in the 1,1th volume element, we introduce a shunt along the axis of the system (i.e., along the left side of the half of the system being modeled) to transport the water entering the 1,1th volume element at the bottom of the aquifer back up to the 1,n_zth volume element at the top. This requires modifying Equation (137) by subtracting a term $(Q_w/4)C_{11}^w$ from the 1,1th equation and adding an identical term to the 1,n_zth equation. The 4 in the denominator is required by the fact that the 1,1th and 1,n_zth volume elements handle only one-fourth of the total flow used to generate the water flow field, since the sink and the source are located on the lower left and upper left corners of these volume elements, respectively.

Material Balance for the Vapor Phase: The Steady-State Approximation for the C_{ij}^g

Construction of a mass balance for vapor phase VOC in the ijth volume element yields

$$(\nu - \omega)l\Delta x\Delta z\frac{dC_{ij}^g}{dt} = (\nu - \omega)l\Delta x\Delta z(K_H C_{ij}^w - C_{ij}^g)$$

$$+ l\Delta xS_z[(i - 1/2)\Delta x, (j - 1)\Delta z] \cdot \frac{P[(j - 3/2)\Delta z]}{P[(j - 1)\Delta z]}$$

$$\cdot C_{i,j-1}^g + l\Delta zS_x[(i - 1)\Delta x, (j - 1/2)\Delta z] \cdot C_{i-1,j}^g$$

$$- l\Delta xS_z[(i - 1/2)\Delta x, j\Delta z] \cdot \frac{P[j\Delta z]}{P[(j - 1/2)\Delta z]} C_{ij}^g$$

$$- l\Delta zS_x[i\Delta x, (j - 1/2)\Delta z]C_{ij}^g \qquad (138)$$

The first term on the right-hand side of Equation (138) corresponds to mass transport of VOC between the vapor and aqueous phases. The next four terms correspond to advective transport of VOC vapor by the sparging gas. It is necessary to include pressure ratio factors in two of these terms to take into account the dilution of VOC in the vapor as it rises into regions of lower pressure; if this is not done, the advective terms do not conserve VOC. The form of the volumetric gas flux here makes it unnecessary to include in these equations the switching terms needed in describing advection in the aqueous phase.

The system of Equations (50), (137), and (138), which have been developed to model DNAPL sparging, is a mathematically stiff set of differential equations. That is, although one can expect experimental runs in the field to require some months, the time increments that one must use in the numerical integration of the model equations must be of the order of ten seconds or less. This leads to excessive computer time requirements. One can avoid this difficulty by noting the fact that, generally, the mass of VOC in the vapor phase is only a very small fraction of the total mass of VOC present in the system. This suggests that one may be able to use the steady-state approximation for the vapor phase concentrations. In this, one sets the left-hand side of Equation (138) equal to zero and then solves the resulting algebraic equation for C^{v}_{ij}, starting with the equation for C^{v}_{11} and going in the directions of increasing i and j. This process converts the stiff differential equations into algebraic equations and thereby permits the use of very much larger values of the time increment Δt in the numerical integrations. A comparison of results of the exact approach and the steady-state approach for a representative set of parameters is given in Table 8.16; we see that the discrepancy is less than 0.1%. In all of the work presented here, the steady-state approximation was used.

One of the points of interest is the extent to which the form of the molar gas flux $q_z(x,z)$ influences the modeling results. This function was chosen somewhat arbitrarily to be given by Equation (93), so it would be helpful

TABLE 8.16. Comparison of the Results of Steady-State and Nonsteady-State Model Calculations.

Time (days)	Total Remaining VOC (kg)	
	Steady State	Nonsteady State
0	1,902.40	1,902.40
1	1,759.39	1,759.51
2	1,657.21	1,657.32
3	1,558.67	1,558.79
4	1,462.09	1,462.19
5	1,367.49	1,367.59
6	1,275.07	1,275.18
7	1,185.04	1,185.14
8	1,097.65	1,097.74
9	1,013.22	1,013.31
10	932.15	932.25
11	854.98	855.09
12	782.44	782.55
13	715.60	715.71
14	656.23	656.35

The parameters used in these calculations are those given in Table 8.17 except that $Q_a = 1.0$ mol/sec and $Q_w = 0$.
In the steady-state runs $\Delta t = 100$ sec. In the nonsteady-state runs $\Delta t = 10$ sec.

if one could show that the calculated cleanup rates were not highly sensitive to the form of Equation (93). We next explore this point.

Let us replace Equation (93) for $q_z(x,z)$ by Equation (139),

$$q_z(x,z) = A(z)[a_0^n z/h - x^n], \quad |x| < a_0(z/h)^{1/n}$$

$$= 0, \quad |x| > a_0(z/h)^{1/n} \tag{139}$$

Note that Equation (93) is a special case of this equation, corresponding simply to the case $n = 2$. To obtain q_x one follows along the lines leading to Equation (97). The requirement that the integral of $q_z(x,z)$ over any plane perpendicular to the z-axis give Q_a, the total molar gas flow rate, yields

$$A(z) = \frac{Q_a(n + 1)}{2nla_0^{n+1}} \cdot (h/z)^{(n+1)/n} \tag{140}$$

The molar gas flux is conservative, so its divergence must vanish. This yields

$$\frac{\partial q_z}{\partial z} = -\frac{\partial q_x}{\partial x} \tag{141}$$

Differentiating the expression for q_z with respect to z and then integrating with respect to x yields an expression for $q_x(x,z)$. Note that the integration constant is easily evaluated from the fact that the symmetry of the problem gives $q_x(0,z) = 0$. The final results for q_z and q_x are

$$q_z = \frac{Q_a(n + 1)}{2nla_0} \cdot (h/z)^{1/n} \cdot [1 - (h/z) \cdot (x/a_0)^n] \tag{142}$$

and

$$q_x = \frac{Q_a(n + 1)}{2n^2lh} \cdot (x/a_0) \cdot (h/z)^{(n+1)/n} \cdot [1 - (h/z) \cdot (x/a_0)^n] \tag{143}$$

The remainder of the analysis follows exactly along the lines described above.

Results: Sparging with a Horizontal Pipe; Inclusion of Water-Vapor Mass Transport Kinetics and Water Circulation

The model was implemented in TurboBASIC, and the runs described below were made on microcomputers equipped with 80286 or 80386-DX microprocessors operating at 12 and 33 MHz, respectively, and equipped with math coprocessors. Typical runs required a half hour or less of computer time.

Default parameters used in the modeling are given in Table 8.17. Where other values of parameters were used, these are indicated in the captions of the figures.

The effect of the parameter controlling mass transport of VOC between the aqueous and vapor phases, λ, on rate of VOC removal is shown in Figure 8.44. For the parameter sets being used, it is evident that this mass transport step is quite significant in controlling the rate of removal. Evidently, one would be well advised to explore methods to increase the amount of air-water interface present during sparging operations, since this should result in increased values of λ. Brown (undated) suggests the use of well screens that generate small bubbles. Use of a pulsed airflow rate might also be helpful in this regard. Another possibility is the injection of air containing a small amount of the vapor of an alcohol, to reduce the surface tension and thereby yield smaller bubbles.

The impact of airflow rate Q_a on VOC removal rate is seen in Figure 8.45. Over the range of airflow rates used, airflow rate is not appreciably limiting. For example, doubling the airflow rate from 1 to 2 mol/sec results in a barely detectable increase in VOC removal rate. In an actual sparging

TABLE 8.17. Default Sparging Model Parameters, DNAPL Runs.

Width of domain of interest	10 m
Thickness of aquifer	5 m
Length of horizontal sparging pipe	20 m
n_x, n_z	5, 5
Width of air sparging pattern at top of aquifer	10 m
Molar gas flow rate of sparging well	4 mol/sec
Air-induced water circulation rate	4 L/sec
Ambient temperature	20°C
Total porosity of aquifer medium	0.4
Water-filled porosity of aquifer medium	0.36
Aquifer medium density	1.7 mg/cm³
Contaminant	Trichloroethylene
Density of contaminant	1.46 gm/cm³
Water solubility of contaminant	1,100 mg/L
Henry's constant of contaminant (dimensionless)	0.20
Diffusivity of contaminant in porous medium	2×10^{-10} m²/sec
Rate constant λ for aqueous phase/vapor transport	0.001 sec⁻¹
Initial DNAPL concentration	2,000 mg/kg
Initial dissolved VOC concentration	1,100 mg/L of water
Width of contaminated zone	8 m
Depth of contaminated zone below water table	3 m
Initial DNAPL droplet diameter	0.1 cm
Δt	100 sec

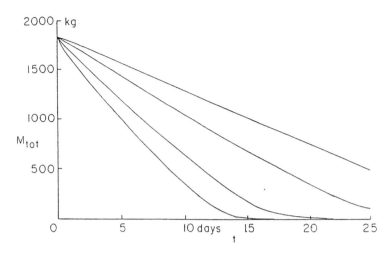

Figure 8.44 Plots of residual mass of TCE versus time; effect of aqueous phase-vapor transport parameter λ. From top to bottom, $\lambda = 1, 2, 5,$ and 10×10^{-4} sec^{-1}. Other parameters as in Table 8.17. [Reprinted from Wilson (1994) p. 80, by courtesy of Marcel Dekker, Inc.]

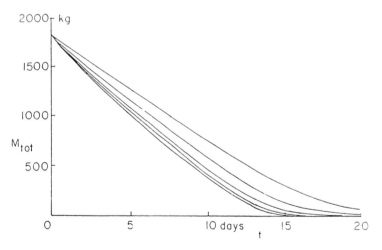

Figure 8.45 Plots of residual mass of TCE versus time; effect of air injection rate Q_a. From top to bottom, $Q_a = 0.125, 0.25, 0.5, 1,$ and 2 mol/sec. Other parameters as in Table 8.17. [Reprinted from Wilson (1994) p. 81, by courtesy of Marcel Dekker, Inc.]

operation, one would probably work at the lower end of the airflow rate range to avoid costs of pumping excessive amounts of air and of treating excessive volumes of highly dilute off-gas if this is being recovered for treatment.

The effect of water circulation rate Q_w on the VOC removal rate is relatively slight, as seen in Figure 8.46. In practice, one is not able to vary this parameter independently, as it is presumably determined by the airflow rate, the well design, and the geological characteristics of the site. For the runs made here, a zero value of Q_w resulted in an unremovable residue of VOC, which was outside the zone of influence of the air. In field operations this parameter would be quite difficult to measure, so it is fortunate that the role it plays is minor.

The parameter controlling the width of influence of the air injection well, a_0, is certainly linked to the geological characteristics of the aquifer, to the well design, and to the rate at which air is being injected. Figure 8.47 shows that it is undesirable to have values of the width parameter that are sufficiently small that portions of the zone of contamination are not aerated. Removal rates are decreased, and we may have a significant increase in tailing if the VOC must be dissolved and then circulated to the aeration zone in order to be removed. Sparging pilot studies should result in site-specific information on the relationship between the width parameter and the air injection rate. This presumably can be obtained by placing piezometer wells close to the water table at various distances from the air

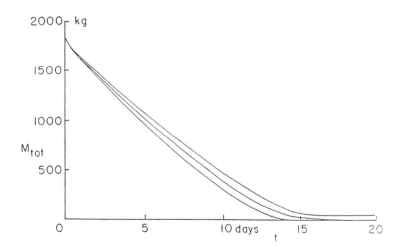

Figure 8.46 Plots of residual mass of TCE versus time; effect of water circulation rate Q_w. From top to bottom, Q_w = 0, 2, and 10 L/sec. Other parameters as in Table 8.17. [Reprinted from Wilson (1994) p. 81, by courtesy of Marcel Dekker, Inc.]

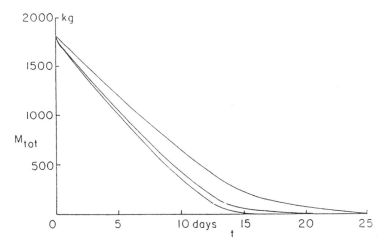

Figure 8.47 Plots of residual mass of TCE versus time; effect of air distribution width parameter a_0. a_0 = 6, 8, and 10 m, top to bottom. Other parameters as in Table 17. [Reprinted from Wilson (1994) p. 82, by courtesy of Marcel Dekker, Inc.]

injection well and monitoring soil gas pressure as a function of air injection rate and distance from the injection well.

The impact of effective DNAPL droplet diameter on removal rate is shown in Figure 8.48. Increasing the effective DNAPL droplet diameter decreases the surface to volume ratio of the DNAPL in the aquifer, thereby decreasing the rate of solution of DNAPL. Initially, these runs all show a rather rapid rate of cleanup, during which dissolved VOC is being removed. Once this has been done, however, the rate of solution of the DNAPL droplets/ganglia/blobs becomes a major factor in controlling the rate of the remediation. Pilot studies must be carried out for a period sufficient to give an indication of the extent to which this type of mass transport will be rate-limiting if they are to be useful in estimating remediation times.

The dependence of removal rate on the form of the air flux function $q_z(x,y)$ is shown in Figure 8.49. In the results shown in this figure, the molar gas fluxes were calculated using Equations (142) and (143) with n = 0.5, 1, 2, 3, 4, and 10. It is seen from Equation (142) that the larger values of n give a flux of sparging gas that is more uniformly distributed across the domain of interest than do the smaller values of n. This is reflected in the more rapid removal rates observed for the runs having the larger values of n. Note, however, that varying n from 10 all the way to 1 results in only about a 30% increase in the time required for cleanup. We therefore conclude that cleanup times are not very sensitive to the precise form of the distribution of the sparging gas. This is fortunate, since this

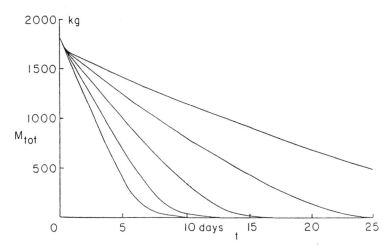

Figure 8.48 Plots of residual mass of TCE versus time; effect of effective DNAPL droplet diameter. Droplet diameter = 0.2, 0.1414, 0.1, 0.0707, and 0.05 cm, top to bottom. Other parameters as in Table 8.17. [Reprinted from Wilson (1994) p. 83, by courtesy of Marcel Dekker, Inc.]

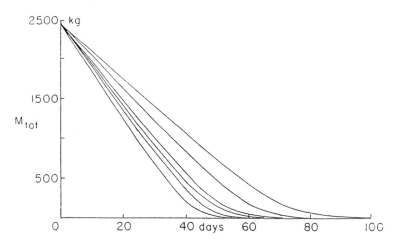

Figure 8.49 Plots of residual mass of TCE versus time; effect of the parameter n in the sparging gas molar flux function $q_z(x,z)$. n = 0.5, 1, 2, 3, 4, 10 from top to bottom; gas flow rate of sparging well = 2 mol/sec; rate constant for aqueous phase/vapor phase mass transport = 0.0001 sec^{-1}. Other parameters as in Table 8.17. [Reprinted from Wilson (1994) p. 83, by courtesy of Marcel Dekker, Inc.]

451

distribution would be quite difficult to calculate and also quite difficult to measure experimentally.

In earlier work (this chapter; Wilson, 1992) we modeled the removal of dissolved VOC by sparging using an approach in which the domain influence of the sparging well was partitioned into two one-dimensional (radial) sets of annular cylindrical domains for computation. One set modeled the upper half of the aquifer, the second, the lower half. For sparging with a horizontal slotted pipe, the computer program used in the present calculations allows us to eliminate this simplification. In Figures 8.50 through 8.54 we present the results of runs simulating the removal of dissolved VOC (in the absence of DNAPL) by sparging with a horizontal slotted pipe. In these calculations the initial DNAPL concentrations were set equal to zero, and the initial dissolved VOC concentrations were set equal to a value below the aqueous solubility of the VOC. Default parameters for these runs are given in Table 8.18.

In Figure 8.50 we see the effect of the mass transfer rate parameter λ for transport of VOC from the aqueous phase to the vapor phase. As expected, the rate of cleanup can be extremely adversely affected if λ is small. The value of λ should increase with increasing air-water interfacial area within the aquifer and with increasing uniformity of distribution of the injected gas flux. These are parameters that would be extremely difficult (if not impossible) to measure or estimate by laboratory experiments, and they may also be rather site-specific. One must therefore plan on experimental estimation of λ during the course of pilot-scale experiments at the site. This,

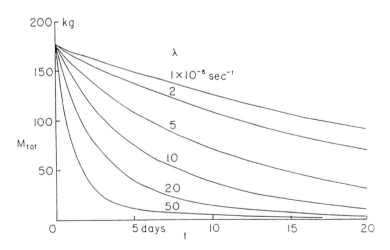

Figure 8.50 Plots of total residual VOC mass versus time (no DNAPL present); effect of aqueous/vapor mass transfer rate parameter. Default parameters are given in Table 8.18. [Reprinted from Wilson (1994) p. 85, by courtesy of Marcel Dekker, Inc.]

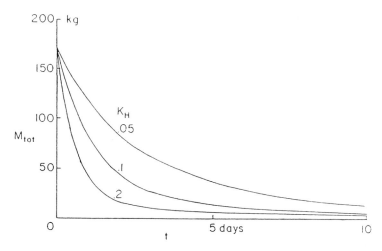

Figure 8.51 Plots of total residual VOC mass versus time (no DNAPL present); effect of Henry's constant K_H. Default parameters are given in Table 8.18. [Reprinted from Wilson (1994) p. 86, by courtesy of Marcel Dekker, Inc.]

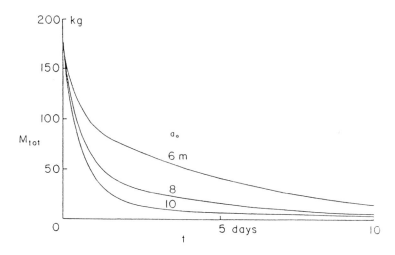

Figure 8.52 Plots of total residual VOC mass versus time (no DNAPL present); effect of the width parameter a_0 of the air distribution. Default parameters as in Table 8.18. [Reprinted from Wilson (1994) p. 87, by courtesy of Marcel Dekker, Inc.]

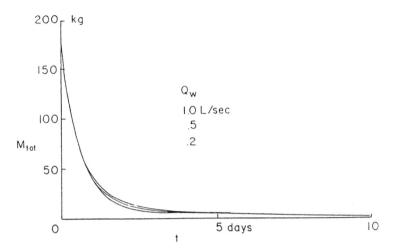

Figure 8.53 Plots of total residual VOC mass versus time (no DNAPL present); effect of the air-induced water circulation rate Q_w. $Q_w = 1.0, 0.5, 0.2$ L/sec, from the top down. Default parameters as in Table 8.18. [Reprinted from Wilson (1994) p. 87, by courtesy of Marcel Dekker, Inc.]

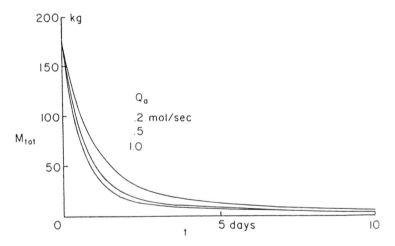

Figure 8.54 Plots of total residual VOC mass versus time (no DNAPL present); effect of the sparging air molar flow rate Q_a. In these runs the aqueous-vapor VOC mass transfer rate parameter λ is held constant at 0.001 sec^{-1}. [Reprinted from Wilson (1994) p. 88, by courtesy of Marcel Dekker, Inc.]

454

TABLE 8.18. Default Sparging Model Parameters, Dissolved VOC Runs.

Width of domain of interest	10 m
Thickness of aquifer	5 m
Length of horizontal sparging pipe	20 m
n_x, n_z	5, 5
Width of air sparging pattern at top of aquifer	10 m
Molar gas flow rate of sparging well	1 mol/sec
Air-induced water circulation rate	1 L/sec
Ambient temperature	20°C
Total porosity of aquifer medium	0.4
Water-filled porosity of aquifer medium	0.36
Aquifer medium density	1.7 mg/cm^3
Contaminant	Trichloroethylene
Density of contaminant	1.46 gm/cm^3
Water solubility of contaminant	1,100 mg/L
Henry's constant of contaminant (dimensionless)	0.20
Diffusivity of contaminant in porous medium	2×10^{-10} m^2/sec
Rate constant λ for aqueous phase/vapor transport	0.001 sec^{-1}
Initial DNAPL concentration	0 mg/kg
Initial dissolved VOC concentration	1,000 mg/L of water
Width of contaminated zone	8 m
Depth of contaminated zone below water table	3 m
Initial DNAPL droplet diameter	0.1 cm
Δt	100 sec

in turn, dictates that the pilot-scale sparging runs be of sufficient duration to allow one to get a reasonably accurate estimate of λ. Initial VOC removal rates (from water lying in the near vicinity of air bubble channels) may be much more rapid than the rate that will be sustained after the easily removable VOC has been sparged out and one is removing VOC that may have to move by aqueous advection or by diffusion into a region from which it may readily be air stripped.

In Figure 8.51 the effect of Henry's constant is shown. Henry's constants are known for all of the environmentally significant VOCs [see Montgomery and Welkom (1990), for example], so the effects of this parameter are readily predicted unless the aquifer medium contains substantial quantities of clay or other material that may sorb the contaminants. If such materials are present, one may expect this to result in the determination of small values of λ during pilot studies, since these will not distinguish between slow aqueous-vapor mass transport kinetics and slow desorption or diffusion kinetics.

The effects of the width parameter of the air distribution, a_0, are shown in Figure 8.52. The effects are not large until the width parameter is sufficiently small that portions of the zone of contamination lie outside the domain through which air is passing. When that occurs, as is the case with the curve for which $a_0 = 6$ m, the dissolved VOC must move by relatively slow advection into the domain through which the air is passing before it can be removed by stripping. This results in severe tailing; evidently, for expeditious cleanup one should design sparging systems in such a fashion that air is delivered to all of the volume of the aquifer that is contaminated.

Figure 8.53 shows that the effect of the rate of air-induced water circulation (Q_w) on the rate of dissolved VOC removal is relatively slight for the system modeled, as was found to be the case when DNAPL was being removed (see Figure 8.46). In these runs, as well as those shown in Figure 8.46, the zone of contamination lies entirely within the domain that is being aerated. One expects that the effect of Q_w on VOC removal rate would be somewhat larger if there were portions of the zone of contamination that were outside of the domain of aeration; this point was explored and found out, in fact, to be the case. The order of the curves appears at first to be counter-intuitive, in that increasing Q_w actually causes a slight *decrease* in the rate of VOC removal. Examination of the distribution of VOC in the simulated aquifer revealed that an increased water circulation rate carried VOC more rapidly out of the domain through which air was moving than was the case at a lower water circulation rate. Once this VOC has left the region in which sparging is actually occurring, it cannot be removed until it is carried back into that domain. Eventually, of course, this does occur, which is why the three curves merge after about four days' sparging.

The impact of sparging airflow rate Q_a on the VOC removal rate is shown in Figure 8.54. The results, however, require some interpretation. If the removal is strictly limited by the ability of the air to carry VOC (i.e., when there is local equilibrium between the vapor and the aqueous phase with respect to VOC transport), the removal rate is directly proportional to the airflow rate. If the airflow rate is large so that removal is strictly limited by the rate of mass transport between the aqueous and vapor phases, one expects the removal rate to be proportional to the air-water interfacial area so that the removal rate would again be proportional to the airflow rate. This dependence would be handled in our model by means of a proportional increase in the value of λ, the aqueous-vapor VOC mass transfer rate parameter. In Figure 8.54, however, the value of λ was held constant, and only Q_a was varied. We therefore observe a substantially weaker dependence of removal rate on Q_a than direct proportionality.

Figure 8.55 shows a set of plots in which Q_a is again varied (as in Figure 8.54); however, the values of λ were chosen to be proportional to the values of Q_a. We see, as anticipated, a much stronger dependence of removal rate on Q_a in Figure 8.55 than was observed in Figure 8.54.

In conclusion, we note that (1) the model presented requires a rather limited number of input parameters, (2) it runs readily on currently

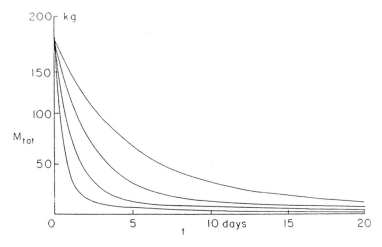

Figure 8.55 Plots of total residual VOC mass versus time (no DNAPL present); effect of the sparging air molar flow rate Q_a. In these runs the aqueous-vapor VOC mass transfer rate parameter is taken proportional to Q_a. Q_a = 0.2, 0.5, 1.0 mol/sec; = 0.0002, 0.0005, 0.001 sec^{-1}, top to bottom. [Reprinted from Wilson (1994) p. 89, by courtesy of Marcel Dekker, Inc.]

available microcomputers, (3) it permits the modeling of DNAPL solution kinetics and aqueous-vapor mass transport kinetics, and (4) the trends shown by the numerical results appear quite reasonable in the light of physical intuition. It is hoped that the model will prove useful as an evaluation and design tool to people involved with the remediation of aquifers contaminated with VOCs and that in the future data from a broad range of sites will provide field validation.

SPARGING WITH A VERTICAL WELL: INCLUSION OF DNAPL SOLUTION KINETICS AND DIFFUSION OF AQUEOUS VOC FROM LOW-PERMEABILITY STRUCTURES

In this section the sparging of aquifers contaminated with VOCs is examined by means of physically rather realistic model, which includes the kinetics of solution of NAPL droplets and diffusion of VOC from porous layers of low permeability. The well configuration is that of a single vertical well screened only near the bottom. The model runs readily on microcomputers equipped with 80486 microprocessors and operating at 50 MHz or faster. This model (Gómez-Lahoz et al., 1994) and a somewhat similar model for sparging of VOCs from aquifers by means of a buried horizontal slotted pipe (Wilson et al., 1994) were developed to provide more realistic approaches to what appears to be a major bottleneck in

SVE, pump-and-treat operations, and sparging-diffusion of VOC from water-saturated porous domains containing dissolved VOC.

In the following sections a somewhat abbreviated version of the analysis leading to the differential equations constituting the model (to avoid excessive repetition of material covered previously) is presented. This is followed by a discussion of the results of calculations done using the model: first, a comparison of sparging by means of a single vertical well with sparging by means of a horizontal slotted pipe. The dependence of cleanup rates on a few of the model parameters is then explored. The section closes with an examination of the rebound of VOC concentration in the mobile aqueous phase when the sparging well is shut down after various periods of operation. A short section on conclusions then completes the discussion.

Theoretical

The Physical System

A schematic diagram of a sparging set-up having the configuration of a single vertical well is shown in Figure 8.56. We assume that the aquifer medium is homogeneous and isotropic, so that the system has axial symmetry and we may work in cylindrical coordinates (r,z). We assume that there are low-permeability porous lenses (clay, till, silt) distributed throughout the aquifer, which may contain VOC either dissolved in the immobile water in the porosity of the lenses or actually present as non-aqueous phase liquid (NAPL) droplets, also distributed in the lenses in this model. Mass transport kinetics limitations can occur via the rate of solution of the droplets and/or the rate of diffusion of dissolved VOC through the immobile water to the mobile water flowing through the more permeable matrix in which the low-permeability lenses are located.

The injected air is assumed to be at local equilibrium with the mobile water with respect to VOC mass transport, and this equilibrium is assumed to be governed by Henry's Law.

The calculation of fluid flow in such multiphasic systems is a formidable task, which we again avoid by postulating the z-components of the air and water fluxes and then calculating the r-components from conservation requirements. The gas flux is handled in the same way as used earlier in this chapter. The equations used here to describe advective transport in the mobile water are slightly more complex than in the previous model and will be treated fully. The equations describing solution of NAPL droplets and diffusion through the immobile water in the clay lenses to the mobile

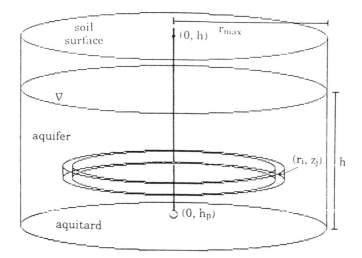

Figure 8.56 Geometry and notation of a single vertical sparging well screened near the bottom. [Reprinted from Gómez-Lahoz et al. (1994) p. 1509, by courtesy of Marcel Dekker, Inc.]

water are virtually identical to those derived in connection with previous models and will be presented without proof.

The Water Circulation Field

We postulate for the z-component of the water superficial velocity $\mathbf{v}$ the expression

$$v_z = Bz(h - z)(b - r) \exp(-cr) \tag{144}$$

where B, b, and c are constants to be determined or assigned. This expression makes v_z positive near the well, negative at distances greater than b from the well, zero at $z = 0$ and at $z = h$, and 0 in the limit of large r. These properties are what one would expect for the water circulation around a sparging well.

It is necessary that the total flux of water through any horizontal plane be equal to zero, since there are no sources or sinks of water in the vicinity. Therefore, we must have

$$\int_0^{2\pi} \int_0^{\infty} v_z r \, dr \, d\theta = 0 \tag{145}$$

Substitution of Equation (144) into Equation (145) yields the requirement that

$$\int_0^\infty (b - r)r \cdot \exp(-cr)dr = 0 \qquad (146)$$

from which we find that c must be given by

$$c = 2/b \qquad (147)$$

and so

$$v_z = Bz(h - z)(b - r) \cdot \exp(-2r/b) \qquad (148)$$

The r-component of the water circulation field is obtained by the method employed above to get q_r for the gas flow field. Since the field is conservative,

$$\nabla \cdot \mathbf{v} = 0 \qquad (149)$$

This gives

$$r^{-1} \frac{\partial}{\partial r}(rv_r) = -\frac{\partial v_z}{\partial z} = -B(h - 2z)(b - r) \cdot \exp(-2r/b) \qquad (150)$$

Multiplication by r, integration from 0 to r, and division by r then yields

$$v_r = -(Bb/2)(h - 2z)r \cdot \exp(-2r/b) \qquad (151)$$

This completes the calculation of the superficial velocity field for the water circulation. A representative set of streamlines for the water circulation is shown in Figure 8.57; the time periods for points on these streamlines to make a complete circuit are indicated in the caption.

Advective Transport of VOC

In sparging, advective transport takes place in both the gaseous and the mobile aqueous phases. The ring-shaped volume elements that will be

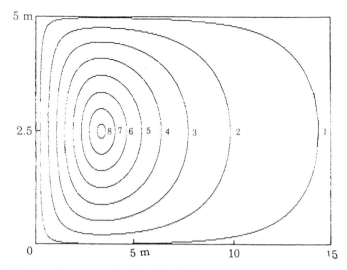

Figure 8.57 Streamlines, water circulation flow field. The times required to make a circuit for each streamline are as follows:

Trajectory Number	Transit Time, Days
1	185.32
2	20.48
3	8.42
4	5.16
5	3.84
6	3.22
7	2.92
8	2.80

The distance b from the axis of the system to the stagnation point is 3.33 m. The water circulation scale factor $B = 0.4/86,400$ m^{-2}sec^{-1}, and $\omega^m = 0.2$. [Reprinted from Gómez-Lahoz et al. (1994) p. 1514, by courtesy of Marcel Dekker, Inc.]

used are shown in Figure 8.56. We assume that the gas phase and the mobile aqueous phases are in local equilibrium, which gives

$$C_{ij}^g = K_H C_{ij}^m \tag{152}$$

Here

C_{ij}^g = gas phase VOC concentration in the ijth volume element, kg/m^3
C_{ij}^m = VOC concentration in the mobile aqueous phase in the ijth volume element, kg/m^3
K_H = Henry's constant for the VOC, dimensionless

The volume of the ijth volume element is given by

$$V_{ij} = (2i - 1)(\Delta r)^2 \Delta z \tag{153}$$

The top, bottom, inner, and outer surfaces of this volume element are given by

$$S_{ij}^T = S_{ij}^B = (2i - 1)(\Delta r)^2 \tag{154}$$

$$S_{ij}^I = 2\Delta(i - 1)\Delta r \Delta z \tag{155}$$

$$S_{ij}^O = 2\Delta i \Delta r \Delta z \tag{156}$$

Let

ω^m = mobile water porosity
ω^i = immobile water porosity

A mass balance on VOC transported in the gas phase then gives

$$\omega^m \Delta V_{ij} \left[\frac{dC_{ij}^m}{dt}\right]_{gas} = K_H S_{ij}^I U_{ij}^I [S(U^I)C_{i-1,j}^m + S(-U^I)C_{ij}^m]$$

$$+ K_H S_{ij}^O U_{ij}^O [-S(-U^O)C_{i+1,j}^m - S(U^O)C_{ij}^m]$$

$$+ K_H S_{ij}^B U_{ij}^B \left[S(U^B)C_{i,j-1}^m \frac{P[(j - 1)\Delta z]}{P[(j - 3/2)\Delta z]} + S(-U^B)C_{ij}^m \frac{P[(j - 1)\Delta z]}{P[(j - 1/2)\Delta z]}\right]$$

$$+ K_H S_{ij}^T U_{ij}^T \left[-S(-U^T)C_{i,j+1}^m \frac{P[j\Delta z]}{P[(j + 1/2)\Delta z]} - S(U^T)C_{ij}^m \frac{P[j\Delta z]}{P[(j - 1/2)\Delta z]}\right]$$

$$\tag{157}$$

where

$$S(u) = 0, u < 0$$
$$= 1, u \geq 0, \text{ and}$$

$$U_{ij}^I = U_r[(i - 1)\Delta r, (j - /2)\Delta z] \tag{158}$$

$$U_{ij}^O = U_r[i\Delta r, (j - 1/2)\Delta z] \tag{159}$$

$$U_{ij}^B = U_z[(i - 1/2)\Delta r, (j - 1)\Delta z] \tag{160}$$

$$U_{ij}^T = U_z[(i - 1/2)\Delta r, j\Delta z] \tag{161}$$

give the gas fluxes at the four surfaces of the volume element. Note that here we have included pressure correction terms to take into account the expansion of the gas as it rises from the bottom surface of the volume element to its center or from the center to the top.

Similarly, a mass balance on VOC transported in the mobile aqueous phase gives

$$
\omega^m \Delta V_{ij} \left[\frac{dC_{ij}^m}{dr}\right]_{water} = S_{ij}^I v_{ij}^I [S(v^I)C_{i-1,j}^m + S(-v^I)C_{ij}^m]
$$

$$
+ S_{ij}^O v_{ij}^O [-S(-v^O)C_{i+1,j}^m - S(v^O)C_{ij}^m] + S_{ij}^B v_{ij}^B [S(v^B)C_{i,j-1}^m
$$

$$
+ S(-v^B)C_{ij}^m] + S_{ij}^T v_{ij}^T [-S(-v^T)C_{i,j+1}^m - S(v^T)C_{ij}^m] \qquad (162)
$$

where

$$
v_{ij}^I = v_r[(i - 1)\Delta r, (j - 1/2\Delta z] \qquad (163)
$$

$$
v_{ij}^O = v_r[i\Delta r, (j - 1/2)\Delta z] \qquad (164)
$$

$$
v_{ij}^B = v_z[(i - 1/2)\Delta r, (j - 1)\Delta z] \qquad (165)
$$

$$
v_{ij}^T = v_z[(i - 1/2)\Delta r, j\Delta z] \qquad (166)
$$

Solution of NAPL Droplets

The solution of NAPL droplets distributed within the low-permeability lenses is handled as in the earlier work. Transcription to the notation used with the cylindrical geometry employed here gives

$$
\frac{dm_{ijk}}{dt} = -\frac{3\Delta V_{ij}C_0^N D(C_{sat} - C_{ijk}^i)(m_{ijk}/m_{ij}^O)^{1/3}}{n_u a_0'^2 \varrho_{voc}[1 - (a_0'/d)(m_{ijk}/m_{ij}^O)^{1/3}]} \qquad (167)
$$

Here

n_u = number of slabs into which the low-permeability clay lenses are partitioned for mathematical analysis

m_{ijk} = mass of NAPL in the kth slab of immobile water in the ijth volume element, kg

$m_{ij}^O = C_0^N V_{ij}/n_u$ = initial mass of NAPL in the kth slab, kg

C_0^N = initial NAPL concentration in the contaminated region, kg/m³

a_0' = initial radius of NAPL droplets, m

$d - a$ = boundary layer thickness around NAPL droplet, m, d estimated from Equation (168)

D = diffusion constant of VOC in the immobile aqueous phase, m²/sec, this includes the molecular diffusivity of the VOC in water, the tortuosity, etc.

C_{sat} = aqueous solubility of VOC, kg/m³

ϱ_{voc} = density of VOC, kg/m³

C_{ijk}^i = VOC concentration in the immobile aqueous phase in the kth slab of the ijth volume element, kg/m³

In Wilson et al. (1994) it was shown that a reasonable value for the outer radius of the boundary layer thickness around an NAPL droplet is given by

$$d = a_0' \left[\frac{\pi \omega \varrho_{voc}}{6 C_0^N} \right]^{1/3} \tag{168}$$

Diffusion of VOC through the Immobile Water in the Clay Lenses

This is handled by partition of the pancake-shaped clay lenses into $2n_u$ slabs and applying Fick's Laws of diffusion to each slab [see Figures 8.58(a) and (b)]. The resulting equations are

$$\frac{dC_{ijk}^i}{dt} = \frac{D}{(\Delta u)^2 \nu_{clay}} \cdot (C_{ij,k+1}^i - 2C_{ijk}^i + C_{ij,k-1}^i)$$

$$- \frac{n_u}{\omega^i \Delta V_{ij}} \cdot \frac{dm_{ijk}}{dt}, \, k = 2,3,\dots,n_u - 1 \tag{169}$$

$$\frac{dC_{ijnu}^i}{dt} = \frac{D}{(\Delta u)^2 \nu_{clay}} \cdot (-C_{ijnu}^i + C_{ij,nu-1}^i) - \frac{n_u}{\omega^i \Delta V_{ij}} \cdot \frac{dm_{ijnu}}{dt} \tag{170}$$

and

$$\frac{dC_{ijl}^i}{dt} = \frac{D}{(\Delta u)^2 \nu_{clay}} \cdot [C_{ij2}^i - C_{ijl}^i + 2(C_{ij}^m - C_{ijl}^i)]$$

$$- \frac{n_u}{\omega^i \Delta V_{ij}} \cdot \frac{dm_{ijl}}{dt} \tag{171}$$

Here

$2l$ = thickness of the clay lenses, m

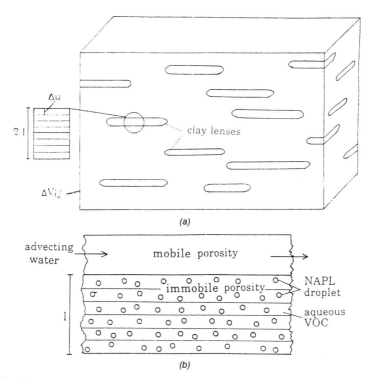

Figure 8.58 (a) Distribution of low-permeability porous lenses in a volume element. (b) Partitioning of half of a lens into slabs, distribution of NAPL droplets within the slabs, and relationship of the lens to the mobile porosity. [Reprinted from Wilson et al. (1994b) p. 1406, by courtesy of Marcel Dekker, Inc.]

Δu = thickness of a slab of immobile water in the clay lenses, m
($\Delta u = l/n_u$)
ν_{clay} = porosity of clay in the low-permeability lenses

Completion of the Mobile Aqueous Phase Material Balance

The term modeling diffusion of VOC from the first immobile aqueous layer in the lenses ($k = 1$) into the mobile aqueous phase is

$$\omega^m \Delta V_{ij}\left[\frac{dC_{ij}^m}{dt}\right]_{diff} = \frac{\omega^i \Delta V_{ij} D}{l\nu_{clay}(\Delta u/2)} \cdot (C_{ijl}^i - C_{ij}^m) \qquad (172)$$

The overall material balance for VOC in the mobile aqueous phase is then given by

$$\frac{dC_{ij}^m}{dt} = \left[\frac{dC_{ij}^m}{dt}\right]_{gas} + \left[\frac{dC_{ij}^m}{dt}\right]_{water} + \left[\frac{dC_{ij}^m}{dt}\right]_{diff} \tag{173}$$

The Model

The model then consists of the following differential equations. The change of mass of NAPL in the kth slab of the ijth volume element is governed by Equation (167). The VOC concentration in the immobile water in the kth slab of the ijth volume element is calculated from Equations (169), (170), or (171). Finally, the VOC concentration in the mobile water in the ijth volume element is calculated by means of Equation (173), which is, in turn, composed of Equations (157), (162), and (172).

Initialization consists of specifying the model parameters and the dimensions of the contaminated zone and the initial VOC concentration in the contaminated zone. The differential equations were then integrated forward in time by the simple Euler formula, since RAM restrictions precluded using a more sophisticated method.

The total mass of residual VOC in the domain of interest at any time t is given by

$$M_{tot} = \sum_{i=1}^{n_r}\sum_{j=1}^{n_z}\left[\Delta V_{ij}\omega^m C_{ij}^m + \sum_{k=1}^{n_u} m_{ijk} + (\omega\Delta V_{ij}/n_u)C_{ijk}^i\right] \tag{174}$$

In the runs presented below a reduced total mass is plotted

$$M' = M_{tot}(t)/M_{tot}(0)$$

We shall also be interested in the rebound of the VOC concentration in the mobile aqueous phase after the sparging well has been shut off. The modeling equations for the system after sparging has been stopped are exactly the same as those given above, except that the gaseous and aqueous advection terms are dropped. One can then plot reduced concentrations at various points within the domain, defined by $C_w' = C_{ij}^m/C_{sat}$, where w is an index indicating the values of i and j being used.

Results

Default parameters for the runs are given in Table 8.19. Variations from this set of values are indicated in the text and figure captions as they occur.

TABLE 8.19. Default Parameters for Sparging Runs, Single Vertical Well.

Thickness of aquifer	5 m
Height of screened section of well above aquitard	0.5 m
Molar airflow rate to well	0.4 mol/sec
(Volumetric airflow rate)	20.7 SCFM
Radius of influence of air at top of aquifer, a_0	5 m
Distance of center of water circulation from well axis, b	3.33 m
Radius of domain of interest, r_{max}	15 m
Scale factor for water circulation, B	0.4/86,400 (1/m^2 sec)
Temperature	25°C
Mobile water-filled porosity, ω^m	0.2
Immobile water-filled porosity, ω^i	0.2
Soil density, ϱ	1.7 gm/cm^3
Identity of VOC	Trichloroethylene, TCE
Density of VOC, ϱ_{voc}	1.46 gm/cm^3
Aqueous solubility of VOC	1,100 mg/L
Henry's constant of VOC (dimensionless)	0.2821
Diffusivity of VOC in immobile water, D	1×10^{-10} m^2/sec
n_r	15
n_z	5
Thickness of clay lenses, $2l$	1 cm
Porosity of clay lenses, ν_{clay}	0.4
Number of slabs into which lenses are partitioned for mathematical analysis	5
Initial total VOC concentration	2,000 mg/kg
Initial NAPL droplet diameter, $2a_0'$	0.1 cm
Radial distance to which contamination extends	4 m
Depth in aquifer to which contamination extends	4 m
Δt	225 sec

Before exploring other aspects of the model, one must determine how large the radius of the domain of interest must be to avoid spurious losses of VOC by the washing of VOC out of the domain of interest by the water circulation. Physically, of course, this VOC is not removed from the aquifer but is only spread out; the domain of interest must be chosen large enough so that this washout effect is negligible. Runs in which the airflow rate is set equal to zero while water circulation takes place allow the investigation of this point. In Figure 8.59 we see that, for runs made with the default parameters, a domain radius of 10 m permits a little washout to occur during a 30-day run but that a domain radius of 15 m shows no evidence of washout; therefore, a domain radius of 15 m was used throughout.

For a horizontal slotted pipe sparging well configuration, we found earlier (Wilson et al., 1994) that the half-width of the domain of interest had to be at least 25 m to avoid significant washout. The difference between the two geometries appears to be associated with the exponential attenuation factors of the water circulation velocity components. For the horizontal

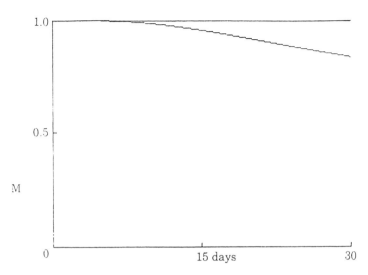

Figure 8.59 Plots of M' versus time; the "wash-out" effect. The airflow rate $Q = 0$; $r_{max} = 15$, 10 m from the top down. Other parameters as in Table 8.19. [Reprinted from Wilson et al. (1994b) p. 1421, by courtesy of Marcel Dekker, Inc.]

slotted pipe system the derived exponential attenuation factor is $\exp(-x/b)$, where b is the distance of the center of circulation from the axis of the system. For the vertical well configuration considered here, one obtains an exponential attenuation factor of $\exp(-2r/b)$, which attenuates with distance from the well substantially more rapidly.

Figure 8.60 shows a comparison of sparging runs made with a single vertical well (v) and a single horizontal well (h) with identical gas flow rates, and values of a_0 (maximum distance to which the gas flux extends laterally from the well), b, etc. Surprisingly, the vertical well performs substantially more efficiently than does the horizontal slotted pipe, even when one takes into account the smaller volume of the domain of interest of the vertical well (201 m³ compared to 320 m³ for the horizontal well). Examination of the mobile aqueous VOC concentration values in the domains of interest provides the explanation for this result. The influence of the water circulation field for the horizontal well extends out laterally much farther than does that of the water circulation field for the vertical well, as a result of the difference in exponential attenuation factors. This results in substantial quantities of VOC being carried far out of the aeration zone of the horizontal well. Before this VOC can be removed, it must be carried by water advection back into the aeration zone, which is a slow process. This is responsible for the extensive tailing shown by the horizontal slotted pipe well results. In soil vapor extraction the relative efficiencies of the two well configurations are just the opposite; in SVE there is no water circulation to spread the VOC out of the aeration zone.

The effect of linked airflow and water circulation rates is shown in Figure 8.61. Here the water circulation rate parameter was taken to be proportional to the airflow rate, on the basis of some quite preliminary data on the aeration of a bench-scale water-saturated sand bed. As one would expect, increased advection results in faster removal. However, this process is eventually limited by the rates of the solution and diffusion processes, so the rate of cleanup is less than proportional to the airflow rate.

The effect of well depth is shown in Figure 8.62. As the well depth is decreased, the expected decrease in removal rate and increase in the extent of tailing toward the end of the cleanup are seen. As was observed with horizontal sparging wells, it is unwise to attempt to economize by drilling shallow wells. Water circulation is not an efficient way to move VOC from the contaminated zone into the zone of aeration.

The effect of the maximum radius of the airflow field is shown in Figure 8.63. A radius that is too small leaves some of the contaminated zone unaerated, so the cleanup rate is reduced somewhat. A radius that is too large results in air being wasted by moving through regions containing no contaminant, so, again, the cleanup rate is reduced.

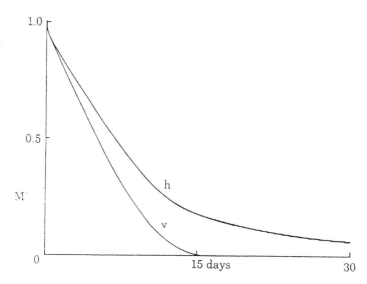

Figure 8.60 Plots of M' versus time; comparison of the performances of a single vertical well v (all parameters as in Table 8.20) and a single horizontal slotted pipe well h (length of pipe $= 10$ m, airflow rate $= 0.4$ mol/sec, water circulation scale factor $B' = 0.05/86,400$ m^{-2}sec^{-1}, other parameters as in Table 8.19). [Reprinted from Gómez-Lahoz et al. (1994) p. 1521, by courtesy of Marcel Dekker, Inc.]

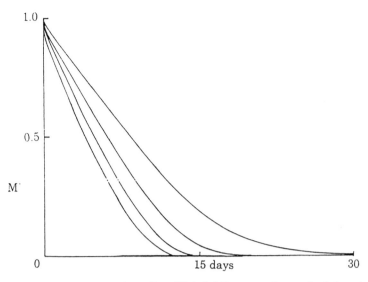

Figure 8.61 Plots of M' versus time; effect of linked airflow rate and water circulation rate. (Q, $86,400B$) = (0.1, 0.1), (0.2, 0.2), (0.4, 0.4), and (1.0 mol/sec, 1.0 m^{-2}sec^{-1}), from the top down. [Reprinted from Gómez-Lahoz et al. (1994) p. 1522, by courtesy of Marcel Dekker, Inc.]

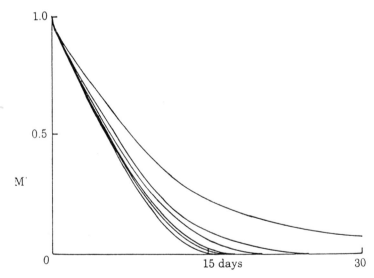

Figure 8.62 Plots of M' versus time; effect of well depth. Height of well above the bottom of the aquifer = 3.0, 2.5, 2.0, 1.5, 1.0, and 0.5 m from the top down. [Reprinted from Gómez-Lahoz et al. (1994) p. 1522, by courtesy of Marcel Dekker, Inc.]

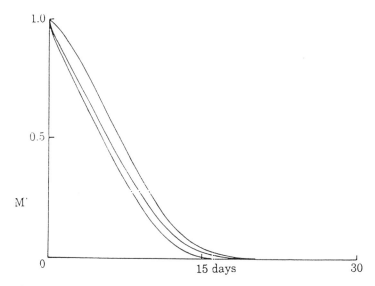

Figure 8.63 Plots of M' versus time; effect of maximum radius of air distribution a_0. $a_0 = 3, 8,$ and 5 m from the top down. [Reprinted from Gómez-Lahoz et al. (1994) p. 1523, by courtesy of Marcel Dekker, Inc.]

Figure 8.64 shows the very large effect that the thickness of the low-permeability lenses has on the rate of cleanup. Initially, there is a period of a few hours during which removal rates are rapid, as dissolved VOC is removed from the mobile water by the sparging air. Shortly thereafter, however, diffusion from the clay lenses becomes the rate-limiting factor, and cleanup rates decrease markedly, depending on the thickness of the clay lenses. The thicknesses of the lenses here range from 0.5 (rapid cleanup) to 3.0 cm (quite slow cleanup). One would be well advised to check the well logs for a site for indications of the presence of such structures.

The effect of NAPL droplet radius a_0' on the cleanup rate is also quite large, with cleanup rates decreasing with increasing droplet size (and correspondingly decreasing NAPL-water interfacial area). This is seen in Figure 8.65. The effects of increasing lens thickness and increasing droplet size are extremely similar, and it would be difficult to distinguish between them on the basis of field pilot studies. However, since the effects are so similar, distinguishing between them is probably unnecessary for practical purposes.

In Figures 8.66–8.69 we examine the phenomenon of rebound—the increase in VOC concentration in groundwater (or soil gas) after a remediation operation (sparging, pump-and-treat, SVE) has been terminated.

Figure 8.66 shows plots of the reduced VOC mass and two reduced

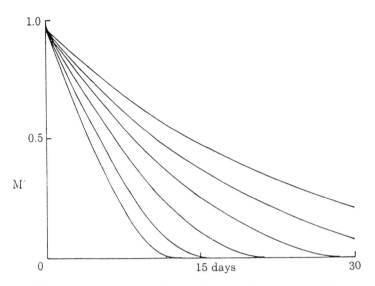

Figure 8.64 Plots of M' versus time; effect of thickness of the low-permeability lenses, $2l$. $2l = 3.0, 2.5, 2.0, 1.5, 1.0,$ and 0.5 cm from the top down. [Reprinted from Gómez-Lahoz (1994) p. 1524, by courtesy of Marcel Dekker, Inc.]

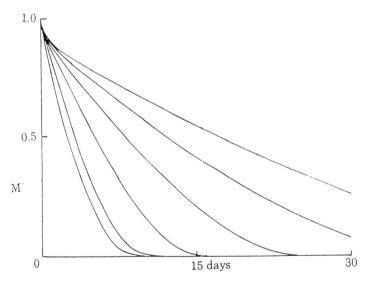

Figure 8.65 Plots of M' versus time; effect of diameter of NAPL droplets, $2a_0'$. $2a_0' = 0.01, 0.02$ (superimposed), $0.05, 0.10, 0.15, 0.20, 0.25$ cm from the bottom up. [Reprinted from Gómez-Lahoz et al. (1994) p. 1524, by courtesy of Marcel Dekker, Inc.]

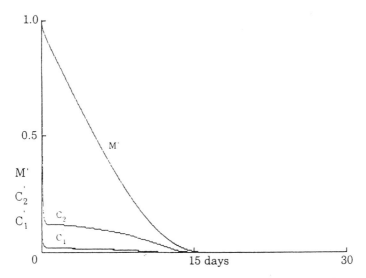

Figure 8.66 Plots of M', C'_1, and C'_2 versus time. C'_1 is the reduced concentration of VOC in the mobile water at the top center of the zone of contamination. C'_2 is the reduced concentration of VOC in the mobile water at the bottom outer edge of the zone of contamination. All parameters as in Table 8.19. [Reprinted from Gómez-Lahoz et al. (1994) p. 1525, by courtesy of Marcel Dekker, Inc.]

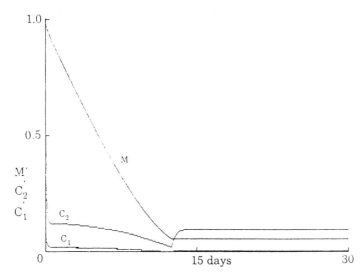

Figure 8.67 Plots of M', C'_1, and C'_2 versus time. The sparging well was shut down after 12.5 days of operation. [Reprinted from Gómez-Lahoz et al. (1994) p. 1526, by courtesy of Marcel Dekker, Inc.]

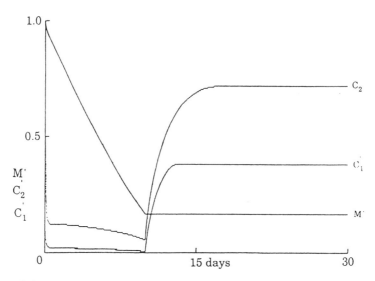

Figure 8.68 Plots of M', C_1', and C_2' versus time. The sparging well was shut down after 10 days of operation. [Reprinted from Gómez-Lahoz et al. (1994) p. 1526, by courtesy of Marcel Dekker, Inc.]

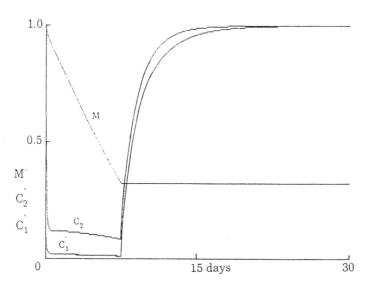

Figure 8.69 Plots of M', C_1', and C_2' versus time. The sparging well was shut down after 7.5 days of operation. [Reprinted from Gómez-Lahoz et al. (1994) p. 1527, by courtesy of Marcel Dekker, Inc.]

mobile water concentrations as functions of time. The concentration C_1' is the concentration of VOC in the mobile water at the top center of the zone of contamination; C_2' is the mobile water VOC concentration at the bottom outer edge of the zone of contamination. In Figure 8.66 the run is carried out for 30 days; cleanup is essentially complete after 15 days. C_1' is always less than C_2', since the top center of the zone of contamination receives a much larger air flux than does the bottom outer edge.

In Figure 8.67 the sparging well is shut off after 12.5 days, shortly before remediation is complete. The rebound of C_1' is negligible, indicating that this portion of the zone of contamination is essentially cleaned up. Rebound of C_2', however, is appreciable, although not large. Evidently, the aquifer near the lower outer edge of the zone of contamination no longer contains NAPL (or rebound would be larger), but there is still dissolved VOC diffusing from the immobile water in the clay lenses.

Rebound is quite a bit larger in Figure 8.68, in which the sparging well is shut down after 10 days of operation. Both C_1' and C_2' show substantial rebound, with C_1' recovering to a value less than that to which C_2' recovers because of the difference in air flux through the two portions of the contaminated zone.

Rebound in Figure 8.69, for which run the sparging well was turned off after 7.5 days of operation, is the maximum possible for both C_1' and C_2', indicating that NAPL is still present in both portions of the contaminated zone.

CONCLUSIONS

The following conclusions can be drawn from the results obtained with this sparging model:

- Somewhat surprisingly, vertical sparging wells appear to be rather more efficient than horizontal slotted pipe sparging wells, in that one does not have nearly as much tailing of the cleanup with the vertical wells as one has with the horizontal wells. This is apparently due to the increased range of the water circulation field for horizontal wells as compared to that for vertical wells.
- Sparging wells should be drilled all the way through the zone of contamination to achieve maximum cleanup rates.
- Increasing airflow rates result in increased cleanup rates, but diffusion and solution mass transfer rates eventually become rate-limiting. At that point further increases in airflow rate are futile.
- NAPL droplet size and low-permeability lens thickness are extremely important in limiting diffusion and solution mass transfer rates.

- Diffusion and solution kinetics result in a rebound in the mobile water VOC concentrations after a sparging well is shut down. The magnitude of this rebound depends on the location at which the water sample is taken and the extent to which the overall cleanup has progressed. Rebound can be expected to be largest in those contaminated regions that receive the lowest flux of air.

REFERENCES

Ahlfeld, D., A. Dahmani, M. Farrell, and W. Ji, 1992, "Detailed Field Measurements of an Air Sparging Pilot Test," *Ground Water* (submitted).

Angell, K. G., 1992, "In situ Remedial Methods: Air Sparging," *National Environmental Journal,* 2:20.

Ardito, C. P. T and J. F. Billings, 1990, "Alternative Remediation Strategies: The Subsurface Volatilization and Ventilation System," *Proc., Conf. on Petroleum Hydrocarbons and Organic Chemicals in Groundwater,* Oct. 31–Nov. 2, p. 281.

Billings, J. F., A. I. Cooley, and G. K. Billings, 1994, "Microbial and Carbon Dioxide Aspects of Operating Air-Sparging Sites," in *Air Sparging for Site Remediation,* R. E. Hinchee, ed., Lewis Publishers, Chelsea, MI, p. 56.

Boersma, P., F. S. Petersen, P. Newman, and R. Huddleston, 1993, "Use of Groundwater Sparging to Effect Hydrocarbon Biodegradation," *Proc., Petroleum Hydrocarbons and Organic Chemicals in Ground Water: Prevention, Detection and Restoration,* Houston, TX, 17:557.

Bohler, U., J. Brauns, H. Hotzl, and M. Hahold, 1990, "Air Injection and Soil Air Extraction as a Combined Method for Cleaning Contaminated Sites–Observations from Test Sites in Sediments and Solid Rocks," in *Contaminated Soil '90,* F. Arendt, M. Hinsenveld, and W. J. van den Brink, eds., Kluwer Academic Publishers, the Netherlands.

Brown, R. A., Undated, "Air Sparging: A Primer for Application and Design," Groundwater Technology, Inc., 310 Horizon Center Dr., Trenton, NJ 08691.

Brown, R. A., 1992, "Air Sparging Projection Delaware," *Underground Tank Technology Update,* 6(3):8.

Brown, R. A. and R. Fraxedas, 1992, "Air Sparging–Extending Volatilization to Contaminated Aquifers," *Symposium on Soil Venting,* April 29–May 1, 1991, Houston, TX, U.S. EPA Report No. EPA/600/R-92/174, p. 249.

Brown, R. A. and F. Jasiulewicz, 1992, "Air Sparging Used to Cut Remediation Costs," *Pollution Engineering* (July):52.

Brown, R. A., C. Herman, and E. Henry, 1991a, "The Use of Aeration in Environmental Cleanups," *Proc., Haztech International Pittsburgh Waste Conference,* Pittsburgh, PA.

Brown, R. A., R. J. Hicks, and P. M. Hicks, 1994, "Use of Air Sparging for in situ Bioremediation," in *Air Sparging for Site Remediation,* R. E. Hinchee, ed., Lewis Publishers, Chelsea, MI, p. 38.

Brown, R. A., R. E. Payne, and P. F. Perlwitz, 1992, "Air Sparging Pilot Testing at a Site Contaminated with Gasoline," *Proc., Petroleum Hydrocarbons and Organic Chemicals in Ground Water: Prevention, Detection and Restoration,* Houston, TX, 14:429.

Brown, R. A., E. Henry, C. Hermann, and W. Leonard, 1991b. "The Use of Aeration in Environmental Cleanups," *Proc., Conf. on Petroleum Hydrocarbons and Organic Chemicals in Groundwater,* Nov. 20–22, p. 265.

Burchfield, S. D. and D. J. Wilson, 1993, "Groundwater Cleanup by in situ Sparging. IV. Removal of DNAPL by Sparging Pipes," *Separ. Sci. Technol.,* 28:2529.

Chandler, P., L. Beabes, and R. Banary, 1992, "Air Sparging Project in Cleveland, Ohio," *Underground Tank Technology Update,* 6(3):6.

Clarke, A. N., R. D. Norris, and D. J. Wilson, 1993, "Saturated Zone Remediation of VOCs through Sparging," in *Hazardous Waste Site Soil Remediation: Theory and Application of Innovative Technologies,* D. J. Wilson and A. N. Clarke, eds., Marcel Dekker, New York, NY.

Conrad, S. H., J. L. Wilson, W. R. Mason, and W. J. Peplinski, 1992, "Visualization of Residual Organic Liquid Trapped in Aquifers," *Water Resources Research,* 28:467.

Fairbanks, P. E., L. Pennington, and J. H. Rabaideau, 1993, "Air Sparging in Reduced Permeability Sediments," *Proc., Petroleum Hydrocarbons and Organic Chemicals in Ground Water: Prevention, Detection and Restoration,* Houston, TX, 17:561.

Feenstra, S. and J. A. Cherry, Undated, "Dense Organic Solvents in Ground Water: An Introduction," in *Dense Chlorinated Solvents in Ground Water,* Institute for Ground Water Research, University of Waterloo, Waterloo, Ont., Progress Report No. 0863985.

Feenstra, S. and J. A. Cherry. 1989, "Subsurface Contamination by Dense Non-Aqueous Phase (DNAPL) Liquid Chemicals," *Proc., International Groundwater Symposium on Hydrogeology of Cold and Temperate Climates and Hydrogeology of Mineralized Zones,* May 1–5, 1988, Halifax, Nova Scotia, p. 61.

Feenstra, S., D. M. Mackay, and J. A. Cherry, 1991, "Method for Assessing Residual NAPL Based on Organic Chemical Concentrations in Soil Samples," *Ground Water Monitoring Review,* 11:128.

Felten, D. W., M. C. Leahy, L. J. Bealer, and B. A. Kline, 1992, "Case Study: Remediation Using Air Sparging and Soil Vapor Extraction," *Proc., Conf. on Petroleum Hydrocarbons and Organic Chemicals in Ground Water: Prevention, Detection and Restoration,* Nov. 4–6, Eastern Regional Ground Water Issues, Houston, TX, 14:395.

Gómez-Lahoz, C., J. M. Rodríguez-Maroto, and D. J. Wilson, 1994, "Groundwater Cleanup by in situ Sparging. VII. VOC Concentration Rebound Caused by Diffusion after Shutdown," *Separ. Sci. Technol.,* 29:1509.

Gudemann, H. and D. Hiller, 1988, "In situ Remediation of VOC Contaminated Soil and Groundwater by Vapor Extraction and Groundwater Aeration," *Proc., Third Annual Haztech International Conference,* Cleveland, OH.

Gvirtzman, H. and S. M. Gorelick, 1992, "The Concept of in-situ Vapor Stripping for Removing VOCs from Groundwater," in *Transport in Porous Media,* Kluwer Academic Publishers, the Netherlands.

Herrling, B. and W. Buerman, 1990, "A New Method for in-situ Remediation of Volatile Contaminants in Groundwater – Numerical Simulation of the Flow Regime," in *Computational Methods in Subsurface Hydrology,* G. Gambolati, A. Rinaldo, C. A. Brebbia, W. G. Gray, and G. F. Pinder, eds., Springer, Berlin, p. 299.

Herrling, B., W. Buermann, and J. Stamm, 1991, "In situ Remediation of Volatile Contaminants in Groundwater by a New System of 'Vacuum-Vaporizer-Wells,' " in *Subsurface Contamination by Immiscible Fluids,* K. U. Weyer, ed., A. A. Balkema, Rotterdam.

Herrling, B., J. Stamm, B. J. Alesi, and P. Brinnel, 1992, "Vacuum-Vaporizer-Wells for in situ Remediation of Volatile and Strippable Contaminants in the Unsaturated and Saturated Zone," *Symposium on Soil Venting*, April 29–May 1, 1991, Houston, TX, U.S. EPA Report No. EPA/600/R-92/174.

Hinchee, R. E., 1994, "Air Sparging State of the Art," in *Air Sparging for Site Remediation*, R. E. Hinchee, ed., Lewis Publishers, Chelsea, MI, p. 56.

Howe, G. B., M. E. Mullins, and T. N. Rogers, 1986, *Evaluation and Prediction of Henry's Law Constants and Aqueous Solubilities for Solvents and Hydrocarbon Fuel Components. Vol. I*. Technical Discussion, USAFESE Report No. ESL-86-66, U.S. Air Force Engineering and Services Center, Tyndall AFB, FL.

Ji, W., A. Dahmani, D. Ahlfeld, J. Lin, and E. Hill, 1993, "Laboratory Study of Air Sparging: Air Flow Visualization," *Ground Water Monitoring Review* (Fall):127.

Johnson, R. L., 1994, "Enhancing Biodegradation with in situ Air Sparging: A Conceptual Model," in *Air Sparging for Site Remediation*, R. E. Hinchee, ed., Lewis Publishers, Chelsea, MI, p. 14.

Johnson, R. L., W. Bagby, M. Perrott, and C. Chen, 1992, "Experimental Examination of Integrated Soil Vapor Extraction Techniques," *Proc., Conf. on Petroleum Hydrocarbons and Organic Chemicals in Ground Water: Prevention, Detection and Restoration*, Nov. 4–6, Eastern Regional Ground Water Issues, Houston, TX, p. 441.

Johnson, R. L., P. C. Johnson, D. B. McWhorter, R. E. Hinchee, and I. Goodman, 1993, "An Overview of in situ Air Sparging," *Ground Water Monitoring Review* (Fall):127.

Johnson, P. C., C. C. Stanley, D. L. Byers, D. A. Benson, and M. A. Acton, 1991, "Soil Venting at a California Site: Field Data Reconciled with Theory," in *Hydrocarbon Contaminated Soils and Groundwater: Analysis, Fate, Environmental and Public Health Effects, and Remediation, Vol. 1*, Lewis Publishers, Chelsea, MI, p. 253.

Kaback, D. S. and B. B. Looney, 1989, "Status of in situ Air Stripping Tests and Proposed Modifications: Horizontal Wells AMH-1 and AMH-2 Savannah River Site," WSRC-RP-89-0544, Westinghouse Savannah River Co., Savannah River Laboratory, Aiken, SC.

Kaback, D. S., B. B. Looney, C. A. Eddy, and T. C. Hazen, 1989a, "Innovative Ground Water and Soil Remediation: In situ Air Stripping Using Horizontal Wells," presented at the *Fifth Outdoor Action Conference on Aquifer Restoration, Ground Water Monitoring and Geophysical Methods*, May 13–16, Las Vegas, NV.

Kaback, D. S., B. B. Looney, J. C. Corey, L. M. Write, and J. L. Steele, 1989b, "Horizontal Wells for in situ Remediation of Groundwater and Soils," *Proc., NWWA 3rd Outdoor Action Conf.*, Orlando, FL.

Kampbell, D., 1993, "U.S. EPA Air Sparging Demonstration at Transverse City, Michigan," in *Environmental Restoration Symposium*, sponsored by the U.S. Air Force Center for Environmental Excellence, Brooks AFB, TX.

Kayano, S. and D. J. Wilson, 1993, "Migration of Pollutants in Groundwater. VI. Flushing of DNAPL Droplets/Ganglia," *Environmental Monitoring and Assessment*, 25:193.

Langseth, D. E., 1990, "Hydraulic Performance of Horizontal Wells," *Superfund '90, Proc., 11th National Conf. of the Hazardous Materials Control Research Institute*, Washington, D.C., p. 398.

Leonard, W. C. and R. A. Brown, 1992, "Air Sparging: An Optimal Solution," *Proc., Conf. on Petroleum Hydrocarbons and Organic Chemicals in Ground Water: Pre-

vention, Detection and Restoration, Nov. 4–6, Eastern Regional Ground Water Issues, Houston, TX, p. 349.

Loden, M. E., 1992, "Technology Assessment of Soil Vapor Extraction and Air Sparging," U.S. EPA, Risk Reduction Engineering Laboratory, report no. EPA 600/R-92/173.

Lundegard, P. D. and G. Anderson, 1993, "Numerical Simulation of Air Sparging Performance," *Proc., Petroleum Hydrocarbons and Organic Chemicals in Ground Water: Prevention, Detection and Restoration,* Houston, TX, 17:461.

Mackay, D. M. and J. A. Cherry. 1989. "Ground Water Contamination: Limitations of Pump and Treat Remediation," *Environ. Sci. Technol.,* 12:630.

Mackay, D. M., W. Y. Shiu, A. Maijanen, and S. Feenstra, 1991, "Dissolution of Non-Aqueous Phase Liquids in Groundwater," *J. Contaminant Hydrology,* 8:23.

Marley, M. C. 1991, "Air Sparging in Conjunction with Vapor Extraction for Source Removal at VOC Spill Sites," *Proc., 5th National Outdoor Action Conf. on Aquifer Restoration, Groundwater Monitoring and Geophysical Methods,* p. 89.

Marley, M. C. and R. Billings, 1992, "Principles of Air Sparging," *Underground Tank Technology Update,* 6(3):1.

Marley, M. C., D. J. Hazebrouck, and M. T. Walsh. 1991, "Air Sparging in Conjunction with Vapor Extraction for Source Removal at VOC Spill Sites," paper presented at *6th Annual Conference on Hydrocarbon Contaminated Soils,* University of Massachusetts, Amherst, MA.

Marley, M. C., D. J. Hazebrouck, and M. T. Walsh, 1992a, "The Application of in-situ Air Sparging as an Innovative Soils and Groundwater Remediation Technology," *Groundwater Monitoring Review,* 12:137.

Marley, M. C., F. Li, and S. Magee, 1992b, "The Application of a 3-D Model in the Design of Air Sparging Systems," *Proc., Conf. on Petroleum Hydrocarbons and Organic Chemicals in Ground Water: Prevention, Detection and Restoration,* Nov. 4–6, Eastern Regional Ground Water Issues, Houston, TX, p. 377.

Marley, M. C., M. T. Walsh, and P. E. Nangeroni, 1990, "A Case Study on the Application of Air Sparging as a Complementary Technology to Vapor Extraction at a Gasoline Spill Site in Rhode Island," *Hydrocarbon Contaminated Soil Conference,* University of Massachusetts, Amherst, MA.

Middleton, A. C. and D. H. Hiller, 1990, "In Situ Aeration of Groundwater, a Technology Overview," *Conf. on Prevention and Treatment of Soil and Groundwater Contamination in the Petroleum Refining and Distribution Industry,* Oct. 16–17, Montreal, Que.

Miller, C. T., M. M. Poirer-McNeill, and A. S. Mayer, 1990, "Dissolution of Trapped Nonaqueous Phase Liquids: Mass Transfer Characteristics," *Water Resources Res.,* 26:2783.

Millington, R. J. and J. M. Quirk, 1961, "Permeability of Porous Solids," *Trans. Faraday Soc.,* 57:1200.

Montgomery, J. H. and L. M. Welkom, 1990, *Groundwater Chemicals Desk Reference, Vols. 1 and 2,* Lewis Publishers, Ann Arbor, MI.

Mutch, R. D., J. I. Scott, and D. J. Wilson, 1993, "Cleanup of Fractured Rock Aquifers: Implications of Matrix Diffusion," *Environmental Monitoring and Assessment,* 24:45.

Neaville, C. C., P. C. Johnson, and R. L. Johnson, 1994, "An Evaluation of in-situ Air-Sparging Technology at a Gasoline UST Site, Hayden Island, Oregon," in *Proc., 1994 Conference on the Environmental Health of Soils,* March 30–April 1, Long Beach, CA.

Norris, R. D. and D. J. Wilson, 1993, "Air Sparging of VOCs from Aquifers: Design and Modeling," *Proc., 6th Ann. Environmental Management and Technology Conf./Central (HAZMAT/CENTRAL 93)*, March 9–11, Rosemont, IL, p. 417.

Ostendorf, D. W., E. E. Moyer, and E. S. Hinlein, 1993, "Petroleum Hydrocarbon Sparging from Intact Core Sleeve Samples," *Proc., Petroleum Hydrocarbons and Organic Chemicals in Ground Water: Prevention, Detection and Restoration*, Houston, TX, 17:415.

Pankow, J. F., R. L. Johnson, and J. A. Cherry, 1993, "Air Sparging in Gate Wells in Cutoff Walls and Trenches for Control of Plumes of Volatile Organic Compounds (VOCs)," *Ground Water*, 31:654.

Powers, S. E., L. M. Abriola, and W. J. Weber, 1992. "An Experimental Investigation of Nonaqueous Phase Liquid Dissolution in Saturated Subsurface Systems," *Water Resources Res.*, 28:2691.

Powers, S. E., C. O. Louriero, L. M. Abriola, aru W. J. Weber, 1991, "Theoretical Study of the Significance of Nonequilibrium Dissolution of Nonaqeous Phase Liquids in Subsurface Systems," *Water Resou,·ces Res.*, 27:463.

Radecki, E. A., C. Matson, and M. R. Brenoel, 1987, "Enhanced Natural Degradation of a Shallow Hydrocarbon Contaminated Aquifer," *Proc., NWWA Focus Conf. on Midwestern Ground Water Issues*, National Water Well Assoc., Dublin, OH, p. 485.

Roberts, L. A. and D. J. Wilson, 1993, "Ground water Cleanup by in situ Sparging. III. Modeling of Dense Nonaqueous Phase Liquid Droplet Removal," *Separ. Sci. Technol.*, 28:1127.

Roberts, L. A., S. Kayano, and D. J. Wilson, 1992, "Modeling of DNAPL Removal from Aquifers by Sparging and Pump-and-Treat Operations," *Proc., Special Symposium: Emerging Technologies for Hazardous Waste Management*, Sept. 21–23, Amer. Chem. Soc., Atlanta, GA, p. 364.

Schwille, F., 1988, *Dense Chlorinated Solvents in Porous and Fractured Media: Model Experiments*, Lewis Publishers, Chelsea, MI.

Sellers, K. L. and R. P. Schreiber, 1992, "Air Sparging Model for Predicting Groundwater Cleanup Rate," *Proc., Conf. on Petroleum Hydrocarbons and Organic Chemicals in Ground Water: Prevention, Detection and Restoration*, Nov. 4–6, Eastern Regional Ground Water Issues, Houston, TX, p. 365.

Underground Tank Technology Update, 1992, "Special Feature on Air Sparging; Principles of Air Sparging, Air Sparging Project in Cleveland, OH, Air Sparging Project in Delaware, Air Sparging and SVE Radius of Influence, Air Sparging/SVE References," *Underground Tank Technology Update*, 6(2):1.

Wehrle, K., 1990, "In-situ Cleaning of CHC Contaminated Sites: Model-Scale Experiments Using the Air Injection (in situ Stripping) Method of Granular Soils," in *Contaminated Soil '90*, F. Arendt et al., eds., Kluwer Academic Publishers, the Netherlands, p. 1061.

Wilson, D. J., 1990, "Soil Cleanup by in-situ Aeration. V. Vapor Stripping from Fractured Bedrock," *Separ. Sci. Technol.*, 25:243.

Wilson, D. J., 1992, "Groundwater Cleanup by in situ Sparging. II. Modeling of Dissolved VOC Removal," *Separ. Sci. Technol.*, 27:1675.

Wilson, D. J., 1993, "Advances in the Modeling of Several Innovative Technologies," *Proc., 6th Ann. Environmental Management and Technology Conf./Central (HAZMAT/CENTRAL 93)*, March 9–11, Rosemont, IL, p. 135.

Wilson, D. J. 1994, "Groundwater Cleanup by in situ Sparging, V. Mass Transport-Limited DNAPL and VOC Removal," *Separ. Sci. Technol.*, 29:71.

Wilson, D. J., C. Gómez-Lahoz, and J. M. Rodríguez-Maroto, 1994a, "Groundwater Cleanup by in situ Sparging. VIII. Effect of Air Channelling on Dissolved VOC Removal Efficiency," *Separ. Sci. Technol.*, 29:2389.

Wilson, D. J., J. M. Rodríguez-Maroto, and C. Gómez-Lahoz, 1994b, "Groundwater Cleanup by in situ Sparging. VI. A Solution/Distributed Diffusion Model for Nonaqueous Phase Liquid Removal," *Separ. Sci. Technol.*, 29:1401.

Wilson, D. J., S. Kayano, R. D. Mutch, and A. N. Clarke, 1992, "Groundwater Cleanup by in situ Sparging. I. Mathematical Modeling," *Separ. Sci. Technol.*, 27:1023.

Wilson, J. L. 1990, "Laboratory Investigation of Residual Liquid Organics from Spills, Leaks, and the Disposal of Hazardous Wastes in Groundwater," U. S. EPA Report No. EPA/600/6-90/004, April. Available from NTIS as PB90-235797.

CHAPTER 9

Bioventing**

INTRODUCTION

R ECENT studies (MacDonald and Rittman, 1993) indicate that bioremediation is one of the fastest growing sectors of the U.S. market for
hazardous waste cleanup; it is expected to become a $500 million per year
industry by the year 2000. On the other hand, soil vapor extraction (SVE)
is already the most used in situ technology. These remediation techniques
can be considered independently when feasibility studies are performed,
but they can also be considered in combination in many situations—a technique known as bioventing. This technique uses the same principles as
SVE to draw air through the contaminated region; contaminants are
removed either by stripping them out of the soil or by biodegradation processes. This makes the technique somewhat more versatile than SVE since
it is not limited only to volatile contaminants but can also address any fractions that are susceptible to biodegradation.

A complete perspective on bioventing was provided recently by Hinchee
(1994), as a section of a *Handbook of Bioremediation*. Here, we will discuss the bioventing technology without entering into the details of the
physical principles involved in drawing air through the unsaturated region,
since these have been discussed earlier in this book. This chapter begins
with an introduction to the bioremediation technologies; then follows a
discussion of the design and operation of bioventing systems, with special
attention to how they differ from SVE systems. It seems to be clear to most
researchers working on the development of in situ bioremediation technologies that one of the most important things required for extensive use of
these technologies is to demonstrate to regulators and clients that it really
works. We have therefore included a section on the monitoring techniques

**This chapter was prepared by J. M. Rodríguez-Maroto and Cesar Gómez-Lahoz, Dpto.
de Ingeniería Química, Universidad de Málaga, 29071-Málaga, Spain.

for in situ bioremediation technologies, with special attention to bioventing. Design, operation, and monitoring require the development of mathematical models to simulate the very complex processes taking place in the subsurface, so that these can be better understood. We have included a one-dimensional column model, which involves a section including rather complex mathematics. This may be skipped if the reader does not feel comfortable with the differential equations. The mathematical section is followed by a section in which results obtained from the model are discussed and compared with the results published by researchers working in the field. We think that the model provides a tool with which the rate-limiting process(es) can be identified. This is very important if one is trying to change the design or operating conditions to optimize the performance of the system. A section on conclusions closes the chapter.

THE BIODEGRADATION PROCESSES

Under natural conditions the microorganisms act on the contaminants and produce other compounds. These processes are known as biotransformations, and the products can be classified in four main groups:

(1) Inorganic salts, H_2O, CO_2 (and CH_4 under anaerobic conditions); processes leading to these products are known as mineralization
(2) Organic compounds of low molecular weight
(3) Humic substances, for which the biodegradation processes continue very slowly (usually no more than 2–5% per year)
(4) New biomass

When one is considering microbiological processes for the removal of the contaminants, it should be noted that success is not defined in terms of the disappearance of the target compound but in terms of detoxification. From the environmental point of view, these processes can be classified as in the schematic representation. Mineralization, formation of humic substances, and growth of new biomass are biotransformations leading to detoxification and are known as biodegradation processes. Some biotransformations producing organic compounds of low molecular weight also lead to nontoxic compounds while others may produce compounds that are either more toxic, more mobile, or both; these processes are known as activation biotransformations. Engineered bioremediations look to establish environmental subsurface conditions that promote inhibition or detoxification biotransformations, so that in a reasonable period of time all the toxics and precursors of toxics are destroyed by the action of microorganisms.

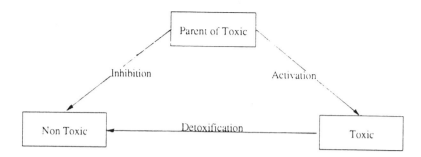

THE CONTAMINANT

The most important critical factor for the selection of any bioremediation technique is the biodegradability of the target contaminant. Although it is not possible to be sure if a contaminant will be biodegraded under field conditions without performing extensive treatability studies (Rawe et al., 1991), the Refractory Index [RI = BOD_5/COD (5-day biological oxygen demand/chemical oxygen demand)] may be used as a preliminary value to establish the likelihood that a compound will undergo aerobic biodegradation. In general terms, most of the organic compounds of frequent use (petroleum hydrocarbons, oxygenated solvents, etc.) are easily biodegraded under aerobic conditions. With respect to the aromatic compounds, as the number of rings increases, the RI decreases sharply.

Other anthropogenic compounds, like most of the chlorinated aliphatics or aromatics, which are not naturally present in the environment, are frequently not used by the microorganisms either as a carbon or as an energy source and can only undergo biotransformation in the presence of other primary carbon and energy sources in what is known as cometabolism (Strand et al., 1988; Semprini and McCarty, 1992; Strandberg et al., 1989; Adriens and Focht, 1990). Also, some of these compounds are found to be more easily biotransformed under anaerobic conditions (Mikesell and Boyd, 1988).

In addition to the nature of the contaminant, its concentration in the aqueous phase is important for the biodegradation processes. If the target compound concentration is too small, the growth of the microorganisms capable of biodegradation may not occur, especially if other substances are present that promote the development of other microorganisms that compete more successfully for the available nutrients and electron acceptors. On the other hand, if the target compound concentration is too high, it can produce toxic effects on the microbial population. Usually, aerobic bacteria are used to biodegrade compounds with concentrations between 50 and 4,000 mg/L; for subsurface in situ applications, concentrations of 10 mg/L are sufficient. Of course, if a biomass population capable of biodegradation is already present, concentrations several orders of magnitude lower than these values can be achieved after biodegradation.

THE MICROORGANISMS

The microorganisms naturally present in soil are able to biodegrade any substance that is synthesized in a natural biological process and are responsible of closing the carbon cycle. If this were not so, nondegradable substances would have accumulated in huge amounts in the soil during the centuries. Nevertheless, there is a large number of anthropogenic substances that are industrially produced today [about 63,000; 500–1,000 new ones are introduced each year (Barcelona et al., 1990)]. Many of these substances are not biodegradable, probably because these new substances have not been present long enough to allow time for the metabolic routes in the microorganisms to be established.

Thus, the second requirement for the feasibility of bioremediation is the presence of the microorganisms capable of biodegrading the target compound. The presence of several kinds of microorganisms with such capabilities in soil is well established (see for instance, Bossert and Bartha, 1984; Lee et al., 1988; Alexander, 1991; Hoeppel and Hinchee, 1994), the bacteria being the most important single group. Usually, indigenous bacteria are used for biodegradation, but in some cases it can be necessary to use foreign microorganisms already adapted to a similar type of contamination at a different site or even to use genetically engineered bacteria. In some cases the development of a biomass population capable of degrading the contaminants requires a period of time during which the biodegradation is almost negligible. This induction period needed for the acclimation of the bacteria is usually not necessary at places that have had a history of pollution.

THE SUBSURFACE MEDIUM

There are several environmental factors that affect the possibilities of success of in situ bioremediation techniques. If these factors are not favorable enough, biodegradation may not occur even if microorganisms able to biodegrade the contaminant are present. Several physico-chemical characteristics of the soil influence biological activity; these include pH, redox potential, texture and porosity, permeability, soil moisture content, temperature, and sorption exchange capacity. Other parameters more related to the biological processes are also very important, such as availability of nutrients and electron acceptor, presence of inhibitory or toxic substances, predators, etc.

Physico-Chemical Parameters

A neutral or slightly alkaline soil is optimum for biodegradation, but a range of pH values between 6 and 8 is considered acceptable. Most natural soils are slightly acidic, and neutralization with lime or other bases may be needed.

Soil temperature is also relevant to the rate of biodegradation, with warmer temperatures usually yielding higher rates of biological activity. Optimum temperature values (20–35°C) are usually far above ambient soil temperatures. Nevertheless, some information indicates that microbiological colonies are adapted to local temperatures and do not succeed when they are transported to warmer regions. In Arctic regions bioremediations of petroleum spills have proved successful in dealing with some well-known accidents [for instance, Pritchard and Costa (1991)]. Also, some soil bioremediation researchers have tried the use of various techniques to increase the soil temperature to enhance bioremediation in Arctic regions (Leeson et al., 1993).

Soil relative moisture is a significant parameter for bioremediation processes in the vadose zone (Harder and Höpner, 1991), such as bioventing. A minimum water content is necessary since microbiological transport processes always take place through aqueous interfaces, even when the bacteria may need some solid support. On the other hand, diffusive transport in the aqueous phase is usually three to four orders of magnitude slower than it is in the gaseous phase, and frequently this diffusion transport controls the rate of supply of electron acceptors and nutrients. Oxygen transport for aerobic biodegradation is more efficient if part of the porous structure of the soil is gas-filled.

The carbon substrate, the electron acceptor, and the nutrients all should be available for the microorganisms, which means that each should maintain an adequate thermodynamic activity in the water solution in the vicinity of the microorganism. Hydrocarbons may sorb onto solids, mainly on humic or clayey material, or in fine-grained soils, the hydrocarbon may need to migrate by aqueous phase diffusion from the pores where the contaminant is confined, which are not directly accessible to microorganisms. These desorption/diffusion processes place an inherent limitation on the rate at which hydrocarbons are biodegraded (Larkin et al., 1991; Coho and Larkin, 1992). Sorption onto solid surfaces may limit the availability of these substances, hence decreasing the rate of biodegradation, but it can also help to reduce some toxic concentrations below their toxic limits (Rijnaarts et al., 1990). Sorption/desorption processes may also act as a reservoir of nutrients, which are released from the sorption sites to replace material removed from the aqueous phase by the microorganisms.

Substances with low aqueous solubility or strongly sorbed onto solid surfaces can be mobilized by some microorganisms that are able to produce surfactants. Some tests have been done to enhance the biodegradation rate by adding surfactants externally, but this has not generally improved the contaminant uptake (Hoeppel et al., 1991; Arthur et al., 1992).

Soil permeability is one of the most important variables for SVE, and it also is important for bioventing to make possible the delivery of oxygen and nutrients to the contaminated regions to be remediated. In some field bioremediation cases a decrease in the soil permeability has been observed due to the growth of biomass around the point at which oxygen and nutrients were supplied. As with some of the previous variables, these conditions are dynamic and may change during the course of the soil remediation process.

Redox potential is easily associated with the aerobic or anaerobic conditions of the soil and with the oxidation states of some metallic ions. Once oxygen is depleted through biological or chemical consumption, if it is not replaced, the microbiological processes that take place change as the environmental conditions move to more reductive potentials. This leads to a preferential sequence of processes in order of redox potential: aerobic biodegradation, denitrification, iron and manganese reduction, sulfate reduction, and methanogenesis (Hoeppel and Hinchee, 1994).

Under anaerobic conditions some contaminants may be transformed more easily when the redox potential reaches some specific values. For instance, chloroform and trichlorethane may not be transformed under denitrifying conditions but will be metabolized when methanogenesis conditions are reached. In a similar fashion, some chlorinated phenols may be degraded by bacteria present during methanogenesis but may be quite resistant to bacteria present under sulfate-reducing conditions.

Redox potential also determines the oxidation state of many inorganic ions, thus changing the adsorption phenomena occurring in the soil, which, in turn, may lead to a change in the availability of some nutrients or a change in the concentrations of some toxic metals. Soils containing organic contaminants usually exhibit reductive potentials that will be changed if aeration is performed.

Biological Parameters

The presence of toxics, such as heavy metals or other substances, may inhibit microbial metabolism. Toxic effects are usually shown above some threshold value of the toxic compound, and even target compounds that may be used as primary substrates by the microorganisms may be toxic for these microorganisms if they are present above that threshold value. Some substances that may be essential for some specific metabolic routes may show toxicity for other routes. For instance, some chlorinated aliphatics that were undergoing biotransformations under anaerobic conditions may no longer be destroyed when aeration of the soil is initiated. This same inhibitory effect may appear after sulfates are introduced into a system in which some methanogenic transformations were already going on.

Growth of new protoplasmic material requires the assimilation of different proportions of several inorganic elements such as nitrogen, phosphorus, potassium, sulfur, etc. Among these, the ones that are more likely to limit the growth of the microorganisms are nitrogen and phosphorus. If the ratios N/C or P/C are below the values required by the microbiological processes, additional nutrients should be supplied. This situation is rather common when petroleum hydrocarbon contamination occurs because, in contrast to most of the natural substrates, the ratios between nutrients and carbon for the hydrocarbons in petroleum are zero. It is estimated that the degradation of 1 L of gasoline would require 44 g of nitrogen and 22 g of phosphorus (Lee et al., 1988) that will have to be supplied as aqueous solution. Typical ratios that have been proposed for CiNiP are 250:10:3 or 100:10:2 (Staps, 1990).

Nutrient supply is difficult to establish without lab studies because the nutrients must not only be present in the soil but must be available to microorganisms where they are needed. This requires not only a quantitative conventional analysis but a chemical speciation to establish if the nutrient will reach the region where it is needed as an aqueous solution or if it will be precipitated or sorbed somewhere close to the point at which it is supplied.

Oxygen is by far the most energy efficient electron acceptor for biological processes. Energy efficiency can be quantitatively measured, for instance, by the number of ATPs (adenosine triphosphates) produced when metabolization of a molecule of substrate is carried out. Thus, when a glucose molecule is metabolized through aerobic metabolism, about thirty-eight ATPs may be formed, whereas the net yield of anaerobic decomposition of glucose by yeast is only two ATP molecules for each glucose molecule degraded (Alexander, 1991). Since protoplasm synthesis requires energy, the growth of biomass in the presence of oxygen is faster and more efficient than it is under anaerobic conditions.

Thus, if oxygen is present, only those microorganisms capable of using aerobic metabolism, which include the facultative bacteria, will succeed. Alternatively, when oxygen is not present, other electron acceptors (NO_3^-, SO_4^{2-}, CO_2 . . .) may be used for the degradation of some substances. Furthermore, some contaminants like carbon tetrachloride, are more easily biotransformed under reductive denitrification conditions.

Oxygen supply is usually the rate-limiting step for the aerobic microbiological processes in the subsurface. Oxygen demand depends on the contaminant and the transformation products. If a significant increase in the microbiological population is produced, oxygen demand is lower than that necessary to achieve complete mineralization of the substrate, as can be seen from the following reaction schemes for a hydrocarbon of composition C_7H_{12}, corresponding to a complete mineralization process and a very efficient system in biomass production:

$$C_7H_{12} + 10\ O_2 \rightarrow 7\ CO_2 + 6\ H_2O$$

$$C_7H_{12} + 5\ O_2 + NH_3 \rightarrow C_5H_7O_2N + 2\ CO_2 + 4\ H_2O$$

Field stoichiometry is more likely to be closer to the first situation if the biomass has achieved a stable population. While a substantial growth of biomass is taking place calculation of the contaminant biodegradation from CO_2 evolution or O_2 uptake will yield conservative results.

If we calculate the amount of oxygen required according to the second stoichiometry, 1,667 kg of oxygen will be required to biodegrade 1,000 kg of hydrocarbon. If this oxygen is supplied by means of an aqueous solution with a concentration of 8 mg/L, i.e., close to equilibrium with air under atmospheric concentrations, 208,000 m^3 of water will have to be circulated through the contaminated region. If the section normal to the flow is 10 m^2 and we assume a superficial velocity of 1–2 m^3/m^2 day, the time needed to achieve remediation will be twenty-five to fifty years. This is obtained by assuming that all the oxygen supplied is utilized, which requires a very efficient contact between the oxygenated solution and the hydrocarbon, so probably less optimistic estimations will result in centuries. Obviously, more efficient ways of providing oxygen are required.

Use of aqueous solutions at equilibrium with pure oxygen at 1 atm will lead to fivefold increases in the oxygen concentration as compared to those in equilibrium with air. This is still very low. It has been suggested (Michelsen and Lotfi, 1991) that aqueous solutions containing stabilized microbubbles of oxygen be used, but, to the authors' knowledge, no field application has used this approach.

Another alternative for the oxygen supply is the use of hydrogen peroxide solutions (Huling et al., 1990; Hinchee et al., 1991a, Aggarwal et al., 1991). Solutions of 3% hydrogen peroxide ($\approx 30,000$ mg/L) are commonly used as a general antiseptic. These solutions can supply about 15,000 mg/L of oxygen. Limitations for this technique are the toxic effect of high concentrations of hydrogen peroxide to microorganisms (bacteria tolerate concentrations up to 500–2,000 mg/L) and the low stability of the hydrogen peroxide solutions, which leads to release of almost all the oxygen close to the injection region. The decomposition of hydrogen peroxide is catalyzed by several inorganic ions, among which Fe^{2+} and Fe^{3+} are particularly effective, and it is also decomposed by some enzymatic processes. While pilot-scale treatability studies will show the catalytic effects of the inorganic salts from the beginning of the experiments, catalysis related to enzymatic processes may arise later in the remediation process. The use of inhibitors of the catalyzed decomposition, such as H_2KPO_4, is recommended in some situations.

Thus, the use of hydrogen peroxide solution may improve significantly the efficiency of oxygen delivery in the aqueous phase, but the mass ratio between the carrier needed to supply the required oxygen and the contaminant is still high, and a substantial amount of additional research work will be needed before one will be able to implement bioremediation systems to demonstrate the efficiency of hydrogen peroxide solutions.

Oxygen may also be supplied in the gaseous phase as it is present in air, both into the saturated zone, as in sparging systems [see, for instance, Brown (1994)] or by injection or extraction of air in the vadose zone (as in SVE). The latter approach is what is known as bioventing, a term generally applied only for in situ applications, although it is sometimes used for excavated piles (Jespersen et al., 1993).

The oxygen content of air is about 233 g/kg of carrier, so air is substantially more concentrated in equivalent oxygen content than are hydrogen peroxide solutions used for bioremediation. Bioventing is certainly the more efficient way to deliver oxygen to the vadose zone. It is also advantageous to remove as much contaminant as possible while it is in the vadose zone and before it reaches the zone of saturation. Among the reasons for this, there is the fact that gas phase diffusivities are three or four orders of magnitude higher than those for the same substance in the aqueous phase. Therefore, if the remediation is rate-limited by diffusion processes, it will be about three or four orders of magnitude faster in the gaseous phase than in the aqueous phase. The likelihood of having diffusion processes controlling the remediation is much greater for the saturated region. Thus, although the scientific literature dealing with bioremediation technologies for the saturated zone is more extensive than the literature on bioventing, the use of air as an efficient oxygen carrier and the advantages related to mass transport processes in the gaseous phase make bioventing one of the most promising bioremediation technologies.

BIOVENTING OPERATION AND DESIGN

As pointed out previously, bioventing consists basically in drawing air into the vadose zone so that biodegradation is enhanced by the augmented oxygen supply. In Dupont's (1993) words, "Bioventing represents a hybrid physical/biological process utilizing venting systems for oxygen transfer, while focusing not on contaminant stripping but rather on in situ aerobic contaminant biodegradation." A conceptual design is shown in Figure 9.1.

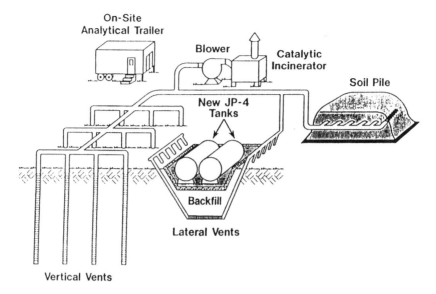

Figure 9.1 Conceptual drawing of the Hill AFB, Utah, field soil venting site [from Dupont et al. (1991)]. (Reprinted by permission from Butterworth-Heinemann; copyright © 1991 by Battelle Memorial Institute.)

Thus, the main difference between SVE and bioventing exists in the principal objective, stripping for SVE, and biodegradation for bioventing. Bioventing equipment is very similar to that used for SVE: horizontal or vertical wells, with the corresponding blowers, pipe and valve systems, etc. Nevertheless, design and operation are different as a result of the different objectives: SVE maximizes the volatilization of the contaminant and is not concerned with enhancing biodegradation, although some biodegradation may occur that will certainly cause no harm. Bioventing, on the other hand, looks for the maximum biological efficiency in the utilization of the oxygen introduced, so that as much contaminant as possible is aerobically degraded; also, in some cases, in bioventing one attempts to minimize the amount of contaminant stripped, to permit the operation of the system with little or no aboveground gas treatment.

At most sites contaminated with petroleum hydrocarbons, when SVE is underway, biodegradation occurs to some extent. It is almost certain that, if oxygen is depleted in the contaminated area before starting the remediation program, this oxygen has been utilized by some microorganisms. Then if aerobic conditions are returned to the soil, significant biological activity will occur. Biodegradation was observed and monitored at some sites without any external actions different from those necessary for SVE [see, for instance, De Paoli et al. (1991) and Urlings et al. (1991)]. As a matter of fact, the bioventing rate is usually reported to be zero order in hydrocarbon, and oxygen supply is the critical factor in the rate of the process.

Both SVE and bioventing are limited to the vadose zone but may act over portions of soil below the original water table if this is lowered by dewatering (Hinchee, 1994). If air is supplied to the saturated region to achieve volatilization of VOCs or biodegradation, the techniques are known as sparging or biosparging and present some particular characteristics that make them different from those of the bioventing systems we are dealing with.

Some contaminants are susceptible to treatment either by stripping or by biodegradation, but while SVE efficiently removes some toxic compounds that are not easy to biodegrade [like chlorinated aliphatics, for which bioventing is still considered as an emerging technology (Wilson, 1994)], some compounds of low volatility, like jet fuels, are more easily removed by biodegradation. Bioventing has also proven to be efficient in biodegrading PAHs at Ironton, Ohio, with zero order rate parameters of oxygen uptake between 0.08 to 0.38% O_2 h^{-1} (Hinchee and Ong, 1992). The removal of part of the contaminants by volatilization may even help the remediation of the other portion by biodegradation through several mechanisms, such as the disappearance of some toxic compounds (chlorinated aliphatics, light aromatic hydrocarbons, etc.) during the remediation, and an increase in the surface/volume ratio of the NAPL droplets, which may help increase contaminant availability of some hydrocarbons of low aqueous solubility (Hoeppel et al., 1991).

Bioventing may be performed by the injection or extraction of air through the soil. For the first approach, if the injected air that has come into contact with the contaminated regions reaches clean soil and is in contact with it for a sufficient period of time, the amount of contaminant that was initially stripped may now be biodegraded in this latter domain of soil. Thus, if the conditions are favorable, biodegradation may account for almost 100% of the contaminant removal. The U.S. EPA Risk Reduction Engineering Laboratory and the U.S. Air Force used this approach at Eielson AFB (Alaska) and demonstrated that concentration of hydrocarbons 2 ft over the soil level during the operation were low (TPH = 61 ppm and benzene = 3.3 ppm). This is a very simple design, but, even if the system is operated at reduced flux rates, the fate of the effluent contaminants must be carefully determined, especially if buildings are present in the vicinity of the well, into which the contaminants can migrate.

If an extraction system is used and the extraction well is placed in the contaminated area, as one will try to do if performing SVE, most of the contaminant may be stripped. This is the type of system that was used at Tyndall AFB, Florida (Hinchee et al., 1992).

Of course, injection (or passive) wells may be used in combination with extraction wells to achieve a better control of the airflow pattern and the fate of the contaminant. Emissions can then be controlled, if high VOC concentrations occur in the off-gas, by reinjecting it back to the soil until the biodegradation occurs or by the use of traditional SVE off-gas treatment units.

The most important variable for the operation of a bioventing system is the air flux, which supplies the oxygen necessary for aerobic biodegradation. Thus, bioventing will only fulfill the promise of the treatability studies if the soil permeability is high enough to ensure an adequate oxygen supply in a reasonable period of time. Permeability values of at least 10^{-4} cm/s (Hoeppel et al., 1991) or 10^{-3} cm/s (van Eyck and Vreeken, 1989) are recommended.

Typical values of the parameters of SVE systems working in continuous operation [1 to 48 pore volumes day (Urlings et al., 1991)] may lead to oxygen supply at rates that are above the rate of oxygen uptake observed in the field: 0.05–1% (vol.) O_2h^{-1} (Hinchee and Ong, 1992). High extraction rates for SVE may also lead to important decreases in the contaminant vapor concentration if the volatilization is rate-limited by the mass transport of the contaminant from the stationary to the mobile gaseous phase. Thus, very high extraction rates may lead to an airstream with an exit concentration of oxygen very close to that for air and a relatively low contaminant concentration, increasing the effluent treatment costs necessary to achieve emission standards, since these costs are typically proportional to the gas volume rather than to the amount of contaminant.

Ely and Heffner (1988), of the Chevron Research Company, presented a patent where they recommend "adjusting the flow rate of oxygen into the hydrocarbon contaminated zone to above the flow rate for maximum hydrocarbon evaporation," because the amount of contaminant biodegraded increases with increasing flow rate. They present graphical results, indicating that this increase takes place for flow rates in the range between 10 and 50 SCFM; no quantitative data are given regarding what happens at higher rates or how many pore volumes per day are equivalent to those flow rates. The values presented correspond to moderate flow rates of typical SVE applications.

Today, almost all the authors working in this field recognize that it can be more convenient to work under such conditions that will avoid large volumes of air that need treatment before discharge to the atmosphere. This can be done by two basic strategies: using reduced flow rates (0.1 to 0.5 pore volumes per day) or by pulse venting (Dupont 1993). The pulsed operation of bioventing columns has been modeled (García-Herruzo et al., 1993).

The equivalent oxygen demand may be estimated from the results obtained with the rapid in situ respirometric test developed by Hinchee and

Ong (1992), which is described in the following section on monitoring technologies. This value can be used to determine the minimum air flux (or the ratio between the nonventing and venting periods for pulse operation). Nevertheless, it should not be forgotten that air transition times through soil are not homogeneous and may vary widely, depending on the system geometry. Thus, if the air flux is calculated to supply enough oxygen to those regions of the soil that are vented by the more rapid streamlines, some other regions may become anaerobic in those sections further from the air entrance as oxygen is depleted before reaching the well. Conversely, if the flux is designed to supply enough oxygen to those regions transited by those streamlines requiring larger periods of time to reach the well, some other regions will be vented at a rate that will not permit the microorganisms to utilize all the oxygen supplied, with a corresponding decrease in the ratio between the amount of contaminant biodegraded and the amount stripped. As a general rule, as one increases the number of wells, the cost increases, but the range between the fastest and the slowest streamlines narrows, so oxygen supply can be more homogeneous.

The second approach, pulse operation, consists of introducing a significant amount of air into the subsurface at a relative large flow rate and then stopping the blower until oxygen is depleted in the subsurface to a level that indicates that the biological activity is being controlled by the oxygen availability. Evidently, careful monitoring may be necessary for this type of operation. It does permit one to minimize the volume of air extracted from the soil and to supply enough oxygen to the soil to optimize biodegradation. Furthermore, if the rate of O_2 uptake decreases as the remediation approaches completion, the periods of time during which the blower is stopped are increased, and the working conditions may easily be kept close to optimum during all the remediation period (Dupont, 1993).

In some cases where minimization of the amount of contaminant stripped was the objective, operating conditions were such that contribution of biodegradation to the removal was close to 100%. The small amount of contaminant stripped was treated by reinjecting the off-gas into clean soil where it was biodegraded to completion (Hinchee et al., 1992).

Some of the variables significant to bioremediation processes are difficult to control, but some may be modified to increase the system efficiency. For instance, two different ways of warming up the soil were investigated at Eielson AFB (Alaska) (Leeson et al., 1993), based on the introduction of warm water or the use of greenhouse systems. Moisture addition enhanced the biological activity at an arid site such as Hill AFB (Utah) (Hinchee and Arthur, 1991), and nutrient addition may be needed at some sites although in many circumstances no improvement was observed after nutrient amendment of soils (as in Hill AFB and Tyndall AFB) (Hinchee et al., 1992).

We have tested the sensitivity of bioventing to different variables by a column model that permitted consideration of different rates of the mass transfer processes between the stationary phases and the mobile gas phase. The results indicated that the mass transfer rate parameters are of high significance and that they determine, in many cases, the relative insensitivity of the system to the other model parameters (Gómez-Lahoz et al., 1994b, 1994c).

Bioventing cost estimation still requires a substantial amount of work because the technique is not yet beyond the experimental stage. Hoeppel et al. (1991) indicate that costs may vary between $26 and $200/ton for a nonresearch project. He also gives another estimation on a different basis as $230–300/gallon of residual oil in soil. Hinchee (1994) indicates that the key costs of bioventing are the monitoring costs and that these are very site-specific. These monitoring costs are proportional to the time needed for remediation, so significant decreases in the removal rate may lead to substantial increases in the remediation costs. This should be considered when optimizing the operating conditions, because if one operates the system at very low flow rates to minimize the volume of gases extracted that need treatment or to achieve complete biodegradation so that the gas treatment system is not necessary, this may lead to increases in the total remediation costs if the time needed to achieve completion is increased substantially. Due to the relatively long time period required for biodegradation, a bioventing site optimized for biodegradation as opposed to volatilization will incur higher monitoring costs. On some sites this may be more than offset by reduced or eliminated gas treatment costs, thus reducing total remediation costs.

MONITORING IN SITU BIOREMEDIATION

To assess the feasibility of a bioremediation project, it is necessary to know how the microorganisms are utilizing the target contaminant, and this must be done for the specific conditions of the site to be remediated. If the project is considering an in situ application of bioremediation, the assessment will be considerably more difficult because of the complexity of the soil environment and the difficulties in obtaining representative data from the subsurface.

Up to the present time several methods have been developed to track and verify in situ bioremediations; these consist of the study of the variations with time of one of the variables significant to bioremediation, such as: a) contaminant concentration, b) oxygen uptake or carbon dioxide evolution, c) temperature, d) number of microorganisms capable of degrading the contaminant, and e) ratio between contaminants that are easily biodegraded to others more difficult to biodegrade (i.e., pristanes or phytanes)

(Heitzer and Sayler, 1993). Even though each of these methods may produce some information to support the fact that in situ bioremediation is really being accomplished, soil heterogeneities and the difficulties in accomplishing complete mass balances make it almost impossible to demonstrate with only one of these monitoring techniques that biological processes are responsible for the contaminant removal and not some other mechanism(s) such as dispersion, volatilization, dissolution, etc.

Thus, it is not enough to demonstrate that the contaminant may be removed by microbiological processes nor even the ability of the site indigenous bacteria to biodegrade the contaminant. The real question is, "Is the contaminant being removed in the actual contaminated soil, with the real soil environmental conditions, by biodegradation?"

As indicated, none of the above techniques alone is capable of answering the previous question on its own. The National Research Council of the National Academy of Sciences has recently published a book entitled *In situ Bioremediation, When Does It Work* (National Research Council, 1993), indicating that it is necessary to use methods based on three types of evidence to demonstrate that bioremediation is really working, i.e., that not only is the contaminant concentration (or, more precisely, the soil toxicity) decreasing but that this decrease is due to the microbiological activity. The three types of evidence required for every well-designed project are

(1) The contaminant concentration is decreasing in the site. This is determined by standardized sampling during the remediation.

(2) Laboratory experiments showing that microorganisms obtained from the site are able to biodegrade the contaminant under conditions as close as possible to those in the field.

(3) One or more kinds of evidence indicate that biodegradation is really taking place in the site.

This third type of evidence is the most difficult to obtain but is also the most important. Three different procedures may be used to obtain this type of information:

(1) Chemical and/or microbiological analysis of soil and groundwater samples to determine bacteria counts or number of protozoans, rates of microbiological activity, electron acceptor concentration, etc.

(2) In situ active experiments to assess the microbiological activity, by addition of various chemicals in a controlled manner to see if their fate is consistent with what should occur during bioremediation, such as the addition of nutrients at subsites, measurement of the electron acceptor uptake rate, and fate of inert tracers or isotopically labeled contaminants.

(3) For the use of models, two different strategies are followed: simulation of the several abiotic phenomena that may lead to a decrease of the contaminant concentration, or simulation of the microbiological processes to establish the rate of biodegradation. For the first approach, if the in situ measured contaminant concentrations are significantly lower than those that may be expected from the abiotic processes as established from the model, then this may be considered as indicative that biodegradation processes are occurring. The second approach follows the biodegradation processes themselves with more direct and detailed information, but a larger volume of quantitative information is required to run such models, and this information is frequently very difficult to obtain.

The models that have been developed may be classified in four groups: saturated flow models, multiphase flow models, geochemical models, and biological reaction rate models, although researchers frequently combine two or more of these kinds of models to obtain a more complete evaluation.

Since none of the three procedures used alone is enough to demonstrate that biodegradation is producing the detoxification of the site at the desired rate, combination of several procedures is usually required. This becomes a difficult task due to the heterogeneous nature of soil and the impossibility of carrying out an analysis program that will indicate without any doubt the amount, nature, and distribution of the contaminant. Furthermore, the skills necessary to gather all this information are not easy to find without the inclusion of a whole team, because such a study requires integrating multidisciplinary concepts and using tools developed to study very different scales of time and space, as for instance, microbiology and hydro-geodynamics.

Uncertainties are minimized as one maximizes the number of samples and the types of analysis performed, but this must be done within the limits of economical feasibility. The utilization of simplified models where the several variables are weighted correctly may help in selecting the best monitoring procedures, which ultimately will need the use of safety factors in the design to account for the uncertainties that are assumed present.

MONITORING TOOLS FOR IN SITU BIOVENTING

The previous paragraphs dealt with the several procedures that are recommended to assess the different in situ bioremediation techniques. This section deals with some of the techniques that may be included in the laboratory scale experiments, soil sampling or in situ experiments that yield specific information about the effectiveness of bioventing.

Thus, among the large number of techniques developed to assess the bioremediation performance with laboratory experiments, Hinchee and Arthur (1991) used soil columns where moisture and/or nutrients were added to establish the feasibility of improving the performance of the biodegradation processes that were already under field application at Hill AFB in Utah. To do this they followed the CO_2 evolved from the columns, as compared with that obtained from sterilized columns under identical conditions. Also, Leeson et al. (1993) used this technique to assess the possibilities of enhancing the performance of bioventing for the remediation of a JP-4 fuel spill at Eielson AFB, Alaska.

Among the soil sampling experiments developed for bioventing experiments, there is one suggested by Aggarwal and Hinchee (1991), and used by them with other coworkers at Hill AFB (Hinchee et al., 1991b), to establish the source of the CO_2 evolved from the soil. The method consists of measuring the ratio between the stable carbon isotopes ($^{13}C/^{12}C$) present in the soil CO_2 mass spectrometrically. The ratio between these two isotopes is related to the source of the CO_2, and differences are found between the $^{13}C/^{12}C$ ratio of CO_2 from fossil sources (petroleum contaminants), the ratio for CO_2 generated by the natural biological processes taking place in the soil (root respiration and mineralization of natural substrates entering the soil) and the ratio for CO_2 arising from other inorganic sources such as carbonate rocks or atmospheric CO_2. At the sites where this test was used, it showed that the CO_2 was not coming from inorganic sources. The analysis is claimed to be economical and the necessary equipment is commercially available.

With respect to the in situ experiments, the most common one (Dupont, 1993) is the analysis of the gases evolved from the soil during the normal operation of a bioventing application, where CO_2, O_2, and VOC concentrations may be tracked to determine the bioremediation progress. O_2 and CO_2 concentrations are compared to those of the incoming air so that mass balances may be done for the estimation of the mean rate of the biological processes, whereas decreases of the VOC concentration with time indicate that the site is being remediated.

Marrin et al. (1991) have compared CO_2 and VOC concentrations measured at the extraction well with others obtained from gas samples of the contaminated region and found differences of one or two orders of magnitude between these values. Thus, rate parameters obtained from the concentrations measured at the well correspond to average values of the actual parameters and may not be used for the estimation of the small-scale biological rate parameters.

A rapid in situ respirometric test has been developed by Hinchee and coworkers (Hinchee and Ong, 1992; Ong et al., 1991; Hinchee et al., 1991b) that permits the determination of the biological activity at a specific location in the contaminated region. The method consists of supplying air to the soil until aerobic conditions are reached, stopping the venting wells, and then following the evolution of O_2 and CO_2 concentrations in the gaseous phase. As indicated by Dupont (1993), "These determinations must be based on a comparison to uncontaminated soil conditions, as only levels of O_2 depletion and CO_2 enrichment in excess of background levels are indicative of increased microbial activity compared to normal basal respiration levels seen in uncontaminated soils at the site."

Biological reaction rates based on oxygen uptake are generally more reliable than those that follow the CO_2 evolution (van Eyck and Vreeken, 1988; Dupont, 1993; and Hinchee and Ong, 1992). Results based on CO_2 evolution can underestimate the microbiological activity, and deviations between calculations of contaminant biodegradation obtained from oxygen uptake measurements and carbon dioxide evolution increase as the soil pH increases; the ratio between the estimations is close to 1 for pH values of 6 and close to zero for a pH of 8.5. This indicates that some CO_2 may undergo acid-base reaction with alkaline species present in the subsoil.

Oxygen uptake for the estimation of the contaminant biodegradation is based on a specific stoichiometry and ratio between the oxygen consumed and the contaminant biodegradation, which may not be easy to know accurately. There may also be other error sources, such as oxygen consumption by the oxidation of inorganic salts (which is not probable if the monitoring experiment is performed after a lengthy period of venting) and those related to the natural respiration processes in uncontaminated soils, as indicated above. Also, oxygen depletion may be due to diffusion into other nonaerated soil regions. Injection of inert gases with higher diffusivities (such as helium) and monitoring of their concentration during the respirometric study have been used to demonstrate that diffusion is not responsible for the decreases in oxygen concentration (Hinchee and Ong, 1992).

Thus, monitoring of bioventing applications will usually require (1) previous determinations of the contaminant, oxygen, and carbon dioxide concentrations; (2) periodic analysis of the VOC, oxygen, and carbon dioxide concentrations in the effluent gases during normal operation; (3) rapid in situ respirometric tests; and (4) probably some other tests such as stable isotope analysis or tests typical of other bioremediation techniques such as bacteria counts. Interpretation of the information obtained from these monitoring experiments will be better understood if it is compared to modeling results, which may be helpful to convince regulators and clients that the bioremediation processes are really occurring in the subsurface.

A MODEL FOR BIOVENTING

The model that is described next is very similar to one described in detail earlier (Gómez-Lahoz, 1994a), but, following Hinchee's suggestions, we have included the presence of NAPL, which permits a more realistic picture of soil contamination. Thus, we have a three-phase system (vapor, aqueous, and NAPL) with contaminant being transferred between the three phases and oxygen being transported from the vapor to the aqueous phase, as schematically represented below. Among the objectives of this work is the development of a tool to permit us to ascertain what is the limiting factor to bioremediation under field conditions. Basically, two major aspects of bioremediation are considered: first, mass transport of the different substances between the three phases present and, second, the biological processes. Thus, two different extreme scenarios are considered: (1) biodegradation is rate limited by the mass transfer processes, i.e., the rate at which the essential substances are made available to microorganisms, and (2) the biological processes are slow compared to the mass transfer kinetics. Obviously, situations with mixed control are also possible. We want to ascertain if it is possible to determine which is the situation from the monitoring tools already available for field experiments. We will simulate the monitoring of O_2 uptake during continuous operation of bioventing and also the fast monitoring tests described previously, under conditions that lead to one of the two scenarios. We will also use the model to study how bioventing efficiency can be enhanced by different strategies, under the two scenarios proposed.

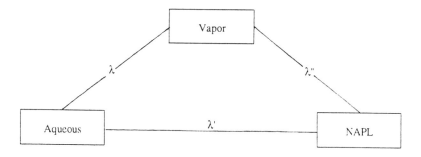

From the biological point of view, the model includes several features, some of which will be simplified later. For instance, it includes the presence of two substrates to allow the study of some common field situations, such as contamination by mixtures of several compounds or the use of cometabolism [see, for instance, Semprini and McCarty (1991, 1992)].

The rather extensive nomenclature used in this section is as follows:

A_j = concentration of jth nutrient (g/cm³ of aqueous phase)
B = biomass concentration (g/cm³ of aqueous phase)
B^* = parameter of the inhibition Equation (12) (g/cm³ of aqueous phase)

C_j^s = contaminant j (j = 1,2) concentration in the aqueous phase (g/cm³ of aqueous phase)

C_j^v = contaminant j (j = 1,2) concentration in vapor phase (g/cm³ of vapor phase)

C_j^L = contaminant j (j = 1,2) concentration in the NAPL phase (g/cm³ of soil)

C_2^* = threshold value of the concentration of contaminant 2 for toxic effects on biomass (g/cm³ of aqueous phase)

F = acute toxicity exponential factor for contaminant 2 (dimensionless)

I = inhibition factor for biological activity (dimensionless)

K_B = die-off coefficient for microorganisms (s⁻¹)

K_{Bo} = die-off coefficient in the absence of toxic compounds (s⁻¹)

K_{cj} = contaminant j (j = 1,2) half saturation constant (g/cm³)

K_D = Darcy's constant, pore basis (cm²/atm s) (volumetric flow rate of air per unit area = $\nu K_D \nabla P$)

K_j = nutrient j half-saturation constant (g/cm³)

K_{hej} = Henry's constant of contaminant j (j = 1,2) (dimensionless)

K_{ho} = Henry's constant of oxygen (dimensionless)

K_{max} = maximum rate of growth of biomass on the best of substrates (s⁻¹)

K_o = oxygen half saturation constant (g/cm³)

L = length of the column (cm)

M_i = mass of contaminant in compartment i (g)

m_i = contaminant concentration in compartment i (g/cm³ of soil)

M_{wj} = molar weight of contaminant j (g/mol)

N = number of compartments into which the column is partitioned (dimensionless)

n_j = stoichiometric coefficient for contaminant j (j = 1,2) (g of new biomass obtained from 1 g of substrate)

$n_{o,j}$ = stoichiometric coefficient for oxygen with contaminant j (j = 1,2) (g of oxygen required for the assimilation of 1 g of contaminant j)

n_o' = stoichiometric coefficient for endogenous respiration (g of oxygen consumed per g of dead biomass)

$n_{z,j}$ = stoichiometric coefficient for nutrient z for contaminant j (j = 1,2) (g of free nutrient z required for the assimilation of 1 g of contaminant j)

n_z' = stoichiometric coefficient for release of nutrient z after biomass die-off (g of nutrient released from 1 g of dead biomass)

O^s = oxygen concentration in aqueous phase (g/cm³ of aqueous phase)

O^v = oxygen concentration in vapor phase (g/cm³ of vapor phase)

P_{in} = column inlet pressure (atm)

P_{out} = column outlet pressure (atm)

P_{vj} = vapor pressure of pure phase (NAPL) of contaminant j at soil temperature (atm)

R = ideal gas constant (atm cm³/mol K)

S_j = solubility of contaminant j in the aqueous phase (g/cm³ of aqueous phase)

T = soil temperature (K)

t = time (s)

V_i = volume of each of the compartments into which the column is partitioned (cm³)

v = pore velocity of air in porous medium (volumetric flow rate = $vA\nu$) (cm/s)

X_j = molar fraction of contaminant j in the NAPL (dimensionless)

x = distance along the column from the gas input (cm)

α_j = relative maximum growth rate of biomass on contaminant j (j = 1,2) relative to the best substrate (dimensionless)

Δx = length of each of the compartments into which the column is partitioned; $\Delta x = L/N$ (cm)

λ_{cj} = mass transfer rate coefficient for contaminant j (j = 1,2) between the aqueous and gaseous phases (s⁻¹)

λ'_{cj} = mass transfer rate coefficient for contaminant j (j = 1,2) between the NAPL and aqueous phases (s⁻¹)

λ''_{cj} = mass transfer rate coefficient for contaminant j (j = 1,2) between the NAPL and gaseous phases (s⁻¹)

λ_o = mass transfer rate coefficient for oxygen between gaseous and aqueous phases (s⁻¹)

ν = air-filled void fraction of soil (dimensionless)

ω = volumetric moisture fraction of soil (dimensionless)

The growth of biomass is described by the usual Monod's kinetic equation, where possible rate limitations may be exerted by low aqueous concentrations of the substrates (C_1 or C_2), an undetermined number of nutrients, A_i, and dissolved oxygen, O^s, where the values are those of the bulk aqueous phase, so the model is considering a fully penetrated biofilm (McCarty et al., 1984). A die-off term is also considered.

$$\frac{\partial B}{\partial t} = K_{max}B \frac{n_1\alpha_1\frac{C_1^s}{K_{c1}} + n_2\alpha_2\frac{C_2^s}{K_{c2}}}{1 + \frac{C_1^s}{K_{c1}} + \frac{C_2^s}{K_{c2}}} \frac{O^s}{K_o + O^s} \prod_{j=1}^{n}\left[\frac{Aj}{K_j + A_j}\right] - K_B B$$

(1)

Nutrients are considered to be present only in the aqueous phase so the rates of change of the nutrients' concentrations are related to that for biomass through the stoichiometric coefficients. A source term is also included, which is related to the biomass mineralization:

$$\frac{\partial A_z}{\partial t} = -K_{max}B \frac{n_{z,1}\alpha_1 \frac{C_1^s}{K_{c1}} + n_{z,2}\alpha_2 \frac{C_2^s}{K_{c2}}}{1 + \frac{C_1^s}{K_{c1}} + \frac{C_2^s}{K_{c2}}} \frac{O^s}{K_o + O^s} \prod_{j=1}^{n} \left[\frac{Az}{K_z + A_z}\right] + n_i' K_B B$$

$$(2)$$

The expression for the rate of change of dissolved oxygen needs the inclusion of oxygen diffusion from the gaseous phase beside the biological terms, which include a term for endogenous respiration:

$$\frac{\partial O^s}{\partial t} = -K_{max}B \frac{n_{o,1}\alpha_1 \frac{C_1^s}{K_{c1}} + n_{o,2}\alpha_2 \frac{C_2^s}{K_{c2}}}{1 + \frac{C_1^s}{K_{c1}} + \frac{C_2^s}{K_{c2}}} \frac{O^s}{K_o + O^s} \prod_{j=1}^{n} \left[\frac{Aj}{K_j + A_j}\right]$$

$$- n_o' K_B B + \lambda_o \left[\frac{O^v}{K_{ho}} - O^s\right] \tag{3}$$

The two substrates have expressions for the rate of change in the aqueous concentration similar to that of dissolved oxygen, but an additional term must be included to describe the rate of dissolution of the NAPL:

$$\frac{\partial C_1^s}{\partial t} = -K_{max}B \frac{\alpha_1 \frac{C_1^s}{K_{c1}}}{1 + \frac{C_1^s}{K_{c1}} + \frac{C_2^s}{K_{c2}}} \frac{O^s}{K_o + O^s} \prod_{j=1}^{n} \left[\frac{Aj}{K_j + A_j}\right]$$

$$+ \lambda_{c1} \left[\frac{C_1^v}{K_{hc1}} - C_1^s\right] + \lambda_{c1}'[S_1 X_1 - C_1^s] \tag{4}$$

$$\frac{\partial C_2^s}{\partial t} = -K_{max}B \frac{\alpha_2 \frac{C_2^s}{K_{c2}}}{1 + \frac{C_1^s}{K_{c1}} + \frac{C_2^s}{K_{c2}}} \frac{O^s}{K_o + O^s} \prod_{j=1}^{n} \left[\frac{Aj}{K_j + A_j}\right]$$

$$+ \lambda_{c2} \left[\frac{C_2^v}{K_{hc2}} - C_2^s\right] + \lambda_{c2}'[S_2 X_2 - C_2^s] \tag{5}$$

For purposes of simplicity, all the mass transport processes are described by means of linear equations with lumped parameters as the proportionality constants. Of course, more refined methods may be included, like that described in Chapter 4 for the dissolution of NAPL droplets distributed in the aqueous phase. That approach indicates that, as the cleanup proceeds, the equivalent rate coefficients of the mass transport processes decrease somewhat.

For the species present in the vapor phase, the rates of change of vapor concentrations require the inclusion of desorption from the aqueous phase, the advection terms, and, for the substrates, NAPL evaporation:

$$\frac{\partial O^v}{\partial t} = -\lambda_o \frac{\omega}{\nu} \left[\frac{O^v}{K_{ho}} - O^s \right] - \frac{\partial (\nu O^v)}{\partial x} \tag{6}$$

$$\frac{\partial C_1^v}{\partial t} = -\lambda_{c1} \frac{\omega}{\nu} \left[\frac{C_1^v}{K_{hc1}} - C_1^s \right] - \frac{\partial (\nu C_1^v)}{\partial x} + \lambda_{c1}'' \left[\frac{M_{w1} P_{v1} X_1}{RT} - C_1^v \right] \tag{7}$$

$$\frac{\partial C_2^v}{\partial t} = -\lambda_{c2} \frac{\omega}{\nu} \left[\frac{C_2^v}{K_{hc2}} - C_2^s \right] - \frac{\partial (\nu C_2^v)}{\partial x} + \lambda_{c2}'' \left[\frac{M_{w2} P_{v2} X_2}{RT} - C_2^v \right] \tag{8}$$

Rates of substrate NAPL concentrations' variation will be given by

$$\frac{\partial C_1^L}{\partial t} = -\lambda_{c1}' \omega [S_1 X_1 - C_1^s] - \lambda_{c1}'' \nu \left[\frac{M_{w1} P_{v1} X_1}{RT} - C_1^v \right] \tag{9}$$

$$\frac{\partial C_2^L}{\partial t} = -\lambda_{c2}' \omega [S_2 X_2 - C_2^s] - \lambda_{c2}'' \nu \left[\frac{M_{w2} P_{v2} X_2}{RT} - C_2^v \right] \tag{10}$$

If one of the substrates (say substrate 2) exhibits toxic effects on the biomass when the concentration exceeds a threshold value, C_2^*, the die-off coefficient may be represented by

$$K_B = K_{Bo} [1 + (C_2^s / C_2^*)] \tag{11}$$

It is clear that, with this model, there is no upper limit for the biomass concentration, so under some situations the biomass may grow up to unrealistic values, where, certainly, several phenomena will inhibit further increases of the biomass growth, such as clogging of the porous structure or, probably sooner, the biofilm will become thick enough to make the assumption of a fully penetrated biofilm incorrect. This phenomenon may be described by means of additional equations representing the diffusion and uptake of nutrients and electron donor and acceptor through the biofilm, leading to some of the biomass having significantly less available concentrations of the several substances necessary for biological activity. A simpler way to describe this phenomenon has been used by Kindred and Celia (1989). They include a biological inhibition factor, I, in the expression for the biomass growth, leading to very similar results. The inhibition factor has a value close to one for very low values of biomass concentration but increases rapidly as the biomass exceeds a value B^*. Our model uses this latter approach, which requires only one additional parameter, B^*; we used $B^* = 1$ mg/L, the value employed by Kindred and Celia:

$$\frac{\partial B}{\partial t} = K_{max}B/I(\cdots), \quad \text{where } I = (1 + B/B^*) \quad (12)$$

DESCRIPTION OF THE SIMPLIFIED MODEL

Several simplifications were used to make the first check of the model by comparison of its results to the available field information, most of which was developed by Hinchee and coworkers. The simplifications are the following:

(1) The contaminant will be considered as if it consists of only one species: ($\alpha_1 = 1$; $C_2 = 0$, subindex 1 dropped in the following).

(2) Transport of contaminant through the NAPL/gas interface is expected to be several orders of magnitude faster than the mass transport through the other two interfaces, especially if the NAPL consists basically of a single component, because there will be no gradient concentrations in the NAPL side of the interface and the higher resistance for the mass flow will be that offered from the gaseous phase. This has two major consequences: first, if this interface exists, VOCs will be very rapidly removed by stripping and this portion of the NAPL will not be significant in controlling the cleanup time. Second, numerical integration involving these much faster processes will require much smaller time increments, making the already very long computer runs not practical for available microcomputers. If needed, algorithms based on the steady-state approximation or on the local equilibrium approximation, similar to those described earlier (Gómez-Lahoz et al., 1994a) can be developed in future work to study the case where a multiple component NAPL presents a gaseous/NAPL interface.

(3) Nutrients are not limiting the growth of biomass, or these limitations are also described properly by the inhibition factor used for the description of the diffusion processes through the biofilms.

The equations for the simplified model are as follows:

$$\frac{\partial B}{\partial t} = nK_{max}B/I \frac{C^s}{K_c + C^s} \frac{O^s}{K_o + O^s} - K_B B \qquad (1 \rightarrow 13)$$

$$\frac{\partial O^s}{\partial t} = -n_o K_{max}B/I \frac{C^s}{K_c + C^s} \frac{O^s}{K_o + O^s} - n'_o K_B B + \lambda_o \left[\frac{O^v}{K_{ho}} - O^s \right]$$

$$(3 \rightarrow 14)$$

$$\frac{\partial C^s}{\partial t} = -K_{max}B/I \frac{C^s}{K_c + C^s} \frac{O^s}{K_o + O^s} + \lambda_c \left[\frac{C^v}{K_{hc}} - C^s \right] + \lambda_c^q [S - C^s]$$

$$(4 \rightarrow 15)$$

$$\frac{\partial O^v}{\partial t} = -\lambda_o \frac{\omega}{\nu} \left[\frac{O^v}{K_{ho}} - O^s \right] - \frac{\partial(\nu O^v)}{\partial x} \qquad (6)$$

$$\frac{\partial C^v}{\partial t} = -\lambda_c \frac{\omega}{\nu} \left[\frac{C^v}{K_{hc}} - C^s \right] - \frac{\partial(\nu C^v)}{\partial x} \qquad (7 \rightarrow 16)$$

$$\frac{\partial C^L}{\partial t} = -\lambda'_c \omega [S - C^s] \qquad (9 \rightarrow 17)$$

Equations (2), (5), (8), and (10) of the rigorous model are not needed for this simplified model. The remaining model equations are coupled and strongly nonlinear, so they must be solved numerically. To do this, the column is partitioned into a number of compartments, *N,* each of which is considered to have homogeneous concentrations. Thus, a mass balance for the contaminant present in compartment i gives

$$M_i = (\nu C_i^v + \omega C_i^s + C_i^L)V_i = m_i V_i \qquad (18)$$

and the velocity of the gas coming into the *i*th compartment is given by

$$v_i = \frac{K_D(P_{in}^2 - P_{out}^2) \left[\sqrt{P_{in}^2 - (P_{in}^2 - P_{at}) \frac{i-1}{N}} \right]^{-1}}{2L} \qquad (19)$$

The runs are initialized, assuming that the initial concentrations are homogeneous throughout the entire column and that all the phases are in equilibrium, so the concentrations are obtained from the initial soil concentration, m_i ($t = 0$) and

$$(t = 0) \quad C_i^L = m_i - (\nu K_{hc} + \omega)S; \quad C_i^s = S; \quad C_i^v = SK_{hc} \quad (20)$$

If $C_i^L < 0$ then $C_i^L = 0$ and $C_i^s = m_i/(\omega + \nu K_{hc})$; $C_i^v = C_i^s K_{hc}$.

As explained above, two different sets of runs have been performed: the first considers continuous operation. Results are plotted showing the mass of the contaminant left in the column and the amount that has been biodegraded versus time. The second set of runs simulates the fast monitoring experiments described previously (measure of oxygen uptake during non-venting periods), so these plots show the oxygen concentration in the vapor phase.

In order to permit easier comparisons between runs simulating continuous and discontinuous (oxygen uptake monitoring) operation, runs were performed assuming that, at time $t = 0$, the biomass had already achieved a stable population under aerobic conditions. The calculation of this stable population is done as follows:

$$\frac{\partial B}{\partial t} = n K_{max} B/I - K_B B = 0$$

$$I = (1 + B/B^*) = n K_{max}/K_B \quad (21)$$

$$B = (n K_{max}/K_B - 1)B^*$$

As long as the biomass concentration stays close to the value given by Equation (21), $B \gg B^*$ and $B/I \to B^*$, and the maximum substrate utilization rate, U, which is one of the most commonly reported quantities in microbiological kinetics, is given by

$$UB = K_{max} B/I \approx B^* K_{max}; \quad U \approx K_{max} B^*/B \quad (22)$$

Equation (22) must be used if one wants to compare our values of K_{max} with typical values in the literature, usually reported as U.

Values for the mass transport coefficients were selected based on the following criteria. (1) The rate of dissolution of the VOC present in the NAPL into the aqueous phase is limited by the resistance of the aqueous phase because there is no gradient in the concentration of the VOC in the NAPL. On the other hand, for most VOCs, the rate of desorption from the aqueous to the gaseous phase will be limited mainly by the transport through the aqueous side of the interface. Therefore, resistivities are similar and the differences between λ_c and λ_c' will be related basically to the differences between the relative amounts of surface shown by the water to the gaseous phase and to the NAPL, which are not likely to be very different, so we have initially assumed that $\lambda_c = \lambda_c'$. (2) The ratio of the rate parameters for mass transfer between the gaseous and the aqueous phases for oxygen and the contaminant (λ_o/λ_c) is taken equal to the ratio of the aqueous diffusivities of oxygen and the contaminant, which is typically close to two.

The model permits us to consider the oxygen utilized in endogenous respiration. Nevertheless, we have used a value for the stoichiometric constant for oxygen uptake equal to the one that would be necessary to achieve mineralization of the contaminant, while the stoichiometric constant for biomass endogenous respiration is set equal to zero. We have assumed a general stoichiometry of the substrate assuming that it is hexane, so

$$C_6H_{14} + \frac{19}{2} O_2 \rightarrow 6CO_2 + 7H_2O$$

Approximately 3.5 g of oxygen are used to mineralize 1 g of substrate.

In addition to the previous considerations, we have selected the mass transfer coefficients and K_{max} to give results similar to those observed by Hinchee and coworkers in their field experiments. All other biological parameters are similar to those used in Gómez-Lahoz et al. (1994a). The values of the parameters are given in Table 9.1.

MODEL RESULTS AND DISCUSSION

From results reported by Hinchee and coworkers [see, for instance Hinchee and Ong (1992); Hinchee (1994)], the values of the kinetic rate parameters as measured from oxygen uptake, in those cases considered to be zero order in oxygen, are typically in the range of 0.1 to 1% O_2 h^{-1} $\sim$4–40 mg (kg soil)$^{-1}$ day^{-1}. As mentioned above, we have postulated that this rate is limited either by the mass transfer processes or the biological processes or both. If the mass transfer processes are limiting,

$$\left[\frac{\partial O^v}{\partial t}\right]_{uptake} = -\lambda_o \frac{\omega}{\nu} \left[\frac{O^v}{K_{ho}} - O^s\right] \tag{23}$$

where if mass transfer is controlling,

$$\left[\frac{O^v}{K_{ho}} - O^s\right] \approx \frac{O^v}{K_{ho}} \approx 9 \text{ mg/L}$$

TABLE 9.1. Values for the Parameters.

Column length (L)	50 cm
Column radius (r)	10 cm
Δx	5 cm
Voids fraction associated with the mobile phase (ν)	0.2
Volumetric moisture content of the soil (ω)	0.2
Inlet pressure (P_{in})	1 atm
Outlet pressure (P_{out})	0.99 to 0.9995 atm
Temperature (T)	15°C
Darcy's constant (K_D)	50 cm²/atm s
Soil density (ϱ)	1.5 g/cm³
Initial contaminant concentration ($M/\varrho V$)	1,000 mg cont./kg soil
Initial biomass concentration (B)	49 to 499 mg/L
Inhibition parameter (B^*)	1 mg/L
Henry's constant of contaminant (Kh_c)	$2 \cdot 10^{-2}$
Henry's constant of oxygen (Kh_o)	30
Stoichiometric coefficient for substrate (n)	0.5 g biomass/g substrate
Stoichiometric coefficient for oxygen (n_o)	3.5 g oxygen/g substrate
Stoichiometric coefficient for endogenous respiration (n'_o)	0 g oxygen/g biomass
Maximum velocity constant (K_{max})	$8 \cdot 10^{-4}$ to $8 \cdot 10^{-3}$ s⁻¹
Michaelis constant of substrate (K_c)	0.1 mg/L
Michaelis constant of oxygen (K_o)	0.1 mg/L
Die-off coefficient of biomass (K_B)	$8 \cdot 10^{-6}$ s⁻¹
NAPL/aqueous phase mass transfer coefficient for contaminant (λ'_c)	$\lambda'_c = \lambda_c$
Aqueous/gaseous phase mass transfer coefficient for contaminant (λ_c)	$\lambda_o = 2\lambda_c$

(assuming this value as the mean value throughout the soil and that dissolved oxygen will almost be exhausted). If $\omega = 0.2$, $\nu = 0.2$; and $\lambda_o = 4 \cdot 10^{-4}$ s⁻¹, this leads to a zero order coefficient for oxygen uptake of approximately 0.9% O_2 h⁻¹.

In a similar fashion, if the biological processes are the ones limiting the rate of oxygen uptake, then,

$$\left[\frac{\partial O_v}{\partial t}\right]_{uptake} = -n_o K_{max} B/I \frac{C^s}{K_c + C^s} \frac{O^s}{K_o + O^s} - n'_o K_B B \quad (24)$$

if mineralization is not consuming oxygen (or if this oxygen demand is included in n_o, as is the case here), and assuming enough availability of oxygen and substrate (mass transfer processes not limiting):

$$\left[\frac{\partial O_v}{\partial t}\right]_{uptake} = -n_o K_{max} B/I \quad (25)$$

with a value of $B^* = 1$ mg/L (Kindred and Celia, 1989), and $n_o = 3.5$ (mg O_2/mg substrate), a value of $K_{max} = 8 \cdot 10^{-4}$ s^{-1} leads to a maximum for the oxygen uptake rate close to 0.7% O_2 h^{-1}.

Figures 9.2 and 9.3 show the results obtained for two runs with the parameter values of Table 9.1, the first one performed under conditions leading to mass transfer limitations, while the second one corresponds to kinetics limited mainly by the biological processes. The initial biomass concentrations were selected to have a stable population, as calculated from Equation (21). The volumetric flow for these runs corresponds to approximately seventeen pore volumes per day, which is a typical SVE rate, and leads to low values of the relative contribution of the biological processes to the cleanup. This relative contribution is expressed as the contaminant mass removed by biodegradation related to the total mass rather than to savings in the time needed to achieve closure standards or to cost reductions attachable to the biological contribution, which are probably more favorable, since SVE efficiency decreases clearly as the cleanup approaches completion, while biodegradation continues to take place at a relatively rapid rate until the substrate is almost exhausted, and so plays an important role in avoiding the tailing effects that would take place if reme-

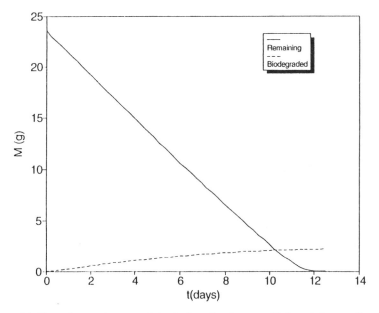

Figure 9.2 Mass of contaminant remaining in the soil and amount biodegraded versus time for $\Delta P = 10^{-2}$ atm; $\lambda_o = 4 \cdot 10^{-4}$ s^{-1}; $K_{max} = 8 \cdot 10^{-3}$ s^{-1}; $B_0 = 499$ mg/L; other parameter values as in Table 9.1.

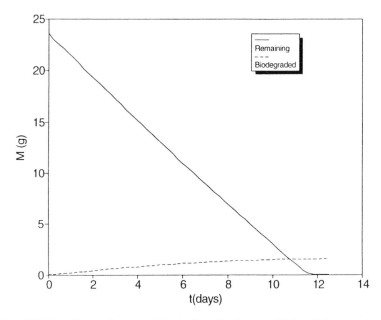

Figure 9.3 Mass of contaminant remaining in the soil and amount biodegraded versus time for $\Delta P = 10^{-2}$ atm; $\lambda_o = 4 \cdot 10^{-3}$ s^{-1}; $K_{max} = 8 \cdot 10^{-4}$ s^{-1}; $B_0 = 49$ mg/L; other parameter values as in Table 9.1.

diation were taking place only by stripping [see Gómez-Lahoz et al., (1994a) on this point].

The zero order kinetic parameter, which is measured in field experiments from O_2 uptake or CO_2 evolution values, can be very easily calculated from the model results tracking these values, too. For the conditions used in these two figures, values of this kinetic constant very close to those calculated from Equations (23) or (25), respectively, are obtained (0.9 and 0.7% O^2 h^{-1}, respectively), with a rather good fit of the values observed at the column exit to this zero order kinetic equation. Thus, there will be no way to discriminate between these two situations if one is working under continuous operation and is following the biological processes by any of the typical procedures that were referred to in the section on monitoring; one will observe zero order kinetics of the biological process no matter which process (mass transfer or biological) is controlling. As a result it will be almost impossible to determine from results obtained during continuous operation (Figures 9.2 and 9.3) if there are mass transfer processes limiting the rate of contaminant biodegradation. Such information is important for making decisions about what to do to improve the efficiency of the system. For instance, if the biological processes are limited by mass transfer kinetics, it will be useless to introduce nutrients into the soil in order to enhance biodegradation.

Figures 9.4 and 9.5 describe runs performed under the same conditions as those of Figures 9.2 (mass transfer rate–limited) and 9.3 (biological activity–limited), respectively, but these runs are following oxygen uptake after the blower is stopped [i.e., these runs are simulating the fast monitoring procedure developed by Hinchee and Ong (1992), described in the section on monitoring]. As can be seen, this kind of procedure leads to very different results for the two situations. Oxygen uptake under mass transfer limited kinetics is first order with respect to oxygen in the vapor phase (the curve fits a logarithmic expression), whereas oxygen uptake for kinetics limited by the biological activity follows a zero order equation until oxygen is close to exhaustion. For these two cases, fits to zero order kinetics of oxygen concentration values obtained for the compartment close to the column exit from the turn-off time until $O^v = 60$ mg/L give $K = 0.454\%$ O_2 h^{-1} (mass transport–limited) and $K = 0.666\%$ O_2 h^{-1} (biologically limited), with correlation coefficients of $r^2 = 0.96$ and $r^2 = 0.99997$, respectively. While the second value is consistent with that calculated from Equation (25) (0.7% O_2 h^{-1}), the first one is about one-half the value obtained from Equation (23) (0.9% O_2 h^{-1}). This discrepancy occurs because the assumption that the oxygen diffusion driving force $(O^v/K_{ho} - O^s) = 9$ mg/L is no longer correct, since this value drops from 9 to 2 mg/L during the portion of the experiment fitted. Nevertheless, the correlation

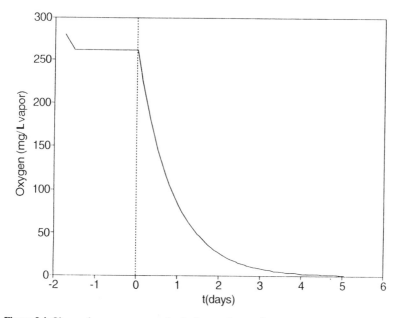

Figure 9.4 Observed oxygen concentration in the gas phase at the column exit versus time. Initial pressure drop $\Delta P = 10^{-2}$ atm; $\Delta P = 0$ atm (no air flow) after 40 h; $\lambda_o = 4 \cdot 10^{-4}$ s^{-1}; $K_{max} = 8 \cdot 10^{-3}$ s^{-1}; $B_0 = 499$ mg/L; other parameter values as in Table 9.1.

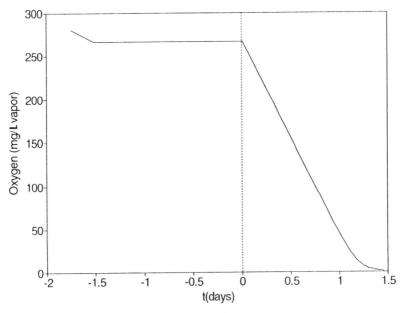

Figure 9.5 Observed oxygen concentration in the gas phase at the column exit versus time. Initial pressure drop $\Delta P = 10^{-2}$ atm; $\Delta P = 0$ atm (no air flow) after 40 h; $\lambda_o = 4 \cdot 10^{-3}$ s^{-1}; $K_{max} = 8 \cdot 10^{-4}$ s^{-1}; $B_0 = 49$ mg/L; other parameter values as in Table 9.1.

coefficient (0.96) could be considered good enough if it were obtained from actual values of a field experiment, so it may prove difficult to establish the order of oxygen uptake, especially if only the initial portion of the oxygen uptake curve is tracked.

If both processes are controlling, as in Figure 9.6, oxygen in the aqueous phase will be far from the value corresponding to equilibrium with the vapor phase, but, at the beginning of the process, the aqueous oxygen concentration will be high enough to exert no limitations on the biological activity, and one will observe an initial linear decrease of the gaseous oxygen concentration being monitored, corresponding to zero order kinetics. Thus, oxygen will be depleted from both phases at a rate that maintains the driving force for oxygen diffusion ($O^v/K_{ho} - O^s$) at a constant level. Later, as O^s approaches the value of the half saturation constant for oxygen ($K_o = 0.1$ mg/L), the biological activity changes to become first order with respect to dissolved oxygen, and the observed vapor oxygen uptake also changes to first order kinetics. Thus, if this change of order occurs at a value, O_c^v, significantly higher than $K_{ho} \cdot K_o$ ($= 30 \cdot 0.1$ mg/L $\approx 0.2\%$ O$_2$ for the values used in this work), it would indicate that the mass transfer of oxygen to the aqueous phase will not be limiting the biological activity as long as $O^v > O^v_c$; the soil may be considered well aerated if this condition is maintained. If working conditions are such that $O^v < O^v_c$, the aqueous phase oxygen concentration is most probably below the half saturation constant for oxygen, and the biological activity will be rate-limited by oxygen availability. The values of O^v_c for the runs presented in Figures 9.4, 9.5, and 9.6 are approximately 250, 10, and 100 mg/L, respectively.

Now that we have a tool to discriminate between the two types of control of the biological activity during bioventing, mass transport–limited and biological rate–limited, a discussion will follow of some of the most typical management modifications for bioventing operations. These include (1) decreases in the airflow rate and (2) increases in the airflow path length such that contaminant-laden air is forced to circulate through noncontaminated areas. Both lead to increases in the residence time of air in the subsoil (Dupont et al., 1991; Miller et al., 1991; Dupont, 1993). Some other alternatives with obvious results are not studied, such as introduction of nutrients (useless if the biological processes are not limiting) or changes in the soil temperature, which will show very little effect on the rate of bioremediation if the process is controlled by mass transfer kinetics. Information on the impacts of such alternatives, however, may also be

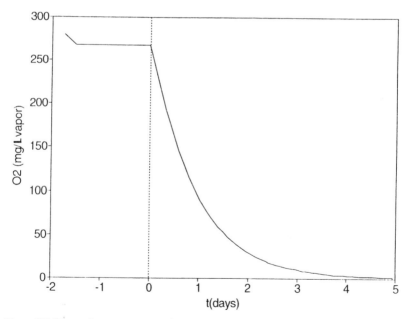

Figure 9.6 Observed oxygen concentration in the gas phase at the column exit versus time. Initial pressure drop $\Delta P = 10^{-2}$ atm; $\Delta P = 0$ atm (no air flow) after 40 h; $\lambda_o = 4 \cdot 10^{-4}$ s^{-1}; $K_{max} = 8 \cdot 10^{-4}$ s^{-1}; $B_0 = 49$ mg/L; other parameter values as in Table 9.1.

derived from the analysis of fast monitoring curves such as are shown in Figures 9.4 to 9.6.

Figures 9.7 and 9.8 show the results corresponding to runs performed under mass transfer–limited conditions and biological airflow rate values similar to those suggested by Dupont (1993) for bioventing operations, i.e., an airflow rate of 0.86 pore volumes/day. Figure 9.7 is for mass transfer limited conditions and Figure 9.8 for biodegradation rate–limited by the biological activity and may be compared to Figures 9.2 and 9.3, respectively, where the flow rate was seventeen pore volumes/day. As can be seen, simple reduction of the volumetric rate leads to significant increases in the biological contribution to the overall cleanup. This may result in significant decreases of the total cleanup cost as the amount of contaminant stripped from the column, which will very likely need further treatment, is decreased significantly. This saving is to be added to those others related to low flux operation of SVE wells such as lower volumes of gas needing treatment for each kg of contaminant stripped, and the improved efficiency of some of the gas-activated carbon (GAC) units (which are able to adsorb larger amounts of contaminants if the residence time in the canister is enough to permit approach to local equilibrium conditions) (Wilson et al., 1994).

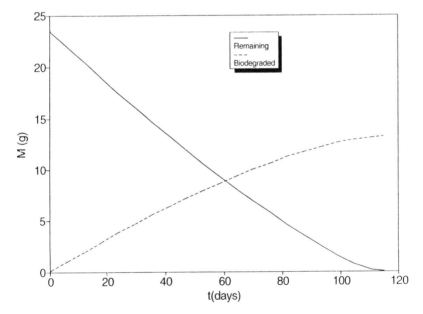

Figure 9.7 Mass of contaminant remaining in the soil and amount biodegraded versus time for $\Delta P = 5 \cdot 10^{-4}$ atm; $\lambda_o = 4 \cdot 10^{-4}$ s^{-1}; $K_{max} = 8 \cdot 10^{-3}$ s^{-1}; $B_0 = 499$ mg/L; other parameter values as in Table 9.1.

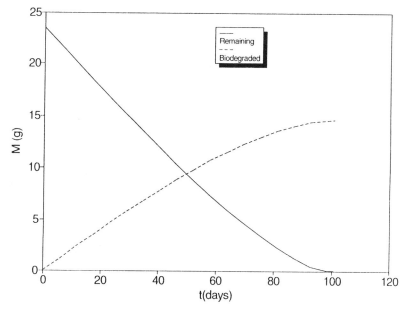

Figure 9.8 Mass of contaminant remaining in the soil and amount biodegraded versus time for $\Delta P = 5 \cdot 10^{-4}$ atm; $\lambda_o = 4 \cdot 10^{-3}$ s^{-1}; $K_{max} = 8 \cdot 10^{-4}$ s^{-1}; $B_0 = 49$ mg/L; other parameter values as in Table 9.1.

The values of the zero order kinetic constants for oxygen uptake, as calculated for the first thirty days, are 0.51 and 0.64% O_2 h^{-1}, respectively. These values are to be compared with 0.91 and 0.68% O_2 h^{-1}, obtained from runs represented in Figures 9.2 and 9.3, corresponding to high values of the flow rate. The values of the vapor phase oxygen concentration at the column exit after this thirty day period are 91.8 and 44.6 mg/L (6.5 and 3.2%, respectively). It is clear that as one decreases the airflow rate, a larger fraction of the oxygen introduced with the airstream is utilized by the microorganisms, so the gas phase oxygen concentration in those regions of the column further from the air entrance decreases with decreasing volumetric air flux. As explained above, microbiological activity is not limited by the oxygen vapor concentration as long as this concentration is above O_c^v (the value for which the rates in the fast monitoring experiments change from zero to first order), and this value is much higher when mass transfer processes are controlling than when microbiological kinetics are the limiting step for the contaminant biodegradation. Thus, a decrease in the volumetric flux may result in a decrease in the microbiological activity in those regions of the column further from the air entrance, which is most important for those situations in which the mass transfer processes are controlling. This will be seen again in more detail in the following paragraphs.

Another series of runs was performed to study the influence of the flow path length in the efficiency of bioremediation. Figure 9.9 presents the results obtained for a run simulating conditions identical to those used for Figure 9.3 (biological activity–limited) but with a column length twice as large, the second half of which initially contains no contamination. The pressure drop was set to keep the volumetric flux rate equal (2,262 mL/h) in both runs; this is at $2 \cdot 10^{-2}$ atm for the run in Figure 9.9 and 10^{-2} atm for that shown in Figure 9.3. Thus, the transition times through the contaminated section are identical in both cases. Under these conditions, the contaminant stripped from the contaminated region of the column reaches an initially clean region, into which it dissolves and is subject to further biodegradation by the microorganisms present in this section of the column. The retention factor introduced by this mechanism of sorption and reaction permits only negligible amounts of contaminant to be stripped from the column during the first two days, during which about seventeen pore volumes of air have been circulated (transit time throughout the entire column = 2.8 h). At the end of this forty-eight hour period, the ox-

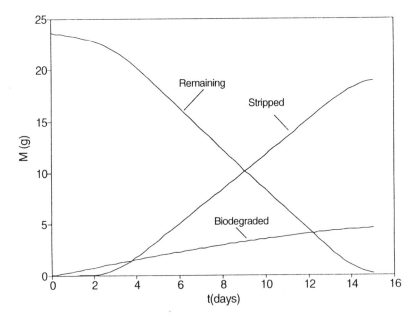

Figure 9.9 Mass of contaminant remaining in the soil, amount biodegraded and amount stripped versus time for $L = 100$ cm (first 50 cm contaminated): $\Delta P = 4 \cdot 10^{-2}$ atm; $\lambda_o = 4 \cdot 10^{-3}$ s^{-1}; $K_{max} = 8 \cdot 10^{-4}$ s^{-1}; $B_0 = 49$ mg/L; other parameter values as in Table 9.1.

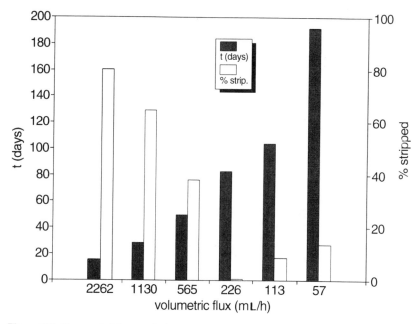

Figure 9.10 Time needed for remediation and amount stripped for different volumetric flow rates; other parameters as those of Figure 9.9.

ygen concentration of the air exiting the column is 254 mg/L, so further decreases in the flow rate may lead to better results in terms of the relative contributions of biodegradation and stripping from the column.

Figure 9.10 shows the time needed for completion of cleanup and the total amount of VOC stripped that was obtained for different volumetric flow rates (pressure drops); all other parameters are as those of Figure 9.9. As expected, as the flow rate decreases, the time needed for complete remediation increases. The amount stripped out of the column is at a minimum value (very close to zero; probably no treatment of the exiting gases would be needed) for a volumetric flux of 226 mL/h (pressure drop 0.002 atm; 0.864 pore volumes per day). Flow rates above this value do not permit the biodegradation of the contaminant in the second part of the column as rapidly as it is supplied from the first section. On the other hand, as the flow decreases to very low values, the oxygen concentration in the vapor phase is almost exhausted due to the biological processes occurring in the first portion of the column, so some of the contaminant stripped will reach the column exit without being biodegraded simply because there is not enough oxygen supplied to the second portion of the column to permit aerobic biodegradation.

Results obtained for similar runs performed under the mass transfer–limited conditions are presented in Figure 9.11. The differences between the two figures are better understood by examining the values obtained for the dissolved and vapor phase oxygen concentrations in the compartment closest to the column exit, when the cleanup is under way— for instance, after 50% of the contaminant has been removed. These values are shown in Table 9.2.

As can be seen, under mass transfer–limited conditions, even for the lower value of the volumetric flux explored, substantial amounts of oxygen reach the column exit because the rate of oxygen uptake decreases linearly with the vapor phase oxygen concentration, as was observed in the fast monitoring experiment simulation. This first order rate is not observed directly from the continuous experiments in any of these runs, not even for those with the lower value of the volumetric flux, as can be seen in Figure 9.12, where results obtained for these conditions are represented. The low availability of oxygen in part of the column could be related to the low values of the slope of the line plotting contaminant removal by biodegradation. Although the linear shape indicates that removal rate does not change with time, this is not an indication of a zero order reaction with respect to oxygen as long as this is supplied at a constant rate.

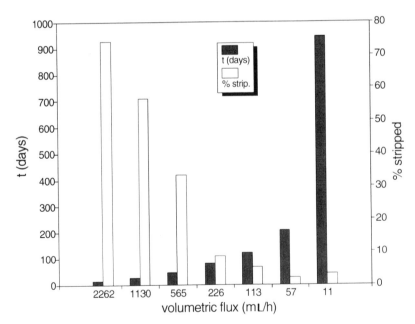

Figure 9.11 Time needed for remediation and amount stripped for different volumetric flow rates; other parameters as those of Figure 9.12.

TABLE 9.2. Aqueous and Vapor Oxygen Concentrations in the Compartment Closest to the Column Exit after 50% of the Contaminant Has Been Removed (Parameters as in Table 9.1).

	Biologically Limited Runs ($K_{max} = 8 \cdot 10^{-4}$ s⁻¹; $\lambda_o = 4 \cdot 10^{-3}$ s⁻¹)		Mass Transfer–Limited Runs ($K_{max} = 8 \cdot 10^{-3}$ s⁻¹; $\lambda_o = 4 \cdot 10^{-4}$ s⁻¹)	
Volumetric Flux (mL/h)	Vapor Phase (mg/L)	Aqueous Phase (mg/L)	Vapor Phase (mg/L)	Aqueous Phase (mg/L)
4,526	261.6	8.042	257.8	0.01416
2,262	257.1	7.892	251.0	0.01381
1,130	239.5	7.309	229.7	0.01252
565	200.4	6.004	184.9	0.00988
226	79.81	2.577	94.41	0.004921
113	0.1166	0.002282	33.77	0.001839
57	0.0857	0.002858	6.477	0.000512

It should also be noticed that, when working under biologically limited conditions, as long as the oxygen concentration in the stream coming out of the soil column is not very low, one can decrease the flow rate and achieve a significant decrease in the amount of contaminant stripped without significant decreases in the rate of biological activity. However, if conditions are those leading to rates of biological activity limited by mass transfer of oxygen from the vapor to the aqueous phase, as the volumetric flux is decreased, the mean value of the rate of biological activity decreases, too, even when relatively large amounts of oxygen reach the column exit with the vapor stream. Thus, decisions related to the optimum volumetric air flux or increases in the flow path length for bioventing should be taken after information obtained from the fast monitoring tests described by Hinchee.

CONCLUSIONS

Bioremediation is one of the most promising technologies for the remediation of contamination in the vadose zone. When feasible, it will probably be the preferred technology, because it may involve costs comparatively smaller than those of other techniques, and it releases the responsible party from later liabilities since it destroys the contaminant rather than just removing it to another site. Nevertheless, one of the most problematic questions related to bioventing, as well as to some other bioremediation technologies, is that of demonstration that it is really working. There are several techniques available that indicate that the bio-

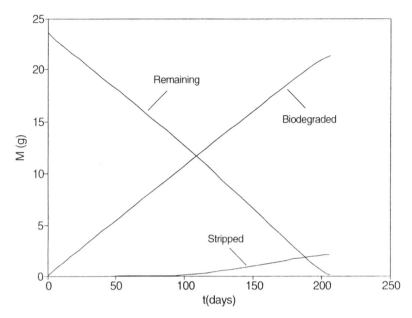

Figure 9.12 Mass of contaminant remaining in the soil, amount biodegraded and amount stripped versus time for $L = 100$ cm (first 50 cm contaminated): $\Delta P = 5 \cdot 10^{-4}$ atm; $\lambda_o = 4 \cdot 10^{-4}$ s^{-1}; $K_{max} = 8 \cdot 10^{-3}$ s^{-1}; $B_0 = 499$ mg/L; other parameter values as in Table 9.1.

degradation is actually taking place, although it seems clear that none of them provides absolute proof, and frequently several would be required to convince the regulatory authorities that the technology is really working.

Among these available monitoring techniques, a simple and reliable one developed for bioventing is the oxygen uptake (and CO_2 evolution from soils of low pH values) during continuous operation of the system as compared to analogous measurements in similar soils outside of the contaminated region. Besides this long-term monitoring, a short-term technique has been developed by Hinchee and coworkers based on gaseous phase oxygen uptake during stops in the venting procedure. Information obtained from these tests not only supplies proof of actual biodegradation and gives an estimate of the rate at which the microorganisms are degrading the contaminants but may also give insight as to the rate-limiting mechanism that is involved in the bioremediation process taking place in the soil.

A model is presented here for the interpretation of the results obtained from the continuous operation monitoring and the short-term monitoring. This model indicates that the rate of the biological processes in the subsurface during the continuous operation of bioventing will probably be constant, with a linear plot no matter which is the rate-limiting process. This should not be interpreted as an indication of zero order dependence on oxygen concentration in the gaseous phase (O^v). As a matter of fact, if the process is rate-limited by the mass transfer processes occurring between the different phases in soil, there will most likely be a first order dependence on O^v as can be seen from the results obtained from the simulation of the short-term monitoring tests. If these mass transfer processes are fast enough to ensure an adequate supply of the different substances essential for microorganisms (basically the electron donor and acceptor), the biological processes will be determining the rate of oxygen uptake. In this case the process will be zero order with respect to O^v, as can be observed from the modeling of the fast monitoring test, down to very low values of O^v. When O^v uptake plots for the fast in situ monitoring tests show a change of order from linear to logarithmic at intermediate values, this may be indicating that there is mixed control of the biodegradation process.

This information should help to improve the management of bioventing systems, indicating the optimum volumetric flow rate, the improvements that may be expected from increases in the flow path length, and the possible advantages of nutrient addition, soil warming, etc.

Bioventing may be performed, either in addition to volatilization or to minimize volatilization. If one is looking to eliminate the usually expensive effluent gas treatment units necessary in most SVE systems, minimization of volatilization will be the option. This can be performed basically by two management techniques: a decrease in the flow rate and an increase in the circulating gas path, both resulting in a larger gas transit time, which may lead to complete biodegradation of the contaminant. This option, in turn, will lead to substantially larger remediation times during which monitoring will be necessary, since the demonstration of no important contaminant concentrations in the effluent during the initial periods is not a warranty that this will continue to be true until the end of the remediation process, as was observed from our column model simulations when the flow path length was increased to include clean portions of soil.

Of course, the best management option will be the one leading to minimization of the remediation costs while achieving good remedial standards. The most important costs involved in the decision about these management options are those related to monitoring and those related to effluent gas treatment. The first costs are basically proportional to the time necessary to reach closure standards, while the second ones are directly proportional to the volume of gas needing treatment (rather than to the amount of contaminant stripped). In previous chapters it has been shown that low extraction rates may have important economical advantages when only SVE is considered, especially if there is no local equilibrium between the contaminant present in the stationary phases and that in the gaseous stream. Our results here indicate that low extraction rates may also help to increase the amount of contaminant biodegraded. Thus, those options looking to optimize the biodegradation processes may also lead to better performance of the SVE system. Optimization of the system will probably require the performance of fast monitoring tests and analysis of the results based on models of characteristics similar to those of the model presented here, so that one will be able to identify the rate-limiting process, which should be helpful in deciding the best way to operate a specific installation.

We believe that there are more similarities than differences between SVE and bioventing and that these technologies should be considered together when feasibility studies are performed. Design of in situ remediation applications is frequently susceptible to errors because, among other reasons, it is difficult to know accurately where the contaminant is. One may design his remediation technology to optimize SVE and, thus, try to place the wells as close as possible to the zone of maximum contamination. If one starts up such an installation and observes that the contaminant concentration increases during the first few days, this may be indicating that the bulk of the contaminant is relatively far from where it was initially thought to be. Larkin et al. (1991) observed such evolution in the contaminant concentration, together with an increase of the CO_2 and a decrease of the O_2 concentrations, indicating that some biological activity was occurring in the subsurface. In such circumstances, a good management option may be to change the operating conditions to enhance the biological processes even if this were not initially considered when the design was performed.

ACKNOWLEDGEMENTS

We acknowledge the helpful suggestions of Dr. Robert Hinchee to obtain a more realistic and complete model. He and Dr. Robert Norris were especially kind in providing much of the scientific literature used in this chapter.

REFERENCES

Adriens, P. and D. D. Focht, 1990, "Continuous Coculture Degradation of Selected Polychlorinated Biphenyl Congeners by Acinetobacter Spp. in an Aerobic Reactor System," *Environ. Sci. Technol.*, 24(7):1042–1049.

Aggarwal, P. K. and R. E. Hinchee, 1991, "Monitoring in situ Biodegradation of Hydrocarbons by Using Stable Carbon Isotopes," *Environ. Sci. Technol.*, 25(6):1178–1180.

Aggarwal, P. K., J. L. Means, D. C. Downey, and R. E. Hinchee, 1991, "Use of Hydrogen Peroxide as an Oxygen Source for in situ Biodegradation, Part II. Laboratory Studies," *J. of Hazardous Materials,* 27:301–314.

Alexander, M., 1991, *Introduction to Soil Microbiology,* Krieger Publishing Company.

Arthur, M. F., G. K. O'Brien, S. S. Marsh, and T. C. Zwick, 1992, "Evaluation of Innovative Approaches to Stimulate Degradation of Jet Fuels in Subsoils and Groundwater," Report NCEL-CR92-004, 35 pp.

Barcelona, M. J., F. Kelly, A. Wehrmann, and W. A. Pettyjohn, 1990, "Contamination of Ground Water: Prevention, Assessment, Restoration," Noyes Data Corp. Pollution Technology Review, 184, no. 2.

Bossert, I. and R. Bartha, 1984, "The Fate of Petroleum in Soil Ecosystems," in *Petroleum Microbiology,* R. M. Atlas, ed., MacMillan Pub., New York, NY.

Brown, R., 1994, "Treatment of Petroleum Hydrocarbon in Ground Water by Air Sparging," in *Handbook of Bioremediation,* Project Officer, J. E. Matthews, Lewis Publishers, Boca Raton, FL.

Coho, J. W. and R. G. Larkin, 1992, "Preliminary Results of a Vapor Extraction Remediation in Low Permeability Soils," *Federal Environment Restoration Conference and Exhibition,* April 15–17, Virginia, pp. 425–430.

De Paoli, D. W., S. E. Herbes, and M. G. Elliot, 1991, "Performance of in situ Soil Venting Systems at Jet Fuel Spill Sites," Appendix H of *Soil Vapor Extraction Technology. Reference Handbook,* T. A. Pedersen and J. T. Curtis, eds., U.S. EPA Report No. EPA/540/2-91/003, Feb.

Dupont, R. R., 1993, "Fundamentals of Bioventing Applied to Fuel Contaminated Sites," *Environmental Progress,* 12(1):45–53.

Dupont, R. R., W. J. Doucette, and R. E. Hinchee, 1991, "Assessment of in situ Bioremediation Potential and Application of Bioventing at a Fuel-Contaminated Site," in *In situ Bioreclamation. Applications and Investigations for Hydrocarbon and Contaminated Site Remediation,* R. E. Hinchee and R. F. Olfenbuttel, eds., Butterworth-Heinemann, Stoneham, MA, pp. 262–282.

Ely, D. L. and D. A. Heffner, 1988, "Process for in situ Biodegradation of Hydrocarbon Contaminated Soil," U.S. Patent 4,765,902, Aug. 23, 1988.

García-Herruzo, F., C. Gómez-Lahoz, J. J. Rodríguez-Jiménez, D. J. Wilson, R. A. García-Delgado, and J. M. Rodríguez-Maroto, 1993, "Soil Clean Up: Pulse Soil Vapor Extraction and Bioremediation," *Proc. 6th Mediterranean Congress on Chemical Engineering, Vol. 2,* pp. 639–640.

Gómez-Lahoz, C., J. M. Rodríguez-Maroto, and D. J. Wilson, 1994a, "Biodegradation Phenomena during Soil Vapor Extraction. A High-Speed Non-Equilibrium Model," *Separ. Sci. Technol.,* 29(4):429–463.

Gómez-Lahoz, C., J. J. Rodríguez, J. M. Rodríguez-Maroto, and D. J. Wilson, 1994b, "Biodegradation Phenomena during Soil Vapor Extraction. Sensitivity Studies for Single Substrate Systems," *Separ. Sci. Technol.,* 29(5):557–578.

Gómez-Lahoz, C., J. J. Rodríguez, J. M. Rodríguez-Maroto, and D. J. Wilson, 1994c, "Biodegradation Phenomena during Soil Vapor Extraction. III Sensitivity Studies for Two Substrates," *Separ. Sci. Technol.,* 29(10):1275–1291.

Harder, H. and Th. Höpner, 1991, "Hydrocarbon Biodegradation in Sediments and Soils. A Systematic Examination of Physical and Chemical Conditions. Part V Moisture," *Hydrocarbon Technology,* 44:329–332.

Heitzer, A. and G. S. Sayler, 1993, "Monitoring the Efficacy of Bioremediation," *Trends in Biotechnology,* 11:334–343.

Hinchee, R. E., 1994, "Bioventing of Petroleum Hydrocarbons," in *Handbook of Bioremediation,* Project Officer, J. E. Matthews, Lewis Publishers, Boca Raton, FL.

Hinchee, R. E. and M. Arthur, 1991, "Bench Scale Studies of the Soil Aeration Process for Bioremediation of Petroleum Hydrocarbons," *J. Appl. Biochem. and Biotech.,* 28/29:901–906.

Hinchee, R. E. and S. K. Ong, 1992, "A Rapid in situ Respiration Test for Measuring Aerobic Biodegradation Rates of Hydrocarbons in Soils," *Journal of the Air & Waste Management Association,* 42(10):1305–1312.

Hinchee, R. E., D. C. Downey, and P. K. Aggarwal, 1991a, "Use of Hydrogen Peroxide as an Oxygen Source for in situ Bioremediation. Part I. Field Studies," *J. Hazardous Materials,* 27:287–299.

Hinchee, R. E., D. C. Downey, R. R. Dupont, P. K. Aggarwal, and R. N. Miller, 1991b, "Enhancing Biodegradation of Petroleum Hydrocarbons through Soil Venting," *J. Hazardous Materials,* 27:315–325.

Hinchee, R. E., S. K. Ong, R. N. Miller, D. C. Downey, and R. Frandt, 1992, "Test Plan and Technical Protocol for a Field Treatability Test for Bioventing," U.S. Air Force Center for Environmental Excellence.

Hoeppel, R. E. and R. E. Hinchee, 1994, "Enhancing Biodegradation for On-Site Remediation of Contaminated Soils and Ground Water," in *Hazardous Waste Site Soil Remediation. Theory and Application of Innovative Technologies,* D. J. Wilson and A. N. Clarke, eds., Marcel Dekker Inc., New York, NY.

Hoeppel, R. E., R. E. Hinchee, and M. F. Arthur, 1991, "Bioventing Soils Contaminated with Petroleum Hydrocarbons," *J. Industrial Microbiology,* 8:141–146.

Huling, S. G., B. E. Bledsoe, and M. V. White, 1990, "Enhanced Bioremediation Utilizing Hydrogen Peroxide as a Supplemental Source of Oxygen. A Laboratory and Field Study," EPA/600/2-90/006, Feb.

Jespersen, C., J. Douglas, and J. Exner, 1993, "Bioremediation Tackles Hazwaste," *Chemical Engineering* (June):116–122.

Kindred, J. S. and M. A. Celia, 1989, "Contaminant Transport and Biodegradation. 2. Conceptual Model and Test Simulations," *Water Resour. Res.,* 25(6):1149–1159.

Larkin, R. G., M. P. Hemingway, and J. W. Coho, 1991, "Results of a Soil Vapor Extraction Pilot Test in Low Permeability Terrace Deposits, Kelly AFB, Texas," *Proc. 1991 Petroleum Hydrocarbons and Organic Chemicals in Ground Water: Prevention, Detection and Restoration, Vol. 8,* Houston, TX, pp. 191–203.

Lee, M. D., J. M. Thomas, R. C. Borden, P. B. Bedient, C. H. Ward, and J. T. Wilson, 1988, "Biorestoration of Aquifers Contaminated with Organic Compounds," *CRC Critical Reviews in Environmental Control,* 18(1):30–89.

Leeson, A., R. E. Hinchee, J. Kittel, G. Sayles, C. M. Vogel, and R. N. Miller, 1993, "Optimizing Bioventing in Shallow Vadose Zones and Cold Climates," *Hydrological Sciences J.,* 38(4):283–295.

MacDonald, J. A. and B. E. Rittman, 1993, "Performance Standards for in situ Bioremediation," *Environ. Sci. Technol.,* 27(10):1974–1979.

Marrin, D. L., J. J. Adriany, and A. J. Bode, 1991, "Results of a Soil Vapor Extraction Pilot Test in Low Permeability Terrace Deposits," *Ground Water Management,* 8:191–203.

McCarty, P. L., B. E. Rittmann, and E. J. Bouwer, 1984, "Microbial Processes Affecting Chemical Transformations in Groundwater," in *Groundwater Pollution Microbiology*, G. Bitton and C. P. Gerba, eds., John Wiley & Sons, Inc., New York, NY.

Michelsen, D. L. and M. Lotfi, 1991, "Oxygen Microbubble Injection for in situ Bioremediation: Possible Field Scenario," in *Innovative Hazardous Waste Treatment Technology Series. Volume 3. Biological Processes*, H. M. Freeman and P. R. Sferra, eds., Technomic. Pub. Co., Inc., Lancaster, PA, pp. 131–142.

Mikesell, M. D. and S. D. Boyd, 1988, "Enhancement of Pentachlorophenol Degradation in Soil Through Induced Anaerobiosis and Bioaugmentation with Anaerobic Sewage Sludge," *Environ. Sci. Technol.*, 22(12):1411–1414.

Miller, R. N., C. C. Vogel, and R. E. Hinchee, 1991, "A Field Scale Investigation of Petroleum Hydrocarbon Biodegradation in the Vadose Zone Enhanced by Soil Venting at Tyndall AFB, Flordia," in *In situ Bioreclamation. Applications and Investigations for Hydrocarbon and Contaminated Site Remediation*, R. E. Hinchee and R. F. Olfenbuttel, eds., Butterworth-Heinemann, Stoneham, MA, pp. 283–302.

National Research Council (Committee on In situ Bioremediation: B. E. Rittman, L. Alvarez-Cohen, P. B. Bedient, F. H. Chapelle, P. K. Kitanidis, E. L. Madsen, W. R. Mahaffey, R. D. Norris, J. P. Salanitro, J. M. Shauver, J. M. Tiedje, J. T. Wilson, and R. S. Wolfe), 1993, *In situ Bioremediation. When Does It Work?*, National Academy Press, Washington, D.C.

Ong, S. K., R. Hinchee, R. Hoeppel, and R. Scholze, 1991, "In situ Respirometry for Determining Aerobic Degradation Rates," in *In situ Bioreclamation*, R. E. Hinchee and R. F. Olfenbuttel, eds., Butterworth-Heinemann, Stoneham, MA.

Pritchard, P. H. and C. F. Costa, 1991, "EPA's Alaska Oil Spill Bioremediation Project," *Environ. Sci. Technol.*, 25(3):372–379.

Rawe, J., et al., 1991, "Guide for Conducting Treatability Studies under CERCLA: Aerobic Biodegradation Remedy Screening. Interim Guidance," U.S. EPA Report No. EPA/540/2-91/013a, July.

Rijnaarts, H. H. M., A. Bachmann, J. C. Jumelet, and A. J. Zehnder, 1990, "Effects of Desorption and Intraparticle Mass Transfer on the Aerobic Biomineralization of α-Hexachlorocyclohexane in a Contaminated Calcareous Soil," *Environ. Sci. Technol.*, 24(9):1349–1354.

Semprini, L. and P. McCarty, 1991, "Comparison between Model Simulations and Field Results for in situ Biorestoration of Chlorinated Aliphatics: Part 1: Biostimulation of Methanotrophic Bacteria," *Ground Water*, 29(3):365–374.

Semprini, L. and P. McCarty, 1992, "Comparison between Model Simulations and Field Results for in situ Biorestoration of Chlorinated Aliphatics: Part 2. Cometabolic Transformations," *Ground Water*, 30(1):37–44.

Staps, S. J. J. M., 1990, "International Evaluation of in situ Biorestoration of Contaminated Soil and Groundwater," EPA/540/2-90/012, Sept.

Strand, S. E., R. M. Seamons, M. D. Bjelland, and H. D. Stransel, 1988, "Kinetics of Methane-Oxidizing Biofilms for Degradation of Toxic Organics," *Wat. Sci. Tech.*, 20(11/12):167—173.

Strandberg, G. W., T. L. Donaldson, and L. L.Farr, 1989, "Degradation of Trichloroethylene and *trans*-1,2-Dichloroethylene by a Methanotrophic Consortium in a Fixed-Film, Packed-Bed Bioreactor," *Environ. Sci. Technol.*, 23(11):1422–1425.

Urlings, L. G. C. M., F. Spuy, S. Coffa, and H. B. R. J. van Vree, 1991, "Soil Vapour Extraction of Hydrocarbons. In situ and On Site Biological Treatment," in *In situ Bioreclamation. Applications and Investigations for Hydrocarbon and Contaminated Site Remediation,* R. E. Hinchee and R. F. Olfenbuttel, eds., Butterworth-Heinemann, Stoneham, MA, pp. 321–336.

van Eyk, J. and C. Vreeken, 1988, "Venting-Mediated Removal of Diesel Oil from Subsurface Soil Strata as a Result of Stimulated Evaporation and Enhanced Biodegradation," *Med. Fac. Landbouww. Rijksuniv. Gent.,* 53(4b).

van Eyk, J. and C. Vreeken, 1989, "Venting-Mediated Removal of Diesel Oil from Subsurface Soil Strata as a Result of Stimulated Evaporation and Enhanced Biodegradation," *Envirotech, Vienna,* pp. 475–485.

Wilson, D. J., T. Cordero, C. Gómez-Lahoz, R. A. García-Delgado, and J. M. Rodríguez-Maroto, 1994, "Soil Cleanup by in situ Aeration. XX. Mass Transport of Volatile Organics in Wet Activated Carbon," *Separ. Sci. Technol.,* 29(16):2079–2101.

Wilson, J. T., 1994, "Bioventing of Chlorinated Solvents for Ground-Water Cleanup through Bioremediation," in *Handbook of Bioremediation,* Project Officer, J. E. Matthews, Lewis Publishers, Boca Raton, FL.

CHAPTER 10

Implementation of Soil Vapor Extraction: Regulatory and Other Issues†

INTRODUCTION

SOIL vapor extraction (SVE) will most likely be the first "innovative" technology to have its status changed to an "established" technology. It has been employed at many sites of varying size, complexity, and degree of site characterization for over a decade. Its popularity, especially in the early days, reflects the relative simplicity of the concept and the fact that, given volatile organic compounds (VOCs) and adequate pneumatic permeability of the soil, the technology will work.

The success of the SVE technology is probably best exemplified by the data presented by the U.S. EPA Technology Innovation Office in their document "Cleanup of the Nation's Waste Sites" (EPA 542-R-92-012, April 1993). This information was derived from Records of Decision (RODs) for fiscal years 1983 through 1991. The number of selections of treatment technologies for Superfund site remediation was 498. Of those, 288, or 58%, were of established technologies such as off- or on-site incineration and solidification/stabilization. Innovative technologies accounted for 42%, or 210 selections. Of those, 83 were SVE. (Note that multiple technologies are often used at one site.) SVE accounts for much of the growth in the selection and use of innovative treatment technologies at Superfund sites, constituting 40% of these applications. For comparison, bioremediation, the next innovative technology after SVE that could have its status changed to "established," makes up 21% of all selections of innovative technologies. Other innovative technologies selected for use at Superfund sites include thermal desorption, soil washing, in situ flushing, dechlorination, in situ vitrification, solvent extraction, air sparging, and in situ steam recovery of oily waste.

---

†This chapter was prepared by Ann N. Clarke, Ph.D., and Jeffrey L. Pintenich, P.E., CHMM, Eckenfelder, Inc., 227 French Landing Drive, Nashville, Tennessee 37288, U.S.A.

529

The wide use of SVE at Superfund sites is understandable given the relative simplicity of the technique and the fact that VOCs are the most commonly identified contaminant group at National Priorities List (NPL) sites. Chlorinated VOCs are by far the most common class of organic contaminants at these sites, followed by nonchlorinated VOCs.

It must be remembered that, as we address the use of SVE at Superfund sites, there are also approximately 5,100 hazardous waste treatment, storage, and disposal facilities (TSDFs) that are potentially subject to corrective action under the Resource Conservation and Recovery Act (RCRA). There are approximately 80,000 pre-existing solid waste management units located at the TSDFs. Similarly, there is the potential for application of SVE for the remediation of soils associated with underground storage tanks that have leaked volatile constituents, especially those compounds not readily amenable to aerobic biodegradation. As of September 1991, the U.S. Department of Defense (DOD) had identified nearly 18,000 potentially contaminated sites and the U.S. Department of Energy (DOE) had responsibility for remediating up to 110 major installations that could include some 4,000 individual contaminated sites.

SVE has become the principal treatment technology for chlorinated and nonchlorinated VOCs in vadose zone soil at Superfund sites; therefore, it is reasonable to believe that, at both NPL sites and other locations throughout the country, the use of SVE technologies will continue at current levels and may even increase. New techniques that modify SVE technology to increase soil permeability and contaminant volatility may lead to further expansion of SVE applications.

SITE CHARACTERIZATION

While the preceding discussion in this chapter is encouraging with respect to the utility of SVE technology, one must acknowledge that an appropriate level of characterization of site conditions is of paramount importance for the logical selection of SVE as the remedial technology of choice, as well as for the proper design of the system. In some cases, SVE will emerge, after thorough site characterization and engineering feasibility studies, as the best or one of the best solutions. At other sites, SVE may be ineffective, or only effective if applied ex situ to soils that have been excavated. This section briefly summarizes the salient issues of site characterization for SVE projects.

NATURE AND EXTENT OF SOIL CONTAMINATION

At the outset, the investigator must determine which chemicals are present, at what concentrations, and in which zones. As discussed in Chapter

2, compounds commonly considered to be VOCs possess the physical and chemical properties that enable them to be potentially amenable to SVE, as do certain semivolatile compounds. However, the concentrations could be so low in some areas that the technology would not be cost-effective. Additionally, some compounds present challenges to the system designer regarding emission control prior to release of the extracted vapor to the atmosphere. Of course, the zones of the site to be treated must be selected.

Subsurface investigations are utilized to help delineate the lateral and vertical extent and chemical characteristics of the contaminant distribution. Most often, these investigations involve the collection of soil samples by boring techniques at a range of depth intervals across the area of concern. In recent years, soil gas probes have also been utilized to help the investigator characterize the site. A reasonable number of sample locations of either type is needed to address variability across and beneath the site; one must recognize that there is an inherent risk of missing zones of contamination in any investigatory program, and there is no hard and fast rule for determining an adequate number of sample locations. Great care must be taken in collecting soil samples for VOC analyses and in maintaining the integrity of the samples prior to analysis (e.g., minimizing headspace in sample jars). Reasonable care should be taken to determine if nonaqueous phase liquids are present in the subsurface (see the following section).

NONAQUEOUS PHASE LIQUIDS

Depending upon the nature, amount, and age of the release, organic chemicals may be present in the subsurface in the form of nonaqueous phase liquids (NAPLs). NAPLs with specific gravities less than that of water are known as light NAPLs or LNAPLs (e.g., toluene), while those denser than water are known as dense NAPLs or DNAPLs [e.g., trichloroethene (TCE)]. NAPLs may be present in the vadose zone and/or in the aquifer. When LNAPL is present in quantities that exceed the saturation capacity of the soil in the vadose zone, to the extent that the localized rate of dissolution into the aquifer limits transport in groundwater, the LNAPL will pool on top of the water table. Its direction of movement is then controlled principally by the horizontal groundwater gradient. DNAPL may occupy interstitial pore spaces in both the vadose and saturated zones and may pool atop the surfaces of confining geologic strata. If present, LNAPL and DNAPL may constitute significant masses of organics to be removed by an SVE system, possibly far outweighing sorbed chemicals in the vadose zone. Failure to account for their presence in the selection and design of SVE has led to unrealistic estimates of cleanup times, inadequate system sizing, and choice of inappropriate emission control equipment at some project sites. In these cases, additional technologies such as recovery of free product may be necessary to remediate the site economically and in a timely fashion.

GEOLOGY AND HYDROGEOLOGY

The horizontal and vertical variability of the site geology and hydrogeology should be understood prior to the selection and design of an SVE system. Cross sections constructed from borings at multiple transects will be helpful to portray these characteristics to the investigator/designer. Especially important are the depth to the water table (the underlying SVE flow boundary), the pneumatic permeability of the soil in the gas flow field of the well(s), and the possible presence of relatively impermeable clay lenses or other subsurface barriers to gas flow.

SURFACE FEATURES

The nature of the land surface may constrain, but not necessarily prohibit, the application of SVE. For example, building floors, paved parking lots, or impervious caps over hazardous waste sites will alter soil gas flow fields around vacuum wells (see Chapter 5). In some cases, passive vent wells may be useful in optimizing the pattern of gas flow to the vacuum wells(s).

EX SITU PROCESS

If evaluations show that an in situ SVE process is unfeasible due to site characteristics, an ex situ process may be worthy of consideration. The contaminated soils could be excavated, perhaps amended with porous inerts, and the vapor extracted in a pug mill, soil aeration cell, or soil pile prior to replacement of the soil into the excavation site. Such an ex situ approach is considerably more cost-effective than thermal treatment if it can be used.

TIMEFRAMES

The investigator and designer must allow sufficient time in project schedules for a reasonable assessment of the issues described above. Even at a small, relatively straightforward site (e.g., a leak from an underground gasoline tank in homogeneous sandy soil), an overall schedule of three months may be needed for the site characterization process. Also, regulatory approval delays, not necessarily attributable to incomplete work, can literally add months to project schedules. At a large site, the process of site characterization and treatability testing could involve, for example, as many as twelve months.

PITFALLS TO IMPLEMENTATION

Although SVE has, for the most part, been successful when employed at a site, there are several issues that could reduce the utility and/or implementability of SVE technology. These "pitfalls" can be encountered even after the two basic requirements for implementation of SVE have been met, namely, that the constituents of interest are indeed volatile and that there is adequate pneumatic permeability in the soil to promote efficient removal. Table 10.1 is a summary listing of various pitfalls to the implementation and success of soil vapor extraction. Each of these items will be discussed briefly. The use of a checklist similar to Table 10.1 prior to any field testing could greatly enhance the performance and timeliness of an SVE-based remediation.

The presence of a high or variable water table can lead to difficulties in the implementation of SVE. A high water table at a given site, resulting in only a modest well depth and screen length, limits the zone of influence achievable from each well. This can increase capital and operating and maintenance (O&M) costs dramatically. A mechanism to circumvent this problem, if feasible is to install horizontal vapor extraction channels (i.e., an array of buried parallel horizontal slotted pipes).

The impacts of a variable water table at a site targeted for SVE can usually be addressed and mitigated, assuming that this condition is known beforehand. If the variable nature of the water table was not addressed, the impacts would be: (1) a well screen filled with water upon the rising of the water table, thus precluding the withdrawal of the VOC laden air stream or (2) failure to remediate the volume of soil exposed by a lowered water table, since the well screen would not have been installed to sufficient depth to draw air efficiently through the entire zone of contamination.

Short circuiting is another potential problem in an SVE-based cleanup. There are several conditions that could cause short circuiting during SVE operations. These include heterogeneous soils; lack of integrity in the "impermeable" cap, if employed; and underground utilities or other structures/debris. When a system short-circuits, a relatively small volume of soil is exhaustively cleaned and recleaned while the other parts of the site remain relatively untouched. It is frequently difficult to know subsurface

TABLE 10.1. Pitfalls to the Implementation/Success of Soil Vapor Extraction.

High water table	Recontamination of soil
Variable water table	Intervening layers/lenses
Short circuiting	Buried materials
Public attitude	Surface construction
Fires	Emission control

characteristics in sufficient detail to preclude this from occurring to some degree during remediation.

For the most part, the public attitude toward SVE is positive. The overall concept is relatively easy and familiar (i.e., not threatening) when parallels are drawn to vacuum cleaners or the use of a straw. Difficulty frequently arises, however, in connection with the question of emission control. There have been instances where SVE had not been able to be employed at a site because of public perceptions about off-gas emission control. Specifically, at one site the public equated the use of catalytic oxidation to incineration, and the resulting objections precluded its use. The use of more traditional activated carbon beds was not an option for this site since vinyl chloride, a known human carcinogen, was present and would not have been adequately sorbed to the carbon to meet local requirements for zero emissions of human carcinogens.

There is another potential difficulty if SVE is used at anaerobic sites at which putrescible material has been disposed. An example of this scenario would be a landfill containing mixed industrial and municipal waste. The potential problem is the in situ generation of methane over the years that, with introduction of air into the anaerobic site, could initiate a fire. Such a site would require extensive monitoring of the off-gas (e.g., opacity from smoke, increased CO_2 levels, temperature elevation) for early indications of any incipient fires, as well as emergency protocols and equipment to break the fire rectangle and stop the fire from spreading.

Another pitfall is the recontamination of the cleaned soil from subsurface dense nonaqueous phase liquid (DNAPL) pools that are not effectively cleaned by soil vapor stripping or from contaminated groundwater passing under the vadose zone. A site study should help define the possibility of DNAPL pools (if not the actual locations), as well as the potential for recontamination from a contaminated aquifer. The latter scenario can be modeled and the projections evaluated. At that point, one needs to evaluate the order and schedule of treatment technologies, e.g., remediation of groundwater prior to or simultaneously with the vapor stripping of the vadose zone soil. Admittedly, the removal of pooled DNAPL at the bottom of an aquifer is an extremely challenging task at best.

The presence of intervening clay or silt layers or lenses can cause several problems in the use of SVE. Specifically, these low-permeability intrusions can lead to bypassing of low-permeability structures by the advecting gas, as well as the continuous contamination of the vadose zone as materials slowly diffuse and desorb from the tight layers, causing a long and inefficient cleanup as a result of unfavorable diffusion/desorption mass transport kinetics. Similarly, buried objects such as drums containing free product residual materials may possibly recontaminate the soil through similar nonequilibrium processes as VOCs slowly evaporate and diffuse from the containers. The openings in the drums, etc., may not expose the contents to the advecting gas stream, so that no enhanced (advective) removal occurs. Also, depending on the types of materials and levels of contamination in the various buried items at a site, there could be different rates of cleanup of different components since the various materials (e.g., wood, paper, rubber, concrete, etc.) would release the VOCs by desorption/diffusion at different rates compared to each other and to the surrounding soil. This could extend the remediation time compared to that of a simpler system of comparable size, as well as make evaluation of the progress of remediation more difficult.

Construction at a site can have both positive and negative impacts on the implementation/success of SVE. Specifically, if a site is heavily paved or built upon, these features can be used as a cap that extends the streamlines and, thus, the zone of influence for each given well. However, breaks in the paving or construction can lead to short circuiting. There are often logistical difficulties in installing the wells at a developed site—both in getting the equipment into the designated areas and in the actual drilling process, where underground utilities can be a problem. Also, daily ongoing activities at the site can impede optimum well placement, as well as the SVE operations themselves. Information about such aspects of the site should be identified early in the site characterization so that provision can be made in the design and engineering plans to mitigate their overall impact.

Emission control warrants special attention. In addition to the impact from the public attitude relating to SVE emission control previously described (which, in ideal circumstances, can be overcome through an educational program), the question of cost needs to be raised. It is not uncommon to have the cost of emission control for the off-gas at an SVE site be equal to 50% of the cost of the remediation. The emission control system will frequently be inefficient after the initial high concentrations of VOCs encountered at the startup have fallen off if appropriate engineering considerations are not afforded to the long-term operation. The lower VOC concentrations in the off-gas, when removal is controlled by nonequilibrium kinetics, can result in the inefficient operation of incineration and catalytic oxidation, as well as activated carbon adsorption.

There are design and operational options that can help create more cost-effective implementation of the various emission control units. These include sequential and joint extraction wells, intermittent operation of the vacuum system (allowing soil gas concentrations to build up during the down time as contaminants are slowly diffused/desorbed into the soil gas), and dilution of the high VOC concentration in the early stages of stripping to permit the use of units better sized to the long "tailing" portion of the removal curve.

The bottom line in addressing the various pitfalls to the implementation/success of soil vapor extraction is that many of the problems can and should be addressed during engineering design, based upon a good working knowledge of existing conditions, and collected during the site characterization and remedial investigation phases. Also the use of mathematical models is recommended to help evaluate the impact of possible options that could be used to mitigate various negative influences without going to the time and expense of elaborate field studies.

REGULATORY ISSUES IN THE UNITED STATES

SVE may be a treatment technology evaluated and selected pursuant to one of the two major federal programs involving site remediation: Superfund and RCRA. SVE might also be employed in regulatory programs involving underground storage tanks (USTs), state Superfund sites, and voluntary actions.

CERCLA OR SUPERFUND PROJECTS

CERCLA projects are conducted under the statutory authority of the Comprehensive Environmental Response, Compensation, and Liability Act (CERCLA), which was enacted in 1980 and amended by the Superfund Amendments and Reauthorization Act (SARA) of 1986. CERCLA establishes a national program for responding to releases of hazardous substances (as defined by statute) into the environment and also addresses the release of hazardous substances from inactive or abandoned waste sites. According to CERCLA terminology, the property affected by a release is referred to as a "Superfund site." The National Contingency Plan of 1990 (NCP; 40 CFR 300) is the regulation that implements CERCLA, and the U.S. EPA is the governing regulatory agency.

The investigations carried out at a Superfund site are conducted by either U.S. EPA (and its contractors) or potentially responsible parties (PRPs) (and their consultants). This includes the Remedial Investigation/Feasibility Study (RI/FS), as well as the remedial design. The remedial action and associated operation and maintenance are conducted by either the U.S. EPA or a PRP contractor. The selected remedy for the site is documented in the Record of Decision (ROD). Funding for the Superfund process is provided either by the PRP(s) for PRP-led sites or via the Hazardous Substance Response Trust Fund (the "Superfund"), for U.S. EPA–led sites. Usually, the Superfund process is overseen by the regional

U.S. EPA offices. The Superfund-cleanup process involves a site discovery or notification step followed by a preliminary assessment/site investigation (PA/SI), which is conducted by the U.S. EPA, the state, or a government contractor. The focus of the PA/SI is the evaluation of the potential for off-site exposure and, in conjunction with this, the collection of risk-related information for use in the site prioritization step, i.e., the Hazard Ranking System (HRS). The HRS is a scoring process used to determine whether a site should be included on the National Priorities List (NPL), a list of sites the U.S. EPA considers the most significant in the United States.

Irrespective of inclusion on the NPL, remedial or investigative actions at a Superfund site may be required via an administrative order, consent order, or a consent decree. Interim measures or removal actions at a Superfund site may also be taken per authority of Section 104 of CERCLA. In-depth site characterization occurs during the remedial investigation (RI). The RI is conducted to characterize site contamination and to obtain the information necessary to identify, evaluate, and select remedial alternatives for the site. SVE treatability studies are typically conducted during the RI.

A baseline risk assessment is conducted during the RI for a Superfund site to quantitatively determine what risks to human health are presented by the site in the absence of any remedial action (i.e., the No Action alternative). The baseline risk assessment includes the identification of constituents of interest, exposure assessment, toxicity assessment, and risk characterization. The results of the baseline risk assessment may indicate that the site poses little or no threat to human health or the environment, and/or it may indicate that certain media, exposure routes, and constituents are not of concern and that others may have the potential to be of concern. The "trigger" within the risk assessment that indicates the need for further evaluation during the feasibility study (FS) is the determination of a carcinogenic risk estimate that exceeds the U.S. EPA's acceptable range of 10^{-6} to 10^{-4} or a hazard index of 1.0 for noncarcinogens. Depending upon the baseline risk results, subsequent activities associated with the FS will be scaled up or down, as appropriate. Also, the baseline risk assessment may be used later in the FS as a benchmark for evaluating the reduction (or increase) of risk with the implementation of various remedial alternatives.

Under current Superfund guidance, numerical preliminary remediation goals (PRGs) may also be calculated using a risk-based process. These PRGs are evaluated, along with applicable or relevant and appropriate requirements (ARARs), and form the basis for cleanup levels. It should be recognized, however, that in the case of SVE and other in situ technologies, final cleanup levels may not be known at the time of remedy selection and may be based upon the capability of the technology as demonstrated during implementation.

The CERCLA program has evolved more extensively than the RCRA program (discussed below) and has more detailed and finalized guidance for use in site evaluations. The CERCLA program is typically a more flexible program within which to conduct a site investigation. The process is geared towards site-specific assessments, including the development of site-specific exposure scenarios for use in the baseline risk assessment. These site-specific exposure scenarios may also be used in the determination of site-specific PRGs. The CERCLA program, however, tends to be more costly and time consuming than RCRA.

RCRA CORRECTIVE ACTION PROJECTS

RCRA projects are conducted under the statutory authority of the Resource Conservation and Recovery Act (RCRA), which was enacted in 1976 and amended by the Hazardous and Solid Waste Amendments (HSWA) of 1984. RCRA (as amended by HSWA) establishes a program for the minimization of waste and the safe management of waste that is generated, transported, or disposed of. The major regulations governing RCRA corrective action projects include: 40 CFR 264.101 (EPA Regulations for Owners and Operators of Permitted Hazardous Waste Facilities); 40 CFR 270.14(d) (EPA Regulations for Federally Administered Hazardous Waste Permit Programs); proposed 40 CFR 264 Subpart S; and proposed changes to 40 CFR 265 (EPA Interim Status Standards for Owners and Operators of Hazardous Facilities) and 40 CFR 270 (EPA Regulations for Federally Administered Hazardous Waste Permit Programs). The corrective action program regulations and guidance are in a state of evolution, with the "RCRA Corrective Action Plan" (CAP, June 1988) currently governing activities. Regulations were proposed in 1990, but, except for small portions thereof, have not yet become final. All guidance documents for the corrective action program are in draft form. The governing authority for RCRA projects is the U.S. EPA, with state agencies having the opportunity to assume primacy. RCRA addresses the release of hazardous wastes and hazardous constituents as specified in Appendix VIII of 40 CFR 261 and Appendix IX of 40 CFR 264. Hazardous wastes are determined either by characteristics (e.g., ignitability, reactivity) or are specifically "listed" in the regulations. RCRA addresses the releases of hazardous wastes and hazardous constituents from active or operating facilities, and the property affected by a release is referred to as a "facility" under RCRA regulations.

The RCRA corrective action program is initiated at a facility holding or seeking a hazardous waste management permit(s). The first step in the RCRA corrective action process is the RCRA Facility Assessment or RFA. The RFA is conducted by either the U.S. EPA or the state prior to the issuance of a RCRA permit or a Section 3008(h) order. The focus of the RFA is the determination of a release or likelihood of a release that poses a threat to human health and the environment (this is in contrast to the Superfund process, which is primarily concerned with off-site exposure). The RFA can serve as a screen, eliminating solid waste management units (SWMUs), environmental media, or entire facilities from consideration if there is no release or likelihood of release at the facility. The RFA process is somewhat similar to the PA/SI and the prioritization step involving the HRS in the Superfund program. Further action and/or interim measures at an RCRA facility maybe initiated via a permit, administrative order, consent order, or court order per Section 3008(h) of RCRA.

The second phase of the corrective action process is the RCRA Facility Investigation (RFI), which is required when the U.S. EPA determines that hazardous waste or hazardous constituents have been, are likely to have been, or are likely to be released from a SWMU or a Corrective Action Management Unit (CAMU) at a permitted facility. The RFI is analogous to the RI at a Superfund site, except that the RFI is generally focused on past and potential releases from specific SWMUs rather than on the presence of constituents across an entire site.

A health and environmental assessment (HEA) is also conducted during the RFI. The HEA, however, unlike the Superfund baseline risk assessment, does not determine site-specific risk estimates. Rather, potential receptors are identified, and a comparison of measured concentrations of site constituents to acceptable levels or "action levels" is performed. Action levels are presumptive risk-based levels, that when exceeded, indicate a closer examination of the site is warranted in the Corrective Measure Study (CMS). The CMS is analogous to the FS in the Superfund process and involves the evaluation of remedial alternatives. Media cleanup standards are specified in the CMS.

The RFI (including the HEA), CMS, and the remedial design are conducted by the owner, operator, or the operator's consultant. The remedial action and associated operation and maintenance are conducted by the owner, operator, or contractor. The selected remedy for the facility is set forth in a Statement of Basis (SOB). Funding for the facility investigation comes from the owner or operator.

The RCRA program is more in a state of flux, especially regarding proposed RCRA Subpart S, than is the CERCLA program. There is less guidance and also less flexibility. The entire process, however, is more streamlined. More states are involved in the oversight process, but regional U.S. EPA oversight is also common. The RCRA risk assessment approach is less flexible yet also less involved (and, thus, less costly) than a CERCLA assessment. Under the RCRA RFI program site risk estimates are not actually generated. Rather, potential receptors are identified, and a comparison of measured concentrations of site constituents to acceptable levels or target concentrations is performed. Abbreviated equations analogous to a PRG calculation are used with default exposure parameters to determine target concentrations. This approach is often more useful in evaluating remedial options because evaluations are completed in terms of concentrations of constituents, rather than risk levels. However, the available RCRA risk assessment guidance is only in draft form and is out of date.

It should be noted that the fact that a site is investigated under the CERCLA program does not preclude regulation under the RCRA program. For example, it is not uncommon to have SWMUs at a CERCLA site.

A brief comparison of the Superfund program and RCRA Corrective Action program is presented in Table 10.2.

UNDERGROUND STORAGE TANKS

Releases from underground storage tanks (USTs) containing petroleum and hazardous substances are regulated by the federal regulations given in 40 CFR 280 and by a variety of individual state programs. In the UST programs, site investigation is undertaken in the course of what is usually termed the "site characterization" or "site assessment," while engineering feasibility and design work is conducted in the course of preparing the "corrective action plan." Many states have established soil and groundwater cleanup standards in their regulations.

STATE AND VOLUNTARY PROGRAMS

SVE may also be a remedial technology employed at state-designated hazardous substance sites or by owners/operators on a voluntary basis. Some of the states have extensive regulatory programs for their hazardous substance sites, while others address sites on a case-by-case basis. Voluntary cleanups often "borrow" approaches from other programs, obviously on an individualized basis.

PERMITS AND APPROVALS

Table 10.3 is a summary of the relationship of SVE to various federal and state regulations as well as voluntary programs. The table summarizes the need for permits under the various regulations relative to the vapor extraction process itself as well as the management of residuals generated during the remediation. The residuals include condensate (which may be classified as groundwater), as well as the residuals from off-gas emissions, such as spent activated carbon. It is suggested that one determine the

TABLE 10.2. Comparison of Superfund Cleanup Program to RCRA Corrective Action Program.

	Superfund Cleanups	RCRA Corrective Actions
Statutory authority	Comprehensive Environmental Response, Compensation, and Liability Act (CERCLA)	Resource Conservation and Recovery Act (RCRA)
Enactment	1980	1984 [For Section 3004(u)]
Major regulations	NCP (40 CFR 300)	40 CFR 264.101 40 CFR 270.14(d) Proposed 40 CFR 264, Subpart S as well as 40 CFR 265 and 40 CFR 270 changes
Regulatory agency	U.S. EPA	U.S. EPA; state agencies (eventually)
Releases covered	Hazardous substances[a]	Hazardous wastes and hazardous constituents[b]
Property affected	"Site"	"Facility"
Program entry	Identification/notification/initial assessment	Facilities holding or seeking hazardous waste management permits
First step	Preliminary Assessment/Site Investigation or PA/SI	RCRA Facility Assessment or RFA
Focus	Off-site exposure	Release
Conducted by	U.S. EPA/state/contractor	U.S. EPA/state/contractor
Prioritization	Hazard Ranking System or HRS	Not applicable
Further action via	Administrative order, consent order, consent decree	Permit or administrative, consent, or court order per Section 3008(h) of RCRA
Interim measures	Per Section 104 of CERLCA	As above
Site characterization	Remedial Investigation or RI	RCRA Facility Investigation or RFI

(continued)

541

TABLE 10.2. (continued).

	Superfund Cleanups	RCRA Corrective Actions
Site risks	Baseline Risk Assessment or BRA	Health and Environmental Assessment or HEA
"Trigger"	Risk (10^{-6})	"Action Levels"
Cleanup evaluations	Feasibility Study or FS	Corrective Measures Study or CMS
Studies by	U.S. EPA contractor or potentially responsible party (PRP) consultant	Owner/operator/consultant
Remedy selection	Record of Decision or ROD	Statement of Basis or SOB
Remedial design	U.S. EPA contractor or PRP contractor	Owner/operator/contractor
Remedial action	U.S. EPA contractor or PRP contractor	Owner/operator/contractor
O&M	U.S. EPA contractor or PRP contractor	Owner/operator/contractor
Funding	Superfund or PRP(s)	Owner/operator

aAs defined by statute.
bIncludes Appendix VIII of 40 CFR 261 and Appendix IX of 40 CFR 264 per U.S. EPA interpretation.

TABLE 10.3. Relationship of SVE to Federal, State, and Voluntary Programs in the United States.

Driving Force	Relationship to SVE	Permit Requirements for Vapor Extraction	Permits or Approval Requirements for Residual Management[a]	
			On Site	Off Site
CERCLA	Cleanup levels evaluated during feasibility study and set forth in an ROD; typically risk-based, considering multiple pathways	None required although substantive requirements of regulations must be met	no	yes
RCRA	Cleanup levels evaluated in corrective measures study, CMS, and set forth in the SOB	No special permit required for activities conducted on site (SVE unit may be temporary unit) other than air emissions	yes	yes
UST	Cleanup level in Corrective Action Plan, i.e., case-by-case basis or set by state regulations	Requirements a function of location; not typically required for vapor extraction itself	yes	yes
State programs	Case-by-case basis; some states do not require state permits but require substantive compliance with regulations	Basically analogous to federal requirements	possibly	yes
Voluntary	Cleanup levels decided upon a case-by-case basis; may borrow from other programs for consistency	Approvals as required on a case-by-case basis	yes	yes

[a]Off-gas emissions/treatment; spent carbon; condensate.

543

regulatory issues applicable to the given type of remediation (in this case, SVE), as well as to the specific location of the site early in the feasibility study.

There is frequently a need for a permit for pilot-scale testing. This includes both mobile and fixed testing scenarios. The former employs a mobile SVE unit brought to the site by a vendor or consultant, while the latter refers to an SVE pilot operation built at the site itself—possibly forming the core of the full-scale unit. There are frequently requirements to control off-gas emission at this level of testing, as well as to monitor the discharge from the selected off-gas treatment technology. Breakthrough limits requiring replacement of the activated carbon, for example, or system shutdown, can also be included in the permit. One should also be apprised of the fact that even simple pneumatic permeability testing might require a permit or at least off-gas emission control and monitoring. Thus, the requirement for permitting pilot-scale testing needs to be determined prior to going on site by contacting the appropriate state regulatory personnel. Some states, such as California, have regional requirements that must also be met—even at the pilot-scale. Several months may be needed before the permits are approved and obtained.

CHAPTER 11

Soil Vapor Extraction Costs[††]

INTRODUCTION

SOIL vapor extraction (SVE) systems have demonstrated an ability to clean up unsaturated zone soil that has been contaminated by a wide variety of volatile and semivolatile organic compounds. Moreover, the costs of SVE systems are often only a fraction of the costs of other competing technologies. In spite of SVE's impressive successes, however, many questions remain unanswered concerning its ability to achieve, in reasonable time frames, specific cleanup levels for different contaminants under varying geological conditions. For example, how effective is SVE in interbedded sands and silts or in high organic content soils? How effective is SVE in removing contaminants from fractured bedrock systems?

The fundamental question that must be answered in connection with the SVE of complex geological regimes is whether or not local equilibrium exists (or is at least approached) between immobile contaminant in the soil and contaminant in the advecting soil gas during SVE system operation. The answer to this question is critical in predicting cleanup times and, therefore, costs. Local equilibrium conditions have typically been assumed in efforts to model the SVE process numerically. However, there would seem to be a variety of conditions where this assumption is not valid. If the soil being vapor stripped is homogeneous and relatively sandy, the local equilibrium assumption is apparently satisfactory (Baehr, et al., 1989). On the other hand, some workers have found that soil gas concentrations of volatile organic compounds (VOCs) increase after the wells have been shut down for a period of time, which is very strong evidence that diffusion and/or desorption kinetics are playing a role and that a nonequilibrium model should be used (Fall et al., 1989; Sterrett, 1989;

[††]This chapter was prepared by Kenton H. Oma, P.E., Eckenfelder, Inc., 227 French Landing Drive, Nashville, Tennessee 37228, U.S.A.

Gómez-Lahoz et al., 1994; Rodríguez-Maroto et al., 1994). DiGiulio (1992) has stressed the importance of these mass transfer effects in SVE, and Gómez-Lahoz et al. (1993) have explored the possibility of cost reductions by intermittent operation of mass transport–limited SVE wells.

Given the dependence of SVE on type of contaminant, site geology nonequilibrium conditions, and other factors, it is difficult to develop cost estimates on a unit volume or unit mass basis, as is commonly done for ex situ technologies such as thermal desorption, soil washing, and incineration, among others. SVE cost estimating for capital equipment items can be performed more easily. Costs for operating, maintenance, and monitoring can be minimized by designing an SVE system that considers the remedial goals of a specific site and that employs equipment that requires little maintenance and has automated features.

In this chapter are presented typical cost components for SVE, review cost estimates for those components and also consideration of the effects of nonequilibrium on predictions of cleanup time and costs of SVE systems.

CONSIDERATIONS IN SVE COST ESTIMATING

The U.S. EPA has developed a methodology for projecting and reporting costs as part of the Superfund Innovative Technology Evaluation (SITE) Program, and this provides a source for "order-of-magnitude" cost estimates for SVE applications (Evans, 1990). An order-of-magnitude cost estimate, as defined by the American Association of Cost Engineers, has an accuracy of $+50\%$ to -30%.

The goal of the SITE cost analysis is to generate (at a minimum) a base-case projection that provides full disclosure of all assumptions and calculations used in the analysis. This is intended to help the end user of the cost analysis to examine the assumptions and determine their appropriateness in the light of the user's own application. Unfortunately, the tendency is for people to focus on a single unit-cost estimate, with the result that the estimates are often taken out of the context of their assumptions and can therefore frequently be in substantial error.

An important consideration in applying SVE cost estimates to a particular site is market pressure. While a "cost" estimate reflects the estimated expenditures to implement the technology, the "price" paid for the implementation includes the seller's profit and is the end result of negotiations between the buyer and seller. It is the action of the market that determines the difference between cost and price. Because of the "negotiated" aspect of establishing price, technology vendors have little incentive to share cost data. Thus, this chapter focuses on cost estimates provided by the SITE program, supplemented by basis engineering cost principles.

TABLE 11.1. Cost-Categories Used for
U.S. EPA Site Program Cost Estimating.[a]

1. Site preparation	7. Utilities
2. Permitting and regulatory requirements	8. Effluent treatment and disposal
3. Capital equipment	9. Residuals/waste shipping/handling
4. Start-up	10. Analytical services
5. Labor	11. Maintenance and modifications
6. Consumables and supplies	12. Demobilization

[a]Evans, 1990.

A set of standard cost categories was developed and provides a common framework for the SITE cost evaluation. The twelve categories are listed in Table 11.1. These categories are intended to encompass a wide range of activities. In reality, cost data for any given category may not apply to a particular type of technology.

REVIEW OF SVE COST ESTIMATES

Cost estimate ranges for SVE have been presented in several publications (U.S. EPA, 1989, 1991a, 1991b, 1993). The more recent publications reference the application analysis report for SVE at the SITE Program's demonstration at a Superfund site in Groveland, Massachusetts (U.S. EPA, 1989). This report also analyzes data from other sites. The Groveland Superfund site contained trichloroethylene (TCE), among other chlorinated solvents. The progress of site cleanup was based on the removal of the TCE to a cleanup level of 0.59 ppm.

The SITE demonstration test run lasted fifty-six days and the total amount of contaminated soil treated was estimated to be 6,000 tons. The technology vendor, Terra Vac, used theoretical models and actual site data to project the time for complete remediation. It was estimated from the models that 150 days would be required to remediate the entire site.

The results of the SITE Project cost estimate are provided in Table 11.2 for the actual fifty-six–day demonstration and for the complete remediation, which was estimated to require 150 days. These costs were published in 1989 and are assumed to be reported in 1988 U.S. dollars. If escalated to 1993, the unit cost of $66/ton becomes $71/ton using an escalation factor of 1.08 (based on averaged ENR construction cost indexes for Means and for Chemical Engineering plant cost). It is possible that the SITE estimated cost is lower than the actual cost would be, given the fact that nonequilibrium effects, if present, can extend the cleanup time beyond what would be projected from the early stages of SVE performance. The off-gas treatment and wastewater disposal costs represent the largest fraction

TABLE 11.2. Estimated Order-of-Magnitude Cost
Estimate from SVE Site Project.[a,b]

	Site Demonstration 56 days, $/ton	Estimated Complete Remediation 150 days, $/ton
1. Site preparation	2.83	2.83
2. Permitting and regulatory	—	—
3. Capital equipment		
Terra Vac, $[c]	50,000	50,000
Contingency (10% of direct costs)	0.28	0.56
4. Start-up and fixed cost		
Operations procedures/training	0.40	0.40
Mobilization and shakedown	0.84	0.84
Depreciation (10% of direct costs)	0.28	0.56
Insurance and taxes (10% of 2 direct costs)	0.28	0.56
5. Labor costs	2.77	6.93
6. Supplies—raw materials	—	—
7. Utilities—electricity	0.40	1.08
8. Effluent treatment	12.83	19.25
9. Residual disposal	7.50	11.25
10. Analytical		
During operation	1.44	3.86
Pre-test and post-test analyses	16.66	16.66
11. Facility modifications		
(10% of direct costs)	0.28	0.56
12. Site demobilization	0.33	0.33
TOTALS	47.12	65.67

[a]U.S. EPA, 1989.
[b]This cost analysis does not include profits of the contractors involved.
[c]Not used directly but is used for the estimate of other costs.

(46%) of the total cost. At an equivalent site where off-gas treatment is not required and wastewater is not generated, the projected unit cost would be $38/ton. At the Groveland site, analytical costs were estimated to be 31% of the total cost for the full remediation. All other costs (total less effluent treatment, residual disposal, and analytical) associated with the SVE site cost estimate were about 22% of the total cost, or about $16/ton (1993 dollars).

Several site-specific factors can have a significant impact on cost. Those factors most affecting cost include

- size of site
- type of soil
- nature of contamination

- level (concentration) of contamination
- requirements for off-gas treatment
- wastewater generation
- nonequilibrium conditions

The size of the remediation project can have a significant impact on the dollars/ton cost of remediation. Equipment costs, analytical costs, labor costs, start-up costs, and fixed costs will decrease on a dollars/ton basis as the sites get larger. For SVE, it is more advantageous to have a deep site rather than one of the same volume that is shallow, since the deeper site requires the less expensive construction of fewer, but deeper, wells.

Sandy soils will typically take less time to remediate than clay soils. Whether the soil is sand or clay, sufficient permeability to air is necessary in the soil matrix to allow a reasonable flow rate for the extraction gas.

The volatility and effective Henry's Law constant (i.e., the partitioning coefficient) of the contaminant have a large effect on the time required to remediate a site. All other conditions being equal, a contaminant with a high Henry's constant will take less time to extract than will one with a low value. If nonequilibrium effects are not dominating, a contaminant with a high Henry's constant will exhibit a higher partial pressure in the effluent soil gas at a given airflow rate than will one with a lower Henry's constant.

A more contaminated site will, naturally, require more time to remediate than a site with a smaller quantity of contaminant, given equal SVE operating parameters and assuming local equilibrium conditions within the soil. In addition, if off-gas treatment is required, this will probably add to the treatment cost of the more contaminated site. (In the past, it has been permitted for some sites to discharge a controlled amount of VOCs directly to the atmosphere, without off-gas treatment.)

Off-gas treatment still may not be required at some sites, depending on the concentration and toxicity of the contaminant. Dispersion stacks may be appropriate in those cases. If off-gas treatment is required, this can amount to 30% of the cost/ton of the remediation. On-site regeneration of activated carbon can reduce carbon costs for off-gas treatment at large sites.

If little or no water is recovered from the vadose zone, the cost per ton could be up to 20% lower than for a site at which a significant quantity of water is collected. Wastewater typically requires analysis and either on-site treatment or off-site disposal.

The significance of nonequilibrium effects is described and an example site is evaluated in the next section.

Based on available data in the SVE SITE report (U.S. EPA, 1989), the cost for remediation by SVE can range from $10/ton to $150/ton or, in 1993 dollars, $11/ton to $162/ton. The lower cost would represent a large remediation project in which no off-gas treatment is required and no wastewater is generated. The higher cost would represent a small urban spill site.

Other references providing cost-related information on SVE include the *Soil Vapor Extraction Reference Handbook* (U.S. EPA, 1991a) and "A Technology Assessment of Soil Vapor Extraction and Air Sparging" (U.S. EPA, 1992). These references both provide vendor cases for system equipment and components such as monitoring equipment, vapor treatment equipment, worker protective equipment, site investigation equipment, chemical analyses, extraction wells, etc. The intent is that the cost estimator can pick and choose the cost components that apply to the site from the vendor cost lists.

EFFECTS OF EQUILIBRIUM ON REMEDIATION COSTS

Geological systems are rarely characterized by high degrees of uniformity. Most geological formations are rather complex, often exhibiting interbedded layers of different textures and permeabilities, as illustrated in Figure 11.1. Figure 11.1 represents an interbedded sand and silt or clay sequence. When SVE is applied to interbedded systems of this type, the advective gas flow paths preferentially follow the more permeable sand beds. Little advection occurs in the finer grained silt or clay beds. Instead, contaminant migration in these beds is controlled more by diffusion as a result of the concentration gradient set up between residual contamination in the silt beds and the more rapidly purged sand beds. It can be expected that, in interbedded formations, the rate-limiting step will often be the diffusion of contaminants from the finer grained beds to the more permeable beds where most of the advection is occurring. A similar phenomenon occurs in fractured rock, in which contaminants tend to diffuse into the relatively immobile pore water or gas within the rock matrix. Using SVE in fractured rock systems sets up a flow of soil gas within the fracture flow system of the rock. However, in fine-grained sedimentary or igneous rock, little advection occurs within the rock matrix itself. As with the above-described silt beds, diffusion of contaminants, in this case from the rock matrix to the adjoining fracture flow systems, becomes the rate-limiting step in the cleanup of sites involving fractured rock.

The overall result of nonequilibrium effects such as those described is that cleanup of such systems can be considerably prolonged, with a concomitant increase in cost.

In order to evaluate the importance of nonequilibrium effects in the design of SVE systems, an evaluation of an example site has been conducted (Oma et al., 1990). The example site is two acres in size and has TCE contamination at a level of 100 ppm from ground surface to the

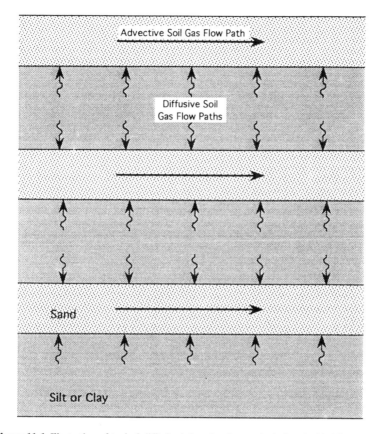

Figure 11.1 Illustration of typical diffusion/advection flow paths in interbedded formations.

groundwater table at a depth of 8 m. Numerical modeling has been conducted to estimate cleanup times for this example site for both equilibrium and nonequilibrium conditions and for a variety of extraction well separation distances. The estimated costs for both capital equipment and operations and maintenance have then been calculated for each of the modeled SVE systems. The cost analysis has considered extracted vapor treatment by granulated activated carbon with off-site regeneration and through use of a regenerative thermal oxidizer.

NUMERICAL MODELING TO PREDICT CLEANUP TIME

Two numerical models were used to estimate SVE cleanup times at the example site for a variety of extraction well spacings. In one of the models, the local equilibrium assumption is made—that is, it is assumed that the

VOC concentration in the soil gas at any point in the domain of interest is at equilibrium with the VOC concentration in the stationary phase(s) at that point. In the second model, a lumped parameter method was used to take into account the kinetic effects of mass transport by diffusion. Both approaches model a single well and assume that the domain being vapor extracted by it is cylindrically symmetrical. The pneumatic permeability of the soil may be anisotropic, and the domain may contain strata of different permeabilities. The local vapor phase VOC concentration is assumed to obey a linear isotherm—that is, under equilibrium conditions it is a linear function of the total local VOC concentration in the stationary phase(s). These include VOC dissolved in soil moisture and VOC held at adsorption sites in the soil. These models have been described in detail in the literature (Wilson et al., 1988; Gannon et al., 1989; Wilson et al., 1989; Wilson and Mutch, 1990; Wilson, 1990) and have been used in feasibility studies and in the design of in situ vapor stripping facilities. A local equilibrium model has also been developed and used extensively by Baehr et al. (1989).

Details of the mathematical equations and assumptions for the equilibrium and nonequilibrium models are presented by Oma et al. (1990). The parameters used in the modeling work are as follows, except as in the figures:

Radius of zone of influence of well	6, 8, 10, 12, 14 m
Depth of water table	8 m
Depth of well	7 m
Screened radius of well	0.3 m
Temperature	14°C
Soil porosity	0.3
Soil volumetric moisture content	0.2
Pneumatic permeability	0.1 m²/atm sec
Effective Henry's constant (dimensionless)	0.005
Wellhead pressure	0.9115 atm
Diffusion/desorption time constant for nonequilibrium model	5.55 hr
Soil density	1.7 gm/cm³

The results of variations in the spacing of the wells in a multiple well array for the local equilibrium model are shown in Figure 11.2. If the wells are placed in a square array, the effective radius of a well is about 0.707 times the well spacing. If the wells are placed on a hexagonal grid (so that each well has six nearest neighbors), the effective radius of a well is essentially one-half the distance between the extraction wells. In these calculations the well depths are all 7 m. Previous modeling has shown that there is a rather rapid decrease in cleanup rates as the effective radius of in-

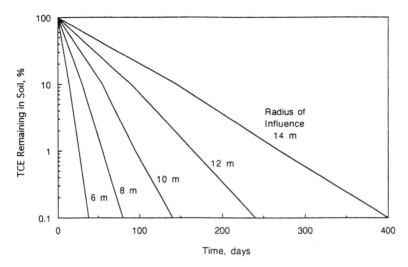

Figure 11.2 Effect of well radius of influence on TCE removal time for local equilibrium case.

fluence of a well is increased much over about 1.5 times the depth of the well.

It is possible to scale the results of calculations with the local equilibrium model to ascertain the effects of changes in several of the parameters in the model. Removal times to reach 90, 99, or 99.9% reductions in total contaminant mass are inversely proportional to the effective Henry's constant. They are inversely proportional to the pneumatic permeability at constant wellhead vacuum or to the molar gas flow rate or (to a good approximation to the wellhead vacuum (the difference between the ambient atmospheric pressure and the wellhead pressure). They are independent of the initial contaminant concentration. (Note, however, that a domain having a very high initial contaminant concentration will typically require a higher percent cleanup to reach target levels.)

The effects of diffusion/desorption kinetics limitation are shown in Figure 11.3, in which two modeling runs with a well radius of influence of 10 m are plotted. The local equilibrium model is used for one, and a time constant of 5.55 hr (a rate constant of 5×10^{-5} sec^{-1}) is used in the nonequilibrium model for the second. Cleanup times for the nonequilibrium case are about five times as long as those for the equilibrium case.

COST ANALYSIS

The comparison of costs for several SVE system configurations has been conducted for the two-acre example site based upon the results of the above-described numerical modeling. The cost analysis has evaluated a va-

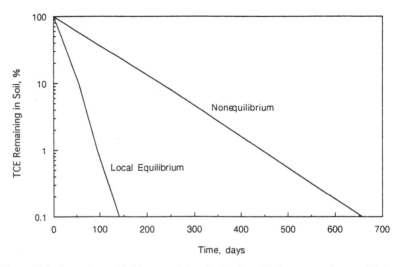

Figure 11.3 Comparison of TCE removal time for local equilibrium case and nonequilibrium case at a radius of influence of 10 m.

riety of extraction well spacings and two vapor treatment methodologies. One methodology analyzed is granulated activated carbon with off-site regeneration of the carbon. The second vapor treatment system evaluated is a regenerative thermal oxidizer and acid gas scrubber. This conceptualized system configuration is shown schematically in Figure 11.4. Air and VOCs are drawn from a series of extraction wells and passed through an induced draft blower and on to either primary or secondary beds of activated carbon or to the regenerative thermal oxidizer. Cost estimates for SVE have been made for both local equilibrium and nonequilibrium conditions and are presented as 1993 U.S. dollars.

Costs are divided into a capital investment component and an operating and maintenance component. The capital cost component consists of direct costs that are: purchased equipment, purchased equipment installation, instrumentation and controls, piping, and electrical. Capital costs are amortized over a five-year period at a conservatively high interest rate of 20%. Indirect costs associated with capital equipment are engineering and supervision, construction expenses, fee, and contingency. The total capital costs were derived by obtaining cost estimates for the purchased equipment and then applying a ratio factor to estimate the cost of the other direct and indirect costs. Table 11.3 lists the capital investment components and their corresponding ratio factors that were used to develop the capital cost estimates. When purchased equipment cost estimates were not available for a specific scale of SVE equipment, the six-tenths power rule was used to approximate the cost of equipment sized for the desired scale (Peters

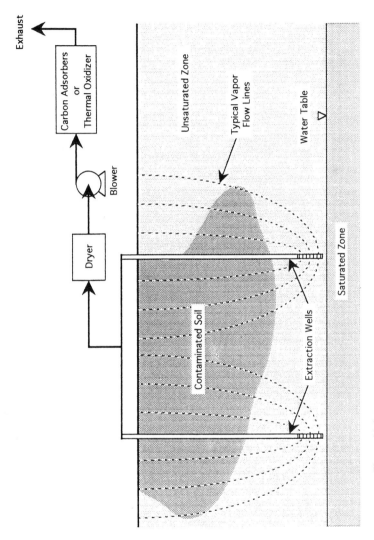

Figure 11.4 SVE concept with off-gas treatment by activated carbon or thermal oxidation.

TABLE 11.3. Capital Investment Components for
Two-Acre Example Site.

Capital Cost Component	Ratio Factor[a]
Direct:	
Purchased equipment	23
Purchased equipment installation	9
Instrumentation and controls (installed)	3
Piping (installed)	9
Electrical (installed)	3
Indirect:	
Engineering and supervision	8
Construction expense	8
Fee	3
Contingency	9

[a]Ratio factors derived from Peters and Timmerhaus (1968, p. 104).

and Timmerhaus, 1968). This rule states that the ratio of costs of two different scales is directly proportional to the ratio of flow capacities for these scales raised to the 0.6 power.

$$\frac{\text{Cost of Scale A}}{\text{Cost of Scale B}} = \left[\frac{\text{Flow Capacity of Scale A}}{\text{Flow Capacity of Scale B}}\right]^{0.6}$$

Assumptions for operating and maintenance costs are shown in Table 11.4. These costs include project labor, chemical analyses, electric power,

TABLE 11.4. Annual Operating and Maintenance Cost
Basis for Two-Acre Example Site.

O&M Cost Component	Cost Bases[a]
Labor:	
Start-up cost (first year)	8% of fixed capital[b]
Maintenance	4% of fixed capital[b]
Operating cost	16 hr/wk @ $53/hr
Carbon changeout	6 hr @ $53/hr per changeout
Analytical	$318 per changeout
Electrical power	Assumes $0.070/k Wh
Activated Carbon:	
Quantity per adsorber	9,000 lb
Purchase cost	$2.00/lb
Disposal/regeneration cost	$0.63/lb
Transportation cost	$0.42/lb

[a]Costs are estimated as 1993 U.S. dollars.
[b]Peters and Timmerhaus (1968, pp. 116 and 134).

and activated carbon. Cost estimates for natural gas were included for the system configuration with regenerative thermal oxidation. Costs related to activated carbon in Table 11.4 are only applied to the cost of the SVE system which employs activated carbon with off-site regeneration. Another cost that is considered separately is the cost to install extraction wells. It is estimated that the costs to install each extraction well will be approximately $3,000; in addition, there will be a mobilization/demobilization fee of $3,200.

The estimated costs for this two-acre example site are for comparative purposes only and are lower than actual SVE remediation costs would be because several cost components have not been included in the estimates. These would be common to all remediation technologies. The costs not included are site investigation, permitting, oversight, administrative, legal, contractor profit, chemical analytical services (other than for carbon disposal), and contingency.

Figure 11.5 shows the costs of SVE as a function of well separation for the local equilibrium cases using off-site regeneration of activated carbon for TCE removal. Figure 11.6 shows similar costs for the system employing the regenerative thermal oxidizer. For both cases, the optimum well separation is about 18 to 22 m. The most cost-effective SVE system employing regenerative thermal oxidation would have a cost of approximately $200,000. The most effective system employing off-site regeneration of granulated activated carbon has a somewhat higher cost of $260,000.

Similar cost estimates have been developed for the nonequilibrium, diffusion-controlled case and are shown in Figures 11.7 and 11.8. Costs for the SVE system using off-site carbon regeneration are substantially higher

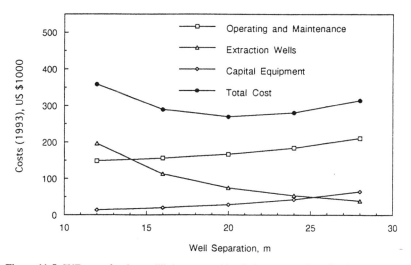

Figure 11.5 SVE costs for the equilibrium case with off-site regeneration of activated carbon.

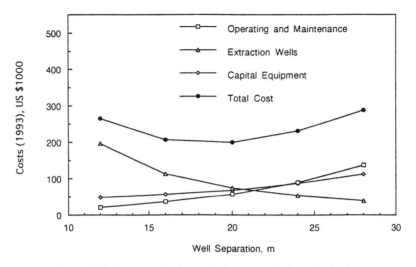

Figure 11.6 SVE costs for the equilibrium case with thermal oxidation.

for the nonequilibrium case than for the local equilibrium case. This is principally due to the longer cleanup time required and the resulting decreased carbon efficiency, which, in turn, results from lower TCE concentrations in the off-gas of the nonequilibrium system. The least cost SVE system for the nonequilibrium case runs about $425,000, using either on-

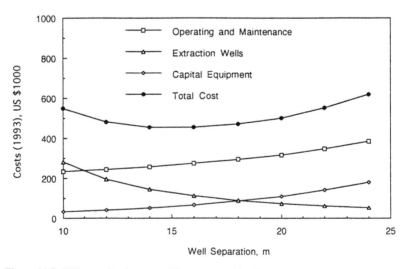

Figure 11.7 SVE costs for the nonequilibrium case with off-site regeneration of activated carbon.

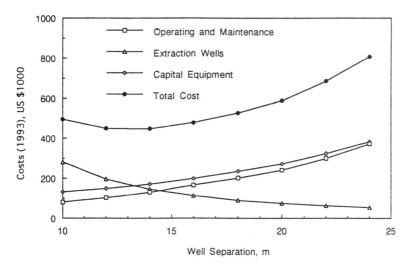

Figure 11.8 SVE costs for the nonequilibrium case with thermal oxidation.

site thermal oxidation or activated carbon with off-site carbon regeneration. It is also noteworthy that the most economical well separation appears to be between 12 and 16 m for the nonequilibrium case. If the well separation suggested by the equilibrium case (18 to 22 m) were employed in the nonequilibrium case, the cost of the example SVE system would be $500,000 to $700,000.

Figure 11.9 compares the total cost estimate for the different cases that were evaluated. This figure shows the significant impact that nonequilibrium conditions can have on site remediation by SVE. The costs for cleanup under nonequilibrium conditions used here are about twice those of the local equilibrium cases.

CONCLUSIONS BASED UPON NONEQUILIBRIUM EFFECTS

A number of conclusions can be drawn from this modeling exercise and the associated cost analyses:

- The analysis reveals that even modest nonequilibrium effects (i.e., a time constant of 5.55 hr) can have significant impact on the rate of soil cleanup and, accordingly, on the overall cost of remediation by SVE.
- The extent to which nonequilibrium conditions affect rates of cleanup has implications with respect to well spacings, pumping rates, and vapor treatment systems. A vapor treatment system that

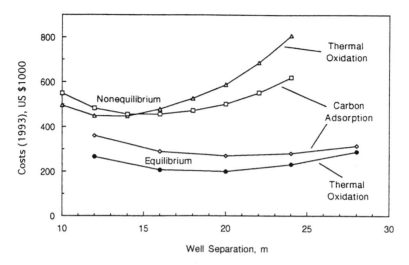

Figure 11.9 Summary of SVE cost estimates for two-acre example site.

may be most cost-effective where local equilibrium conditions exist may become less desirable than another system if the cleanup is prolonged due to nonequilibrium effects.

- Careful consideration of the potential for nonequilibrium effects should be given in the evaluation and design of SVE systems. Pilot-scale studies should be designed to estimate the rate constant for diffusion-controlled release of VOCs. This can often be accomplished by intermittent pumping or by repeated injection/extraction cycles.

REFERENCES

Baehr, A. L., G. E. Hoag, and M. C. Marley, 1989, "Removing Volatile Contaminants from the Unsaturated Zone by Inducing Advective Air-Phase Transport," *J. Contaminant Hydrology,* 4:1.

DiGiulio, D. C., 1992, "Evaluation of Soil Venting Applications," *J. Hazardous Materials,* 32:279.

Evans, G. M., 1990, "Estimating Innovative Technology Costs for the SITE Program," *J. Air Waste Management Association,* 40:1047.

Fall, E. W., et al., 1989, "In situ Hydrocarbon Extraction: A Case Study," as summarized in *The Hazardous Waste Consultant* (Jan./Feb.): 1-1–1-4.

Gannon, K., D. J. Wilson, A. N. Clarke, R. D. Mutch, Jr., and J. H. Clarke, 1989, "Soil Cleanup by in situ Aeration. II. Effects of Impermeable Caps, Soil Permeability, and Evaporative Cooling," *Separ. Sci. Technol.,* 24:831.

Gómez-Lahoz, C., J. M. Rodríguez-Maroto, and D. J. Wilson, 1994, "Soil Cleanup by in situ Aeration. XVII. Field Scale Model with Distributed Diffusion," *Separ. Sci. Technol.*, 29:1251.

Gómez-Lahoz, C, R. A. García Delgado, F. García-Herruzo, J. M. Rodríguez-Maroto, and D. J. Wilson, 1993, "Extraccción a Vacio de Contaminantes Orgánicos del Suelo. Fenómenos de No-Equilibrio," *III Congreso de Ingeniería Ambiental*, March 24–26, Bilbao, Spain.

Oma, K. H., D. J. Wilson, and R. D. Mutch, Jr., 1990, "In situ Vapor Stripping: The Importance of Nonequilibrium Effects in Predicting Cleanup Time and Cost," *Proc., 8th Ann. Hazardous Materials Management Conf. International*, June 5–7, Atlantic City, New Jersey, p. 330.

Peters, M. S. and K. D. Timmerhaus, 1968, *First Design and Economics for Chemical Engineers*, 2nd ed., McGraw-Hill, New York, NY.

Rodríguez-Maroto, J. M., C. Gómez-Lahoz, and D. J. Wilson, 1994, "Soil Cleanup by in situ Aeration. XVIII. Field Scale Models with Diffusion from Clay Structures," *Separ. Sci. Technol.*, 29:1367.

Sterrett, R. J., 1989, "Analysis of in situ Soil Air Stripping Data," presented at the *Workshop on Soil Vacuum Extraction*, April 27–28, Robert S. Kerr Environmental Research Laboratory, Ada, OK.

U.S. EPA. 1989, "Terra Vac in situ Vacuum Extraction System Application Analysis Report," EPA/540/A5-89/003, July.

U.S. EPA, 1991a, *Soil Vapor Extraction Technology Reference Handbook*, EPA/540/2-91/003, Feb.

U.S. EPA, 1991b, "Engineering Bulletin in situ Vapor Extraction Treatment, "EPA/540/2-91/006, May.

U.S. EPA, 1992. "A Technology Assessment of Soil Vapor Extraction and Air Sparging," EPA/600/R-92/173, Sept.

U.S. EPA, 1993, "Superfund Innovative Technology Evaluation Program Technology Profiles," EPA/540/R-93/526, Nov.

Wilson, D. J., 1990, "Soil Cleanup by in situ Aeration. V. Vapor Stripping from Fractured Bedrock," *Separ. Sci. Technol.*, 25:243.

Wilson, D. J. and R. D. Mutch, Jr., 1990, "Soil Cleanup by in situ Aeration. IV. Anistropic Permeabilities," *Separ. Sci. Technol.*, 25:1.

Wilson, D. J., A. N. Clarke, and J. H. Clarke, 1988, "Soil Cleanup by in situ Aeration. I. Mathematical Modeling," *Separ. Sci. Technol.*, 23:991.

Wilson, D. J., A. N. Clarke, and R. D. Mutch, Jr., 1989, "Soil Cleanup by in situ Aeration. III. Passive Vent Wells, Recontamination, and Removal of Underlying Nonaqueous Phase Liquid," *Separ. Sci. Technol.*, 24:939.

Index

563